D1737606

THEORY AND APPLICATIONS OF

Step Motors

THEORY AND APPLICATIONS OF

Step Motors

BENJAMIN C. KUO
University of Illinois at Urbana-Champaign

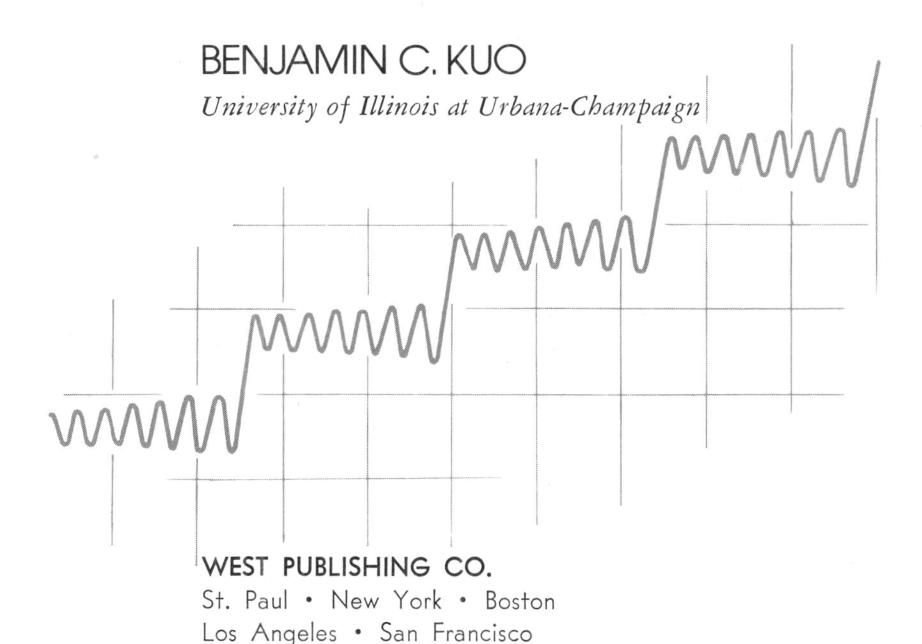

WEST PUBLISHING CO.
St. Paul • New York • Boston
Los Angeles • San Francisco

Printed in the United States of America
Library of Congress Catalog Card Number: 74-4508
ISBN: 0-8299-0015-2
Kuo-Theory & App. Of Step Motors

PREFACE

The developments of digital and incremental control, the advancement of electronic packaging, and the evolution of business machine peripherals are now combining to establish a growing interest in the technology of step motors. Since 1972, an annual Symposium on Incremental Motion Control Systems and Devices has been held at the University of Illinois at Urbana-Champaign. The Symposium is cosponsored by the Electrical Engineering Department of the University of Illinois, the Warner Electric Brake and Clutch Company, Beloit, Wisconsin, and Westool Ltd., Durham, England. The main theme of the Symposium has been incremental motion control by step motors, clutch-brakes, and d-c motors. The Proceedings of these past Symposia have been in great demand.

The objective of this book is to bring some of the papers on step motors in the 1972 and 1973 Proceedings under one cover with additional editing. Because of limitation of space, several excellent papers had to be left out.

The contents of this book may be divided into three major parts: The first part, consisting of the first seven chapters, deals with the description, modeling, and design consideration of step motors; the second part, Chapters 8 through 13, deals with the control of step motors; and the third part, covers special types of step motors such as the linear step motors and the electrohydraulic step motor.

The Symposium and this book were made possible only through the cooperative efforts of many hard working people. The author appreciates the support of the University of Illinois at Urbana-Champaign, Warner Electric Brake and Clutch Company, and Westool Ltd. The encouragement of Dr. E. C. Jordan, Head of the Department of Electrical Engineering, and Mr. W. W. Keefer, President, Warner Electric Brake and Clutch Company, is greatly appreciated. Special thanks go to all the contributing authors, and to the publication staff of the Department of Electrical Engineering who produced the original Symposium manuscript.

The editor is grateful to Dr. G. Singh for his valuable suggestions, and to Jane Braun, his secretary, for her assistance and superb handling of this manuscript.

March, 1974
Urbana, Illinois

B. C. Kuo

CONTENTS

†

1 STEP MOTORS AS CONTROL DEVICES

B. C. Kuo
Department of Electrical Engineering
University of Illinois at Urbana-Champaign
Urbana, Illinois

1-1 INTRODUCTION

A step motor is an electromagnetic incremental actuator which converts digital pulse inputs to analog output shaft motion. In a rotary step motor, the output shaft of the motor rotates in equal increments in response to a train of input pulses. When properly controlled, the output steps of a stepping motor are always equal in number to the number of input pulses.

The idea of using stepping devices in control systems has been in existence for quite some time. For instance, as early as the 1930's a remote positioning system[1,2*] used by the British Royal Navy for transmitting shaft rotations contains a bidirectional step motor. The system was later adopted by the U. S. Navy during World War II. Since that time, step motors have been used in a wide range of applications in control systems. Today we find stepping motors in practically all types of computer peripheral equipment, such as, printers, tape drives, capstan drives, memory access mechanisms, and machine tool systems, process control systems. Typical applications are illustrated in Figures 1-1 through 1-4.

Figure 1-1 illustrates the application of two step motors on an X-Y image selector. The objective of the drive system is to move a matrix of fiche of 256 1/4" images simultaneously along either of two Cartesian coordinate axes in order to position a desired image over a projection lens. In this case the step motors are directly coupled to lead screws with the nuts fastened to the matrix. The selector has a maximum access time of 0.2 second and the position error is ±0.002 inch.

Step motors are natural for this type of application, since the positional error does not accumulate, and no feedback is necessary.

Figure 1-2 illustrates the use of a step motor in the open-loop control of a numerical machine tool system. However, conventional electric step motors are

---

*Superscript numbers refer to References at the end of the chapter.

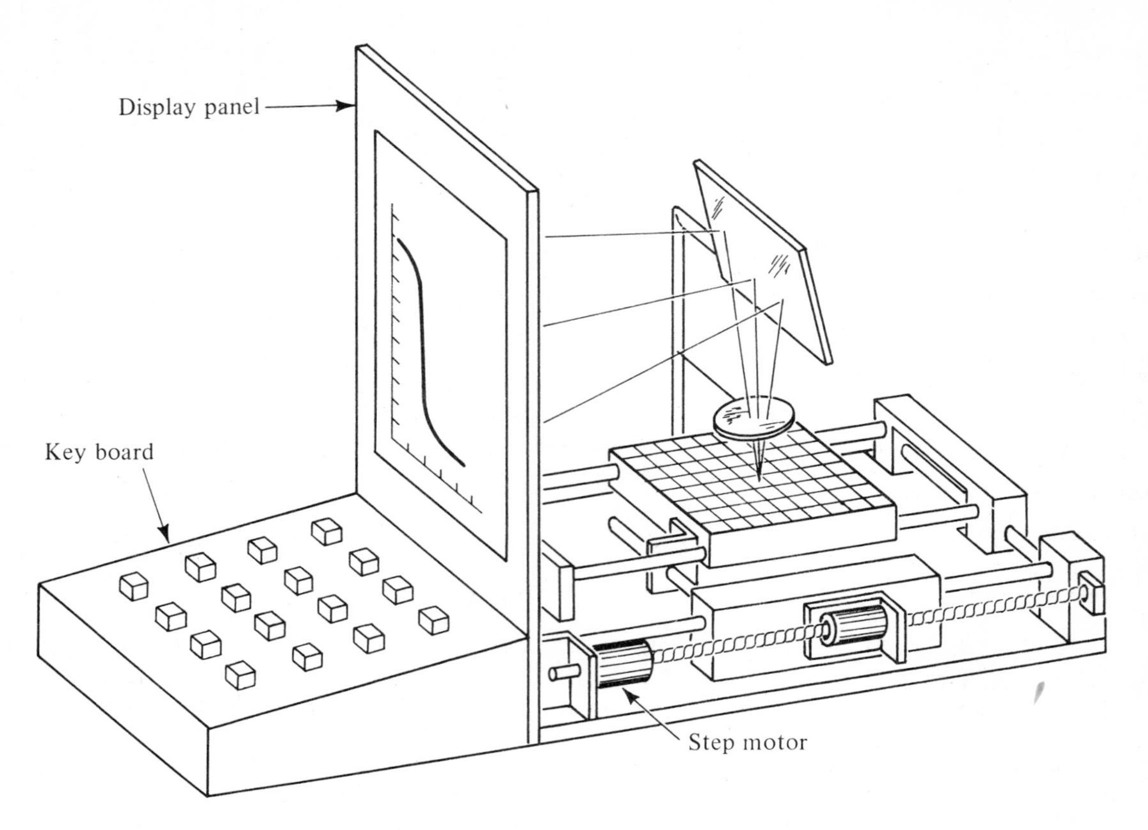

Figure 1-1. An X-Y image selector with step motors.

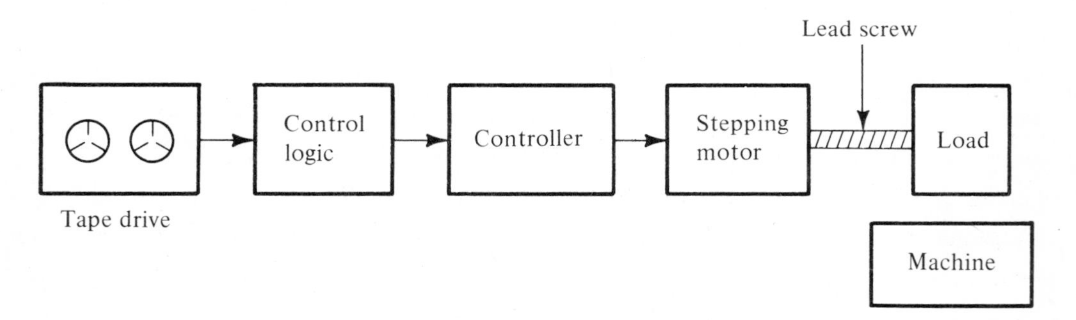

Figure 1-2. An open-loop control system using a step motor in a numerically controlled machine tool.

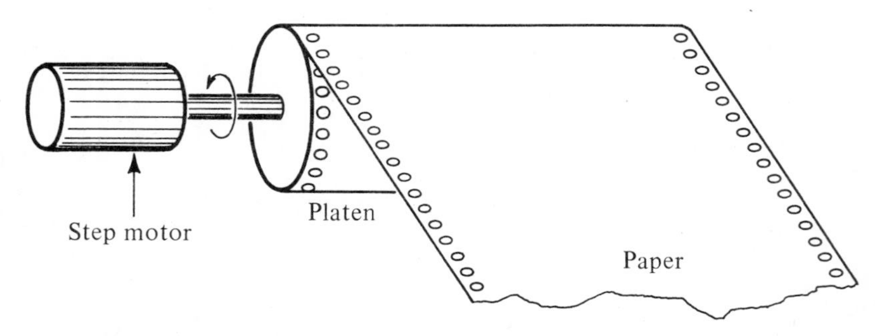

Figure 1-3. A paper-drive mechanism using a step motor.

limited in torque output to about 1500 oz-in. For heavy machine tool systems electrohydraulic step motors[6] which go up to thousands of lb-in in torque are used.

Figure 1-3 illustrates a paper-drive mechanism of a line printer using a step motor. The motor displacement can be programmed to give different printing intervals on the paper. For instance, it may be desired to print either 6 lines-per-inch or 8 lines-per-inch increments. Then, for a given step motor, we may use, say, 4 motor steps for each line for a 6 line-per-inch print, and a 3 motor steps per line for an 8 line-per-inch print interval. Paper drives probably represent one of the most popular applications for step motors. High-speed printers of up to 3000 lines per minute can be driven satisfactorily with step motors.

Figure 1-4 shows an example of a typical digital control system with a step motor. In this case, feedback is used to monitor the position of the output and the stepping rate for high-performance control.

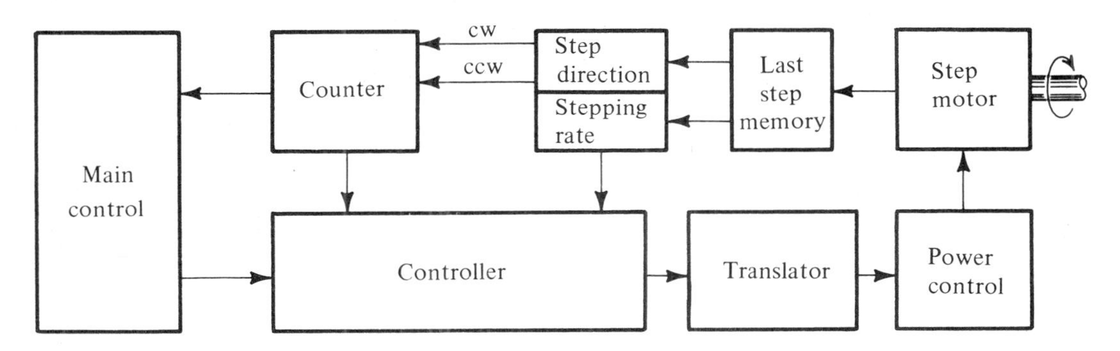

Figure 1-4. A typical digital control system using a step motor.

Recently, increased effort has been spent on the development of more improved step motors in size, speed, accuracy and resolution. It is predicted that we will see much wider use of this type of motors in control systems in the years to come.

Although control systems using conventional a-c or d-c motors and prime movers can be designed to control incremental motion satisfactorily, a system with step motor offers at least the following advantages:

(1) A step motor is inherently a discrete motion device. It is more compatible with modern digital control techniques. It is more easily adaptable for interfacing with other digital components.

(2) The positional error in a step motor is non-cumulative.

(3) It is possible to achieve accurate position and speed control with a step motor in an open-loop system, thus avoiding ordinary instability problems. Other transducers such as tachometer, gear train and feedback transducers can often be eliminated.

(4) Power consumption for intermittent operations is reduced during quiescent periods for a permanent-magnet type step motor.

(5) Design procedure is simpler for a step motor control system.

There are five basic types of step motors, classified according to the principle of operation and construction. These are: solenoid and ratchet type, variable-reluctance type, permanent-magnet rotor type, harmonic-drive type, and the phase-pulsed synchronous type. These motors will be discussed in various detail in the ensuing sections.

1-2 SOLENOID AND RATCHET TYPE STEP MOTORS

There are many different versions of switches and actuators which give stepping motion through the principle of solenoid action. These devices are not truly step motors because they do not operate on the principle of electromagnetic such as a motor does. A solenoid-ratchet type step motor[4,7] usually consists of a spring-return solenoid connected to a shaft through a pawl and ratchet arrangement. A simplified schematic diagram of such a device is shown in Figure 1-5. Each time the solenoid is de-energized, the spring advances the shaft by one step. A mechanical detent is necessary to hold the motor at the indexed position. The motor shown in Figure 1-5 can rotate in both directions. However, many solenoid type motors can step only in one direction due to their simple construction. Furthermore, the physi-

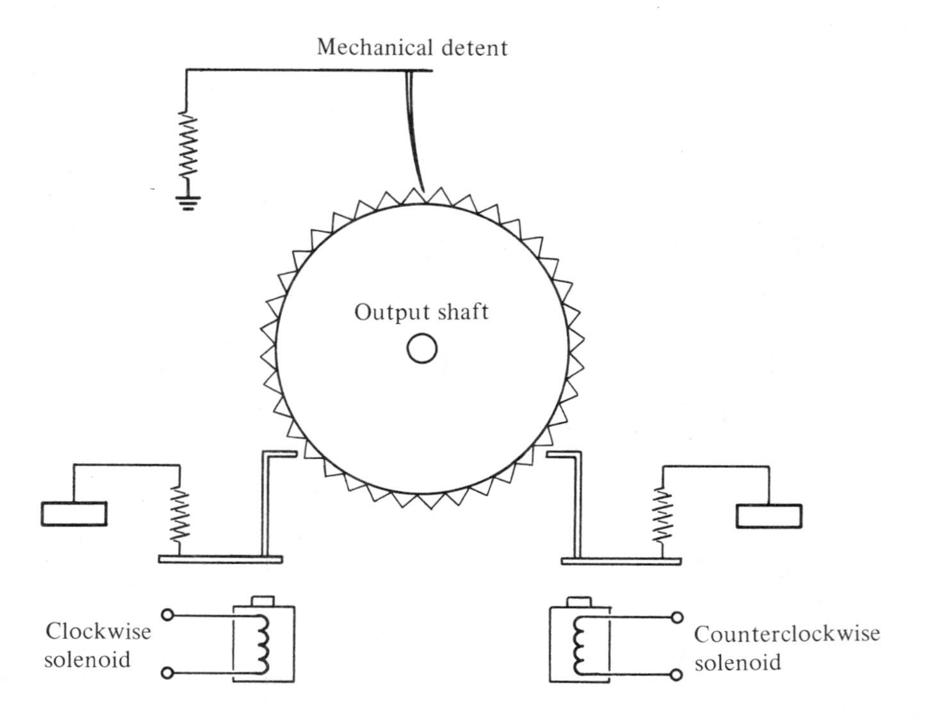

Figure 1-5. A Simplified Schematic Diagram of a Solenoid-ratchet type step motor.

cal limitations of the mechanical parts and the response of the solenoid prevent the motor from stepping at a high rate. Therefore, these devices can be used only in applications where the performance requirements are not so critical. The advantages and disadvantages of this type of step motor are summarized below:

Advantages	Disadvantages
High holding torque	Slow stepping rates
Freedom from oscillation and overshoot	Limited life due to mechanical wear
Low cost	

1-3 HARMONIC DRIVE STEP MOTOR

The harmonic-drive type of step motor is also known as the Responsyn step motor. It is a special adaptation of harmonic drive,[5] which is a unique power transmission system. Figure 1-6 shows the basic elements of a Responsyn step motor: A rigid, internally splined circular spline; a non-rigid externally splined flexspline; and a wave generator. A cut-out view of the motor is shown in Figure 1-7. The non-rigid spline is smaller than the circular spline and has fewer teeth. The wave generator provides a radial deflecting force to the flexspline. In a practical step motor the wave generator is replaced by electromagnetic poles. These poles, when energized in sequence will produce a rotating field which causes a rotating wave in the flexspline.

As the rotating magnetic field makes one complete revolution in one direction, the flexspline will rotate in the opposite direction through a distance equal to the difference between the circumference of the circular spline and that of the flex-

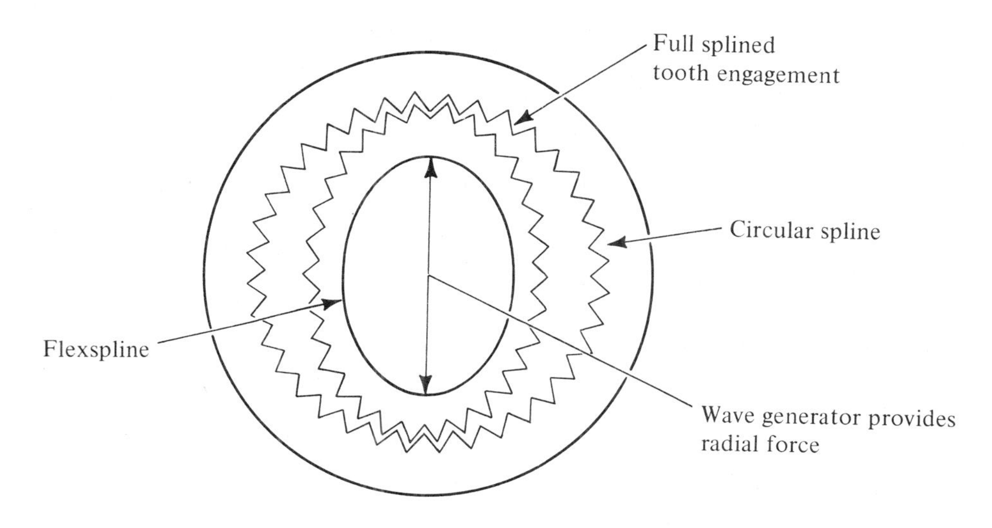

Figure 1-6. A Simplified Schematic Diagram of a Harmonic Drive Motor.

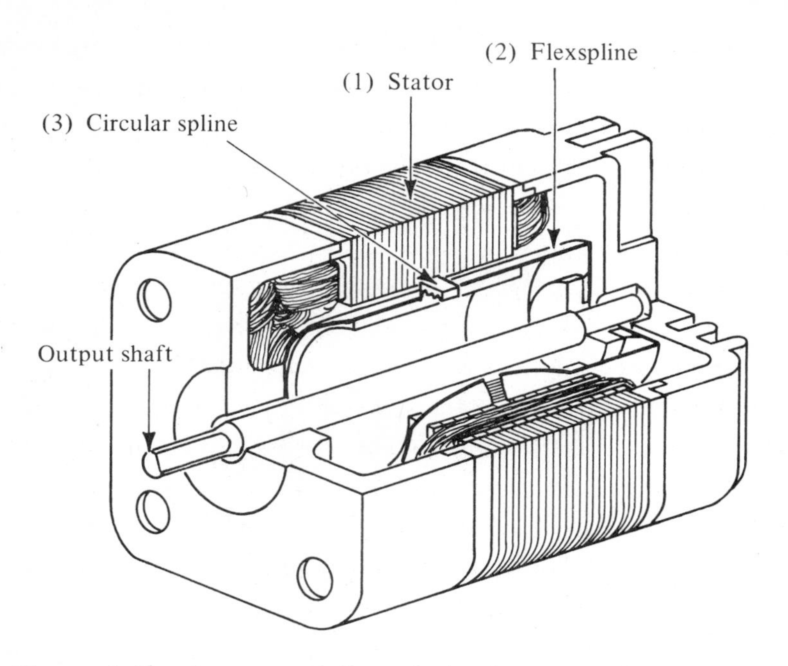

Figure 1-7. A cut-out view of the Responsyn Step Motor.

spline. Thus, the rotation of the magnetic field is directly converted to a rotary motion of the output shaft in very small steps. Typical steps resolutions of the Responsyn motors are in the range of 480 to 2000 steps per revolution.

The principal advantages and disadvantages of the Responsyn step motor are listed as follows:

Advantages	Disadvantages
Extremely high torque to inertia ratio	Slow output speed due to small step increments
Extremely small step increments	Could have limited life because of gearing
Capable of high stepping rates	Requires periodic lubrication
Less oscillation in step response than magnetic detent step motors	
Low inertia, therefore easy to start, stop, and reverse direction of rotation in very short period of time	

1-4 PHASE-PULSED SYNCHRONOUS TYPE STEP MOTOR

The phase-pulsed synchronous type step motor[4] has a permanent-magnet stator. It is also known as the cyclonome step motor. As shown in Figure 1-8, two L-shaped Alnico V permanent magnets of high coercive force are located in the stator. The flux from these magnets passes through the soft-iron toothed pole pieces which react with a toothed soft-iron rotor. The rotor has no windings and is of low inertia.

Figure 1-8 shows a Cyclonome motor with ten rotor teeth. Therefore, the rotor tooth pitch is 36 degrees. The stator has two salient pole groups, A and B. Each group has three stator teeth also with a tooth pitch of 36 degrees. These two groups of stator teeth are arranged so that when one group is aligned with the rotor teeth (minimum reluctance) the other group is aligned with the spaces between the rotor teeth (maximum reluctance). For reasons which will become clear later these two groups of stator teeth are called the driving teeth.

Connected between the two permanent magnets is another stator pole group, C, which contains the detent teeth. Figure 1-8 shows that there are six such detent teeth at 18-degree tooth pitch. Because of the permanent magnet action, when the stator winding is not energized, the rotor teeth will seek the minimum reluctance path and line up with the detent teeth. Figure 1-8 illustrates the case where the rotor teeth are in line with the stator detent teeth 2, 4, and 6.

In order to establish a specific direction of rotation, the detent teeth 1 and 6 are placed at approximately 40 1/2 degrees and 31 1/2 degrees away from their nearest driving teeth, respectively. This way, when the rotor teeth are in perfect alignment with the detent teeth, they would be approximately 4 1/2 degrees away from

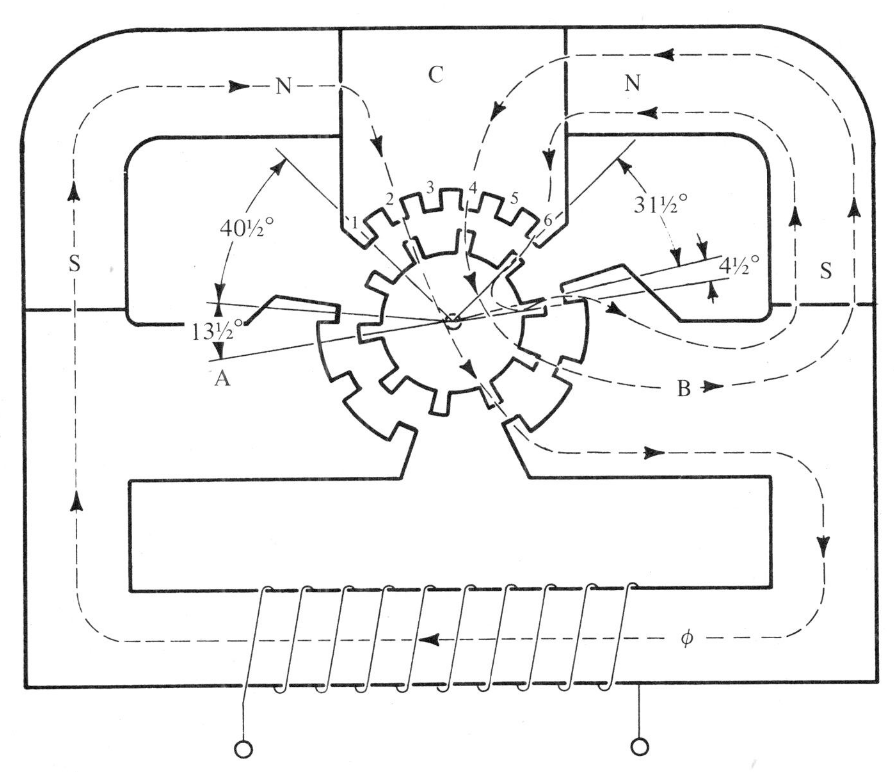

Figure 1-8. Schematic diagram of a Cyclonome step motor. No current in stator windings.

the alignment with the nearest driving teeth group. For instance, in Figure 1-8 the rotor teeth are lined up with the detent teeth 2, 4, and 6. But the three rotor teeth that are nearest to the group B driving teeth are 4 1/2 degrees away from perfect alignment. The flux paths set up by the permanent magnet are shown in Figure 1-8, with most of the flux passing through section B. If we regard this rotor-stator position as the reference, the idealized static reluctance torque of the motor as a function of rotor displacement is as shown in Figure 1-11. With the rotor position as shown in Figure 1-8, in order to make the motor rotate one step in the clockwise direction the stator windings are energized with the current direction as shown in Figure 1-9. The magnetic field set up by this current will oppose the flux from the magnet, thus advancing the rotor to the position shown in Figure 1-9. However, with the stator windings still energized the rotor would have travelled only approximately 13 1/2 degrees from the original position of Figure 1-8.

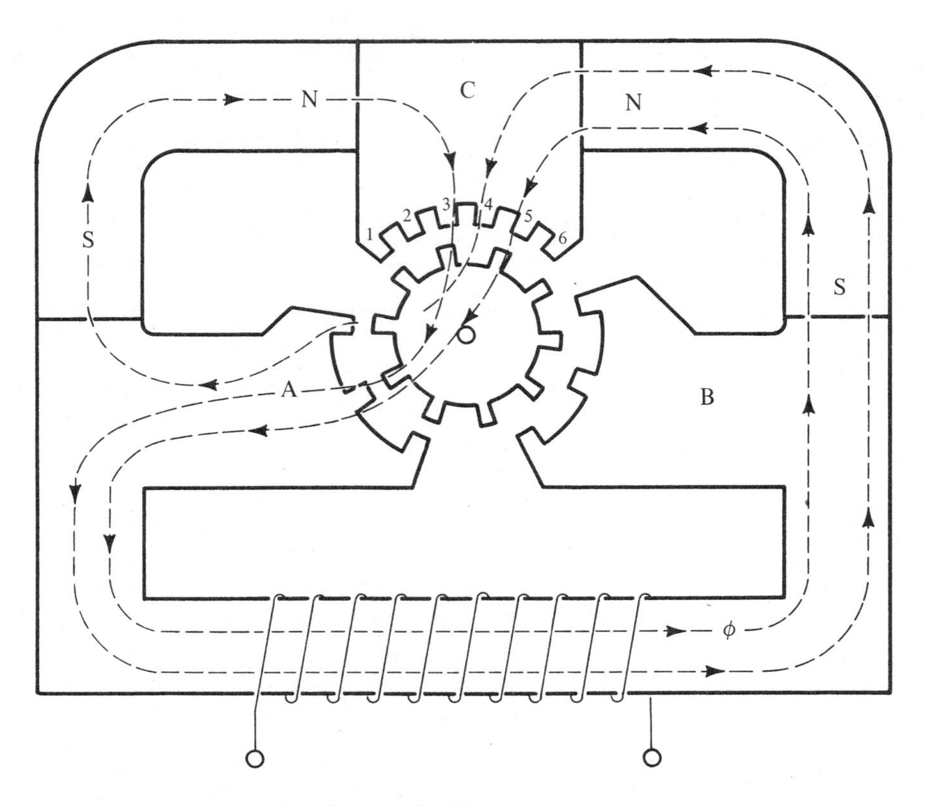

Figure 1-9. Schematic diagram of a Cyclonome step motor. Current in stator windings as shown. Rotor position is 13½ degrees clockwise from that shown in Figure 1-8.

The complete step of 18 degrees is made by de-energizing the stator, as shown in Figure 1-10. Now in order to advance another step, clockwise, from the position of Figure 1-10, a current opposite to that of the case in Figure 1-9 should be sent through the stator windings.

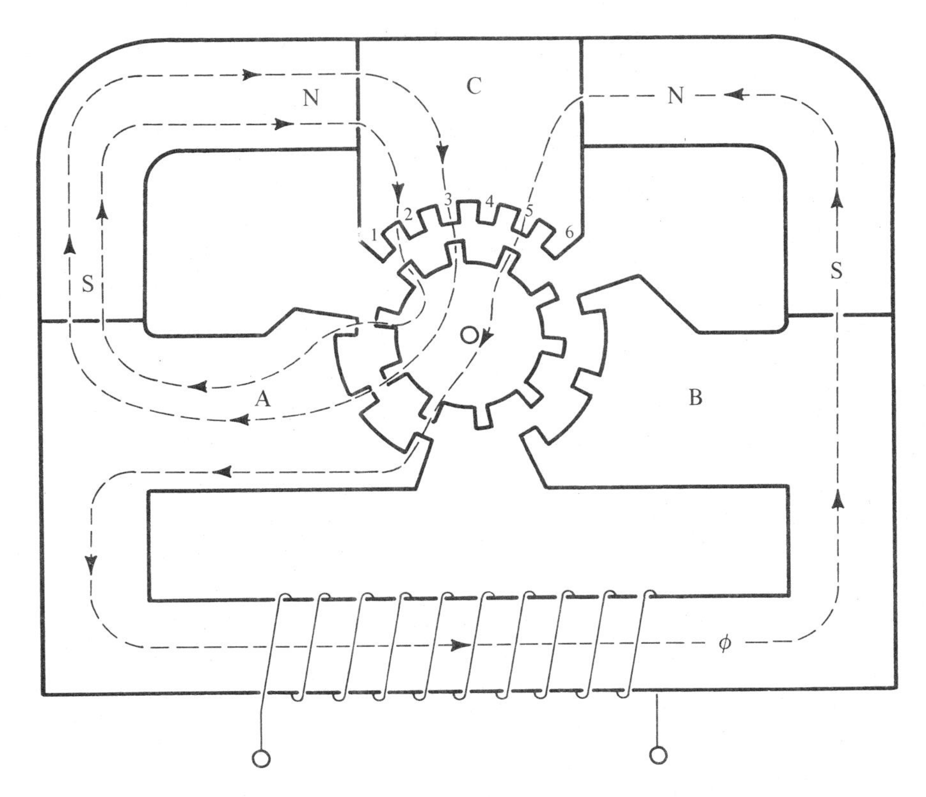

Figure 1-10. Schematic diagram of a Cyclonome step motor. No current in stator windings. Rotor position is displaced 4½ degrees and 18 degrees, clockwise, respectively, from those of Figures 1-9 and 1-8.

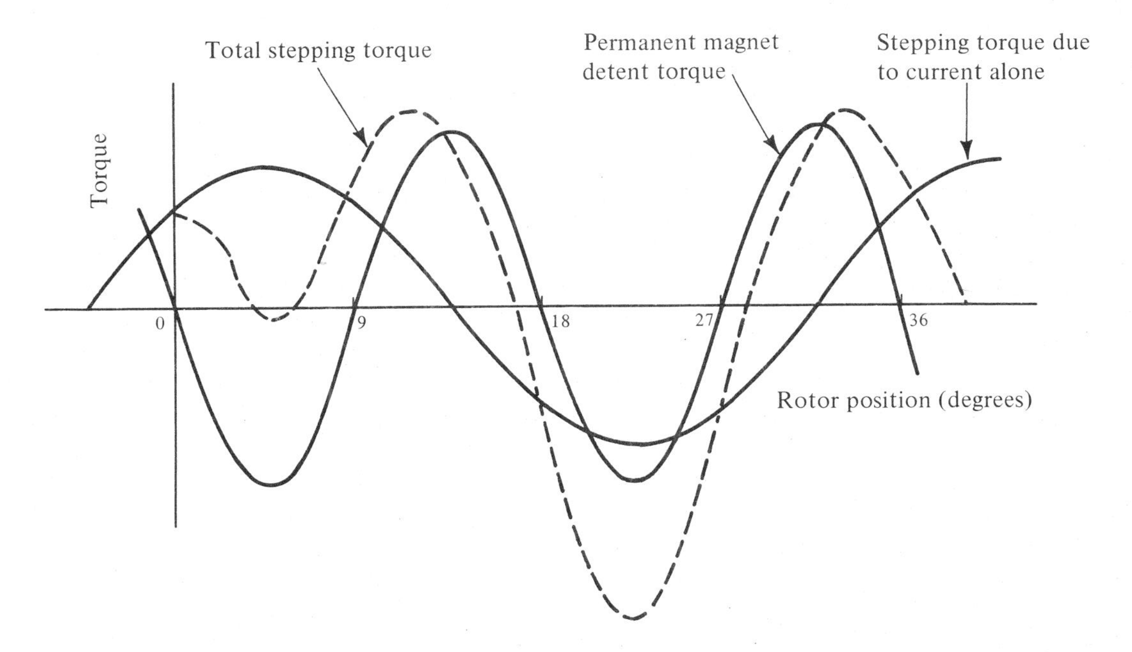

Figure 1-11. Typical torque curves developed in a Cyclonome step motor.

Since the detent teeth and the driving teeth are of different tooth pitch, and the three groups are not uniformly spaced around the rotor, in general, the exact torque relationship of the motor is quite difficult to determine analytically. Figure 1-11 shows the idealized static torque curve of the driving teeth when a "positive" direction of current is in the stator windings, as shown in Figure 1-9. The sum of the torques due to the detent teeth and the driving teeth gives the total torque as shown in the figure. A "negative" current applied to the stator when the rotor is in the position of Figure 1-10 will produce an identical torque curve but shifted to the right by 18 degrees. Therefore, the two torque curves for opposite excitations will intersect at intervals of 13.5 degrees. This means that disregarding the transient in the electrical circuit and the motor speed, for continuous stepping, the ideal switching point from a "positive" to a "negative" current, and back, should be at multiples of 13.5 degrees from the zero reference. The torque curves also indicate that the single stepping characteristics of the motor will be somewhat oscillatory. Another important feature is that the motor can only rotate in one direction. For bidirectional operation, two motors are needed.

The Cyclonome step motor is manufactured by the Sigma Instruments Inc. of Braintree, Massachusetts. Typical stepping rates are from 20 to 350 steps per second at 18-degree intervals. The holding torque range is from 80 to 375 gram-centimeters.

The principal advantages and disadvantages of the Cyclonome step motor are listed below.

Advantages	Disadvantages
Provides high holding torque with winding de-energized	Will not free wheel; cannot be driven without excessive power loss
Mechanically simple	Oscillation and overshoot in step response. Needs damping
Long life	Permanent magnet may change in strength due to excessive excitation
Relatively fast step response	Relatively slow stepping rate
High torque to inertia ratio	Unidirectional unless two motors are combined

1-5 THE VARIABLE-RELUCTANCE STEP MOTOR

Because of its simplicity in construction and ruggedness the variable reluctance (VR) step motor is one of the most popular stepping motors in use today. The motor operates on the variable reluctance principle, and works somewhat like a solenoid. The advantages and disadvantages of the motor are tabulated below:

Advantages	Disadvantages
High torque to inertia ratio	No holding torque when windings are not excited
High stepping rates	Oscillations and overshoot in step response
Fast step response	
Ability to free wheel; the motor is completely free to rotate with no current applied to the windings	
Well suited for electronic damping	
Mechanically simple with long life	
Bidirectional rotation	

Construction and Principle of Operation

The VR stepping motor has a wound stator and an unexcited rotor. The motor can be of the single-stack type or the multiple-stack type. In the multiple-stack

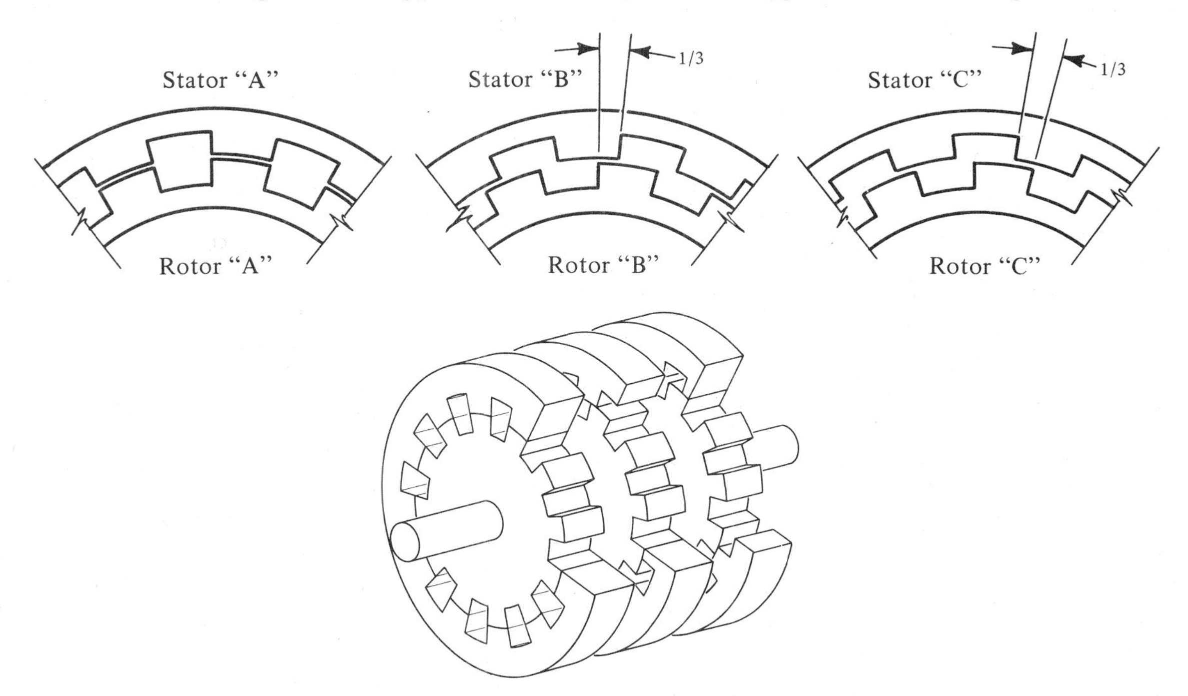

Figure 1-12. A schematic diagram of the rotor and stator teeth arrangement of a multiple-stack, three-phase VR step motor. The motor is shown to have 12 teeth on each stack or phase.

motor, the stator and the rotor consist of three or more separate sets of teeth. The separate sets of teeth on the rotor, usually laminated, are mounted on the same shaft. The teeth on all portions of the rotor are perfectly aligned. Figure 1-12 shows a typical rotor-stator model layout of the VR motor which has three separate sections on the rotor. It is also known as a three-phase motor. A VR motor must have at least three phases in order to have directional control. The three sets of rotor teeth are magnetically independent, and are assembled to one shaft which is supported by two bearings. Arranged around each rotor section is a stator core with windings. Each stator section is supported in a common housing. A schematic diagram showing the windings on the stator is given in Figure 1-13. The end view of the stator of one phase, and the rotor, of a practical VR motor is shown in Figure

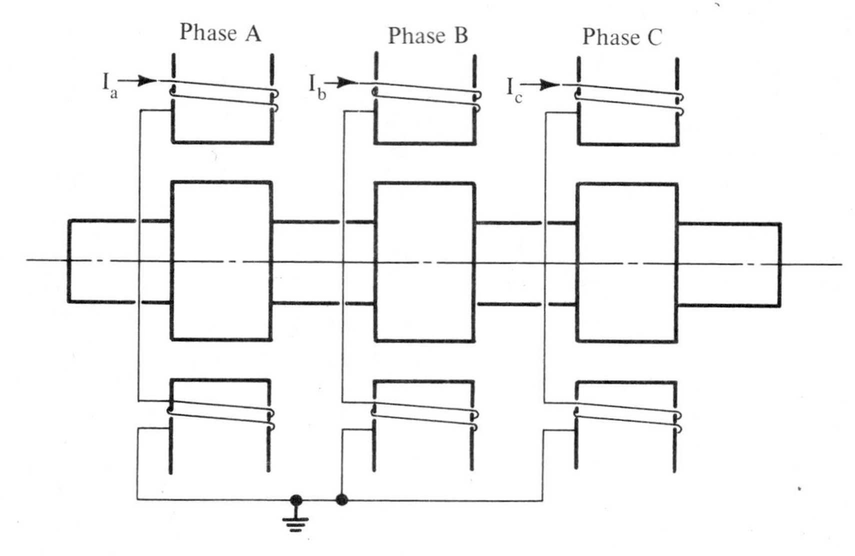

Figure 1-13. A schematic diagram of a multiple-stack, three-phase variable-reluctance step motor.

1-14. In this case the rotor teeth are shown to be in alignment with that of the particular phase of the stator.

The rotor and stator each have the same number of teeth, which means that the tooth pitch on the stator and the rotor are the same. In order to make the motor rotate, the stator sections for the three-phase motor are indexed 1/3 of a tooth pitch in the same direction. Figure 1-15 shows this arrangement with a ten-teeth rotor. Therefore, the teeth on one stator phase are displaced 12 degrees with respect to the stator phase. Here the teeth of phase C of the stator are shown to be aligned with the corresponding rotor teeth. The teeth of phase A of the stator are displaced clockwise by 12 degrees with respect to the teeth of phase C. The teeth of phase B of the stator are displaced 12 degrees clockwise with respect to those of phase A, or 12 degrees counterclockwise with respect to those of phase C. It is

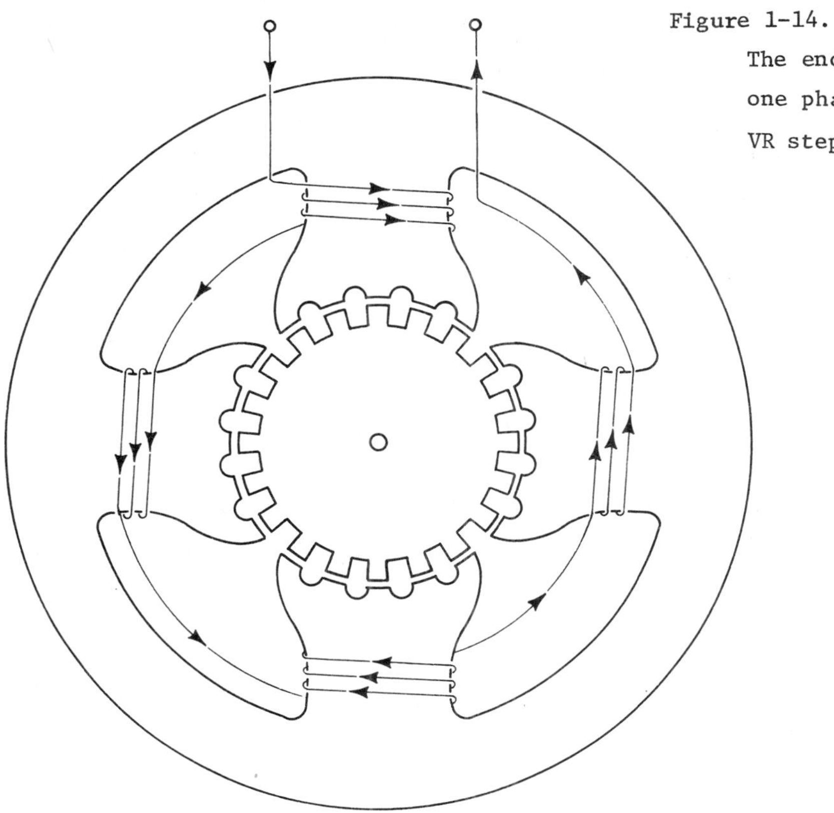

Figure 1-14.
The end view of the stator of one phase of a multiple-stack VR step motor.

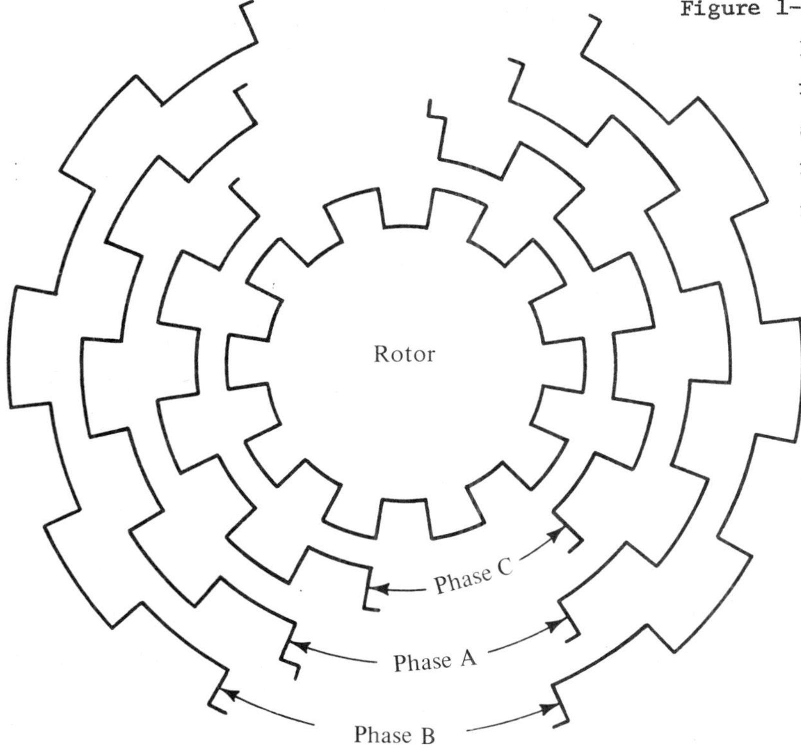

Figure 1-15.
Rotor and stator teeth arrangements of a multiple-stack, three-phase variable-reluctance step motor. The rotor has ten teeth.

easy to see that a minimum of three phases are necessary to give directional control. Four- and five-phase motors are also common, and motors with as many as eight phases are available commercially. For an n-phase motor, the stator teeth are displaced by 1/n of a tooth pitch from section to section. There is, of course, an upper limit to how many phases which can be built into one motor because the motor may become too long and the unsupported distance between the bearings becomes excessive, allowing the rotor shaft to flex.

The step angle of a multiple-stack VR step motor is determined by the number of teeth on the rotor and the stator, as well as the number of phases. The following equations give the relations between the number of steps per revolution, number of phases, number of teeth in each phase, and the step angle (resolution):

$$N = Tn \tag{1-1}$$

where N = number of steps per revolution

T = number of teeth per phase

n = number of phases

$$R = \frac{360^\circ}{Tn} \tag{1-2}$$

where R = step angle in degrees or resolution

For instance, the motor illustrated in Figure 1-12 has 12 teeth on the rotor and has 3 phases. The step angle is 10°, or 36 steps per revolution.

The operating principle of the VR motor just described is rather straightforward. Let one phase of the windings be energized by a constant voltage. The magnetomotive force set up by the current will position the rotor in such a way that the teeth of the rotor section under the excited phase are lined up opposite the teeth on the same phase of the stator. This is the position of minimum reluctance and the motor is in a stable equilibrium. For instance, if phase C is energized in Figure 1-15, the rotor would be positioned as shown in that figure. It can be visualized from the same figure that if the voltage is switched from phase C to phase A, the rotor will rotate by 12 degrees, clockwise, and the rotor teeth will be aligned opposite the teeth of phase A of the stator. Continuing in the same way, the input sequence of CABCAB will rotate the motor clockwise a total of five steps in 12-degree steps. Reversing the input sequence to CBACB . . . will cause the motor to rotate in the counterclockwise direction.

When the step motor is energized and with its rotor at the equilibrium position, no torque is developed on the rotor shaft. When the rotor is displaced from the equilibrium position, a restoring torque is developed which tends to restore the rotor to its stable equilibrium position. This restoring torque is referred to as the static holding torque. A typical static torque curve of one phase of a 12-degree-

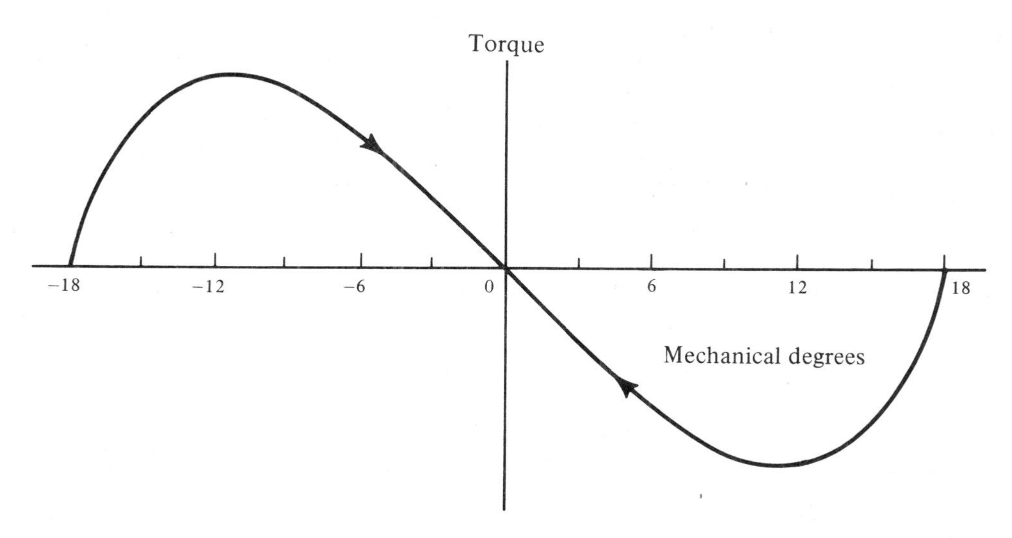

Figure 1-16. Typical static torque of a variable-reluctance step motor. The step angle is 12 degrees.

per-step motor is shown in Figure 1-16. The zero-degree position shown on the torque curve represents the stable equilibrium position of the rotor, or the center axis of any tooth of the energized stator phase. For the motor under consideration, the nearest rotor tooth axis will always lie within 18 degrees on either side of the zero-degree axis indicated in Figure 1-16.

Let us assume that phase C of the motor has been excited for a long time and the rotor position with respect to the stator teeth is as shown in Figure 1-15. If phase A is now energized, with C de-energized, and Figure 1-16 represents the torque variation of a tooth of phase A, the initial position of the rotor will be at -12 degrees in Figure 1-16. As soon as phase A is energized the rotor will proceed

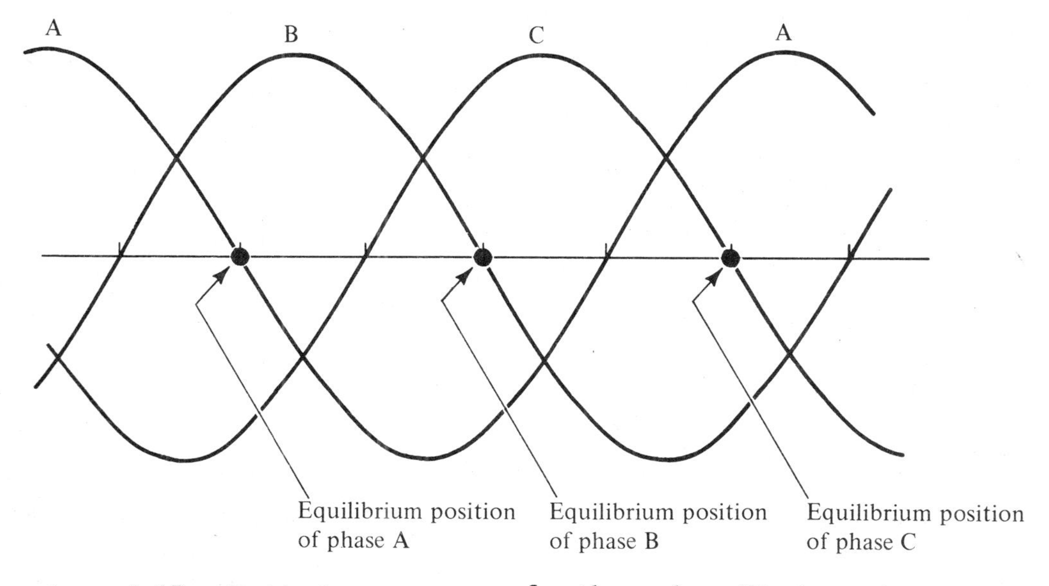

Figure 1-17. Static torque curves of a three-phase VR step motor.

toward the stable equilibrium point (0 degree) where the rotor will finally settle after some oscillations, assuming, of course, there is enough friction to prevent overshoot to carry the rotor beyond the 18-degree point.

It may be noticed that the position, ±18 degrees, also represents an equilibrium point. This is because the deflecting torque is zero at that position. However, this is a position of unstable equilibrium, since the slightest shift for this position will send the motor straight to the next equilibrium point on either side.

Figure 1-17 illustrates typical static torque curves of a three-phase VR step motor which has a 12-degrees-per-step resolution. These curves are quite useful in the determination of the best control scheme for the motor.

In general, the shape of the holding torque curves depend on the construction of the motor as well as the way the motor is excited. For instance, a step motor may be excited by a constant voltage or a dual-voltage control in which a high voltage is first applied to the motor for a short time interval and is then switched to a lower level to maintain rated current in the motor. The switching from the high voltage to the low voltage is accomplished either by measuring time or by sensing the current level. In general, dual-voltage control is used when fast response is desired.

A step motor may also be driven with one phase on at a time or more than one phase on simultaneously. For instance, for a three-phase motor the phase switchings can be C-A-B-C . . . which is one phase at a time, or CA-AB-BC-CA . . . , that is, two phases are excited simultaneously. The step angle is identical for both schemes, although the exact detent positions are offset by one-half of the step angle. The important point is that the static torque curves due to the two types of phase switchings are different, so that the responses will be different. For a single-stack VR motor, because of mutual coupling between the phases, the torque characteristics and the motor responses due to the two types of phase switchings can be drastically different. It will be shown later that a single-stack VR motor usually exhibits better damping characteristics when two phases of the motor are excited simultaneously.

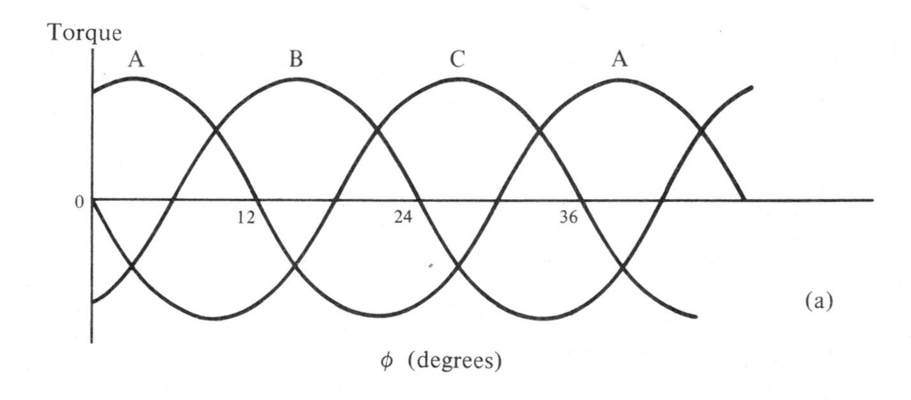

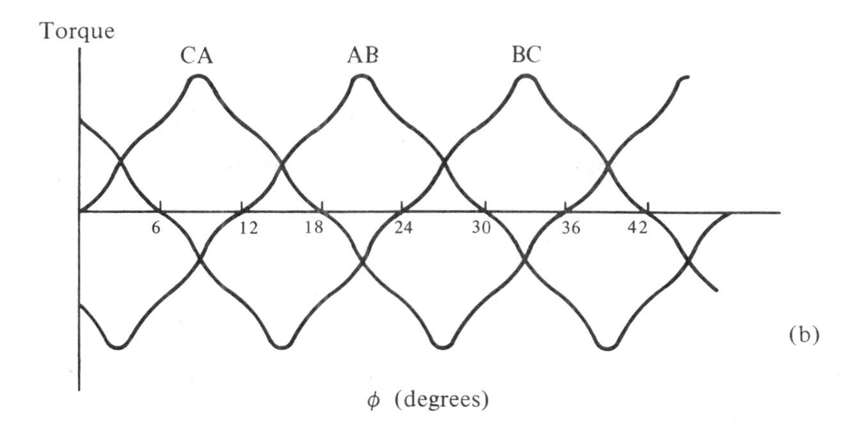

Figure 1-18. Typical static torque curves of a multiple-stack VR step motor.

Figure 1-18 illustrates typical static torque curves of a multiple-stack VR motor with one-phase excitation and two-phase excitation.

A step motor usually generates its highest output torque at standstill. As the input pulse rate is increased, the motor inductance prevents phase currents from attaining their steady-state values, and the motor torque decreases. In general, however, the torque-speed characteristic of a step motor depends on how the motor is controlled. For instance, the output torque of a step motor drops rapidly as a function of speed when the motor is driven by a sequence of input pulses under the open-loop condition. Theoretically, when the motor stalls, the torque drops to zero. The output torque can be improved by the use of a mechanical damper, or a dual-voltage control, or a closed-loop control scheme. Figure 1-19 illustrates typical torque-speed curves of a step motor.

The single-stack VR step motor offers simplicity in construction, although only certain step resolutions can be obtained with this arrangement. In this type of motor the rotor and the stator contain only one group of teeth. However, the number of teeth on the rotor and the stator are necessarily different. For example, Figure

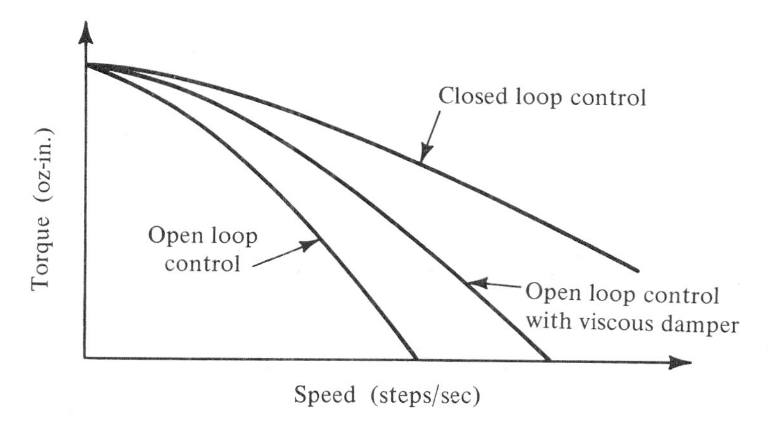

Figure 1-19. Typical torque-speed curves of a step motor.

1-20 illustrates the schematic diagram of the teeth configuration of a single-stack, three-phase, VR motor with twelve stator teeth and eight rotor teeth. There are four teeth per phase on the stator, and the distribution of the teeth for each phase is indicated in the figure. The rotor is shown to be at the detent position when phase

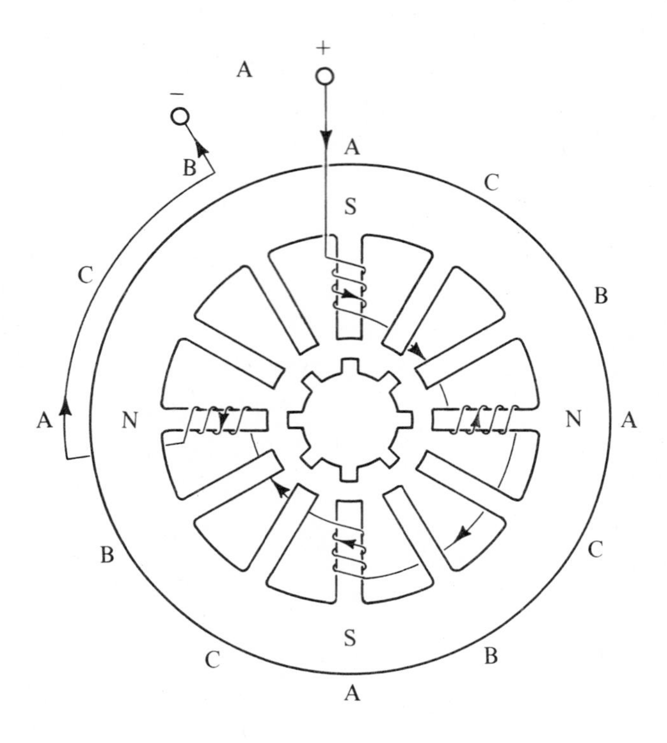

Figure 1-20. Schematic diagram of a single-stack, three-phase, VR step motor.

A is excited with a d.c. voltage. If the d.c. excitation is shifted to the windings of phase B, the motor will make a clockwise rotation of 15 degrees which is the difference between the stator and the rotor tooth pitch. If, instead, phase C is energized, the motor will make a 15-degree step in the counterclockwise direction. Therefore, this is a 24-step-per-revolution motor.

In general, the numbers of teeth on the stator and the rotor are related through

$$N_s = N_r \pm p \tag{1-3}$$

where

N_s = number of teeth on the stator

N_r = number of teeth on the rotor

p = number of stator teeth per phase

The step angle R in degrees is given by

$$R = \frac{360}{n} \text{ degrees} \tag{1-4}$$

where

n = number of steps per revolution

$$= \frac{1}{\frac{1}{N_r} - \frac{1}{N_s}} \tag{1-5}$$

If N designates the number of phases, $N_s = pN$. The minimum value of N is again 3 for the single-stack motors.

As an illustrative example Figure 1-21 shows the layout of a 24-step-per-revolution single-stack step motor that has four phases. In this case the stator is shown with eight teeth and the rotor has six teeth. We can show that for the 15-

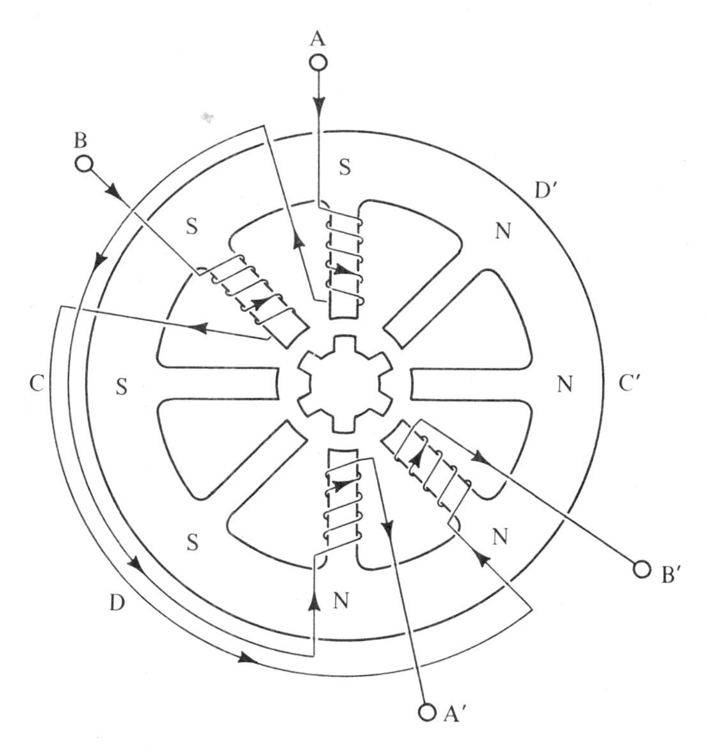

Figure 1-21. Schematic diagram of a single-stack, four-phase, VR step motor.

degree step angle, and the 4-phase configuration, the stator-rotor teeth combination is the only possible one. Using Eqs. (1-3), (1-5), and $N_s = pN$, we have

$$p = \frac{n}{N(N \pm 1)} \tag{1-6}$$

where the minus or plus sign refers to those of Eq. (1-3). For the present case, $n = 24$, $N = 4$. It is apparent that p is an integer; only the minus sign can be used in Eq. (1-6). Therefore,

$$p = \frac{24}{12} = 2 \tag{1-7}$$

and

$$N_s = pN = 8 \tag{1-8}$$

$$N_r = N_s - p = 6 \tag{1-9}$$

It can be shown by using Eq. (1-6) that the maximum number of phases that a 15-degree single-stack step motor can have is four. Since $n = 24$, for p to be a positive integer N can be only three or four.

Using the plus sign in Eq. (1-6), with $n = 24$, we have

$$p = \frac{24}{N(N+1)} \tag{1-10}$$

Then, when $N = 3$, $p = 2$.

$$N_s = pN = 6 \tag{1-11}$$

and

$$N_r = N_s + p = 8 \tag{1-12}$$

In this case the rotor will have more teeth than the stator.

Possible combinations of phase number, resolution, and teeth combinations are computed using the equations derived above, and are tabulated in Table 1-1.

The limitation on the number of phase places certain disadvantages on the single-stack step motor. For the 15° motor shown in Figure 1-21, the polarities of the phases are assumed to be as shown. However, because of the mutual coupling between the phases, unsymmetry in the flux will occur when the motor is switched continuously, A, B, C, D, A, B, ... As shown in Figure 1-21, because the poles are wound as N, N, N, N, S, S, S, S, the phase switchings will produce unsymmetrical flux patterns in the air gap due to the mutual coupling between phases. Therefore, the step responses at positions around the rotor will not be uniform. Since the motor has only two teeth per phase we cannot wind the poles with alternate north and south poles unless two motor phases are excited simultaneously. In practice, however, a 15-degree step motor is quite popular; and for four phases, the poles are arranged to be N, S, N, S . . . , and the control is designed so that two phases are excited at a given time.

TABLE 1-1.

n No. of Steps per Revolution	Step Angle	N No. of Phases	p No. of Stator Teeth per Phase	$N_s = pN$ No. of Stator Teeth	N_r No. of Rotor Teeth
12	30°	3	2	6	4
18	20°	3	3	9	6
24	15°	3	4	12	8
24	15°	4	2	8	6
24	15°	3	2	6	8
30	12°	3	5	15	10
36	10°	3	6	18	12
36	10°	4	3	12	9
36	10°	3	3	9	12
40	9°	5	2	10	8
40	9°	4	2	8	10
48	7.5°	3	8	24	16
48	7.5°	4	4	16	12
48	7.5°	3	4	12	16
54	6.667°	3	9	27	18
60	6°	3	10	30	20
60	6°	4	5	20	15
60	6°	5	3	15	12
60	6°	6	2	12	10
60	6°	3	5	15	20
60	6°	4	3	12	15
60	6°	5	2	10	12
66	5.45°	3	11	33	22
72	5°	3	12	36	24

TABLE 1-1 (Continued)

n No. of Steps per Revolution	Step Angle	N No. of Phases	p No. of Stator Teeth per Phase	$N_s = pN$ No. of Stator Teeth	N_r No. of Rotor Teeth
72	5°	4	6	24	18
72	5°	3	6	18	24
84	4.28°	3	14	42	28
84	4.28°	4	7	28	21
90	4°	3	15	45	30
90	4°	6	3	18	15
90	4°	5	3	15	18
100	3.6°	5	5	25	20
100	3.6°	4	5	20	25
120	3°	3	20	60	40
120	3°	4	10	40	30
120	3°	5	6	30	24
120	3°	6	4	24	20
120	3°	3	10	30	40
120	3°	4	6	24	30
120	3°	5	4	20	24
180	2°	3	30	90	60
180	2°	4	15	60	45
180	2°	5	9	45	36
180	2°	6	6	36	30
180	2°	3	15	45	60
180	2°	4	9	36	45
180	2°	5	6	30	36
200	1.8°	5	10	50	40
200	1.8°	4	10	40	50

The effects of the mutual coupling between phases of a single-stack motor will be discussed in detail in the next chapter. However, it should be kept in mind of the limitations in the step angle, number of phases, as well as the stator winding configuration when designing a single-stack VR step motor.

1-6 SINGLE-STEP RESPONSE OF THE VR STEP MOTOR

One of the disadvantages of the VR step motor is the oscillatory nature of its step response. When the motor is not subject to external damping, it is not uncommon to witness an overshoot of as high as 80 percent in the single-step response. Figure 1-22 illustrates a typical single-step response of a VR motor without damping. In

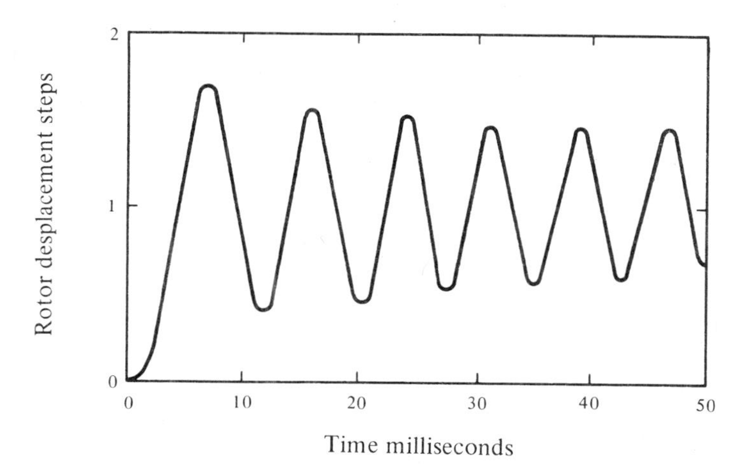

Figure 1-22. Typical single-step response of a variable-reluctance step motor (no damping and no load).

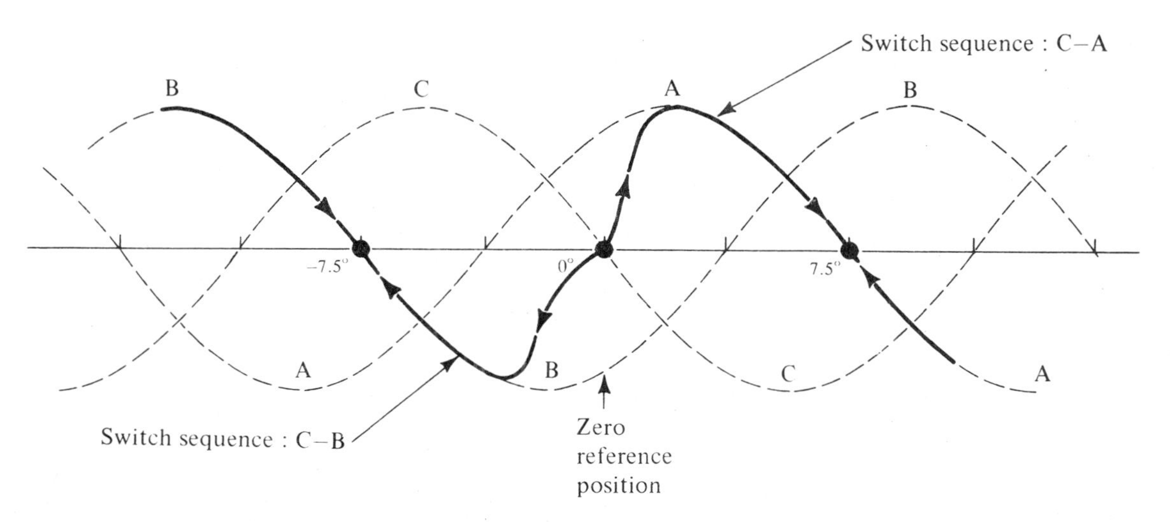

Figure 1-23. Static torque curves of a three-phase VR motor. Switching trajectories for single-step responses are shown.

this case the rotor shaft overshoots the steady-state position by approximately 70 percent and goes through a period of decaying oscillation that lasts almost 150 milliseconds. In order to control the step response of the motor properly we must first understand the relation between the motor torque and the current pulses which excite the motor. Figure 1-23 shows the idealized (sinusoidal) static torque curves of all phases of a three-phase VR motor. It is assumed to be a 7.5-degree-per-step motor (48 steps per revolution) so the torque curves are displaced by 7.5 degrees along the horizontal axis which denotes rotor displacement. The dynamic torque of the motor is usually difficult to determine, but is known to be a function of time, and the position and velocity of the rotor. It is a function of the rotor position since there is a torque present on the rotor shaft only when the teeth of the rotor and the stator are misaligned. The dynamic torque is also a function of time since a finite amount of time is required for the current to build up in the windings; and it is a function of velocity since the induced back emf on the windings depends on the position and velocity. Therefore, to determine the exact value of torque present on the rotor at any given instant is a very difficult task. However, for practical purposes, the static torque curves, idealized as shown in Figure 1-23, can be used in a qualitative way to predict the performance of the stepping motor. These static torque curves differ from the dynamic curves in that they represent the maximum torque which can exist at a given rotor position.

The reason for the long period of sustained oscillation in the step response can now be explained using the static torque curves of Figure 1-23. When the rotor shaft is stationary at the zero reference position, phase C of the motor is excited. The motor torque is zero. When the current pulse is switched from phase C to phase A, there will be a positive torque on the shaft which tends to move it until it passes through its equilibrium position at 7.5 degrees, after which there will be a negative torque on the shaft. Since there is very little friction on the motor bearing, the rotor will oscillate about the 7.5-degree equilibrium point for a period of time.

Similarly, if the current pulse had been switched to phase B, there would have been an initial negative torque (see Figure 1-23), and the rotor would have eventually stopped at the -7.5-degree position.

It has been customary to approximate the single-step response by a linear second-order system. In other words, the response shown in Figure 1-22 may be considered as the unit-step response of a linear system whose transfer function is

$$\frac{\theta(s)}{R(s)} = \frac{\omega_n^2}{s^2 + 2\zeta\omega_n s + \omega_n^2} \tag{1-13}$$

where

ζ = damping ratio

ω_n = natural undamped frequency

and R(s) is the Laplace transform of a step input.

However, it should be pointed out that this linear approximation is good only for a single step operation. Since step motor operations usually involve multiple stepping or slewing, the linear model of Eq. (1-13) is of little value in practice.

1-7 PERMANENT-MAGNET STEP MOTOR

A permanent-magnet step motor, which is also known as a synchronous inductor motor, is described in this section. This type of motor which was originally designed to operate as a low-speed synchronous motor, also has the ability to step when the phase windings are excited properly.

The stator of the motor usually contains a multiphase winding, and the rotor has a permanent magnet or a unidirectional field which is produced by a separate d.c. source.

Figure 1-24 illustrates the axial view of the permanent-magnet stepping motor. In Figure 1-25 the unidirectional field is produced by a circular magnetizing coil. The flux of the unidirectional field in each case is shown by the dotted lines.

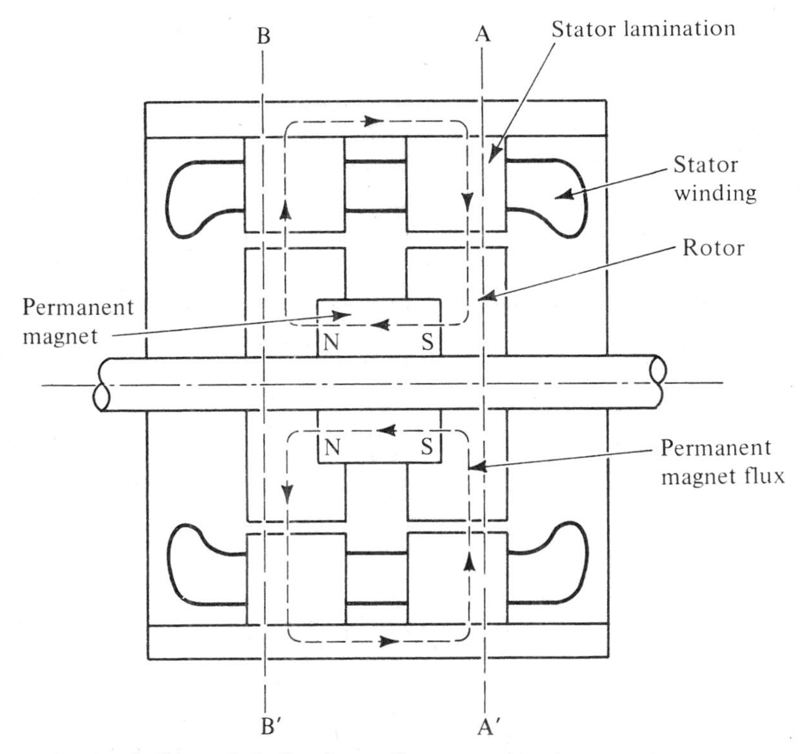

Figure 1-24. Axial view of a permanent-magnet step motor.

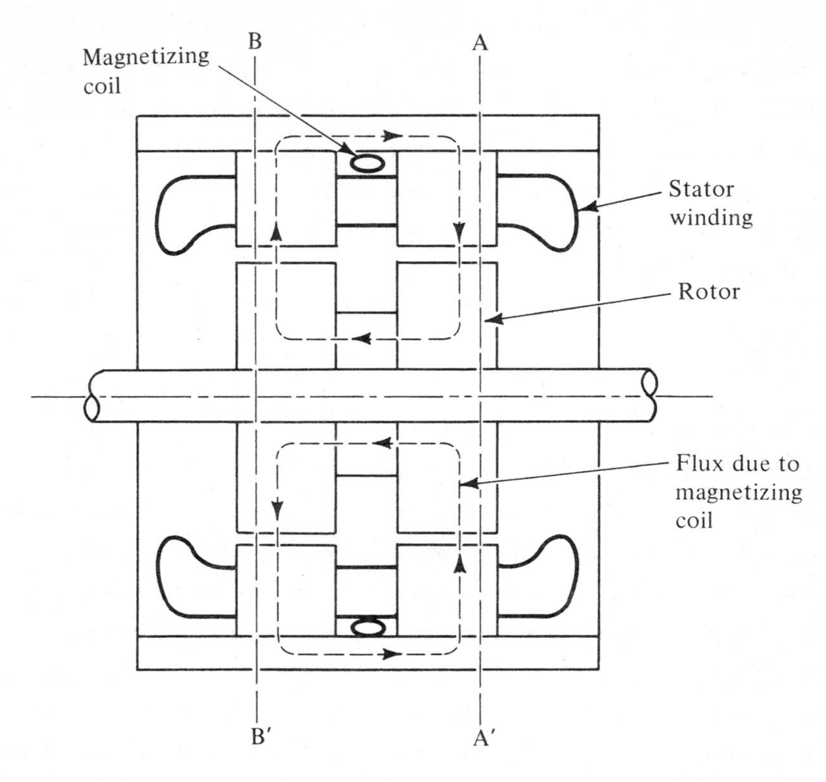

Figure 1-25. Axial view of a permanent-magnet step motor with a magnetizing coil replacing the permanent magnet.

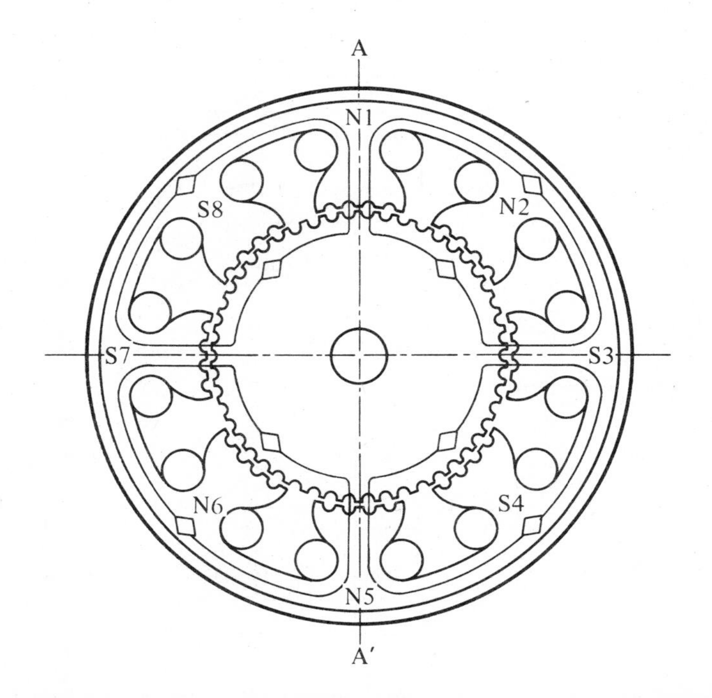

Figure 1-26. Cross section of permanent-magnet step motor perpendicular to shaft. Cross section A-A'.

Figure 1-26 shows the cross-sectional view at Section A-A' of Figure 1-24 or Figure 1-25. The stator shown in this case has eight salient poles with a two-phase four-pole winding. The stator teeth are at a pitch of 48 teeth for one complete cycle, although there are only 40 actual stator teeth. One tooth per pole is being omitted to allow space for the windings. The rotor is shown to have a total of 50 teeth. In general, the stator winding is of p poles per phase, where p is related to the number of stator teeth, N_s, and the number of rotor teeth, N_r, by

$$N_r = mN_s \pm \frac{p}{2} \qquad (1\text{-}14)$$

where m is a positive integer. In the case illustrated in Figure 1-26 $N_r = 50$, $N_s = 48$, $m = 1$, and $p = 4$, and the plus sign is used in Eq. (1-14).

As shown in Figure 1-26, one of the teeth on the rotor is aligned with the center tooth of salient pole N1. The same is true for salient pole N5. The center teeth of salient poles S3 and S7 are one-half of a rotor tooth pitch from the nearest rotor teeth. Therefore, these teeth are at the maximum reluctance position.

Figure 1-27 illustrates the cross-sectional view of the motor at Section B-B'. Notice that the two rotor sections A and B in Figure 1-24 are displaced rotationally by one-half rotor tooth pitch, while the two stator sections are lined up. Therefore, in Figure 1-27, the center teeth of salient poles N1 and N5 are at the maximum

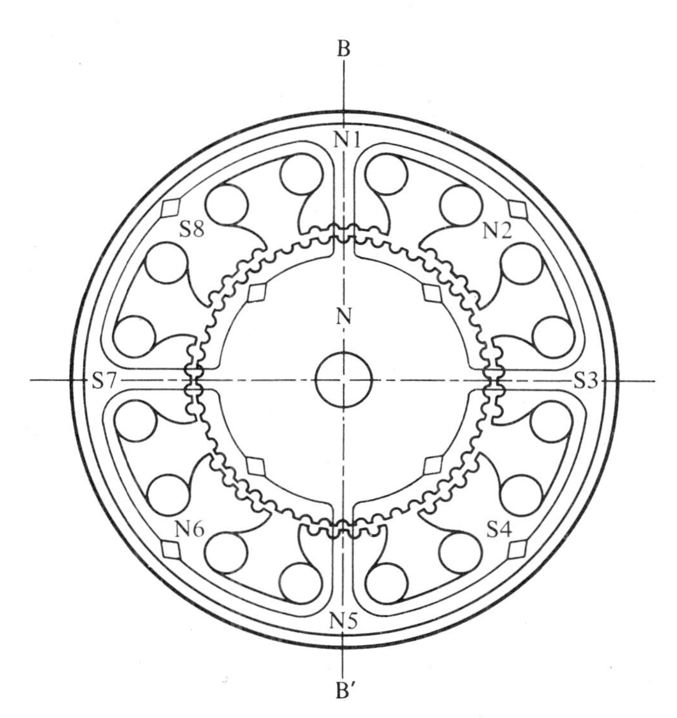

Figure 1-27. Cross section of permanent-magnet step motor perpendicular to shaft. Cross section B-B'.

reluctance position with respect to the rotor teeth, whereas those of S3 and S7 are at the maximum reluctance position.

The paths of the alternating flux produced by the stator windings are shown in Figures 1-26 and 1-27. The d.c. flux and the a.c. flux combine and interact in the air gap. Since the a.c. flux does not flow through the permanent magnet there is no demagnetization effect.

As a slow-speed synchronous motor, the motor is self starting and the synchronous speed is given by

$$S = \frac{60f}{N_r} \tag{1-15}$$

where

S = motor speed in rpm

f = frequency of supply in Hz

N_r = number of rotor teeth

Therefore, the motor illustrated in Figure 1-26 has a synchronous speed of 72 rpm.

In order to illustrate the principle of operation of the synchronous inductor motor as a step motor let us refer to the simplified layout shown in Figure 1-28. The stator has a two-phase four-salient pole winding configuration. The rotor has five teeth on each of its two sections. The teeth on the two rotor sections are offset by one-half of the rotor tooth pitch which is 72 degrees. When the two phases of the stator windings are energized in a proper sequential fashion, the

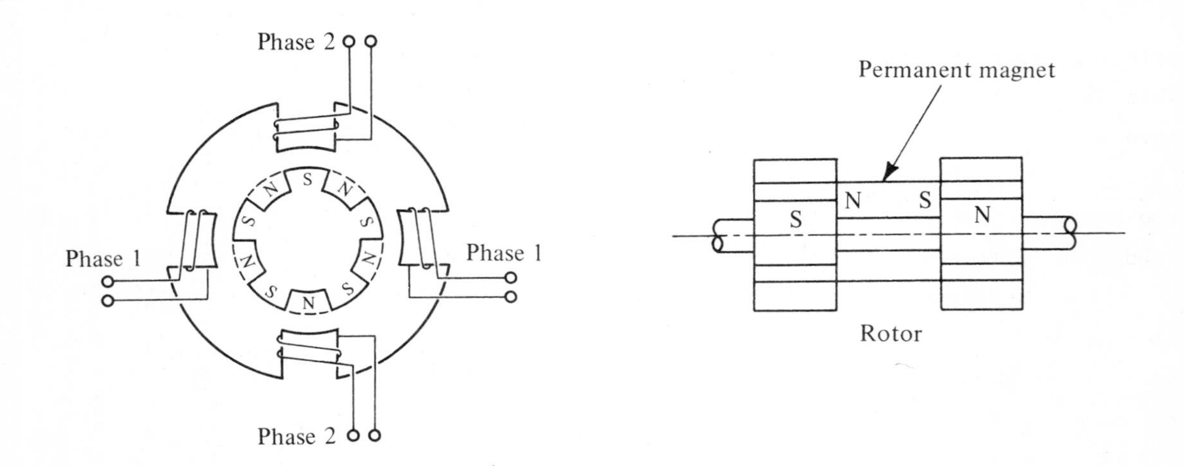

Figure 1-28. A synchronous inductor motor with five teeth on each rotor section.

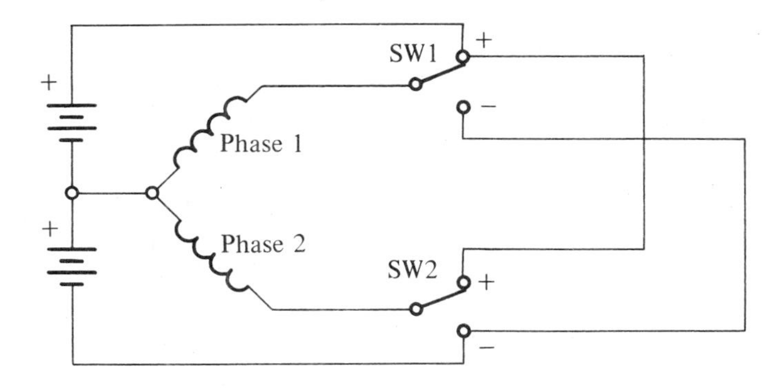

Step	CW Rotation		CCW Rotation	
	SW1	SW2	SW1	SW2
1	+	+	+	+
2	−	+	+	−
3	−	−	−	−
4	+	−	−	+

Figure 1-29. Connection of stator windings of a synchronous inductor motor as a permanent-magnet step motor and the combinations of switching sequences.

motor is capable of making 18-degree steps. Figure 1-29 shows the basic connection of the two phase windings and the required switching cycle for clockwise and counter-clockwise rotation. Each switching operation corresponds to a step angle which is equal to 1/4 of the rotor tooth pitch. Therefore, four switching operations are required to advance the rotor one tooth pitch. With five teeth on each section on the rotor, 20 switchings will cause one revolution. Thus this is a 20-step-per-revolution step motor.

The development of step torque in a synchronous inductor motor is based on the principle that when the stator windings are energized a magnetic flux pattern is set which interacts with the permanent-magnet field in such a way that the rotor will move to line up the two fields.

Figure 1-30 illustrates how the rotor seeks to line up the magnetic fields as the excitation to the stator windings are switched according to the possible sequences tabulated in Figure 1-29 for clockwise rotation.

Let us assume that the stator windings are wound in such a way that when SW1 is at + and SW2 is at - as shown in step 4 of Figure 1-29, the stator poles are magnetized as in Figure 1-30a. Now if the current in phase 1 is unchanged while that in phase 2 is reversed, we have the situation of step 1 with SW1 at + and SW2 at + in Figure 1-29. The stator poles are now magnetized as shown in Figure 1-30b, and the rotor will rotate clockwise 18 degrees to the new equilibrium position. As the next

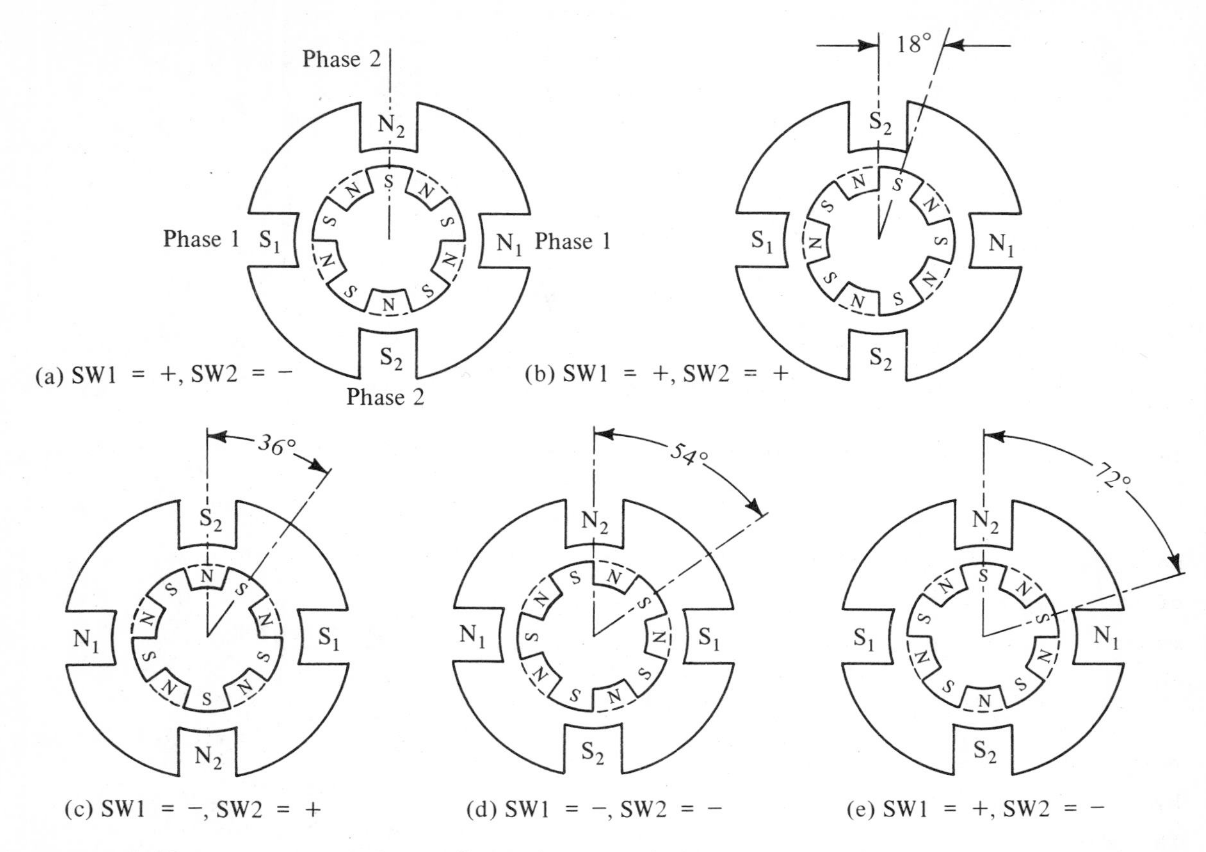

Figure 1-30. Rotor positions of synchronous inductor motor due to sequential switchings of stator excitations.

step, we proceed to switch SW1 to - and SW2 to + as in step 2 of Figure 1-29. The stator poles are now magnetized as shown in Figure 1-30c, with the rotor position now displaced another 18 degrees, clockwise, from that of Figure 1-30b. Continuing with the process, Figures 1-30d and 1-30e show the complete sequence of switchings which causes the motor to step one rotor tooth pitch, or 72 degrees in the present case. We can easily show that reversing the switching sequence would reverse the direction of rotation. The switching control of the permanent-magnet step motor as illustrated in Figure 1-29 requires double-ended power supply that has both polarities. A step motor with bifilar windings would simplify the power requirement to a single-ended supply. Bifilar windings represent windings which are wound in the opposite directions on the same pole. Therefore, instead of reversing the direction of the current in a winding, current in the bifilar windings are switched on and off. Figure 1-31 illustrates the connections of a permanent-magnet step motor with bifilar windings. However, because only half of the windings are used at a given time, the motor of the same size with bifilar windings will not produce as much torque as one that has a single winding.

Step	CW Rotation		CCW Rotation	
	SW1	SW2	SW1	SW2
1	1A	2A	1A	2A
2	1B	2A	1A	2B
3	1B	2B	1B	2B
4	1A	2B	1B	2A

Figure 1-31. Connection of stator windings of a synchronous inductor motor with bifilar windings, and the combinations of switching sequences.

1-8 ELECTROHYDRAULIC STEP MOTOR

The greatest single disadvantage of the current generation of electric step motors is the limited mechanical power output of these motors. Because the control of step motors essentially involves the transferring of currents in and out of the motor during each step operation of the motor, for large motors which require high currents, the control can be quite cumbersome and inefficient.

The electrohydraulic step motor is an answer to large step motors. It is essentially a torque amplifier which utilizes a regular electric step motor and a hydraulic torque amplifier. In other words, the torque developed by the step motor is amplified by the hydraulic motor. This made possible the use of step motors in heavy applications such as machine tools, material handling systems, welder, etc.

Figure 1-32 illustrates a block diagram of the elements of an electrohydraulic step motor. The reduction gear train reduces the angular displacement of the electric step motor due to a single pulse input. The speed at the output side of the gear train is also reduced. The pilot valve is a four-way valve, consisting of a rotary-linear spool and sleeve, which controls the flow of oil proportional to the relative deviation of the spool and sleeve.

The hydraulic motor is of the fixed-displacement axial type. The speed of the hydraulic motor is proportional to the rate of flow of oil from the pilot valve. The rotation of the oil motor shaft is translated through a nut fixed to the shaft

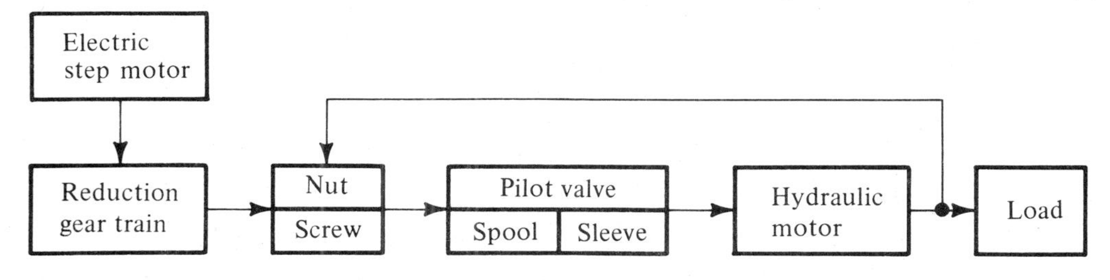

Figure 1-32. A block diagram illustrating the elements of an electrohydraulic step motor.

to a screw fixed to the valve spool, thus making a feedback in the direction to reduce the relative deviation of the spool and sleeve.

The electric step motor used in an electrohydraulic motor is similar to the variable-reluctance type described earlier. The motor usually has three or more phases. The control of the electric step motor also follows the principle of the control of VR motors.

REFERENCES

1. Porter, J., "Stepping Motors Move In.," Prod. Engrg., Vol. 34, February, 1963, pp. 74-78.

2. Kieburtz, B. R., "The Step Motor--The Next Advance in Control Systems," IEEE Trans. on Automatic Control, January, 1964.

3. Bailey, S. J., "Incremental Servos, Part III--How They've Been Used," Control Engineering, January, 1961, pp. 85-88.

4. Bailey, S. J., "Incremental Servos, Part II--Operation and Analysis," Control Engineering, December, 1960, pp. 79-102.

5. Musser, C. W., "Breakthrough in Mechanical Drive Design: The Harmonic Drive," Machine Design, Vol. 32, No. 8, April, 1960, pp. 160-173.

6. Budzilovich, P. N., "Use Electrohydraulic Stepping Motors for All-Digital Drives," Control Engineering, January, 1970, pp. 82-88.

7. Hong, K. Y., "A Dynamic and Static Analysis of a Solenoid Actuated Stepping Motor," IEEE Trans. on Industrial Electronics and Control Instrumentation, August, 1971, pp. 106-109.

8. Snowdon, A. E., E. W. Madsen, "Characteristics of a Synchronous Inductor Motor," AIEE Transactions, Application and Industry, March, 1962.

2 MATHEMATICAL MODELING OF STEP MOTORS

G. Singh
Department of Electrical Engineering
University of Illinois at Urbana-Champaign
Urbana, Illinois

2-1 INTRODUCTION

In this paper mathematical models which describe the dynamic characteristics of step motors are developed. The objective is to obtain a system of ordinary differential equations which represents the behavior of a step motor under all operating conditions. Two different types of step motors are considered; the multiple-stack variable-reluctance type and the permanent-magnet type.

The basic theory underlying the operation of step motors is essentially the same as that of any rotating electromagnetic energy convertor. Four fundamental laws are involved in the operation of such devices; the law of conservation of energy, the laws of magnetic fields, the laws of electric circuits, and the laws of Newtonian mechanics.

As is the case in any physical device, nonlinearities and stray losses complicate the modeling process considerably. In the case of step motors and other rotary electromagnetic energy convertors, the presence of iron parts in the magnetic structure produces nonlinearities due to saturation of the flux density and core losses due to eddy currents and magnetic hysteresis. In addition, there are resistance losses in the windings and friction and windage losses due to the rotation of mechanical parts. Although the winding losses and mechanical losses are generally easy to represent, magnetic nonlinearities and core losses are considerably more difficult to model accurately. Often, because of the simplicity of the resulting relations, magnetic saturation and core losses in the iron are neglected during the analyses of practical devices. The results of such approximate analyses can, if necessary, be corrected for the effects of these neglected factors by semiempirical methods. Final justification for the use of this or any other approach must of course be the pragmatic one given by experimental verification of the predicted results.

In the development of the mathematical model for step motors, first the general equations which describe the behavior of synchronous reluctance type machines are derived. A general N-phase machine is considered, without rotor windings, but with possibly a permanent magnet on the rotor or field coils on the stator to magnetize all rotor teeth to the same polarity as in permanent magnet step motors. Hysteresis and eddy currents are neglected to simplify the analysis. This assumption is realistic if the machine has laminated cores in which case the iron losses are indeed very small. The general derivation includes saturation effects in iron parts and, at each stage of the development, the simplification obtained by assuming a linear magnetic circuit is also derived.

In the subsequent sections the system of general equations is applied to the multiple-stack variable-reluctance and the permanent-magnet type step motors. The modeling of the single-stack variable reluctance step motor is treated in Chapter 3. These three types of step motors are the most important and most widely used step motors in practice. The models obtained in all three cases are nonlinear for saturated magnetic circuits as well as for unsaturated magnetic circuits. Except for analysis of the motors under a few operating conditions, at the moment there is little that can be done with these models on a purely analytical basis. They can, however, be readily implemented on a digital computer for all operating conditions and at present their maximum utility is in the area of performance prediction of step motors by digital simulations.

Frequently, some of the parameters of a motor are obtained with greater ease and more accuracy from experimental test data rather than from design data. This test data, which is generally in the form of inductance and static torque measurements, should therefore be used to derive the actual parameters and improve the accuracy of the model. This approach is discussed in the section dealing with multiple-stack variable-reluctance step motors.

Although the use of experimental data, in order to determine the parameters of a model, leads to a more accurate representation, it limits the use of such models to step motors which have already been constructed and also tested to some extent. Ideally, it is desirable to have an accurate representation of a device using only the design data and predicting its response at the design stage. This would then result in a significant improvement in the design of step motors. However, with the present state of the art it is not possible to accurately model step motors using design data alone; in fact, in some cases, for example where solid rotors are used and core losses are excessive, it is difficult to accurately predict the response even by using all the available experimental data. In these cases the modeling process yields approximate results and the experience of the engineer plays a major role in deciphering the information.

2-2 GENERAL DYNAMIC EQUATIONS OF STEP MOTORS

Consider a step motor with N-phase windings on the stator and no windings on the rotor. The rotor may, however, be magnetized by a permanent magnet located in it or by one or more field coils on the stator carrying constant current such that all the teeth of the rotor are magnetized to the same polarity. Hysteresis and eddy currents are assumed negligible. The mks system of units is used consistently unless otherwise mentioned. Figure 2-1 shows a schematic diagram of the system under consideration.

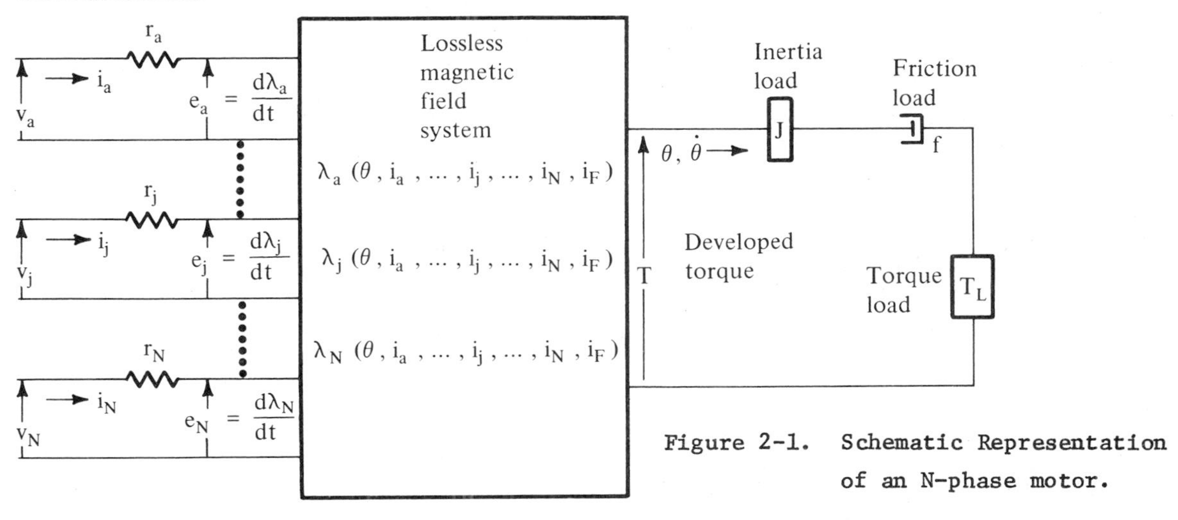

Figure 2-1. Schematic Representation of an N-phase motor.

Voltage equations. Using Kirchhoff's voltage law and Faraday's law of electromagnetic induction, the voltage equations of each winding are

$$v_a = i_a r_a + \frac{d}{dt}(\lambda_a)$$

$$v_b = i_b r_b + \frac{d}{dt}(\lambda_b)$$

$$\cdots\cdots\cdots$$

$$v_j = i_j r_j + \frac{d}{dt}(\lambda_j)$$

$$\cdots\cdots\cdots$$

$$v_N = i_N r_N + \frac{d}{dt}(\lambda_N) \qquad (2\text{-}1)$$

where v_a is the voltage applied to phase a,

i_a is the current in phase a,

r_a is the resistance of phase a,

λ_a denotes the flux linkages of phase a,

$i_a r_a$ is the voltage drop in phase a due to r_a,

and $e_a = \frac{d\lambda_a}{dt}$ is the voltage induced in phase a.

Similar statements can be made for the voltages, currents, etc., in phases b, c, . . . , N.

Flux linkage relations. The flux linkages, $\lambda_a, \lambda_b, \ldots, \lambda_N$, are functions of all the phase currents, the rotor angle, and the magnetomotive force of the permanent magnet or currents in the field coils. The magnetomotive force developed by the permanent magnet of field coils is represented by

$$\mathcal{F}_F = N_F i_F \tag{2-2}$$

where $\mathcal{F}_F$, N_F, and i_F are constants.

In the case of the field coils on the stator,

$\mathcal{F}_F$ is the developed magnetomotive force,

i_F is the current in the coils,

N_F is the effective number of turns,

and in the case of a permanent magnet,

$\mathcal{F}_F$ is the developed magnetomotive force

i_F and N_F represent equivalent currents and effective turns, respectively, such that Eq. (2-2) is satisfied.

Thus, the permanent magnet is represented by an equivalent current source.

The flux linkages of each winding can be written as

$$\lambda_a = \lambda_a(\theta;\ i_a, i_b, \ldots, i_j, \ldots, i_N;\ i_F)$$

$$\lambda_b = \lambda_b(\theta;\ i_a, i_b, \ldots, i_j, \ldots, i_N;\ i_F)$$

$$\cdots\cdots\cdots\cdots\cdots\cdots\cdots$$

$$\lambda_j = \lambda_j(\theta;\ i_a, i_b, \ldots, i_j, \ldots, i_N;\ i_F)$$

$$\cdots\cdots\cdots\cdots\cdots\cdots\cdots$$

$$\lambda_N = \lambda_N(\theta;\ i_a, i_b, \ldots, i_j, \ldots, i_N;\ i_F)$$

$$\lambda_F = \lambda_F(\theta;\ i_a, i_b, \ldots, i_j, \ldots, i_N;\ i_F) \tag{2-3}$$

where λ_F denotes the flux linkages due to the field coils with current i_F.

When there is no saturation, the flux linkages may be written as

$$\lambda_a = L_{aa}\, i_a + L_{ab}\, i_b + \cdots + L_{aj}\, i_j + \cdots + L_{aN}\, i_N + L_{aF}\, i_F$$

$$\lambda_b = L_{ba}\, i_a + L_{bb}\, i_b + \cdots + L_{bj}\, i_j + \cdots + L_{bN}\, i_N + L_{bF}\, i_F$$

$$\cdots\cdots\cdots\cdots\cdots\cdots\cdots\cdots\cdots\cdots$$

$$\lambda_j = L_{ja}\, i_a + L_{jb}\, i_b + \cdots + L_{jj}\, i_j + \cdots + L_{jN}\, i_N + L_{jF}\, i_F$$

$$\cdots\cdots\cdots\cdots\cdots\cdots\cdots\cdots\cdots\cdots$$

$$\lambda_N = L_{Na}\, i_a + L_{Nb}\, i_b + \cdots + L_{Nj}\, i_j + \cdots + L_{NN}\, i_N + L_{NF}\, i_F$$

$$\lambda_F = L_{Fa}\, i_a + L_{Fb}\, i_b + \cdots + L_{Fj}\, i_j + \cdots + L_{FN}\, i_N + L_{FF}\, i_F \qquad (2\text{-}4)$$

where L_{kk} is the self-inductance of phase k; k = a, b, . . . , N. $L_{\ell k} = L_{k\ell}$ is the mutual inductance between phases ℓ and k; $\ell \neq k$; ℓ = a, b, . . . , N; k = a, b, . . . , N.

In the case of a permanent magnet the quantity $L_{kF}\, i_F$, k = a, b, . . . , N, represents the flux linkages of phase k due to the magnetomotive force of the permanent magnet. A similar statement is made for the quantities $L_{Fk}\, i_k$; k = a, b, . . . , N. The inductances in Eq. (2-4) are independent of currents and depend only on the rotor position, θ. Substituting the flux linkage relations of Eq. (2-3) into the voltage relations of Eq. (2-1) yields

$$v_a = i_a r_a + \sum_{k=a}^{N} \frac{\partial \lambda_a}{\partial i_k}\frac{di_k}{dt} + \frac{\partial \lambda_a}{\partial \theta}\frac{d\theta}{dt}$$

$$v_b = i_b r_b + \sum_{k=a}^{N} \frac{\partial \lambda_b}{\partial i_k}\frac{di_k}{dt} + \frac{\partial \lambda_b}{\partial \theta}\frac{d\theta}{dt}$$

$$\cdots\cdots\cdots\cdots\cdots$$

$$v_j = i_j r_j + \sum_{k=a}^{N} \frac{\partial \lambda_j}{\partial i_k}\frac{di_k}{dt} + \frac{\partial \lambda_j}{\partial \theta}\frac{d\theta}{dt}$$

$$\cdots\cdots\cdots\cdots\cdots$$

$$v_N = i_N r_N + \sum_{k=a}^{N} \frac{\partial \lambda_N}{\partial i_k}\frac{di_k}{dt} + \frac{\partial \lambda_N}{\partial \theta}\frac{d\theta}{dt} \;. \qquad (2\text{-}5)$$

In the above system of equations the second term on the right-hand side of each equation, $\sum_{k=a}^{N} \frac{\partial \lambda_j}{\partial i_k}\frac{di_k}{dt}$; j = a, b, . . . , N is the transformer voltage induced in each winding and the last term on the right-hand side of each equation, $\frac{\partial \lambda}{\partial \theta}\frac{d\theta}{dt}$; j = a, b, . . . , N represents the speed voltage in each winding.

When the magnetic circuit is linear, the relations in Eq. (2-4) are used instead of Eq. (2-3) to give

$$v_a = i_a r_a + \sum_{k=a}^{N} L_{ak} \frac{di_k}{dt} + \frac{d\theta}{dt} \left[\sum_{k=a}^{N} i_k \frac{dL_{ak}}{d\theta} + i_F \frac{dL_{aF}}{d\theta} \right]$$

$$v_b = i_b r_b + \sum_{k=a}^{N} L_{bk} \frac{di_k}{dt} + \frac{d\theta}{dt} \left[\sum_{k=a}^{N} i_k \frac{dL_{bk}}{d\theta} + i_F \frac{dL_{bF}}{d\theta} \right]$$

. . . .

$$v_j = i_j r_j + \sum_{k=a}^{N} L_{jk} \frac{di_k}{dt} + \frac{d\theta}{dt} \left[\sum_{k=a}^{N} i_k \frac{dL_{jk}}{d\theta} + i_F \frac{dL_{jF}}{d\theta} \right]$$

. . . .

$$v_N = i_N r_N + \sum_{k=a}^{N} L_{Nk} \frac{di_k}{dt} + \frac{d\theta}{dt} \left[\sum_{k=a}^{N} i_k \frac{dL_{Nk}}{d\theta} + i_F \frac{dL_{NF}}{d\theta} \right] . \qquad (2\text{-}6)$$

As before, in Eqs. (2-6) above the second term on the right-hand side of each equation, $\sum_{k=a}^{N} L_{jk} \frac{d_{ik}}{dt}$; j = a, b, . . . , N represents transformer voltages and the last two terms on the right-hand of each equation,

$$\frac{d\theta}{dt} \left[\sum_{k=a}^{N} i_k \frac{dL_{jk}}{d\theta} + i_F \frac{dL_{jF}}{d\theta} \right]$$

$$j = a, b, \ldots, N,$$

represent speed voltages.

Equations (2-5) and (2-6) are the circuit equations which represent the dynamics of each phase winding. Either Eq. (2-5) or Eq. (2-6) could be used depending on whether saturation is being represented or neglected, respectively. These equations are differential equations for the phase currents. The flux linkage expressions of Eq. (2-3) or Eq. (2-4) are assumed known when these equations are used. The phase voltages v_j and the rotor position θ and rotor velocity $\frac{d\theta}{dt}$ are the possible time-dependent quantities needed to form the forcing functions of these circuit equations.

The flux linkage expressions of Eqs. (2-3) and (2-4) have to be always derived individually for each new device. When the three different types of step motors are considered later, these expressions will be explicitly found for each type of step motor. In the linear case this will merely involve the determination of expressions for self and mutual inductances as functions of rotor position. The phase voltages are inputs for the systems of Eq. (2-5) and (2-6), and the rotor variables θ and $\dot{\theta}$ are coupling variables to be obtained from the equation of motion of the rotor.

2-3 ELECTROMAGNETIC TORQUE EXPRESSIONS

At this stage, it is necessary to determine the expressions for electromagnetic torque developed by the rotor of the N-phase step motor. Torque expressions are best obtained through energy considerations and although it is possible to directly use several such expressions it will be worthwhile to digress from the development of the general equations for step motors and consider the details of torque development in electromagnetic devices. For simplicity, a singly excited system is first considered; the results for N-phase step motors are obtained as straightforward extensions of the results for singly excited magnetic circuits.

To begin the analysis it is necessary to start with the fact that, in an electromagnetic energy conversion device such as the step motor, four forms of energy are involved:

(1) electrical energy input,

(2) energy stored in the magnetic field coupling the stator and the rotor,

(3) energy converted to heat,

(4) mechanical energy output.

The law of conservation of energy states that item one listed above is equal to the sum of the other three. The energy conversion relation in the motor can be represented by the following equation and Figure 2-2.

$$dW_e = dW_f + dW_m \tag{2-7}$$

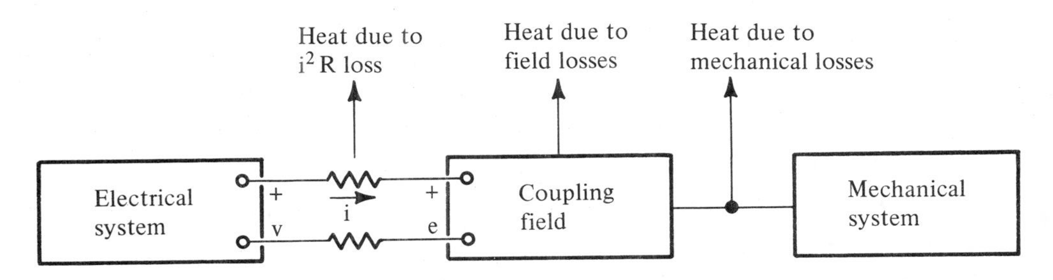

Figure 2-2. Energy relation in an electromagnetic energy conversion device.

where

dW_e = eidt = net differential electrical energy input to coupling device after i^2R losses;

dW_f = total differential energy stored in coupling field plus field losses;

dW_m = mechanical differential energy including mechanical losses.

The conversion of energy to heat is always present, but it is not essential to the process. The two phenomena which are basic to the process are, first, the coupling field must produce a reaction in the electrical circuit for the device to absorb energy from the circuit, i.e., the induced voltage e; and second, energy in the coupling field can be converted to mechanical energy in the device only if relative motion of its parts is possible and if such motion influences the energy in the coupling fields. Thus, the coupling field acts as a reservoir of energy, releasing energy to the output when mechanical motion occurs and being replenished through reaction of the field on the electrical input system.

The induced voltage e is given by

$$e = \frac{d\lambda}{dt} \tag{2-8}$$

where λ is the flux linkage with the circuit, and the average flux ϕ is given by

$$\phi = \frac{\lambda}{N} \tag{2-9}$$

and N is the effective number of turns in the circuit. ϕ is then the flux linking each of these N turns.

Using Eqs. (2-8) and (2-9), the differential electrical energy supplied by the source after i^2R losses is written

$$dW_e = ei\,dt = Ni\,d\phi = id\lambda = F\,d\phi \tag{2-10}$$

where F denotes the magnetomotive force (mmf) defined as $F = Ni$.

When the rotor is restrained, $dW_m = 0$, and the differential energy absorbed in the field is

$$dW_f = dW_e = Fd\phi = id\lambda\ . \tag{2-11}$$

Starting from zero initial flux, the total energy absorbed in the field is

$$W_f = \int_o^{\phi} F(\phi)\,d\phi = \int_o^{\lambda} i(\lambda)\,d\lambda\ . \tag{2-12}$$

In these equations, the mmf is a function of the flux, the relation between them depending on the geometry of the system and the magnetic properties of the core material. In a ferromagnetic system this relationship is non-linear. If eddy currents and hysteresis were not neglected, the relationship between flux and mmf would not be single-valued. The flux-mmf relationship and the quantities in Eq. (2-12) are portrayed in Figure 2-3. The cross-hatched area is the energy absorbed by the coupling field in building up a flux ϕ_1. Since hysteresis and eddy currents are neglected, all this energy is stored in the magnetic field.

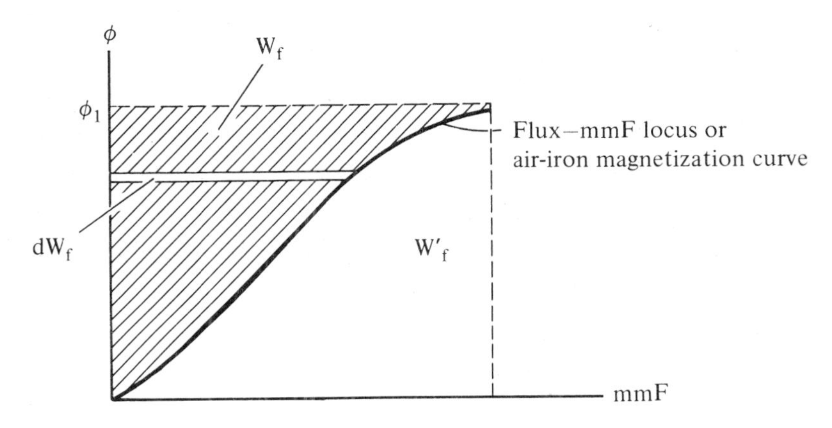

Figure 2-3. Total energy absorbed in the magnetic field.

In Eq. (2-12) the field energy is explicitly expressed in terms of flux linkage or flux as the variable of integration. It is convenient to also define the quantity

$$W_f' = \int_o^i \lambda(i)\, di = \int_o^F \phi(F)\, dF \tag{2-13}$$

which is called the co-energy and is expressed explicitly in terms of current or mmf as the variable of integration. The co-energy is shown in Figure 2-3 as the area under the magnetization curve.

When there is no saturation, the flux and mmf are linearly related as in air, and the cross-hatched area in Figure 2-3 becomes a triangle. Then, the energy and co-energy are equal, and Eqs. (2-12) and (2-13) become

$$W_f = W_f' = 1/2\ \lambda i = 1/2\ \phi F\ . \tag{2-14}$$

The energy and co-energy in this case may also be expressed as

$$W_f = W_f' = 1/2\ R\ \phi^2 = 1/2\ PF^2 = 1/2\ L\ i^2 \tag{2-15}$$

where

$\mathcal{R}$ is the reluctance of the magnetic circuit and is defined as

$$\mathcal{R} = \frac{\mathcal{F}}{\phi} ; \tag{2-16}$$

$\mathcal{P}$ is the permeance of the magnetic circuit and is defined as

$$\mathcal{P} = \frac{\phi}{\mathcal{F}} = \frac{1}{\mathcal{R}} ; \tag{2-17}$$

and

L is the inductance of the coil and is defined as

$$L = \frac{\lambda}{i} = \frac{N\phi}{i} = N^2\mathcal{P} \tag{2-18}$$

all three quantities being constants for a linear lossless magnetic circuit.

The torque developed by the rotor can be calculated by using the principle of virtual work. This involves investigation of the energy balance when the rotor rotates by an infinitesimal amount in the direction of the torque acting on it.

When the rotor rotates by an infinitesimal amount, $d\theta$, in the direction of torque T, mechanical work $Td\theta$ is performed. If the flux ϕ is kept constant during this change, then from Eq. (2-10) the differential electrical energy input $dW_e = 0$. The work is thus performed by drawing energy from the coupling field.

The energy balance Eq. (2-7) becomes

$$dW_e = 0 = dW_f\ (\phi, \theta) + Td\theta \tag{2-19}$$

with constant ϕ where the functional notation $W_f(\phi, \theta)$ is used here to emphasize the fact that field energy must be explicitly expressed in terms of flux ϕ (or flux linkage λ) and θ.

In the limit as $d\theta$ goes to zero, Eq. (2-19) becomes

$$T = -\frac{\partial W_f}{\partial \theta}(\phi, \theta) = -\frac{\partial W_f}{\partial \theta}(\lambda, \theta) \tag{2-20}$$

with ϕ or λ constant. Thus, torque is the rate, with respect to motion, at which energy can be extracted from the magnetic field at constant flux. The developed torque can also be expressed in terms of the co-energy W_f'. In this case it can be shown that

$$T = \frac{\partial W_f'}{\partial \theta}\ (\mathcal{F}, \theta) = \frac{\partial W_f'}{\partial \theta}\ (i, \theta) \tag{2-21}$$

with $\mathcal{F}$ or i constant. Here the co-energy must be expressed explicitly in terms of the mmf, $\mathcal{F}$ (or current i), and θ.

When the flux and mmf are linearly related Eq. (2-21) may be written interchangeably with stored energy if the proper variables are used; thus,

$$T = \frac{\partial W_f}{\partial \theta} (\mathcal{F}, \theta) = \frac{\partial W_f}{\partial \theta} (i, \theta) \tag{2-22}$$

with $\mathcal{F}$ or i constant.

In the linear case, useful expressions for torque may be obtained by differentiating Eq. (2-15) according to Eq. (2-20) or Eq. (2-22); thus

$$T = - 1/2\ \phi^2 \frac{d\mathcal{R}}{d\theta} = 1/2\ \mathcal{F}^2 \frac{d\mathcal{P}}{d\theta} = 1/2\ i^2 \frac{dL}{d\theta} \tag{2-23}$$

where the reluctance $\mathcal{R}$, the permeance $\mathcal{P}$, and the inductance L are functions of θ.

The torque is thus seen to act in a direction to tend to decrease the stored energy at constant flux, to increase the stored energy and co-energy at constant mmf, to decrease the reluctance and to increase the permeance and inductance.

The torque and energy expressions developed for a single winding can now be generalized for N phases with possible rotor magnetization. The field energy and co-energy in the N-phase step motor can be written in the functional form

$$W_f = W_f\ (\phi_a, \phi_b, \ldots, \phi_N; \phi_F; \theta)$$

$$= W_f\ (\lambda_a, \lambda_b, \ldots, \lambda_N; \lambda_F; \theta) \tag{2-24}$$

$$W_f' = W_f'\ (\mathcal{F}_a, \mathcal{F}_b, \ldots, \mathcal{F}_N; \mathcal{F}_F; \theta)$$

$$= W_f'\ (i_a, i_b, \ldots, i_N; i_F; \theta) \tag{2-25}$$

where all the variables have been defined earlier. The general expression for field energy may be derived analogously to Eq. (2-12). Assuming that the rotor is restrained, the electrical input and field energy are equal; thus,

$$W_f = \int_o^{\phi_a} \mathcal{F}_a\, d\phi_a + \int_o^{\phi_b} \mathcal{F}_b\, d\phi_b + \ldots + \int_o^{\phi_N} \mathcal{F}_N\, d\phi_N + \int_o^{\phi_F} \mathcal{F}_F\, d\phi_F$$

$$= \int_o^{\lambda_a} i_a\, d\lambda_a + \int_o^{\lambda_b} i_b\, d\lambda_b + \ldots + \int_o^{\lambda_N} i_N\, d\lambda_N + \int_o^{\lambda_F} i_F\, d\lambda_F \tag{2-26}$$

and similarly for the co-energy

$$W_f' = \int_0^{i_a} \lambda_a \, di_a + \int_0^{i_b} \lambda_b \, di_b + \ldots + \int_0^{i_N} \lambda_N \, di_N + \int_0^{i_F} \lambda_F \, di_F$$

$$= \int_0^{F_a} \phi_a \, dF_a + \int_0^{F_b} \phi_b \, dF_b + \ldots + \int_0^{F_N} \phi_N \, dF_N + \int_0^{F_F} \phi_F \, dF_F \tag{2-27}$$

In the linear case the most useful form of the field energy is in terms of self and mutual inductances of the individual windings. Using the inductances defined in Eq. (2-4) the field energy of the N-phase step motor is

$$W_f = \frac{1}{2} \sum_{k=a}^{N} L_{kk} \, i_k^2 + \frac{1}{2} L_{FF} \, i_F^2 + \sum_{j=a}^{N} \sum_{k=a}^{j-1} L_{jk} \, i_j \, i_k + \sum_{k=a}^{N} L_{kF} \, i_k \, i_F \tag{2-28}$$

where all the inductances are functions of the rotor position θ.

The developed torque is obtained by generalizations of the expressions of Eqs. (2-20) and (2-21); thus,

$$T = -\frac{\partial W_f}{\partial \theta} (\phi_a, \phi_b, \ldots, \phi_N; \phi_F; \theta)$$

$$= -\frac{\partial W_F}{\partial \theta} (\lambda_a, \lambda_b, \ldots, \lambda_N; \lambda_F; \theta) \tag{2-29}$$

$$T = \frac{\partial W_f'}{\partial \theta} (F_a, F_b, \ldots, F_N; F_F \; \theta) \; ;$$

$$= \frac{\partial W_f'}{\partial \theta} (i_a, i_b, \ldots, i_N; i_F; \theta) \; . \tag{2-30}$$

As in the single-winding case, the energy and co-energy in Eqs. (2-29) and (2-30) must be explicitly expressed in terms of the indicated variables.

When the magnetic circuit is linear, the torque expression of Eq. (2-30) can be written interchangeably with stored energy if the proper variables are used,

$$T = \frac{\partial W_f}{\partial \theta} (i_a, i_b, \ldots, i_N, i_F, \theta) \; . \tag{2-31}$$

Equation (2-31), when used with the energy expressed in Eq. (2-28), yields a very convenient expression for torque in the linear case. Differentiating Eq. (2-28) according to Eq. (2-31) gives

$$T = \frac{1}{2} \sum_{k=a}^{N} i_k^2 \frac{dL_{kk}}{d\theta} + \frac{1}{2} i_F^2 \frac{dL_{FF}}{d\theta} + \sum_{j=a}^{N} \sum_{k=a}^{j-1} \frac{dL_{jk}}{d\theta} i_j i_k + \sum_{k=a}^{N} \frac{dL_{kF}}{d\theta} i_k i_F . \quad (2\text{-}32)$$

This completes the development of torque expressions for the N-phase step motor.

2-4 EQUATIONS OF ROTATING MASS

The dynamic equations for the motion of the rotor are now developed. Let T be the developed torque, J the inertia of the rotor, B the coefficient of viscous friction, T_F the coulomb frictional torque, θ the rotor position, and ω the rotor velocity, then the dynamics of the rotor can be expressed by the following differential equation

$$T = J \frac{d^2\theta}{dt^2} + B \frac{d\theta}{dt} + T_F . \quad (2\text{-}33)$$

In state variable form, with θ and ω as the states, Eq. (2-33) becomes

$$\dot{\theta} = \omega \quad (2\text{-}34)$$

$$\dot{\omega} = \frac{T}{J} - \frac{B}{J} \omega - \frac{T_F}{J} . \quad (2\text{-}35)$$

The derivation of the general system of equations for an N-phase step motor with possible rotor magnetization is now complete. In summary, it is to be noted that four distinct relations are needed to adequately model any step motor. These are

(1) the voltage equations of the electrical circuits,

(2) the flux linkage relations including the definition of inductances in the case of a linear magnetic circuit,

(3) the expressions for developed torque including expressions for stored energy and co-energy,

(4) the dynamic equations of the rotor.

In the following sections these four types of relations will be defined for two different types of step motors and the respective mathematical models developed. Any other type of step motor can be similarly modeled by appropriately specializing the general expressions derived in this chapter.

2-5 DYNAMIC MODELS OF MULTIPLE-STACK VARIABLE-RELUCTANCE STEP MOTORS

In this section the modeling of the multiple-stack variable-reluctance step motor is considered in detail. The construction and principle of operation of this type of step motor is described in reference 1. The model is developed for a motor with N stator phases and Z rotor teeth. This corresponds to a step size of $\frac{360}{ZN}$ degrees. A linear model is first derived and this is later modified to include the nonlinear effects of saturation in the iron parts. As an illustration, a typical multiple-stack variable-reluctance step motor, the SM060AB step motor manufactured by the Warner Electric Brake and Clutch Company, is also modeled. The SM060AB is a three-phase VR motor with a resolution of sixty steps per revolution. The tooth pitch of each of the three stacks of the rotor is 18 degrees since there are twenty teeth on each rotor stack. The step angle is 6 degrees.

Considering the case of a linear magnetic circuit first and following the procedure described in Section 2-2, the voltage relations for the N phases can be written as in Eq. (2-1) which are repeated as follows.

$$v_a = i_a r_a + \frac{d\lambda_a}{dt}$$

$$v_b = i_b r_b + \frac{d\lambda_b}{dt}$$

$$\cdots\cdots\cdots$$

$$v_N = i_N r_N + \frac{d\lambda_N}{dt} \quad . \qquad (2\text{-}1)$$

The flux linkages of the phase windings in the linear case are defined in Eq. (2-4). Since there is negligible mutual coupling between phases in a multiple-stack variable-reluctance step motor, all mutual inductances are zero. The flux linkages can, therefore, be written as

$$\lambda_a = L_a \, i_a$$

$$\lambda_b = L_b \, i_b$$

$$\cdots\cdots$$

$$\lambda_N = L_N \, i_N \quad . \qquad (2\text{-}36)$$

In Eq. (2-36) the quantities L_a, L_b . . . , L_N are the self inductances of the phase windings and are functions of the rotor position θ. In order to determine these self inductances, consider the kth phase of the step motor. The permeance of each coil group oscillates with the motion of the rotor, being a maximum when the rotor and stator teeth are fully aligned and a minimum when the rotor and stator teeth are fully misaligned. The frequency of this oscillation is equal to the number of rotor teeth per revolution.

If all harmonics higher than the first are neglected, the permeance of each stator pole can be written as

$$P = P_o + P_1 \cos Z\theta \tag{2-37}$$

where the reference, $\theta = 0$, is chosen where the rotor teeth are fully aligned with the stator teeth. If P_{max} and P_{min} are the maximum and minimum permeances of each pole, respectively, then

$$P_o = \frac{P_{max} + P_{min}}{2}$$

and

$$P_1 = \frac{P_{max} - P_{min}}{2} \tag{2-38}$$

The flux linkages of the phase under consideration are

$$\lambda_k = \frac{N_k^2\ i_k}{p_k} (P_o + P_1 \cos Z\theta) \tag{2-39}$$

where N_k is the number of turns of phase k and p_k is the number of poles on the stator stack of phase k.

Using Eq. (2-36), the self inductance of phase k is

$$L_k = L_{ko} + L_{k1} \cos Z\theta \tag{2-40}$$

with

$$L_{ko} = \frac{N_k^2 P_o}{p_k}$$

and

$$L_{k1} = \frac{N_k^2 P_1}{p_k} \ . \tag{2-41}$$

The self inductances of all the other phases will also have the same form as Eq. (2-40) with some appropriate phase shifts. If all phases of the motor are identical in construction with the same number of turns and the same number of poles, then the quantities L_{ko} and L_{k1}; k = a, b, . . . , N, will be independent of the individual phases. For notational simplicity let

$$L_{ko} = L_o; \; k = a, b, \ldots, N$$

and

$$L_{k1} = L_1; \; k = a, b, \ldots, N \,. \tag{2-42}$$

Choosing phase a as reference and assuming that the sequence a, b, c, . . . represents positive motion of the rotor, the inductances of the individual phases are

$$\begin{aligned}
L_a &= L_o + L_1 \cos(Z\theta) \\
L_b &= L_o + L_1 \cos(Z\theta - 2\pi/N) \\
L_c &= L_o + L_1 \cos(Z\theta - 4\pi/N) \\
L_k &= L_o + L_1 \cos(Z\theta - 2\pi(K-1)/N) \\
L_N &= L_o + L_1 \cos(Z\theta - 2\pi(N-1)/N) \,.
\end{aligned} \tag{2-43}$$

Although, in practice, the permeance of each stator pole and, consequently, the inductance of each phase is not a pure cosine function of θ, the use of the inductance expressions of Eq. (2-43) generally gives satisfactory results. This is due to the fact that the fundamental component of the permeance variation does dominate and the effects of the higher harmonics are secondary. If more accuracy is desired, it is possible to include some of the higher harmonics in the model. This is, however, rarely necessary.

The values of L_o and L_1 can be obtained from the design data of the motor by using Eqs. (2-38) and (2-41). If design data are not available, then L_o and L_1 can be obtained by measuring the inductance of the phase winding as a function of the rotor position. In the linear case, the measurement of inductance has to be performed with zero d.c. bias to get the unsaturated values of L_o and L_1. If L_{max} and L_{min} are the maximum and minimum values of inductance, respectively, for the motion of the rotor over each tooth pitch, then

$$L_o = \frac{L_{max} + L_{min}}{2} \tag{2-44}$$

$$L_1 = \frac{L_{max} - L_{min}}{2} \quad . \tag{2-45}$$

In the linear case the measurement of inductance of any one phase is sufficient. Figure 2-4 shows the variation of inductance of one phase of the motor assuming that the fundamental component is dominant. Figure 2-5 shows that variation of inductance of the SM060AB motor as obtained by inductance measurements. If the Eqs. (2-44) and (2-45) are used, we have

$$L_o = 55.5 \text{ millihenries and } L_1 = 30.9 \text{ millihenries}$$

and the inductances of the three phases of the SM060AB motor are

$$L_a = (55.5 + 30.9 \cos (20\theta)) \times 10^{-3}$$

$$L_b = (55.5 + 30.9 \cos (20\theta - 2\pi/3)) \times 10^{-3}$$

$$L_c = (55.5 + 30.9 \cos (20\theta - 4\pi/3)) \times 10^{-3} \tag{2-46}$$

all values being in henries.

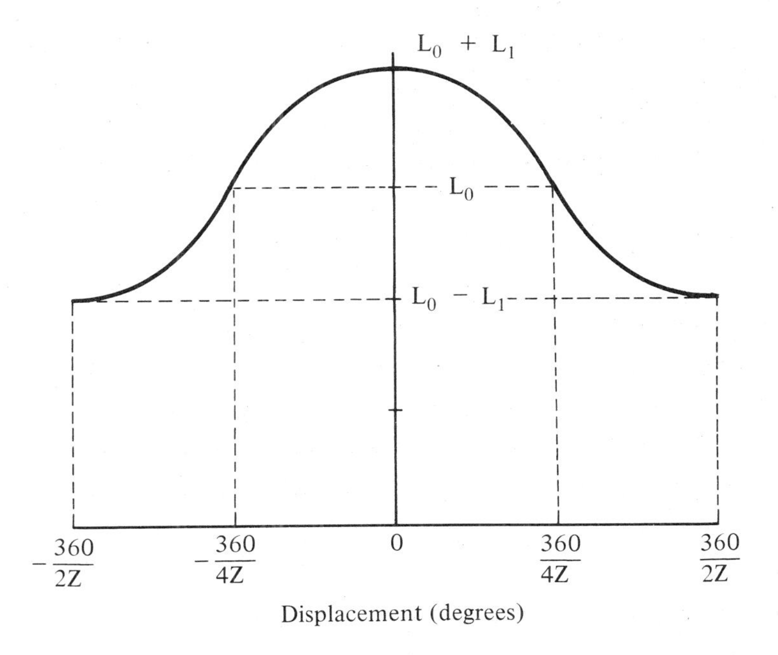

Figure 2-4. The idealized variation of inductance with displacement.

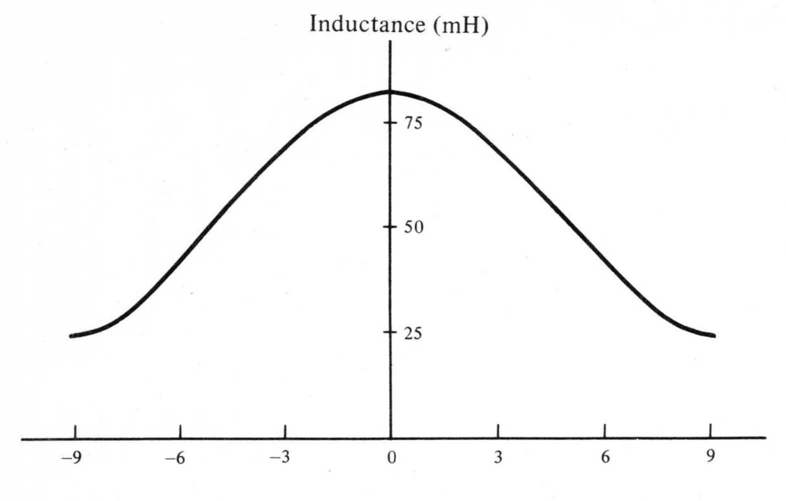

Figure 2-5.
The unsaturated inductance of one phase of the SM060AB step motor.

Substituting the flux linkage expressions of Eq. (2-36) into the voltage expressions of Eq. (2-1) gives, as in Eq. (2-6),

$$v_a = i_a r_a + L_a \frac{di_a}{dt} + i_a \frac{dL_a}{d\theta} \frac{d\theta}{dt}$$

$$v_b = i_b r_b + L_b \frac{di_b}{dt} + i_b \frac{dL_b}{d\theta} \frac{d\theta}{dt}$$

$$\cdots\cdots\cdots\cdots$$

$$v_k = i_k r_k + L_k \frac{di_k}{dt} + i_k \frac{dL_k}{d\theta} \frac{d\theta}{dt}$$

$$\cdots\cdots\cdots\cdots$$

$$v_N = i_N r_N + L_N \frac{di_N}{dt} + i_N \frac{dL_N}{d\theta} \frac{d\theta}{dt} \tag{2-47}$$

If the differentiation of the inductances is performed in Eq. (2-47), according to Eq. (2-43), the complete set of voltage equations becomes

$$v_a = i_a r_a + [L_o + L_1 \cos(Z\theta)] \frac{di_a}{dt} - ZL_1 \sin(Z\theta)\, i_a \frac{d\theta}{dt}$$

$$v_b = i_b r_b + [L_o + L_1 \cos(Z\theta - 2\pi/N)] \frac{di_b}{dt} - ZL_1 \sin(Z\theta - 2\pi/N)\, i_b \frac{d\theta}{dt}$$

$$\cdots\cdots\cdots\cdots$$

$$v_k = i_k r_k + [L_o + L_1 \cos(Z\theta - 2\pi(k-1)/N)] \frac{di_k}{dt} - ZL_1 \sin[Z\theta - 2\pi(k-1)/N]\, i_k \frac{d\theta}{dt}$$

$$\cdots\cdots\cdots\cdots$$

$$v_N = i_N r_N + [L_o + L_1 \cos(Z\theta - 2\pi(N-1)/N)] \frac{di_N}{dt}$$

$$- ZL_1 \sin[Z\theta - 2\pi(N-1)/N]\, i_N \frac{d\theta}{dt} \tag{2-48}$$

Equations (2-48) can be rewritten as differential equations for the currents

$$\frac{di_a}{dt} = \frac{1}{L_o + L_1 \cos (Z\theta)} \left[v_a - i_a r_a + ZL_1 \sin (Z\theta)\, i_a \frac{d\theta}{dt} \right]$$

$$\frac{di_b}{dt} = \frac{1}{L_o + L_1 \cos (Z\theta - 2\pi/N)} \left[v_b - i_b r_b + ZL_1 \sin (Z\theta - 2\pi/N) \cdot i_b \frac{d\theta}{dt} \right]$$

. . . .

$$\frac{di_k}{dt} = \frac{1}{L_o + L_1 \cos (Z\theta - 2\pi(k-1)/N)} \left[v_k - i_k r_k + ZL_1 \sin (Z\theta - 2\pi(k-1)/N)\, i_k \frac{d\theta}{dt} \right]$$

. . . .

$$\frac{di_N}{dt} = \frac{1}{L_o + L_1 \cos (Z\theta - 2\pi(N-1)/N)} \left[v_N - i_N r_N + ZL_1 \sin (Z\theta - 2\pi(N-1)/N)\, i_N \frac{d\theta}{dt} \right] . \qquad (2\text{-}49)$$

For the SM060AB motor the corresponding differential equations are

$$\frac{di_a}{dt} = \frac{10^3}{55.5 + 30.9 \cos (20\theta)} \left[v_a - i_a r_a + 0.618 \sin (20\theta)\, i_a \frac{d\theta}{dt} \right]$$

$$\frac{di_b}{dt} = \frac{10^3}{55.5 + 30.9 \cos (20\theta - 2\pi/3)} \left[v_b - i_b r_b + 0.618 \sin (20\theta - 2\pi/3)\, i_b \frac{d\theta}{dt} \right]$$

$$\frac{di_c}{dt} = \frac{10^3}{55.5 + 30.9 \cos (20\theta - 4\pi/3)} \left[v_c - i_c r_c + 0.618 \sin (20\theta - 4\pi/3)\, i_c \frac{d\theta}{dt} \right] . \qquad (2\text{-}50)$$

The torque developed by the multiple-stack variable-reluctance step motor is easily obtained by using Eq. (2-32) and recognizing that all the mutual inductances are zero. Thus,

$$T = \frac{1}{2} i_a^2 \frac{dL_a}{d\theta} + \frac{1}{2} i_b^2 \frac{dL_b}{d\theta} + \ldots + \frac{1}{2} i_k^2 \frac{dL_k}{d\theta} + \ldots + \frac{1}{2} \cdot i_N^2 \frac{di_N}{d\theta} . \qquad (2\text{-}51)$$

Using Eqs. (2-43) for the inductances in Eq. (2-51) gives

$$T = -\frac{ZL_1}{2}\Big[i_a^2 \sin(Z\theta) + i_b^2 \sin(Z\theta - 2\pi/N) + \ldots$$

$$+ i_k^2 \sin(Z\theta - 2\pi \cdot (k-1)/N) + \ldots + i_N^2 \sin(Z\theta - 2\pi(N-1)/N\Big] \qquad (2\text{-}52)$$

Equation (2-52) can also be written as

$$T = T_a + T_b + \ldots + T_k + \ldots + T_N \qquad (2\text{-}53)$$

where $T_k = -\frac{ZL_1}{2} \sin(Z\theta - 2\pi(k-1)/N)\, i_k^2$ is the torque developed by the kth phase, $k = a, b, c, \ldots, N$.

For the SM060AB motor, the torque is

$$T = -.309\Big[i_a^2 \sin(20\theta) + i_b^2 \sin(20\theta - 2\pi/3) + i_c^2 \sin(20\theta - 4\pi/3)\Big] . \qquad (2\text{-}54)$$

If the mks system of units is used, the torque in Eq. (2-54) is in Newton meters.

The dynamics of the rotor for the multiple-stack variable-reluctance motor are as in Eqs. (2-34) and (2-35),

$$\frac{d\theta}{dt} = \omega$$

$$\frac{d\omega}{dt} = \frac{1}{J}(T - B\omega - T_F) . \qquad (2\text{-}55)$$

Thus, the complete model of the multiple-stack variable-reluctance motor with a linear magnetic circuit consists of

(1) the differential equations for current, Eq. (2-49);

(2) the torque expression, Eq. (2-52);

(3) the rotor dynamic equations, Eq. (2-55).

Figure 2-6 shows the block diagram of the model for the N-phase motor and Figure 2-7 shows the same for the SM060AB motor. The inputs to the model consist of the phase voltages v_a, v_b, . . . , v_N, the initial conditions for the currents, $i_a(0)$, $i_b(0)$, . . . , $i_N(0)$, and the initial conditions for the rotor position, $\theta(0)$, and the velocity $\dot{\theta}(0)$. With the appropriate inputs any operation of the motor can be simulated with this model.

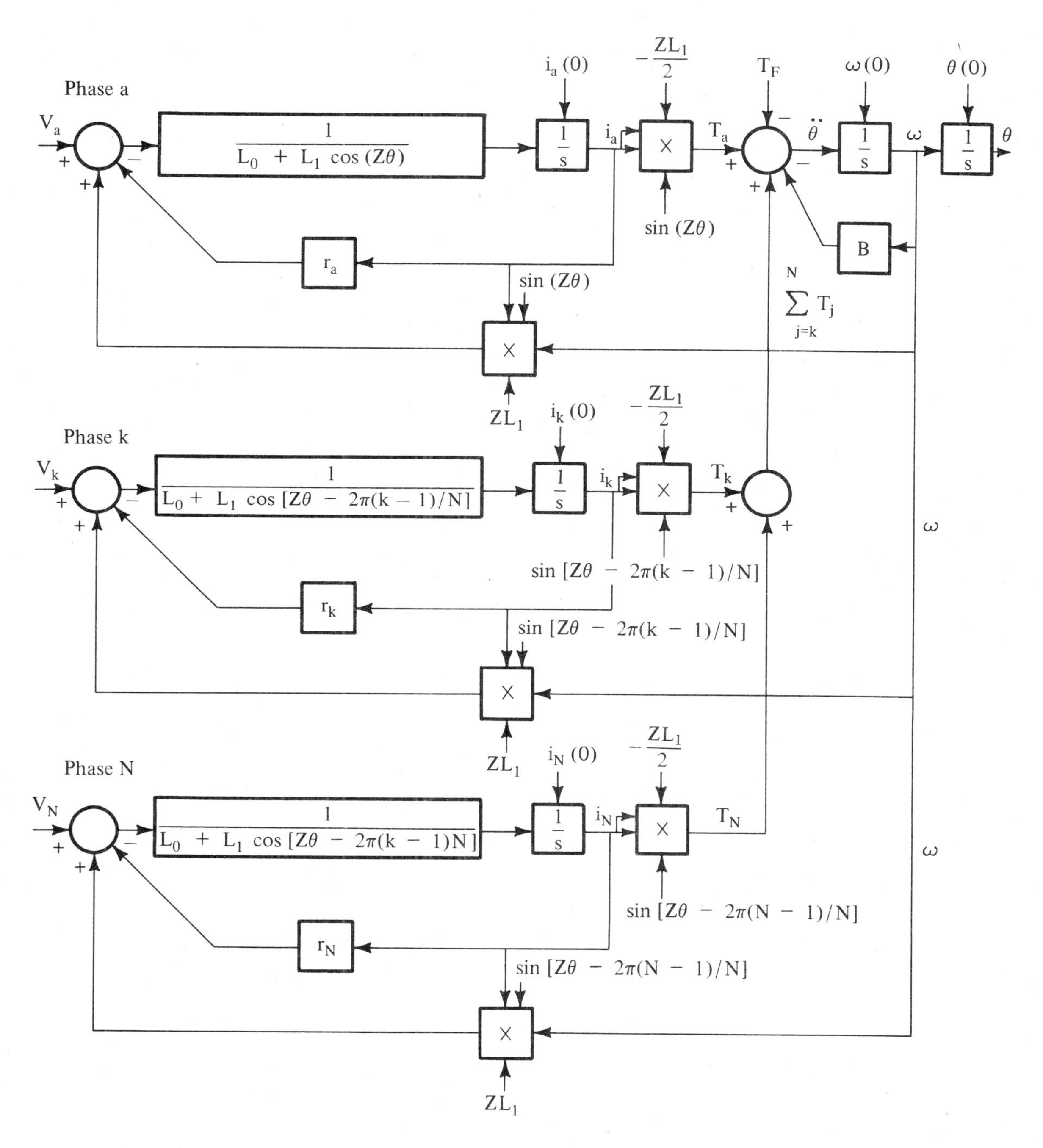

Figure 2-6. Block diagram for an N phase multiple-stack variable-reluctance step motor in the case of a linear magnetic circuit.

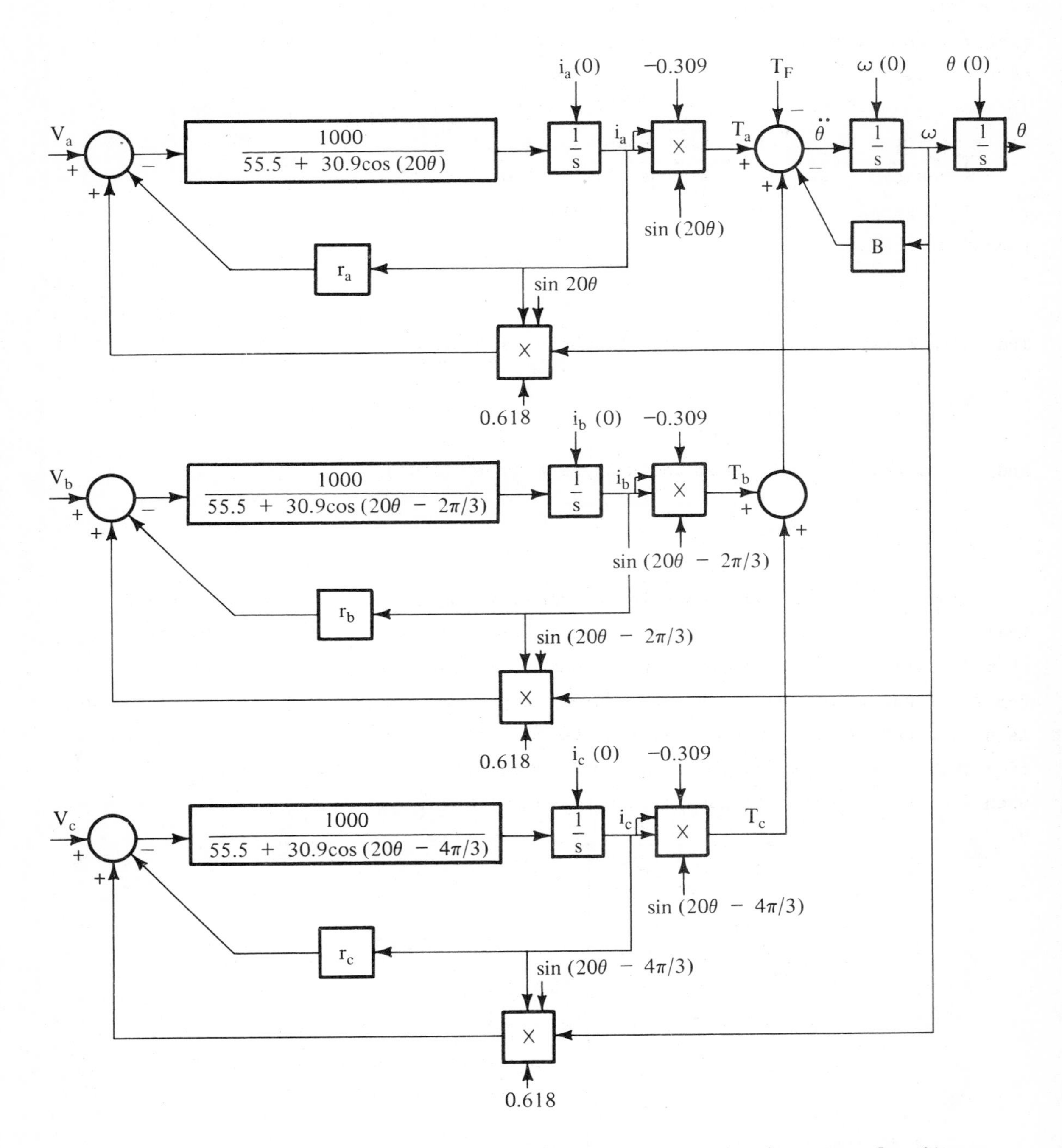

Figure 2-7. Block diagram for the SM060AB step motor in the case of a linear magnetic circuit.

The model derived above can now be modified to account for nonlinear effects of saturation in the magnetic circuit. Basically, two quantities have to be redefined: the inductance of each phase, and the torque developed by each phase of the motor. Because of symmetry, it is sufficient to derive the new expressions for the inductance and torque for just one phase of the motor. These can then be appropriately modified for all the other phases.

Consider then, the kth phase of the motor. When saturation is present, the flux linkage of this phase is a nonlinear function of the phase current and the rotor position; thus,

$$\lambda_k = \lambda_k (i_k, \theta) . \tag{2-56}$$

The incremental inductance of the kth phase is defined as

$$\ell_k(i_k, \theta) = \frac{\partial \lambda_k}{\partial i_k} (i_k, \theta) \tag{2-57}$$

and the average inductance of the kth phase is defined as

$$L_k(i_k, \theta) = \frac{\lambda_k(i_k, \theta)}{i_k} . \tag{2-58}$$

The incremental inductance is the slope of the λ_k versus i_k characteristic (magnetization characteristic) for fixed θ, and the average inductance is the ratio of the ordinate to the abscissa of the same characteristic, also for fixed θ. Figure 2-8 shows the relationships of Eqs. (2-57) and (2-58) in graphical form. When there is no saturation, the incremental and average inductances are equal and independent of current. When saturation is present, these inductances are different, except when $i_k = 0$, and are functions of the phase current i_k as well as the rotor position θ.

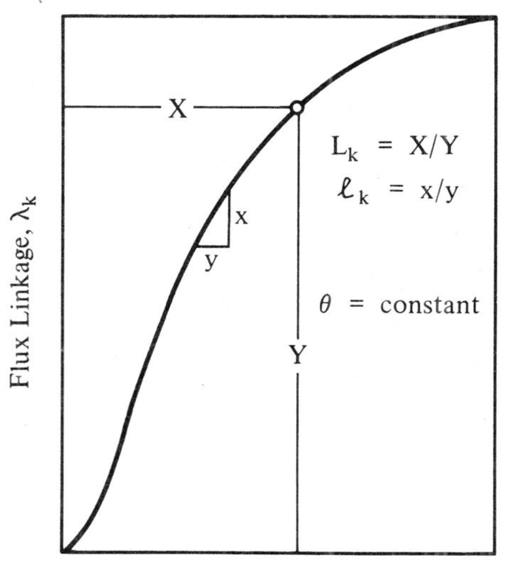

Figure 2-8. Graphical representation of incremental and average inductance of a coil with a saturated magnetic circuit.

It is possible to measure the incremental inductance $\ell_k(i_k, \theta)$ by keeping the rotor fixed and using an incremental inductance bridge. Figures 2-9 and 2-10 show the data obtained for the Warner Electric SM060AB step motor. If the incremental inductance $\ell_k(i_k, \theta)$ which is measured is integrated with respect to i_k at constant θ, the result is the flux linkage $\lambda_k(i_k, \theta)$ for that value of θ. Figures 2-11 and 2-12 show the results of integrating $\ell_k(i_k, \theta)$ for the SM060AB step motor. The average inductance $L_k(i_k, \theta)$ can be calculated by the use of Eq. (2-58) with θ constant. Figures 2-13 and 2-14 show the variation of the average inductance for the SM060AB step motor.

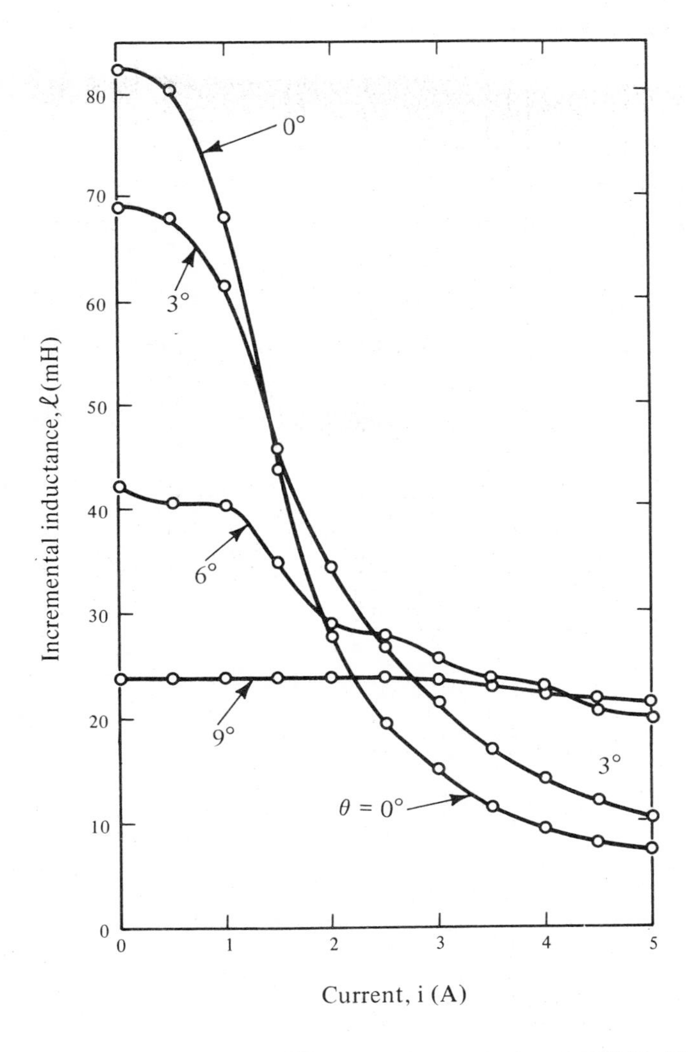

Figure 2-9. Incremental inductance of one phase of the SM060AB VR motor of Warner Electric Brake and Clutch Company.

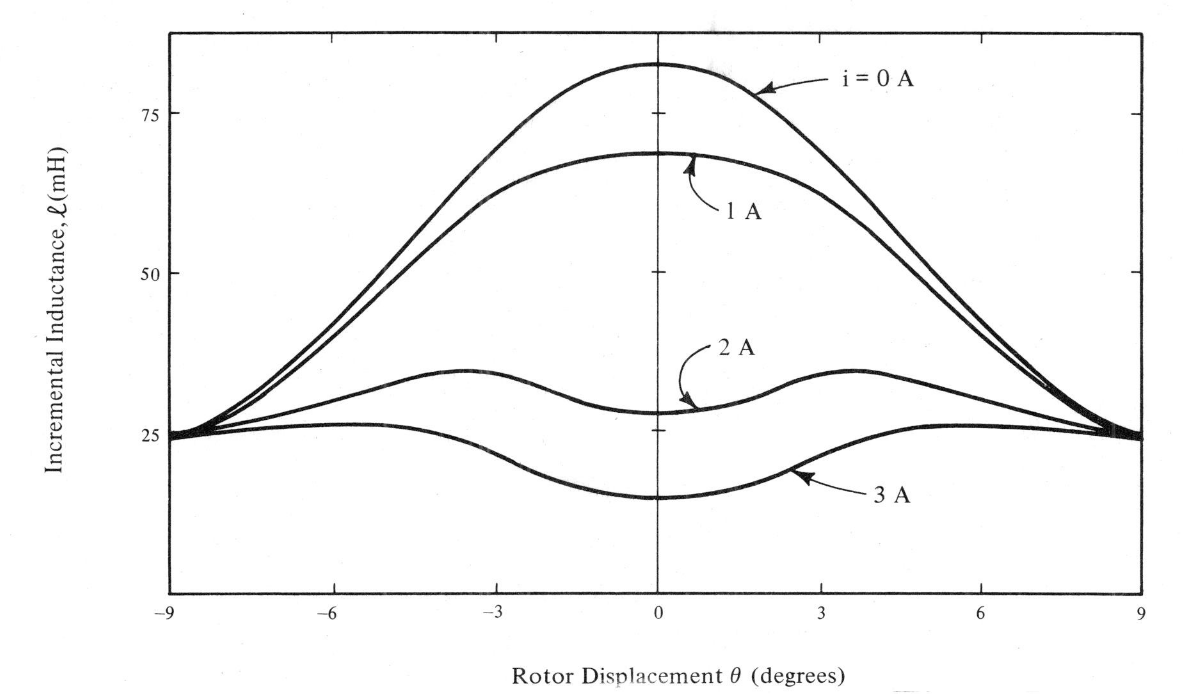

Figure 2-10. Incremental inductance of phase a of the SM060AB step motor of the Warner Electric Brake and Clutch Company.

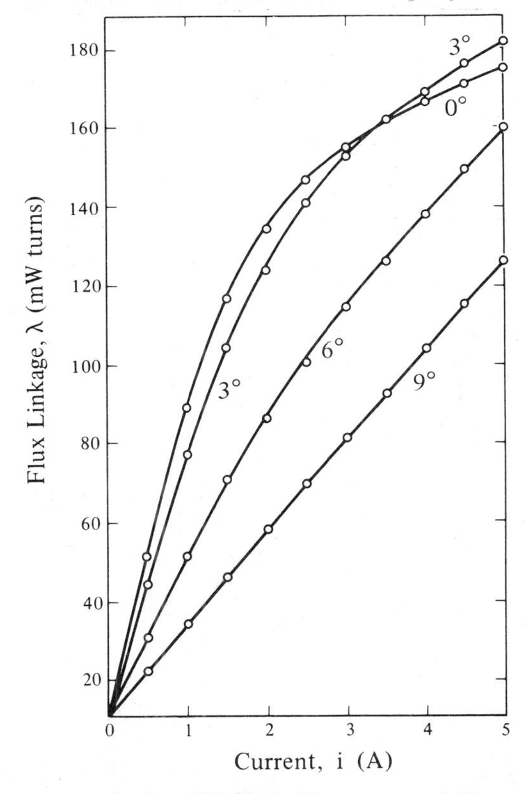

Figure 2-11. Flux curves of the SM060AB VR motor of Warner Electric Brake and Clutch Company.

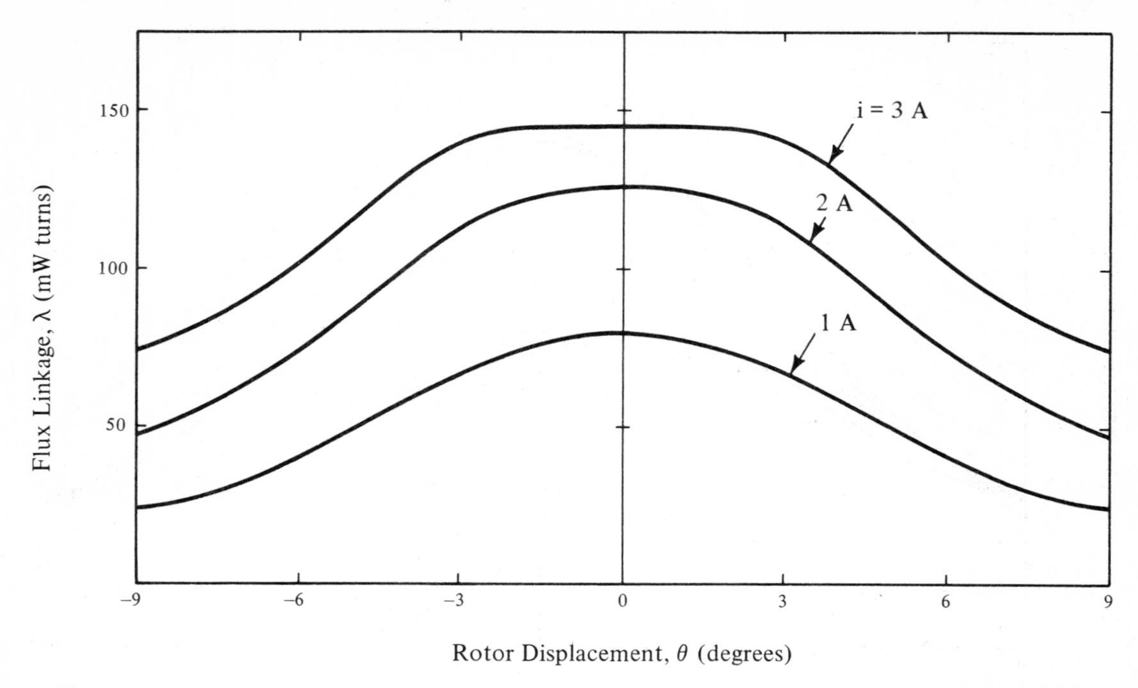

Figure 2-12. Flux linkage of phase a of the SM060AB step motor of Warner Electric Brake and Clutch Company.

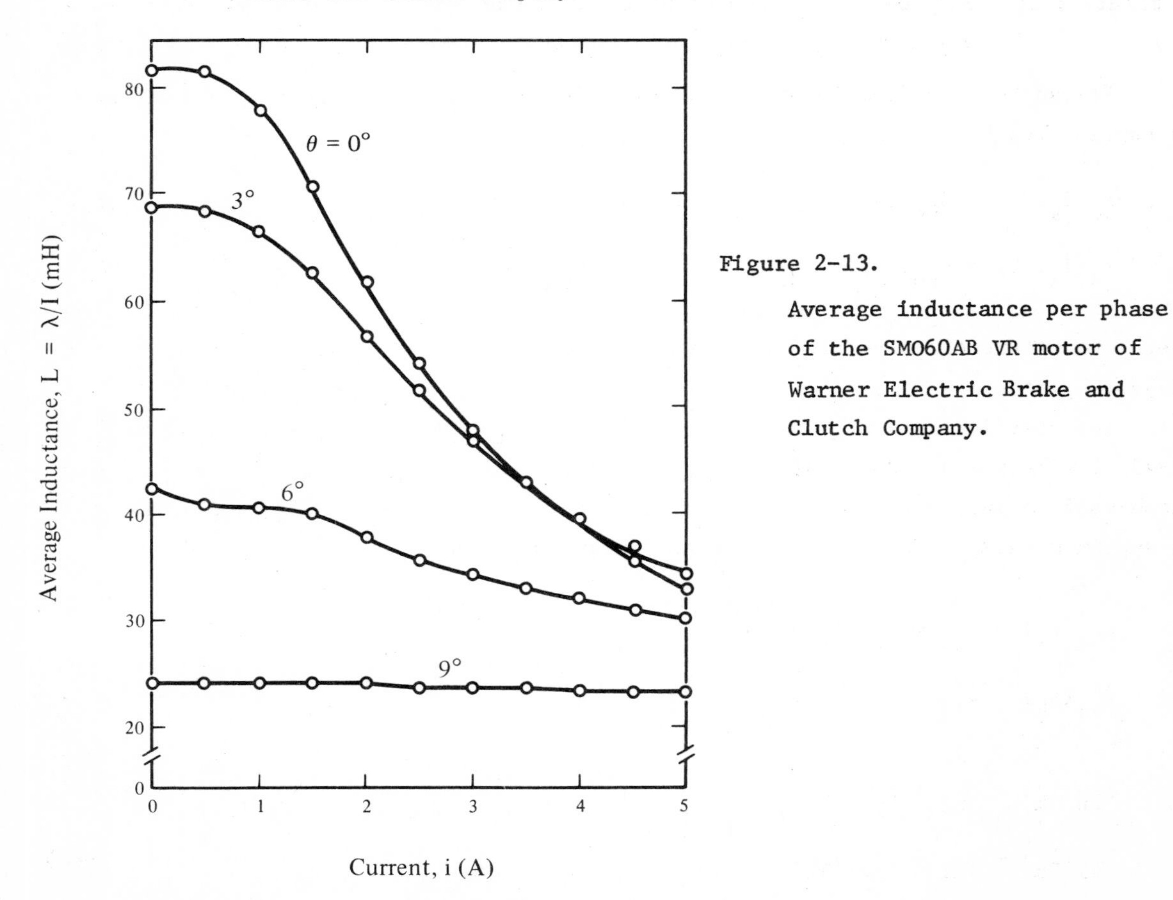

Figure 2-13.

Average inductance per phase of the SM060AB VR motor of Warner Electric Brake and Clutch Company.

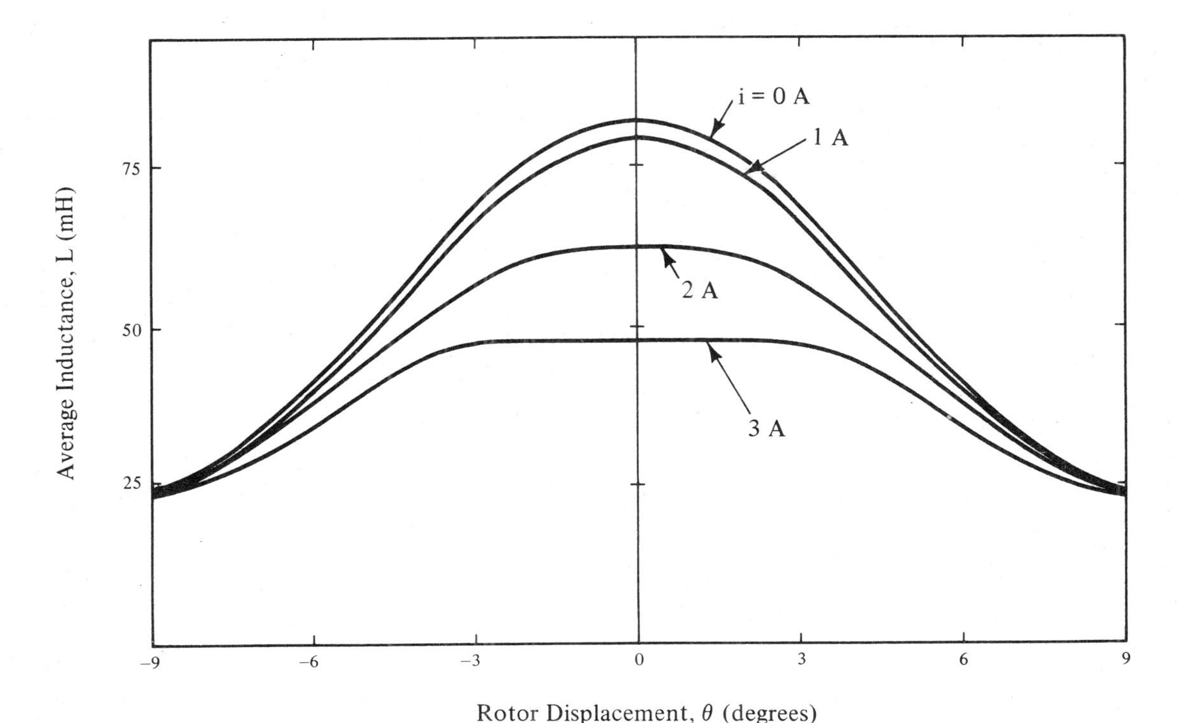

Figure 2-14. Average inductance of phase a of the SM060AB step motor of Warner Electric Brake and Clutch Company.

It can be seen from Figures 2-10 and 2-14 that the inductances ℓ_k and L_k may be represented by

$$\ell_k(i_k, \theta) = \ell_{ko}(i_k) + \ell_{k1}(i_k) \cos [Z\theta - 2\pi(k-1)/N] \tag{2-59}$$

$$L_k(i_k, \theta) = L_{ko}(i_k) + L_{k1}(i_k) \cos [Z\theta - 2\pi(k-1)/N] \tag{2-60}$$

where the reference value of θ is at an equilibrium position of phase a, as in Eq. (2-46).

The coefficients, ℓ_{ko}, ℓ_{k1}, L_{ko}, L_{k1}, are obtained by using Eqs. (2-44) and (2-45) for each individual value of i_k. Figures 2-15 and 2-16 show the variation of these coefficients with respect to the phase current for the SM060AB step motor. If we approximate ℓ_{ko}, ℓ_{k1}, L_{ko} and L_{k1} as linear functions of i_k, we have

$$\begin{aligned} \ell_{ko}(i_k) &= \ell_{o1} + \ell_{o2}\, i_k \\ \ell_{k1}(i_k) &= \ell_{11} + \ell_{12}\, i_k \, , \end{aligned} \tag{2-61}$$

$$\begin{aligned} L_{ko}(i_k) &= L_{o1} + L_{o2}\, i_k \\ L_{k1}(i_k) &= L_{11} + L_{12}\, i_k \, . \end{aligned} \tag{2-62}$$

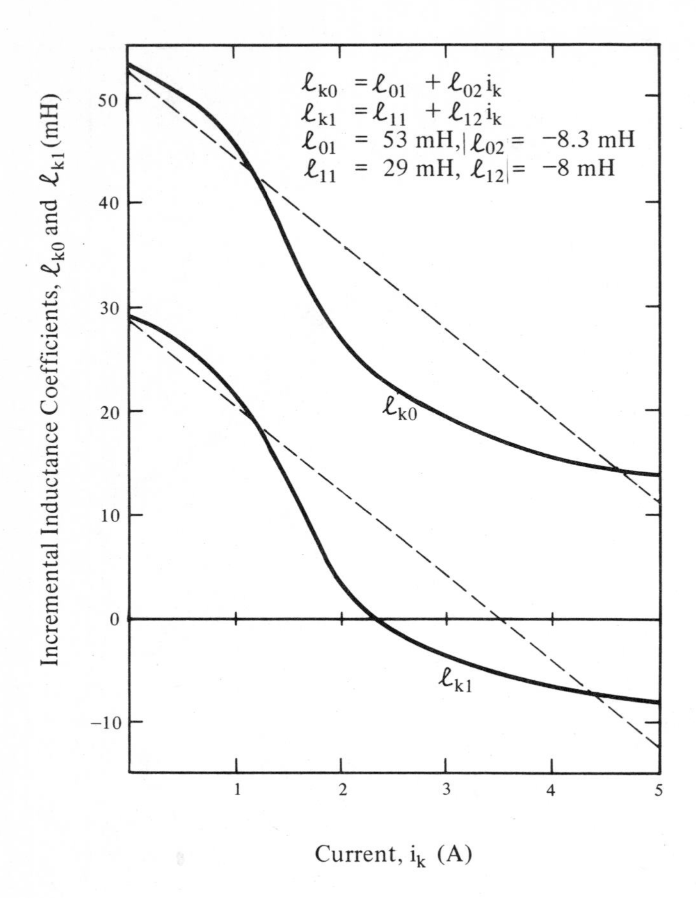

Figure 2-15.

The variation of the incremental inductance coefficients ℓ_{ko}, ℓ_{k1} with the phase current i_k for the SM060AB motor of the Warner Electric Brake and Clutch Company.

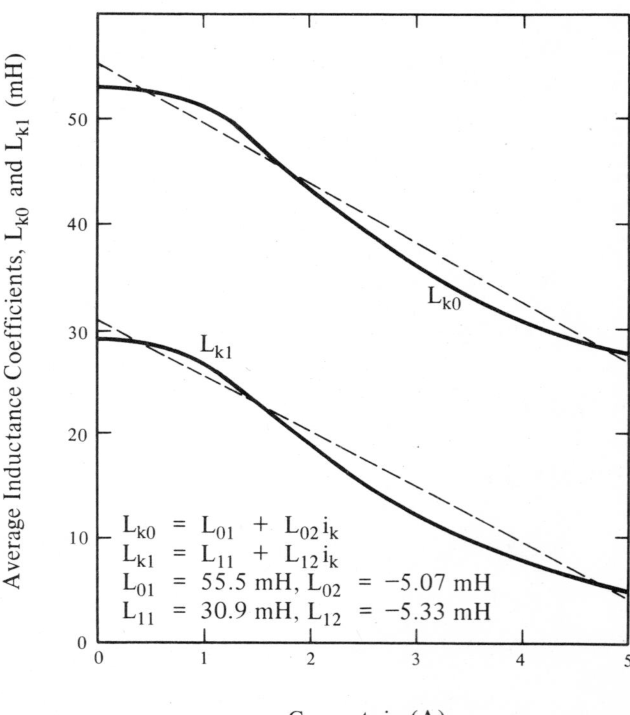

Figure 2-16.

The variation of the average inductance coefficients L_{ko}, L_{k1} with the phase current i_k for the SM060AB step motor of the Warner Electric Brake and Clutch Company.

The coefficients, ℓ_{o1}, ℓ_{o2}, ℓ_{11}, ℓ_{12}, L_{o1}, L_{o2}, L_{11}, L_{12} are calculated by using linear approximations to the curves of ℓ_{ko}, ℓ_{k1}, L_{ko}, L_{k1} versus i_k and these coefficients are independent of the individual phases. For the SM060AB step motor, the values are

$$\ell_{o1} = 53. \ , \ \ell_{o2} = 8.25 \ , \ \ell_{11} = 29 \ , \ \ell_{12} = 8.0 \ ,$$

$$L_{o1} = 55.5 \ , \ L_{o2} = -5.67 \ , \ L_{11} = 30.9 \ , \ L_{12} = -5.33 \ ,$$

all quantities being in millihenries.

The incremental and average inductances of the N-phase step motor therefore become

$$\ell_a = (\ell_{o1} + \ell_{o2} i_a) + (\ell_{11} + \ell_{12} i_a) \cos (Z\theta)$$

$$\ell_b = (\ell_{o1} + \ell_{o2} i_b) + (\ell_{11} + \ell_{12} i_b) \cos (Z\theta - \frac{2\pi}{N})$$

$$\cdots\cdots\cdots\cdots\cdots\cdots$$

$$\ell_k = (\ell_{o1} + \ell_{o2} i_k) + (\ell_{11} + \ell_{12} i_k) \cos [Z\theta - \frac{2\pi(k-1)}{N}]$$

$$\cdots\cdots\cdots\cdots\cdots\cdots$$

$$\ell_N = (\ell_{o1} + \ell_{o2} i_N) + (\ell_{11} + \ell_{12} i_N) \cos [Z\theta - \frac{2\pi(N-1)}{N}] \ , \quad (2\text{-}63)$$

$$L_a = (L_{o1} + L_{o2} i_a) + (L_{11} + L_{12} i_a) \cos (Z\theta)$$

$$L_b = (L_{o1} + L_{o2} i_b) + (L_{11} + L_{12} i_b) \cos (Z\theta - \frac{2\pi}{N})$$

$$\cdots\cdots\cdots\cdots\cdots\cdots$$

$$L_k = (L_{o1} + L_{o2} i_k) + (L_{11} + L_{12} i_k) \cos [Z\theta - \frac{2\pi(k-1)}{N}]$$

$$\cdots\cdots\cdots\cdots\cdots\cdots$$

$$L_N = (L_{o1} + L_{o2} i_N) + (L_{11} + L_{12} i_N) \cos [Z\theta - \frac{2\pi(N-1)}{N}] \ . \quad (2\text{-}64)$$

For the SM060AB step motor these are

$$\ell_a = [(53 - 8.25\ i_a) + (29 - 8\ i_a) \cos (20\theta)]$$

$$\ell_b = [(53 - 8.25\ i_b) + (29 - 8\ i_b) \cos (20\theta - \frac{2\pi}{3})]$$

$$\ell_c = [(53 - 8.25\ i_c) + (29 - 8\ i_c) \cos (20\theta - \frac{4\pi}{3})] \ , \quad (2\text{-}65)$$

$$L_a = [(55.5 - 5.67\, i_a) + (30.9 - 5.33\, i_a) \cos (20\theta)] \times 10^{-3}$$

$$L_b = [(55.5 - 5.67\, i_b) + (30.9 - 5.33\, i_b) \cos (20\theta - \frac{2\pi}{3})] \times 10^{-3}$$

$$L_c = [(55.5 - 5.67\, i_c) + (30.9 - 5.33\, i_c) \cos (20\theta - \frac{4\pi}{3})] \times 10^{-3}. \qquad (2\text{-}66)$$

The use of the representation of Eqs. (2-63) and (2-64) will generally yield results of satisfactory accuracy. If greater accuracy is desired, higher harmonics could be introduced into Eqs. (2-59) and (2-60), and higher order polynomials could be used in Eqs. (2-61) and (2-62). This, however, requires a significant increase in effort and is generally not necessary.

The voltage equation for phase k in the case when saturation is present, from Eq. (2-5), is

$$v_k = i_k r_k + \frac{\partial \lambda_k}{\partial i_k} \frac{di_k}{dt} + \frac{\partial \lambda_k}{\partial \theta} \frac{d\theta}{dt}\, . \qquad (2\text{-}67)$$

Using Eqs. (2-57) and (2-58), Eq. (2-67) becomes

$$v_k = i_k r_k + \ell_k \frac{di_k}{dt} + i_k \frac{dL_k}{d\theta} \frac{d\theta}{dt}\, . \qquad (2\text{-}68)$$

Substituting for ℓ_k and L_k from Eqs. (2-59) and (2-60), the last equation becomes

$$v_k = i_k r_k + \{\ell_{ko} + \ell_{k1} \cos [Z\theta - \frac{2\pi(k-1)}{N}]\} \frac{di_k}{dt}$$

$$- i_k ZL_{k1} \sin [Z\theta - \frac{2\pi(k-1)}{N}] \frac{d\theta}{dt} \qquad (2\text{-}69)$$

where ℓ_{ko}, ℓ_{k1}, and L_{k1} are defined in Eqs. (2-61) and (2-62).

The differential equation for the phase current, i_k, is easily obtained from Eq. (2-69)

$$\frac{di_k}{dt} = \frac{1}{\ell_{ko} + \ell_{k1} \cos [Z\theta - \frac{2\pi(k-1)}{N}]} \{v_k - i_k r_k$$

$$+ i_k ZL_{k1} \sin [Z\theta - \frac{2\pi(k-1)}{N}] \frac{d\theta}{dt} \}\, . \qquad (2\text{-}70)$$

With k = a, b, c, . . . , N, Eq. (2-70) represents the modified form of Eq. (2-49) when saturation is present.

For the SM060AB step motor, Eq. (2-70) becomes

$$\frac{di_a}{dt} = \frac{1000}{(53 - 8.25\, i_a) + (29 - 8\, i_a)\cos(20\theta)} [v_a - i_a r_a +$$

$$i_a\, (.618 - .1066\, i_a) \sin (20\theta) \frac{d\theta}{dt}]$$

$$\frac{di_b}{dt} = \frac{1000}{(53 - 8.25\, i_b) + (29 - 8\, i_b)\cos(20\theta - \frac{2\pi}{3}} [v_b - i_b r_b +$$

$$i_b\, (.618 - .1066\, i_b) \sin (20\theta - \frac{2\pi}{3}) \frac{d\theta}{dt}]$$

$$\frac{di_c}{dt} = \frac{1000}{(53 - 8.25\, i_c) + (29 - 8\, i_c)\cos(20\theta - \frac{4\pi}{3})} [v_c - i_c r_c +$$

$$i_c\, (.618 - .1066\, i_c) \sin (20\theta - \frac{4\pi}{3}) \frac{d\theta}{dt}] \; .$$

In summary, the procedure for obtaining the modified differential equations for the currents in the phase windings is

(1) measure the incremental inductance, ℓ_k, of one phase of the motor as a function of the phase current i_k and rotor position θ;

(2) integrate $\ell_k(i_k, \theta)$ with constant θ to get $\lambda_k(i_k, \theta)$;

(3) use Eq. (2-58) to obtain the average inductance $L_k(i_k, \theta)$;

(4) calculate the coefficients $\ell_{ko}(i_k)$, $\ell_{k1}(i_k)$, $L_{ko}(i_k)$, $L_{k1}(i_k)$ as defined in Eqs. (2-59) and (2-60) by the use of Eqs. (2-44) and (2-45), respectively, for individual values of i_k;

(5) approximate ℓ_{ko}, ℓ_{k1}, L_{ko} and L_{k1} as linear functions of i_k, and calculate ℓ_{o1}, ℓ_{o2}, ℓ_{11}, ℓ_{12}, L_{o1}, L_{o2}, L_{11} and L_{12};

(6) substitute ℓ_{o1}, ℓ_{o2}, ℓ_{11}, ℓ_{12}, L_{o1}, L_{o2}, L_{11} and L_{12} appropriately into Eq. (2-70) and obtain the differential equations for all the phase currents.

The modified expression for torque is also best obtained from experimental data. When saturation is present, the torque is no longer given by Eq. (2-51). As in the case of inductance, the dependence of torque on rotor position is unchanged from the unsaturated case, but the dependence on current is affected significantly. For phase k, the torque can now be written as

$$T_k = -K_t(i_k) \sin [Z\theta - \frac{2\pi(k-1)}{N}] \; . \tag{2-71}$$

To simplify the form of $K_t(i_k)$, it is generally satisfactory to take the linear approximation

$$K_t(i_k) = K_{ok} \left| i_k \right| \tag{2-72}$$

where K_{ok} is independent of i_k. The absolute value sign in Eq. (2-72) is necessary to insure that the torque is in the proper direction.

The value of K_{ok} is obtained by dividing the peak holding torque of phase k at rated current, $T_{k\ max}$, by the ratio current of phase k; thus,

$$K_{ok} = \frac{T_{k\ max}}{i_k \text{ rated}} \ . \tag{2-73}$$

Using Eqs. (2-72) and (2-73) in Eq. (2-71), the torque developed by phase k is

$$T_k = - \frac{T_{k\ max} |i_k|}{i_k \text{ rated}} \sin \left[Z\theta - \frac{2\pi(k-1)}{N} \right] \ . \tag{2-74}$$

Generally, all phases of a step motor are identical and we can write

$$T_{a\ max} = T_{b\ max} = \ . \ . \ . = T_{k\ max} = \ . \ . \ . = T_{N\ max} \tag{2-75}$$

$$i_a \text{ rated} = i_b \text{ rated} = \ . \ . \ . = i_c \text{ rated} = \ . \ . \ . \ i_N \text{ rated} \ . \tag{2-76}$$

Using these last two equations, Eq. (2-73) now gives

$$K_{oa} = K_{ob} = K_{ok} = \ . \ . \ . = \ . \ . \ . = K_{oN} = K_o \ . \tag{2-77}$$

Thus, the torque constant is the same for all the phases and may be calculated by using Eq. (2-73) for any value of k. The total torque is now given by

$$T = - K_o \left[|i_a| \sin Z\theta + |i_b| \sin (Z\theta - \frac{2\pi}{N}) + \ . \ . \ . \right.$$
$$\left. + |i_k| \sin (Z\theta - \frac{2\pi(k-1)}{N}) + \ . \ . \ . + |i_N| \sin (Z\theta - \frac{2\pi(N-1)}{N}) \right] \ . \tag{2-78}$$

For the SM060AB step motor, the torque is

$$T = - .882 \left[|i_a| \sin (20\theta) + |i_b| \sin (20\theta - \frac{2\pi}{3}) \right.$$
$$\left. + |i_c| \sin (20\theta - \frac{4\pi}{3}) \right] \tag{2-79}$$

where the currents are in amperes, and the torque is in Newton-meters.

Equations (2-78) and (2-79) have been found to be consistent with experimental torque data. If a more accurate representation is desired, a higher order polynomial approximation of $K_t(i_k)$ in Eq. (2-72) could be used, although this is usually not needed.

The complete model of the multiple-stack variable-reluctance step motor with saturation present, therefore, consists of

(1) the modified differential equations for the phase currents, Eq. (2-70), with the current dependent inductances, Eq. (2-64);

(2) the modified expression for torque, Eq. (2-78);

(3) the rotor dynamic equations, (2-55), which are unaffected by saturation.

The use of this model has been found to give good results in all the operating modes and with a large variety of multiple-stack variable-reluctance step motors. It can be used adequately to the extent that all the approximations make pragmatic sense. As in any modeling process, the final check has to be with experimental observations and a model is valid only so long as it yields results which corroborate satisfactorily with experimental data.

We end this section by deriving an expression for approximating the developed torque of the multiple-stack variable-reluctance step motor using only the design data. This expression is derived from Eq. (2-23) and can be used to determine K_o in Eq. (2-78) or to estimate the peak static torque of a step motor during the design stages.

Let us consider that one phase of the motor is approximated by the construction shown in Figure 2-17. Neglecting fringing flux, it is shown in the literature[1] that the permeance between two concentric cylinders is given by

$$P = \frac{\mu_o \text{ x area of flux path}}{\text{length of flux path}}$$

$$= \frac{2\pi\mu_o L(r + \frac{g}{2})}{g} \tag{2-80}$$

where r = radius of inner cylinder (rotor)

L = length of rotor

g = length of air gap.

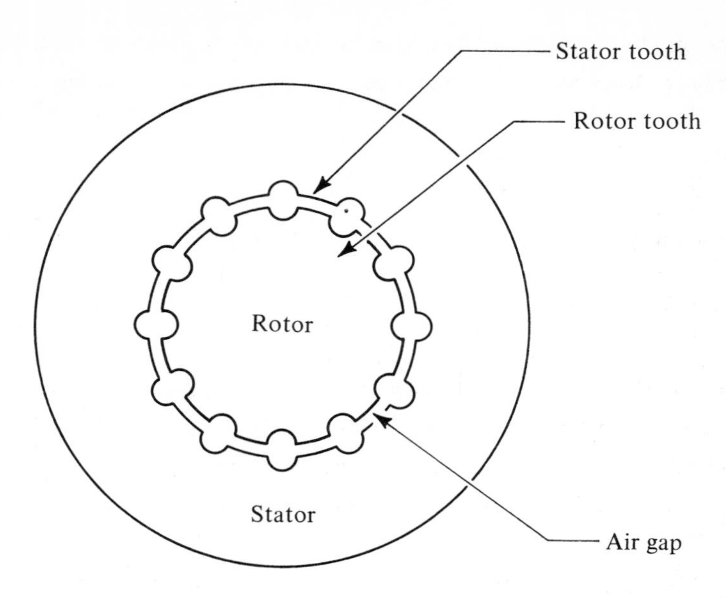

Figure 2-17. Simplified Diagram of One Phase of Stepping Motor.

However, in the situation illustrated in Figure 2-17, the area of flux path is not the area of a complete cylinder which is given by $2\pi L(r + g/2)$. If the rotor teeth are in complete alignment with the stator teeth, the effective area will be $\pi L(r + g/2)$. If the rotor is moved from this position by an angle θ, the effective area will be

$$\text{area of flux path} = \frac{\theta_t - \theta}{\theta_t} \pi L(r + g/2) \tag{2-81}$$

for $\theta_t \geq |\theta|$, but $\theta_t = \pi/N$, where N is the number of teeth in parallel. Therefore,

$$\text{area of flux path} = (1 - \frac{\theta N}{\pi}) \pi L(r + g/2) \tag{2-82}$$

and

$$P = (1 - \frac{\theta N}{\pi}) \frac{\mu_o \pi L(r + g/2)}{2g} . \tag{2-83}$$

$\theta_t \geq |\theta|$ where the length of the flux path is the two air gaps in series.

Now, differentiating both sides of (2-83) with respect to θ, we get

$$\frac{dP}{d\theta} = \frac{-\mu_o NL(r + g/2)}{2g} , \theta_t \geq |\theta| . \tag{2-84}$$

Substituting Eq. (2-84) into Eq. (2-23) yields

$$T = \frac{-\mu_o NL(r + g/2)\ F^2}{4g} , \theta_t \geq |\theta| . \tag{2-85}$$

The negative sign in the last equation simply indicates that the torque is in the direction to increase permeance or decrease θ. Using the relation

$$B = \mu_o H = \frac{\mu_o \cdot F}{2g}, \tag{2-86}$$

Equation (2-49) becomes

$$T = \frac{- NLg(r + g/2)\ B^2}{\mu_o}. \tag{2-87}$$

This can be interpreted as the externally applied torque which would be necessary to displace the rotor of the motor, when one phase is energized, from its aligned or reference position. The torque equation in Eq. (2-87) may be rewritten to include the necessary constants for a convenient set of units.

$$T = \frac{NL(r + g/2)\ F^2}{14.15 \times 10^6\ g} \tag{2-88}$$

where

T = torque in lb-in,

N = number of teeth per phase,

L = axial length of rotor in inches,

g = rotor-to-stator radial air gap in inches,

r = rotor radius in inches,

F = ampere turns developed across the two air gaps in series through which a line of flux must pass in one phase,

or $F = 2(0.3133)\ Bg^2$.

As an illustrative example on the application of Eq. (2-88), let us consider the following set of motor data:

N = 16 teeth per phase (48 steps per revolution for a three-phase motor),

g = 0.0025 in,

r = 0.509 in,

F = 720 ampere-turns,

L = 0.25 in,

then
$$T = \frac{(16)(0.25)(0.509 + 0.00125)(720)^2}{14.15\ (10)^6 (0.0025)}$$
$$= 30 \text{ lb-in}. \tag{2-89}$$

2-6 DYNAMIC MODELING OF PERMANENT MAGNET STEP MOTORS

In this section the detailed modeling of the last of the two most important types of step motors, the permanent-magnet or synchronous-inductor type, is considered. The construction and principles of operation of this type of step motor have been discussed in Chapter 1. As explained there, it is not necessary that a permanent magnet be located in the rotor, the same effect can be achieved by using one or two field coils on the stator, appropriately placed and carrying a constant current. What is essential to the process is that all the rotor teeth on each section of the rotor are magnetized to the same polarity. Since a permanent magnet and a field coil carrying constant current are interchangeable mathematically, the modeling process is identical for both types of rotor magnetization. In the present development, a step motor with a permanent magnet in its rotor is considered. The motor has N_r rotor teeth, N_s stator teeth, p poles per phase on the stator, and two rotor sections. The two rotor sections are offset by one-half of the rotor tooth pitch, resulting in a step angle of $\frac{360}{4N_r}$ degrees. This makes the motor a $4N_r$ steps-per-revolution step motor. Figure 1-28 shows such a motor with $N_r = 5$, $N_s = 4$, and $p = 2$.

The permanent magnet is represented by an equivalent coil having N_F turns and carrying a current of i_F amperes. The mmf developed by the permanent magnet is denoted by $2\mathcal{F}_F$; thus

$$2\mathcal{F}_F = N_F i_F \ . \tag{2-90}$$

In the derivation of the model, we assume that:

i) The magnetic circuit is linear.

ii) The permeance of the kth stator tooth on Section I of the motor is

$$\mathcal{P}_k = \mathcal{P}_f + \mathcal{P}_v \cos N_r(\theta - \theta_{sk}) \tag{2-91}$$

and the permeance of the kth stator tooth on Section II of the motor is

$$\bar{\mathcal{P}}_k = \mathcal{P}_f - \mathcal{P}_v \cos N_r(\theta - \theta_{sk}) \tag{2-92}$$

where

P_k, $\bar{P}_k$	are the permeances of the kth stator tooth on Sections I and II, respectively;
P_f	is the average value of these permeances;
P_v	is the peak magnitude of the variation of these permeances due to rotor position;
N_r	is the number of rotor teeth;
θ_{sk}	is the position of the kth stator tooth from a fixed reference axis;
θ	is the position of the rotor from a fixed reference axis.

The variation of P and $\bar{P}_k$ with rotor position is shown in Figure 2-18.

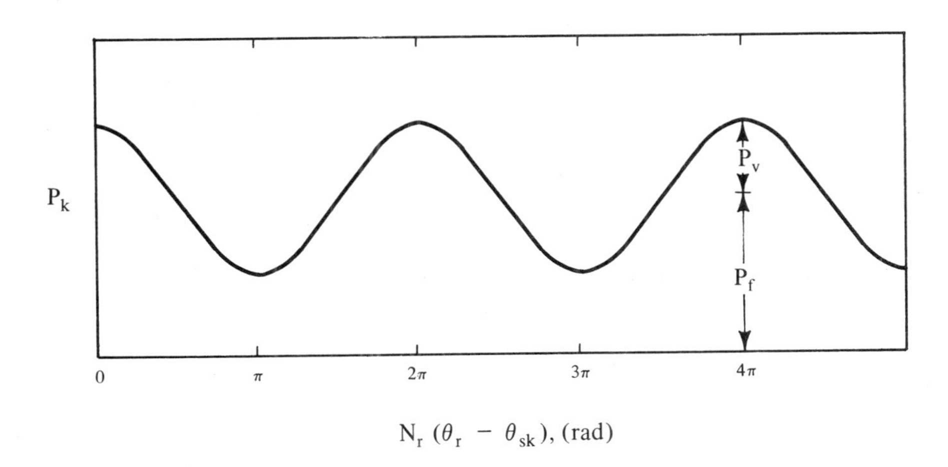

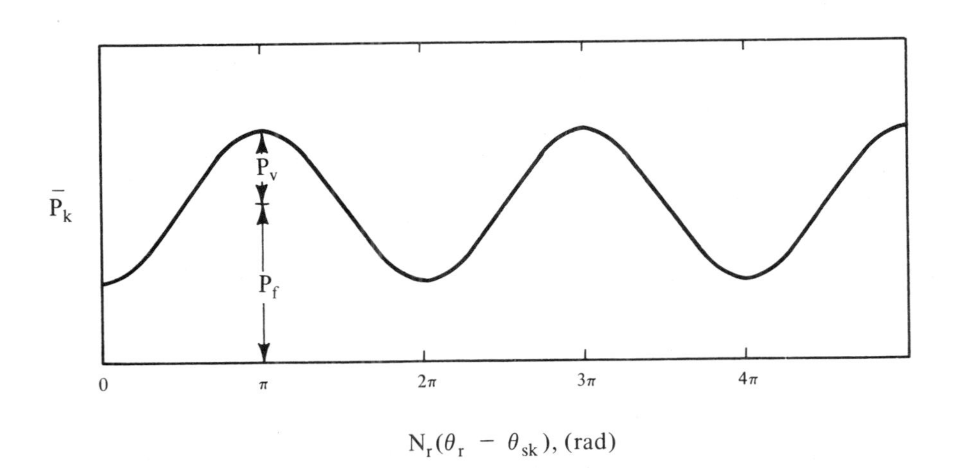

Figure 2-18. Permeances of stator teeth at Section I and Section II of rotor.

Without loss of generality let each pole have an odd number of teeth, and let the axis of the center tooth of a north pole of phase a be the reference axis for the stator; if this tooth is denoted as tooth number 0, with all the other teeth being numbered consecutively as 1, 2, 3, . . . , $N_s - 1$ in the counterclockwise sense, then θ_{sk} becomes

$$\theta_{sk} = \frac{2\pi k}{N_s} . \tag{2-93}$$

If each pole has an even number of teeth, the reference axis is chosen as the center line of a north pole of phase a and θ_{sk} is shifted by half a stator tooth pitch, but the final results are unchanged. The reference axis of the rotor, $\theta = 0$, is chosen as the equilibrium position of the rotor when phase a is energized.

iii) The phase windings are sinusoidally distributed along the periphery of the stator and the number of turns linking the kth stator tooth due to the phase windings are

$$N_{ka} = N_p \cos\left(\frac{p}{2}\theta_{sk}\right) \tag{2-94}$$

$$N_{kb} = N_p \cos\left(\frac{p}{2}\theta_{sk} - \frac{\pi}{2}\right) \tag{2-95}$$

where

N_{ka}, N_{kb} are the number of turns linking the kth stator tooth due to the windings of phase a and phase b, respectively;

N_p is the effective peak number of turns per pole of each phase.

To derive the model, we start by writing the voltage equation of the phase windings; using Eq. (2-1) we have

$$v_a = i_a r_a + \frac{d\lambda_a}{dt}$$

$$v_b = i_b r_b + \frac{d\lambda_b}{dt} . \tag{2-96}$$

The flux linkages of the two phases and the permanent magnet can be written by using Eq. (2-4),

$$\lambda_a = L_{aa}\, i_a + L_{ab}\, i_b + L_{aF}\, i_F$$

$$\lambda_b = L_{ba}\, i_a + L_{bb}\, i_b + L_{bF}\, i_F$$

$$\lambda_F = L_{Fa}\, i_a + L_{Fb}\, i_b + L_{FF}\, i_F . \tag{2-97}$$

The inductances have to be calculated next. This is done, as in the earlier sections, by finding the flux linkages of the respective windings as functions of motor parameters and winding currents.

The total mmf at the kth stator tooth on Section I is

$$\mathcal{F}_k = \mathcal{F}_F + N_p i_a \cos \frac{p\theta_{sk}}{2} + N_p i_b \cos \left(\frac{p\theta_{sk}}{2} - \frac{\pi}{2}\right) , \tag{2-98}$$

and the total mmf at the kth stator tooth on Section II is

$$\bar{\mathcal{F}}_k = -\mathcal{F}_F + N_p i_a \cos \frac{p\theta_{sk}}{2} + N_p i_b \cos \left(\frac{p\theta_{sk}}{2} - \frac{\pi}{2}\right) . \tag{2-99}$$

The net flux entering the kth stator tooth from both sections is

$$\phi_k = \mathcal{F}_k \mathcal{P}_k + \bar{\mathcal{F}}_k \bar{\mathcal{P}}_k . \tag{2-100}$$

Using Eqs. (2-91), (2-92) and Eqs. (2-98), (2-99), Eq. (2-100) becomes

$$\phi_k = \left[\mathcal{F}_F + N_p i_a \cos \frac{p\theta_{sk}}{2} + N_p i_b \cos \left(\frac{p\theta_{sk}}{2} - \frac{\pi}{2}\right)\right] \left[\mathcal{P}_f + \mathcal{P}_v \cos N_r(\theta - \theta_{sk})\right]$$

$$+ \left[-\mathcal{F}_F + N_p i_a \cos \frac{p\theta_{sk}}{2} + N_p i_b \left(\frac{p\theta_{sk}}{2} - \frac{\pi}{2}\right)\right] \left[\mathcal{P}_f - \mathcal{P}_v \cos N_r(\theta - \theta_r)\right] \tag{2-101}$$

or

$$\phi_k = 2\mathcal{P}_f N_p i_a \cos \frac{p\theta_{sk}}{2} + 2\mathcal{P}_f N_p i_b \cos \left(\frac{p\theta_{sk}}{2} - \frac{\pi}{2}\right) + 2\mathcal{P}_v \mathcal{F}_F \cos N_r(\theta - \theta_{sk}) . \tag{2-102}$$

The total flux linkage with winding a is

$$\lambda_a = \sum_{k=0}^{N_s - 1} N_{ka} \phi_k = \sum_{k=0}^{N_s - 1} N_p \cos \left(\frac{p\theta_{sk}}{2}\right) \phi_k$$

$$= \sum_{k=0}^{N_s - 1} 2 N_p \cos \frac{p\theta_{sk}}{2} [N_p i_a \cos \frac{p\theta_{sk}}{2} + \mathcal{P}_f N_p i_b \cos \left(\frac{p\theta_{sk}}{2} - \frac{\pi}{2}\right)$$

$$+ \mathcal{P}_v \mathcal{F}_F \cos N_r (\theta - \theta_{sk})] . \tag{2-103}$$

Expanding the right side of Eq. (2-103) and using Eq. (2-93) yields

$$\lambda_a = \sum_{k=0}^{N_s-1} 2N_p^2 \mathcal{P}_f i_a \cos^2 \frac{p}{2}\left(\frac{2\pi k}{N_s}\right) + \sum_{k=0}^{N_s-1} 2N_p^2 \mathcal{P}_f i_b \cos \frac{p}{2}\left(\frac{2\pi k}{N_s}\right)$$

$$\cdot \cos\left(\frac{p}{2}\frac{2\pi k}{N_s} - \frac{\pi}{2}\right) + \sum_{k=0}^{N_s-1} 2\mathcal{F}_F \mathcal{P}_v N_p \cos \frac{p}{2}\left(\frac{2\pi k}{N_s}\right) \cos N_r\left(\theta - \frac{2\pi k}{N_s}\right) . \qquad (2\text{-}104)$$

Since, if N_s is not equal to p,

$$\sum_{k=0}^{N_s-1} \cos^2\left(\frac{\pi kp}{N_s}\right) = \sum_{k=0}^{N_s-1} \frac{1}{2}\left(1 + \cos \frac{2\pi kp}{N_s}\right) = \frac{N_s}{2} \qquad (2\text{-}105)$$

$$\sum_{k=0}^{N_s-1} \cos\left(\frac{\pi kp}{N_s}\right) \cos\left(\frac{\pi kp}{N_s} - \frac{\pi}{2}\right) = - \sum_{k=0}^{N_s-1} \cos\left(\frac{\pi kp}{N_s}\right) \sin\left(\frac{\pi kp}{N_s}\right)$$

$$= -\frac{1}{2} \sum_{k=0}^{N_s-1} \sin\left(\frac{2\pi kp}{N_s}\right) = 0 \qquad (2\text{-}106)$$

and

$$\sum_{k=0}^{N_s-1} \cos\left(\frac{\pi kp}{N_s}\right) \cos N_r\left(\theta - \frac{2\pi k}{N_s}\right) = \frac{N_s}{2} \cos N_r\theta . \qquad (2\text{-}107)$$

Then Eq. (2-104) becomes

$$\lambda_a = N_p^2 \mathcal{P}_f N_s i_a + N_p \mathcal{F}_F \mathcal{P}_v N_s \cos(N_r\theta)\, i_F$$

$$= N_p^2 \mathcal{P}_f N_s i_a + N_p N_F \mathcal{P}_v N_s \cos(N_r\theta)\, i_F . \qquad (2\text{-}108)$$

Similarly, the total flux linkage of phase b is

$$\lambda_b = \sum_{k=0}^{N_s-1} N_{kb} \phi_k = \sum_{k=0}^{N_s-1} N_p \cos\left(\frac{p\theta_{sk}}{2} - \frac{\pi}{2}\right) \phi_k$$

$$= \sum_{k=0}^{N_s-1} 2N_p \sin\left(\frac{p\theta_{sk}}{2}\right) [\mathcal{P}_f N_p i_a \cos \frac{p\theta_{sk}}{2}$$

$$+ \mathcal{P}_f N_p i_b \cos\left(\frac{p\theta_{sk}}{2} - \frac{\pi}{2}\right) + \mathcal{F}_F \mathcal{P}_v \cos N_r (\theta - \theta_{sk})] \qquad (2\text{-}109)$$

which reduces to

$$\lambda_b = N_p^2 \mathcal{P}_f N_s i_b + \mathcal{F}_F \mathcal{P}_v N_p N_s \sin N_r\theta$$

$$= N_p^2 \mathcal{P}_f N_s i_b + N_p N_F \mathcal{P}_v N_s \sin (N_r\theta)\, i_F \,. \tag{2-110}$$

The flux linkage of the coil representing the permanent magnet due to its own current i_F is

$$\lambda_{FF} = N_f \sum_{k=0}^{N_s-1} \mathcal{F}_F \mathcal{P}_k = N_F \mathcal{F}_F \sum_{k=0}^{N_s-1} (\mathcal{P}_f + \mathcal{P}_v \cos N_r(\theta - \theta_{sk})) \tag{2-111}$$

which reduces to

$$\lambda_{FF} = N_F \mathcal{F}_F \mathcal{P}_f N_s$$

$$= N_F^2 \mathcal{P}_f N_s i_F \,. \tag{2-112}$$

Using Eqs. (2-108, (2-110) and (2-112), the inductances of the motor can be written as

$$L_{aa} = N_p^2 \mathcal{P}_f N_s$$

$$L_{bb} = N_p^2 \mathcal{P}_f N_s$$

$$L_{FF} = N_F^2 \mathcal{P}_f N_s$$

$$L_{ab} = L_{ba} = 0$$

$$L_{aF} = L_{Fa} = N_p \mathcal{P}_v N_s \cos N_r\theta$$

$$L_{bF} = L_{Fb} = N_p \mathcal{P}_v N_s \sin N_r\theta \,. \tag{2-113}$$

Using the inductance expressions of Eq. (2-113) and the flux linkage expressions of Eq. (2-97), the voltage equations of the phase windings, Eq. (2-96) becomes

$$v_a = i_a r_a + N_p^2 \mathcal{P}_f N_s \frac{di_a}{dt} - N_p \mathcal{F}_F \mathcal{P}_v N_s N_r \sin (N_r\theta) \frac{d\theta}{dt}$$

$$v_b = i_b r_b + N_p^2 \mathcal{P}_f N_s \frac{di_b}{dt} + N_p \mathcal{F}_F \mathcal{P}_v N_s N_r \sin (N_r\theta) \frac{d\theta}{dt} \,. \tag{2-114}$$

From Eq. (2-114) the differential equations for the phase currents are written as

$$\frac{di_a}{dt} = \frac{1}{N_p^2 P_f N_s} [v_a - i_a r_a + N_p F_F P_v N_s N_r \sin (N_r\theta) \frac{d\theta}{dt}]$$

$$\frac{di_b}{dt} = \frac{1}{N_p^2 P_f N_s} [v_b - i_b r_b - N_p F_F P_v N_s N_r \cos (N_r\theta) \frac{d\theta}{dt}] \quad . \tag{2-115}$$

The torque developed by the permanent magnet step motor can be obtained by using Eq. (2-32), which is rewritten as

$$T = \frac{1}{2} i_a^2 \frac{dL_{aa}}{d\theta} + \frac{1}{2} i_b^2 \frac{dL_{bb}}{d\theta} + \frac{1}{2} i_F^2 \frac{dL_{FF}}{d\theta}$$

$$+ i_a i_b \frac{dL_{ab}}{d\theta} + i_a i_F \frac{dL_{aF}}{d\theta} + i_b i_F \frac{dL_{bF}}{d\theta} \quad . \tag{2-116}$$

Substituting for the inductances from (2-113) gives

$$T = -i_a i_F N_p N_F P_v N_s N_r \sin N_r\theta + i_b i_F N_p N_F P_v N_s N_r \cos N_r\theta \tag{2-117}$$

or

$$T = N_p F_F P_v N_s N_r (-i_a \sin N_r\theta + i_b \cos N_r\theta) \quad . \tag{2-118}$$

The dynamic equations of the rotor are unchanged from the previous section and are given in Eqs. (2-34) and (2-35).

Thus, the complete model of the permanent-magnet step motor consists of

(1) the differential equations for current, Eq. (2-115),

(2) the torque expression, Eq. (2-118),

(3) the rotor dynamics, Eq. (2-35).

This completes the modeling of the two-phase permanent-magnet step motor.

For a three-phase motor, it can be shown that the voltage equations for the three phases are

$$v_a = r_a i_a + N_p^2 P_f N_s \frac{di_a}{dt} - \frac{1}{2} N_p^2 P_f N_s \frac{di_b}{dt} - \frac{1}{2} N_p^2 P_f N_s \frac{di_c}{dt}$$

$$- N_p F_F P_v N_s N_r \sin (N_r\theta) \frac{d\theta}{dt}$$

$$v_b = r_b i_b + N_p^2 P_f N_s \frac{di_b}{dt} - \frac{1}{2} N_p^2 P_f N_s \frac{di_a}{dt} - \frac{1}{2} N_p^2 P_f N_s \frac{di_c}{dt}$$

$$- N_p F_F P_v N_s N_r \sin (N_r\theta - \frac{2\pi}{3}) \frac{d\theta}{dt}$$

$$v_c = r_c i_c + N_p^2 P_f N_s \frac{di_c}{dt} - \frac{1}{2} N_p^2 P_f N_s \frac{di_a}{dt} - \frac{1}{2} N_p^2 P_f N_s \frac{di_b}{dt}$$

$$- N_p F_F P_v N_s N_r \sin (N_r\theta + \frac{2\pi}{3}) \frac{d\theta}{dt} \ , \tag{2-119}$$

and the total instantaneous torque is given by

$$T = - N_p F_F P_v N_s N_r [i_a \sin N_r\theta + i_b \sin (N_r\theta - \frac{2\pi}{3}) + i_c \sin (N_r\theta + \frac{2\pi}{3})] . \qquad (2\text{-}120)$$

REFERENCES

1. A. Fitzgerald and C. Kingley, Jr., Electric Machinery, McGraw Hill, New York, 1961.

3 SOME ASPECTS OF MODELING AND DYNAMIC SIMULATION OF VARIABLE-RELUCTANCE STEP MOTORS

G. Singh and S. N. Chen
Department of Electrical Engineering
University of Illinois at Urbana-Champaign
Urbana, Illinois

3-1 INTRODUCTION

This chapter is concerned with the development of a mathematical model for the single-stack variable-reluctance (VR) step motor. The objective is to obtain a system of differential equations which represents the dynamic behavior of the motor under all operating conditions. A comprehensive model for the multiple-stack VR step motor is described in Chapter 2.

The exact modeling of the single-stack VR step motor is considerably more complex than that of the multiple-stack VR step motor. This is because all the windings in the single-stack motor share a common magnetic circuit, and the effects of saturation and mutual inductances are difficult to predict and represent accurately. All the self and mutual inductances and self and mutual torques are influenced to some extent by all the phase currents.

Although a comprehensive mathematical model for a single-stack VR step motor is not available at this time, some results have been obtained which provide a substantial improvement over the model which assumed a linear magnetic circuit. The new model utilizes more experimental data than the multiple-stack model. The model developed in this paper is for a particular step motor--the SM024-0018-FE (made by the Warner Electric Brake and Clutch Company). This is a four-phase step motor, with 8 stator teeth and 6 rotor teeth. The step angle is 15 degrees which corresponds to 24 steps per revolution. However, the method of this paper is applicable to any single-stack VR step motor. Only the expressions for the various parameters have to be appropriately modified.

The basic expressions for a single-stack VR step motor are first presented in the next section by assuming a linear magnetic circuit and sinusoidal winding distributions. These expressions are then modified in the subsequent section to account for the effects of saturation and non-sinusoidal winding distributions. The procedure for experimentally determining the necessary parameters is explained. The use of these expressions for simulating the dynamic response of the motor on a digital computer is also discussed, and examples of typical simulation responses are presented.

3-2 THE BASIC MODEL FOR A SINGLE-STACK VR STEP MOTOR

In this section the basic expressions for modeling a single-stack VR step motor are presented. It is assumed that

(1) there is no saturation of the iron parts,

(2) the permeance of the stator teeth varies sinusoidally,

(3) the windings on the stator are sinusoidally distributed,

(4) hysteresis and eddy currents are neglected.

A four-phase step motor--the SM024-0018-FE (made by Warner Electric Brake and Clutch Company) is considered here. Figure 3-1 shows the schematic diagram of this motor. The step angle is 15 degrees which corresponds to 24 steps per revolution. The expressions presented here are for this motor while the expressions for a general single-stack motor are derived in Chapter 2.

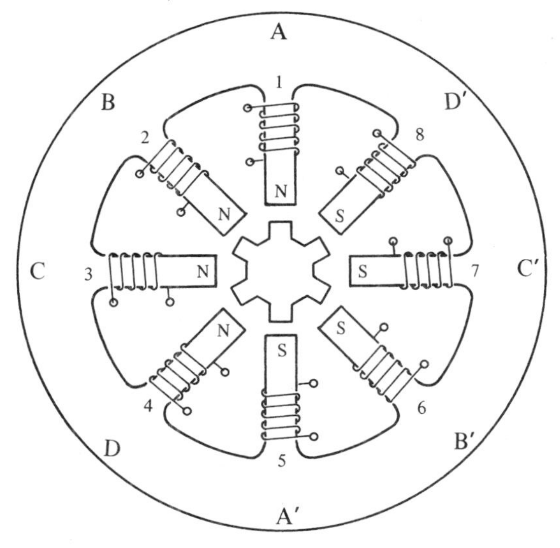

Figure 3-1. Schematic Diagram of a 4 phase step motor with 8 stator teeth and 6 rotor teeth.

Voltage Equations

$$v_a = i_a r_a + \frac{d}{dt}(\lambda_a)$$

$$v_b = i_b r_b + \frac{d}{dt}(\lambda_b)$$

$$v_c = i_c r_c + \frac{d}{dt}(\lambda_c)$$

$$v_d = i_d r_d + \frac{d}{dt}(\lambda_d) \qquad (3\text{-}1)$$

where v_a is the voltage applied to phase a

i_a is the current in phase a

r_a is the resistance of phase a

and λ_a denotes the flux linkages of phase a.

Similar statements can be made for the voltage and current, etc., in phases b, c and d.

Flux Linkages and Inductances

Since the magnetic circuit is assumed to be linear, the flux linkages are written as

$$\lambda_a = L_{aa}i_a + L_{ab}i_b + L_{ac}i_c + L_{ad}i_d$$

$$\lambda_b = L_{ba}i_a + L_{bb}i_b + L_{bc}i_c + L_{bd}i_d$$

$$\lambda_c = L_{ca}i_a + L_{cb}i_b + L_{cc}i_c + L_{cd}i_d$$

$$\lambda_d = L_{da}i_a + L_{db}i_b + L_{dc}i_c + L_{dd}i_d \tag{3-2}$$

where L_{jj} is the self inductance of phase j, j = a, b, c, d, and L_{jk} is the mutual inductance between phase j and phase k, with $L_{jk} = L_{kj}$, j,k = a, b, c, d, j ≠ k.

The self and mutual inductances are a function of the rotor position θ. In matrix notation, Eq. (3-2) becomes

$$\Lambda = L(\theta)I \tag{3-3}$$

where

$$\Lambda = \begin{bmatrix} \lambda_a \\ \lambda_b \\ \lambda_c \\ \lambda_d \end{bmatrix}, \quad I = \begin{bmatrix} i_a \\ i_b \\ i_c \\ i_d \end{bmatrix}, \quad L = \begin{bmatrix} L_{aa} & L_{ab} & L_{ac} & L_{ad} \\ L_{ba} & L_{bb} & L_{bc} & L_{bd} \\ L_{ca} & L_{cb} & L_{cc} & L_{cd} \\ L_{da} & L_{db} & L_{dc} & L_{dd} \end{bmatrix}$$

The expressions for inductances for the motor under consideration are

$$L_{aa} = L_1 + L_2 \cos(6\theta)$$

$$L_{bb} = L_1 + L_2 \cos(6\theta + \pi/2)$$

$$L_{cc} = L_1 + L_2 \cos(6\theta + \pi)$$

$$L_{dd} = L_1 + L_2 \cos(6\theta + 3\pi/2)$$

$$L_{ba} = L_{ab} = L_3 + L_2 \cos(6\theta + \pi/4)$$

$$L_{ca} = L_{ac} = L_2 \cos(6\theta + \pi/2)$$

$$L_{da} = L_{ad} = -L_3 + L_2 \cos(6\theta + \frac{3\pi}{4})$$

$$L_{bc} = L_{cb} = L_3 + L_2 \cos(6\theta + \frac{3\pi}{4})$$

$$L_{bd} = L_{db} = L_2 \cos(6\theta + \pi)$$

$$L_{cd} = L_{dc} = L_3 + L_2 \cos(6\theta + \frac{5\pi}{4}) \qquad (3\text{-}4)$$

where the quantities, L_1, L_2 and L_3 are defined in Chapter 2 and satisfy the relation

$$L_1 - L_0 = \sqrt{2}\ L_3 \qquad (3\text{-}5)$$

and L_0 is the leakage inductance of each phase. Using Eq. (3-3) with Eq. (3-1) yields

$$V_R = L\frac{dI}{dt} + \frac{d\theta}{dt}\frac{dL}{d\theta}\ I \qquad (3\text{-}6)$$

where

$$V_R = \begin{bmatrix} v_a - i_a r_a \\ v_b - i_b r_b \\ v_c - i_c r_c \\ v_d - i_d r_d \end{bmatrix}, \quad \frac{dL}{d\theta} = -[6L_2]$$

$$\times \begin{bmatrix} \sin 6\theta & \sin(6\theta + \pi/4) & \sin(6\theta + \pi/2) & \sin(6\theta + \frac{3\pi}{4}) \\ \sin(6\theta + \pi/4) & \sin(6\theta + \pi/2) & \sin(6\theta + \frac{3\pi}{4}) & \sin(6\theta + \pi) \\ \sin(6\theta + \pi/2) & \sin(6\theta + \frac{3\pi}{4}) & \sin(6\theta + \pi) & \sin(6\theta + \frac{5\pi}{4}) \\ \sin(6\theta + 3\pi/4) & \sin(6\theta + \pi) & \sin(6\theta + \frac{5\pi}{4}) & \sin(6\theta + \frac{3\pi}{2}) \end{bmatrix}$$

and $\frac{d\theta}{dt} = \omega$ is the angular velocity of the rotor.

For purposes of computer simulation, a more suitable form of Eq. (3-6) is

$$\frac{dI}{dt} = L^{-1}[V_R - \frac{d\theta}{dt}\frac{dL}{d\theta} I] \tag{3-7}$$

The inverse of the inductance matrix will generally exist if leakage is not neglected.

Torque Developed and Rotor Dynamics

The torque developed by a single-stack step motor is written as

$$\begin{aligned} T &= \frac{1}{2} I' \frac{dL}{d\theta} I \\ &= \frac{1}{2}\sum_j \sum_k i_j i_k \frac{dL_{jk}}{d\theta} \end{aligned} \tag{3-8}$$

where the summation is over all the phases. In the case of the four-phase motor under consideration, Eq. (3-8) becomes

$$\begin{aligned} T &= \frac{1}{2} I' \frac{dL}{d\theta} I \\ &= -3L_2[i_a^2 \sin 6\theta + i_b^2 \sin(6\theta + \pi/2) + i_c^2 \sin(6\theta + \pi) \\ &\quad + i_d^2 \sin(6\theta + \frac{3\pi}{2}) + 2\, i_a i_b \sin(6\theta + \pi/4) \\ &\quad + 2\, i_a i_c \sin(6\theta + \pi/2) + 2\, i_a i_d \sin(6\theta + 3\pi/4) \\ &\quad + 2\, i_b i_c \sin(6\theta + 3\pi/4) + 2\, i_b i_d \sin(6\theta + \pi) \\ &\quad + 2\, i_c i_d \sin(6\theta + \frac{5\pi}{4})] \end{aligned} \tag{3-9}$$

The rotor dynamics, in state variable form, are

$$\begin{aligned} \frac{d\theta}{dt} &= \omega \\ \frac{d\omega}{dt} &= \frac{T}{J} - \frac{B}{J}\omega - \frac{T_F}{J} \end{aligned} \tag{3-10}$$

where T is the developed torque

B is the coefficient of viscous friction

T_F is the coulomb frictional torque

J is the inertia of the rotor

θ is the rotor position

ω is the angular velocity of the rotor.

Equations (3-7), (3-9), and (3-10) represent the complete performance equations of the single-stack step motor, under the assumptions 1 through 4.

3-3 EFFECTS OF SATURATION AND NONSINUSOIDAL DISTRIBUTIONS

The performance equations derived in the last section will be modified here to account for the effects of saturation and nonsinusoidal distributions. The expressions for flux linkage, inductance and developed torque will be modified while the rotor dynamics remain unchanged.

Flux Linkages and Inductances

In order to describe the effect of saturation and nonsinusoidal winding distributions it is useful to redefine the flux linkages as

$$\Lambda(\theta,I) = L(\theta,I)I \tag{3-11}$$

where $L(\theta,I)$ is referred to as the average inductance matrix, and its elements are the average self and mutual inductances. The dependence of the average inductance on current indicates that its magnitude will change with changing current levels, due to saturation in the magnetic core.

It is also useful to define the incremental-inductance matrix as

$$L_1(\theta,I) = \frac{\partial\Lambda(\theta,I)}{\partial I} = \begin{bmatrix} \ell_{aa} & \ell_{ab} & \ell_{ac} & \ell_{ad} \\ \ell_{ba} & \ell_{bb} & \ell_{bc} & \ell_{bd} \\ \ell_{ca} & \ell_{cb} & \ell_{cc} & \ell_{cd} \\ \ell_{da} & \ell_{db} & \ell_{dc} & \ell_{dd} \end{bmatrix} \tag{3-12}$$

where the elements of $L_1(\theta,I)$, the ℓ_{ij}'s, are functions of θ and I, and are referred to as the incremental self and mutual inductances. When the magnetic circuit is linear, the incremental and average inductances are equal and are independent of the phase currents.

When saturation is present, the incremental and average inductances depend, in general, on all the phase currents. Thus, the incremental and average inductances are written as

$$
\begin{aligned}
\ell_{jk} &= \ell_{jk}(\theta, i_a, i_b, i_c, i_d) \\
L_{jk} &= L_{jk}(\theta, i_a, i_b, i_c, i_d) \qquad \text{(3-13)} \\
j,k &= a, b, c, d
\end{aligned}
$$

In order to simplify the analysis, it is assumed that the self inductance of a phase winding depends only on the current in that particular phase; thus,

$$\ell_{jj} = \ell_{jj}(\theta, i_j) \qquad \text{(3-14a)}$$

$$L_{jj} = L_{jj}(\theta, i_j), \qquad \text{(3-14b)}$$

$$j = a, b, c, d$$

and it is also assumed that the mutual inductance between two phase windings depends only on the average of the currents in these two phases; thus,

$$\ell_{jk} = \ell_{jk}(\theta, i_j, i_k) = \ell_{jk}(\theta, i_{jk}) \qquad \text{(3-15a)}$$

$$L_{jk} = L_{jk}(\theta, i_j, i_k) = L_{jk}(\theta, i_{jk}) \qquad \text{(3-15b)}$$

$$j \neq k, \quad j, k = a, b, c, d$$

where $i_{jk} = \dfrac{i_j + i_k}{2}$.

The assumptions leading to Eqs. (3-14) and (3-15) have been made in order that the resulting model is of a manageable complexity. However, it is expected that the use of this model will provide an improvement over the earlier case when saturation was completely neglected. Ultimately, of course, it is desirable to include the effects of all the currents in the self and mutual inductances.

It is possible to measure the incremental self and mutual inductances by means of an incremental inductance bridge. With θ held constant (rotor fixed), ℓ_{jj} is measured as a function of i_j, and ℓ_{jk} is measured as a function of i_{jk}. If these measurements are repeated for different rotor positions, the incremental self and mutual inductances of Eqs. (3-14a) and (3-15a) are completely defined. These can now be utilized to determine the average self and mutual inductances as defined in Eqs. (3-14b) and (3-15b).

First, it is useful to write Eq. (3-11) in component form,

$$\lambda_k = \sum_j \psi_{kj} = \sum_j L_{kj}\, i_j \tag{3-16}$$

$k = a, b, c, d$

where the summation is performed over the four phases, $j = a, b, c, d$.

The term, ψ_{kj}, which represents the component of flux in phase k due to current in phase j can be written as

$$\psi_{kj} = L_{kj} i_j \tag{3-17}$$

Since ℓ_{jj} is known as a function of θ and i_j, it can be integrated for each θ, with respect to i_j, to yield $\psi_{jj}(\theta, i_j)$. Then, L_{jj} can be determined from

$$L_{jj}(\theta, i_j) = \frac{\psi_{jj}(\theta, i_j)}{i_j} \tag{3-18}$$

Similarly, the average mutual inductance L_{kj} can be determined by integrating ℓ_{kj} with respect to i_{jk}, for each θ, to yield $\psi_{jk}(\theta, i_{jk})$, and then using

$$L_{jk}(\theta, i_{jk}) = \frac{\psi_{jk}(\theta, i_{jk})}{i_{jk}} \tag{3-19}$$

Figure 3-2 shows all the inductances of the motor as a function of θ with no bias current. Since this represents unsaturated values, the average and incremental values are the same. Figures 3-3 through 3-6 show some of the incremental inductances of the motor under consideration for different current levels as a function of θ. Figures 3-7 through 3-10 show the average inductances for the same phases and same currents as a function of θ. The curves in Figures 3-7 through 3-10 are obtained by integrating the data in Figures 3-3 through 3-6 with respect to the current, for each θ, and then using Equations (3-18) and (3-19).

Figures 3-2 through 3-10 show that, to a first approximation, the dependence of the incremental and average inductances on θ can be assumed to be sinusoidal, as in the basic model of the last section. However, due to the nonsinusoidal distribution of the windings in the motor, the phase relationships between the mutual inductances are not the same as those in Eq. (3-4). The phase relationships of all the mutual inductances between two adjacent phases (i.e., ℓ_{ab}, ℓ_{bc}, etc.) are shifted by 180 degrees with respect to the ones in Eq. (3-4).

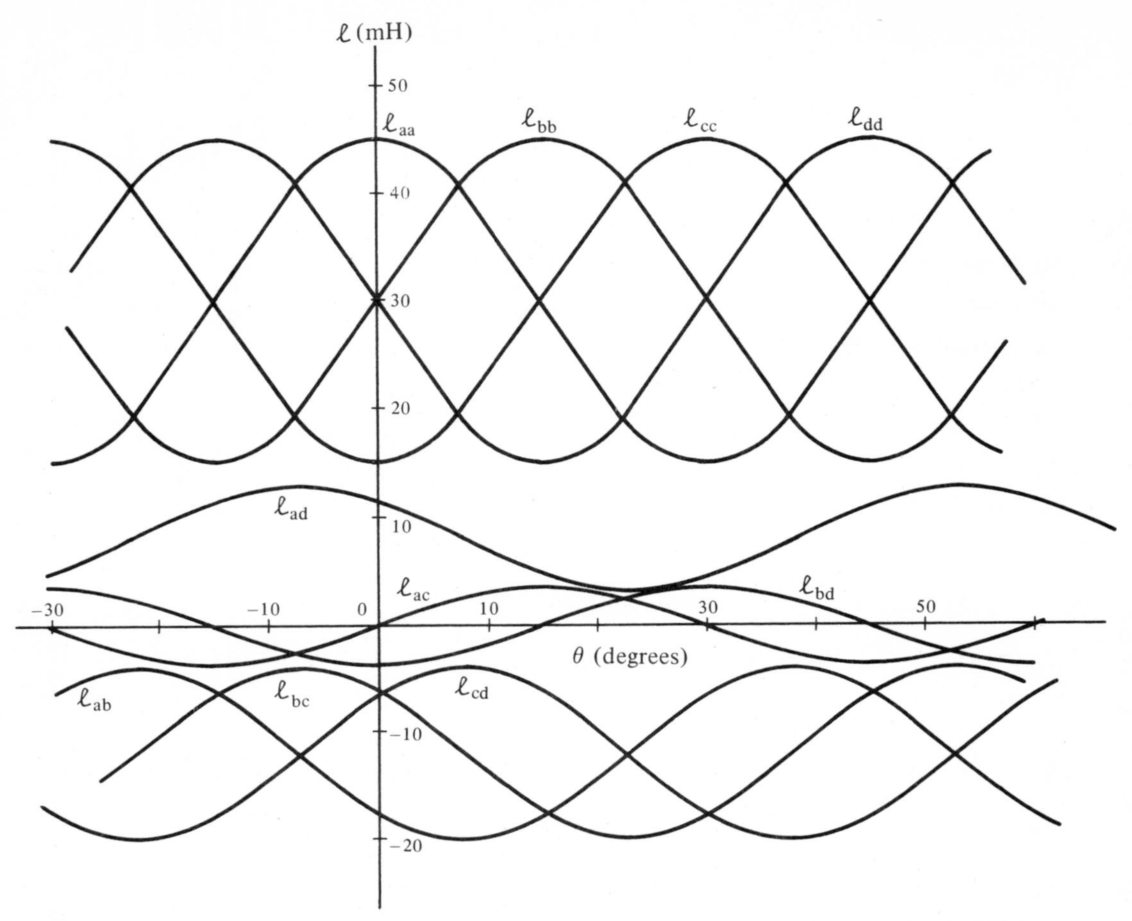

Figure 3-2. Unsaturated self and mutual inductances of the SM024-0018-FE Step Motor (made by Warner Electric Brake and Clutch Company).

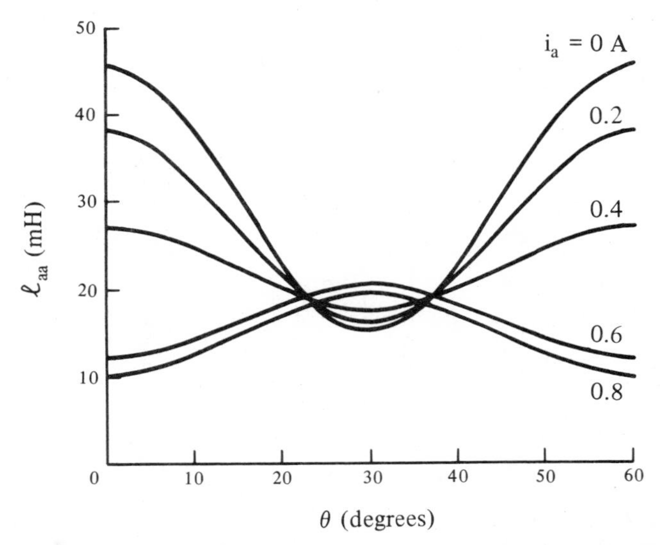

Figure 3-3. Incremental self inductance of phase A for the SM024-0018-FE step motor.

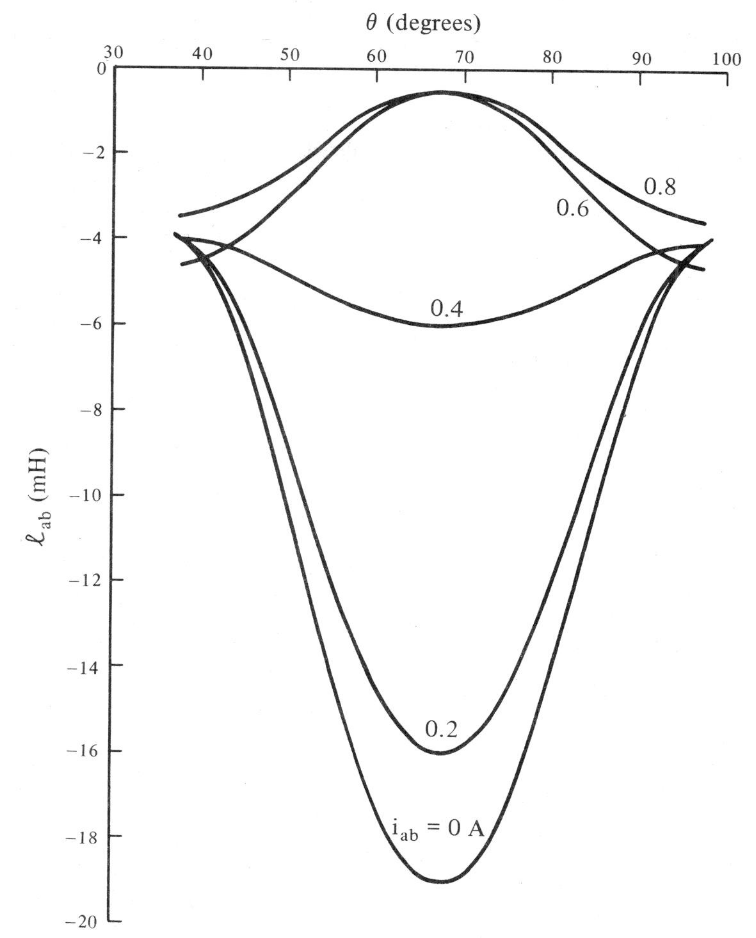

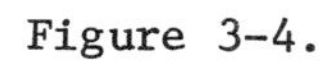

Figure 3-4.
Incremental mutual inductance, ℓ_{ab}, of the SM024-0018-FE step motor.

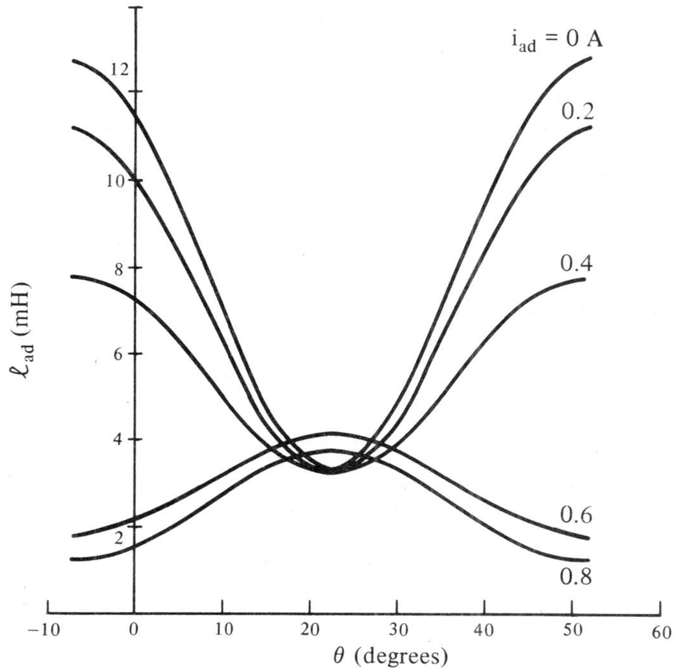

Figure 3-5.
Incremental mutual inductance, ℓ_{ad}, of the SM024-0018-FE step motor.

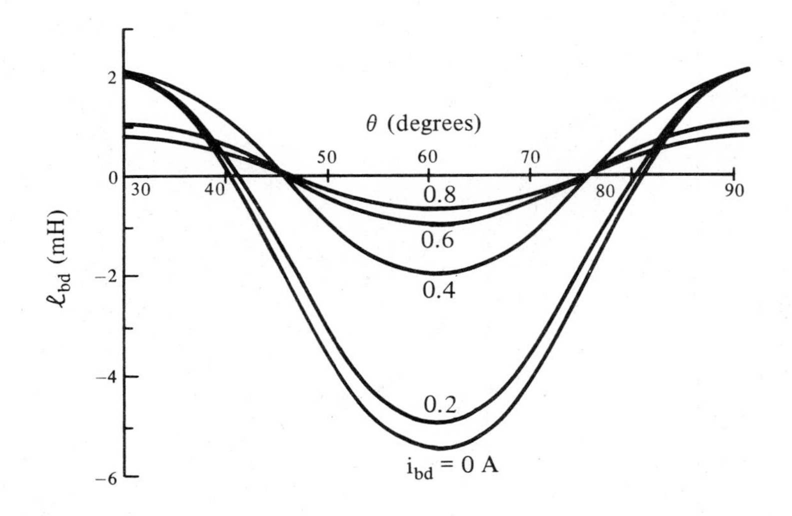

Figure 3-6. Incremental mutual inductance, ℓ_{bd}, of the SM024-0018-FE step motor.

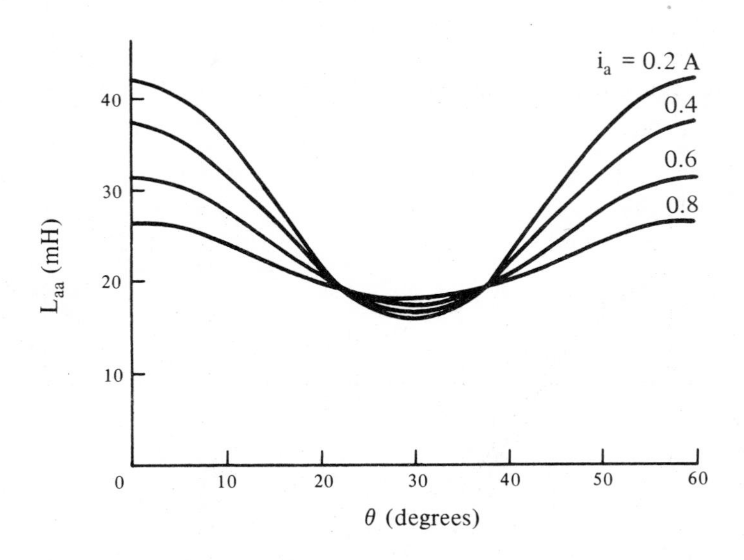

Figure 3-7. Average self inductance of phase A for the SM024-0018-FE step motor.

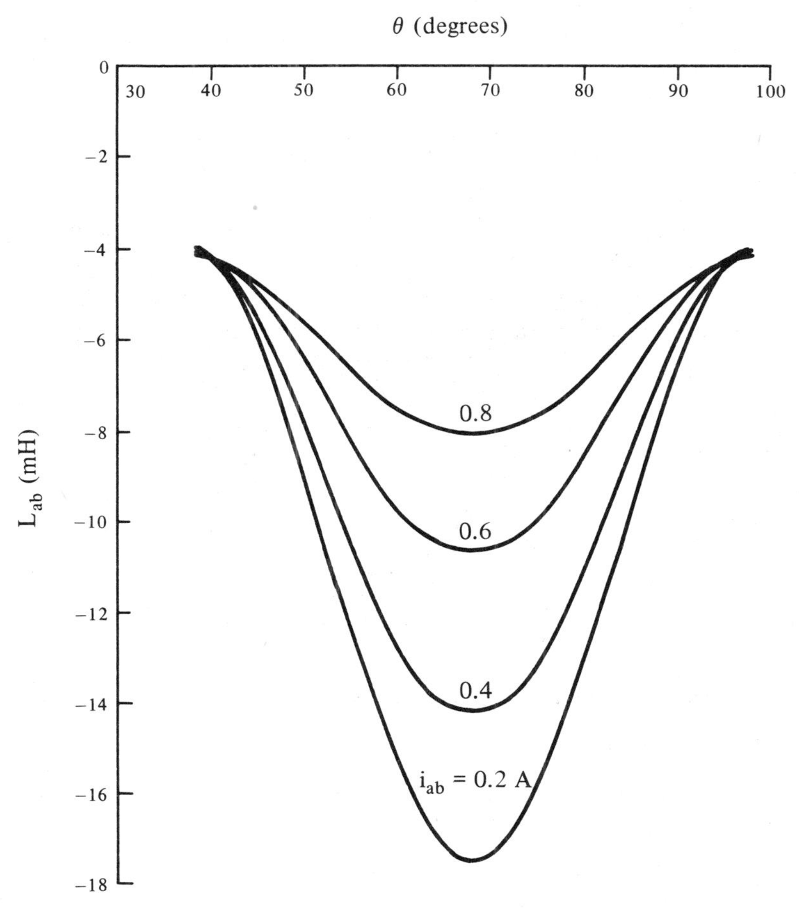

Figure 3-8. Average mutual inductance, L_{ab}, of the SM024-0018-FE step motor.

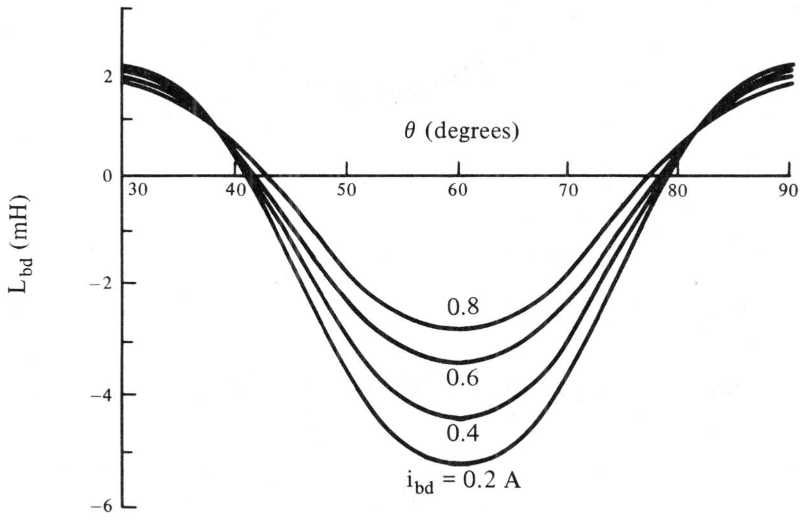

Figure 3-9. Average mutual inductance, L_{bd}, of the SM024-0018-FE step motor.

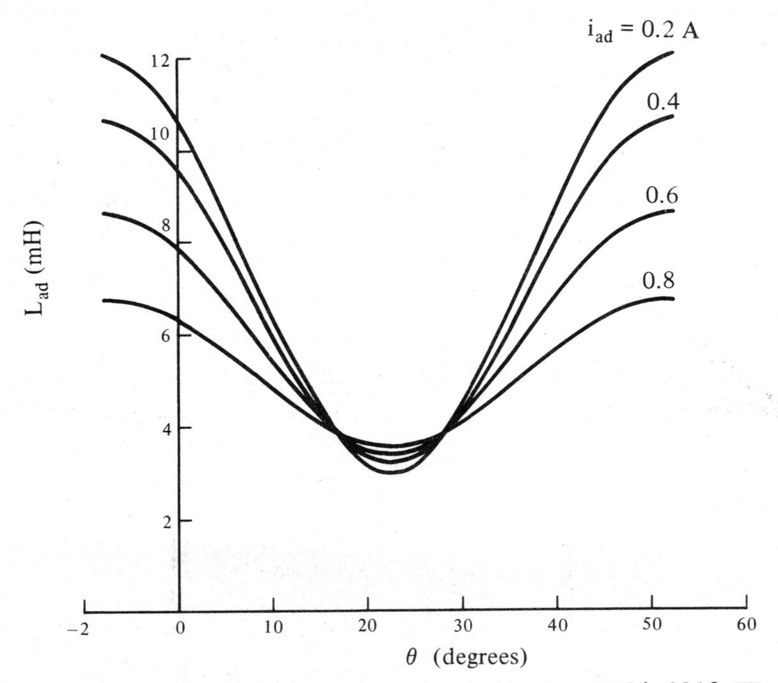

Figure 3-10. Average mutual inductance, L_{ad}, of the SM024-0018-FE step motor.

Thus, the modified incremental self and mutual inductances can be written as

$$\ell_{aa} = \ell_{aa1} + \ell_{aa2} \cos(6\theta)$$

$$\ell_{bb} = \ell_{bb1} + \ell_{bb2} \cos(6\theta + \pi/2)$$

$$\ell_{cc} = \ell_{cc1} + \ell_{cc2} \cos(6\theta + \pi)$$

$$\ell_{dd} = \ell_{dd1} + \ell_{dd2} \cos(6\theta + 3\pi/2)$$

$$\ell_{ab} = \ell_{ab1} + \ell_{ab2} \cos(6\theta - 3\pi/4)$$

$$\ell_{ac} = \ell_{ac1} + \ell_{ac2} \cos(6\theta + \pi/2)$$

$$\ell_{ad} = \ell_{ad1} + \ell_{ad2} \cos(6\theta - \pi/4)$$

$$\ell_{bc} = \ell_{bc1} + \ell_{bc2} \cos(6\theta - \pi/4)$$

$$\ell_{bd} = \ell_{bd1} + \ell_{bd2} \cos(6\theta + \pi)$$

$$\ell_{cd} = \ell_{cd1} + \ell_{cd2} \cos(6\theta + \pi/4) \qquad (3\text{-}20)$$

The expressions for the average self and mutual inductances are similar to those in Eq. (3-20).

The coefficients, ℓ_{aa1}, ℓ_{aa2}, ℓ_{bb1} and ℓ_{bb2}, etc., are functions of the phase currents, i_a and i_b, etc. Tables 3-1 and 3-2 list these coefficients for the incremental and average inductances, respectively. The coefficients are listed for five different values of the current. The current associated with a particular self or mutual inductance corresponds to the dependencies in Eqs. (3-14) and (3-15).

Table 3-1. The incremental self and mutual inductance coefficients for the SM024-0018-FE step motor (made by Warner Electric Brake and Clutch Company).

Current A	ℓ_{aa1}, ℓ_{bb1} ℓ_{cc1}, ℓ_{dd1} mH	ℓ_{aa2}, ℓ_{bb2} ℓ_{cc2}, ℓ_{dd2} mH	ℓ_{ab1}, ℓ_{bc1} ℓ_{cd1} mH	ℓ_{ab2}, ℓ_{bc2} ℓ_{cd2} mH	ℓ_{ac1} ℓ_{bd1} mH	ℓ_{ac2} ℓ_{bd2} mH	ℓ_{ad1} mH	ℓ_{ad2} mH
0	30	15	-11.5	7.5	-1.7	3.7	8	4.8
.2	27	11	-10.0	6	-1.5	3.5	7.2	4
.4	22	5	- 5	1	0	2.0	5.5	2.3
.6	16	- 4	- 2.5	- 2	0	1.8	3.0	- 1.2
.8	14.5	- 4.5	- 2	- 1.5	0	0.8	2.5	- 1.3

Table 3-2. The average self and mutual inductance coefficients for the SMO24-0018-FE step motor.

Current A	L_{aa1}, L_{bb1}, L_{cc1}, L_{dd1} mH	L_{aa2}, L_{bb2}, L_{cc2}, L_{dd2} mH	L_{ab1}, L_{bc1}, L_{cd1} mH	L_{ab2}, L_{bc2}, L_{cd2} mH	L_{ac1}, L_{bd1} mH	L_{ac2}, L_{bd2} mH	L_{ad1} mH	L_{ad2} mH
0.0	30	15	-11.5	7.5	- 1.7	3.7	8	4.8
0.2	28.5	13	-10.8	6.8	- 1.7	3.7	7.5	4.5
0.4	26.5	10.5	- 9.2	5.2	- 1.2	3.2	7.1	3.8
0.6	24	7.3	- 7.5	3.3	- 0.8	2.6	6.1	2.8
0.8	21.9	4.4	- 6.1	2.0	- 0.6	2.2	5.3	1.8

It is interesting to note that the coefficients listed in Tables 3-1 and 3-2 do not have the same relationships as in Eq. (3-5). In Eq. (3-5) the mean values of inductances, L_1 and L_3, have a very simple relationship, while in Tables 3-1 and 3-2 the relations are more complex, and very different from the earlier ones. Also, the amplitude of the inductance variation, L_2 in Eqs. (3-4) and (3-5), is the same for all self and mutual inductances, but in Tables 3-1 and 3-2 it has four distinct values.

In order to incorporate the effect of current on the inductances, a linear dependence will be assumed. This is a justifiable first approximation and should provide considerable improvement over the basic model of the last section. Again, the ultimate aim may be to incorporate higher-order dependencies if they prove necessary.

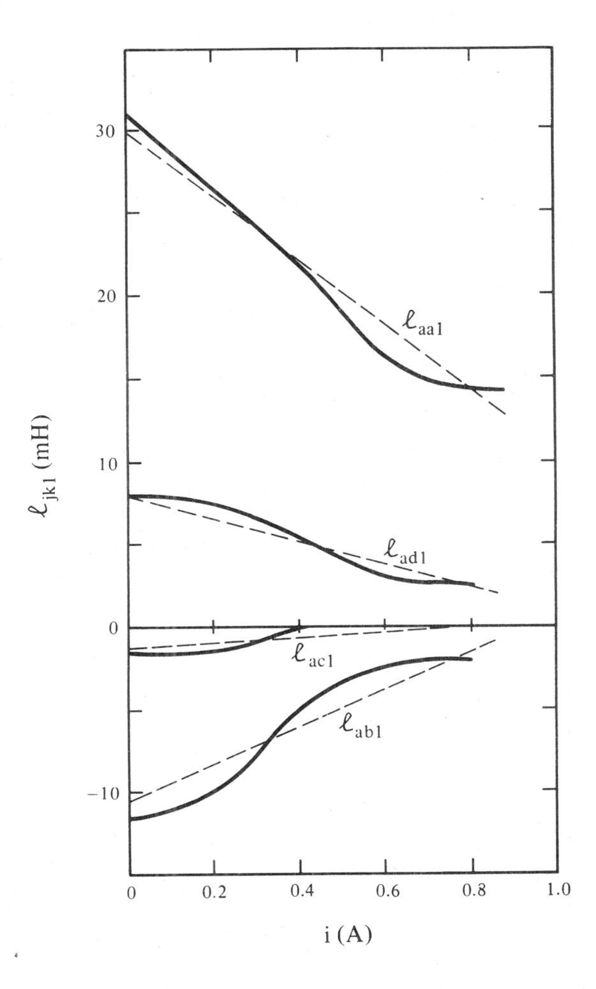

Figure 3-11. Inductance coefficients, ℓ_{aa1}, ℓ_{ab1}, ℓ_{ac1}, ℓ_{ad1}, and their linearization. SM024-0018-FE step motor.

Figure 3-11 shows a plot of the coefficients ℓ_{aa1}, ℓ_{ab1}, ℓ_{ac1} and ℓ_{ad1}, with respect to their current dependencies. The linearized approximations of these coefficients are shown by dotted lines on the same figure. All the coefficients have to be linearized in this way. The incremental and average inductances which are then obtained are written as

$$\ell_{aa} = [30. - 19.4\,|i_a|] + [13.5 - 23.1\,|i_a|]\cos(6\theta)$$

$$\ell_{bb} = [30. - 19.4\,|i_b|] + [13.5 - 23.1\,|i_b|]\cos(6\theta + \frac{\pi}{2})$$

$$\ell_{cc} = [30. - 19.4\,|i_c|] + [13.5 - 23.1\,|i_c|]\cos(6\theta + \pi)$$

$$\ell_{dd} = [30. - 19.4\,|i_d|] + [13.5 - 23.1\,|i_d|]\cos(6\theta + \frac{3\pi}{2})$$

$$\ell_{ab} = \left[-10.6 + 11.4\,\frac{|i_a| + |i_b|}{2}\right] + \left[7.0 - 11.3\,\frac{|i_a| + |i_b|}{2}\right]\cos(6\theta - \frac{3\pi}{4})$$

$$\ell_{ac} = \left[-1.4 + 2.0\,\frac{|i_a| + |i_c|}{2}\right] + \left[3.7 - 3.6\,\frac{|i_a| + |i_c|}{2}\right]\cos(6\theta + \frac{\pi}{2})$$

$$\ell_{ad} = \left[8.1 - 7.1\,\frac{|i_a| + |i_d|}{2}\right] + \left[4.5 - 7.6\,\frac{|i_a| + |i_d|}{2}\right]\cos(6\theta - \frac{\pi}{4})$$

$$\ell_{bc} = \left[-10.6 + 11.4\,\frac{|i_b| + |i_c|}{2}\right] + \left[7.0 - 11.3\,\frac{|i_b| + |i_c|}{2}\right]\cos(6\theta - \frac{\pi}{4})$$

$$\ell_{bd} = \left[-1.4 + 2.0\,\frac{|i_b| + |i_d|}{2}\right] + \left[3.7 - 3.6\,\frac{|i_b| + |i_d|}{2}\right]\cos(6\theta + \pi)$$

$$\ell_{cd} = \left[-10.6 + 11.4\,\frac{|i_c| + |i_d|}{2}\right] + \left[7.0 - 11.3\,\frac{|i_c| + |i_d|}{2}\right]\cos(6\theta + \frac{\pi}{4})$$

(3-21)

$$L_{aa} = [30. - 10.\ |i_a|] + [16. - 14.\ |i_a|]\cos(6\theta)$$

$$L_{bb} = [30. - 10.\ |i_b|] + [16. - 14.\ |i_b|]\cos(6\theta + \frac{\pi}{2})$$

$$L_{cc} = [30. - 10.\ |i_c|] + [16. - 14.\ |i_c|]\cos(6\theta + \pi)$$

$$L_{dd} = [30. - 10.\ |i_d|] + [16. - 14.\ |i_d|]\cos(6\theta + \frac{3\pi}{2})$$

$$L_{ab} = \left[-12. + 7.5\ \frac{|i_a| + |i_b|}{2}\right] + \left[8. - 7.5\ \frac{|i_a| + |i_b|}{2}\right]\cos(6\theta - \frac{3\pi}{4})$$

$$L_{ac} = \left[-\ 2 + 1.7\ \frac{|i_a| + |i_c|}{2}\right] + \left[4. - 2.2\ \frac{|i_a| + |i_c|}{2}\right]\cos(6\theta + \frac{\pi}{2})$$

$$L_{ad} = \left[8. - 3.1\ \frac{|i_a| + |i_d|}{2}\right] + \left[5. - 3.9\ \frac{|i_a| + |i_d|}{2}\right]\cos(6\theta - \frac{\pi}{4})$$

$$L_{bc} = \left[-12. + 7.5\ \frac{|i_b| + |i_c|}{2}\right] + \left[8. - 7.5\ \frac{|i_b| + |i_c|}{2}\right]\cos(6\theta - \frac{\pi}{4})$$

$$L_{bd} = \left[-\ 2. + 1.7\ \frac{|i_b| + |i_d|}{2}\right] + \left[4. - 2.2\ \frac{|i_b| + |i_d|}{2}\right]\cos(6\theta + \pi)$$

$$L_{cd} = \left[-12. + 7.5\ \frac{|i_c| + |i_d|}{2}\right] + \left[8. - 7.5\ \frac{|i_c| + |i_d|}{2}\right]\cos(6\theta + \frac{\pi}{4})$$

(3-22)

With Λ as in Eq. (3-11), the modified form of the voltage equations (3-1) and (3-6) is

$$V_R = \frac{d\Lambda}{dt} = \frac{\partial \Lambda}{\partial I}\frac{dI}{dt} + \frac{\partial \Lambda}{\partial \theta}\frac{d\theta}{dt} = L_1 \frac{dI}{dt} + \frac{dL}{d\theta} I \frac{d\theta}{dt} \qquad (3\text{-}23)$$

The differential equation form of Eq. (3-23), suitable for simulation purposes, is

$$\frac{dI}{dt} = L_1^{-1}\ [V_R - \frac{d\theta}{dt}\frac{dL}{d\theta} I] \qquad (3\text{-}24)$$

Comparison of Eq. (3-24) with Eq. (3-7) shows that the inverse of the incremental inductance matrix L_1 must now be taken.

Torque Developed

The presence of saturation results in the developed torque being different from the expressions of Eqs. (3-8) and (3-9). It is known that the developed torque is a function of the rotor position θ and the phase currents I. Thus

$$T = T(\theta, I) \tag{3-25}$$

It is assumed that, as in Eq. (3-8), the total developed torque is composed of self and mutual components. Using the same subscript convention as is used for the self and mutual inductances, the total developed torque is written as

$$T = T_{aa} + T_{bb} + T_{cc} = T_{dd} + T_{ab} + T_{ac} + T_{ad} + T_{bc} + T_{bd} + T_{cd} =$$

$$\sum_i \sum_j T_{ij} \qquad i,j = a, b, c, d \tag{3-26}$$

As in the case of the inductances, it is assumed that the self torque is a function of only one current; that is,

$$T_{jj} = T_{jj}\,(\theta, i_j)$$

$$j = a, b, c, d \tag{3-27}$$

and the mutual torque is a function of the geometric mean of the two phase currents. Thus,

$$T_{jk} = T_{jk}(\theta, i_j, i_k) = T_{jk}(\theta, \hat{i}_{jk}) \tag{3-28}$$

where $\hat{i}_{jk} = \sqrt{i_j i_k}$

The geometric mean is used here since the torque without saturation in Eq. (3-8) has a product-type dependency $((\sqrt{i_j i_k})^2)$.

It is possible to determine the nature of the current and θ dependency in Eqs. (3-27) and (2-38) by use of experimental torque measurements. First, the self torques are determined as a function of the rotor position θ for a given value of phase current. This is done by using a torque transducer, exciting one phase of the motor, and measuring the torque, while slowly changing the rotor position θ. Figure 3-12 shows this for the four phases of the SM024-0018-FE motor for one value of phase current. Note that all four self torques are similar with appropriate phase shifts. Figure 3-13 shows the torque of one phase, T_{aa}, as a function of θ, for different values of phase current, i_a. As expected, the other self torques also exhibit a similar behavior with different phase currents.

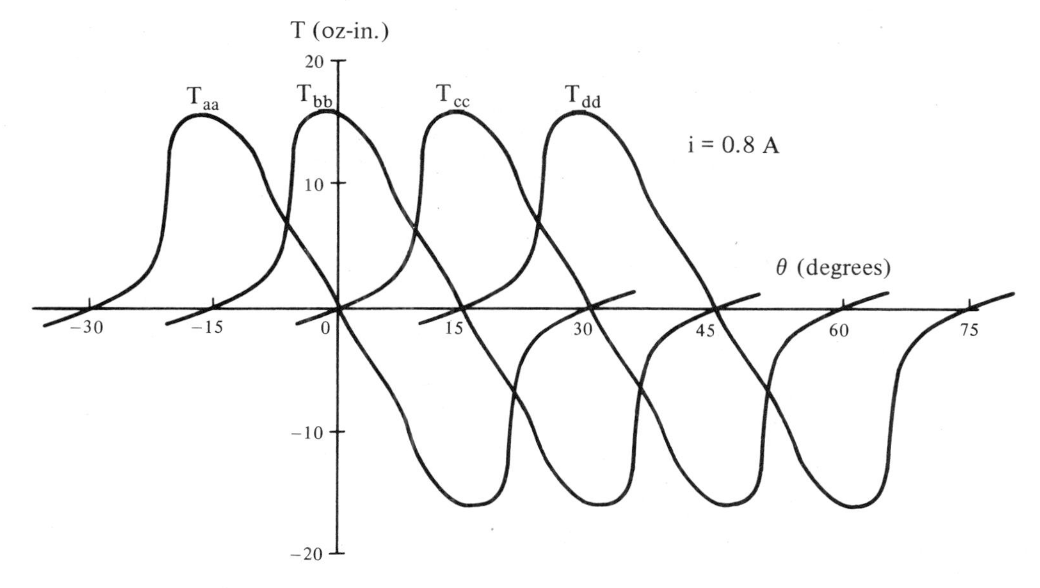

Figure 3-12. Static self torque curves for all the phases of the SM024-0018-FE step motor.

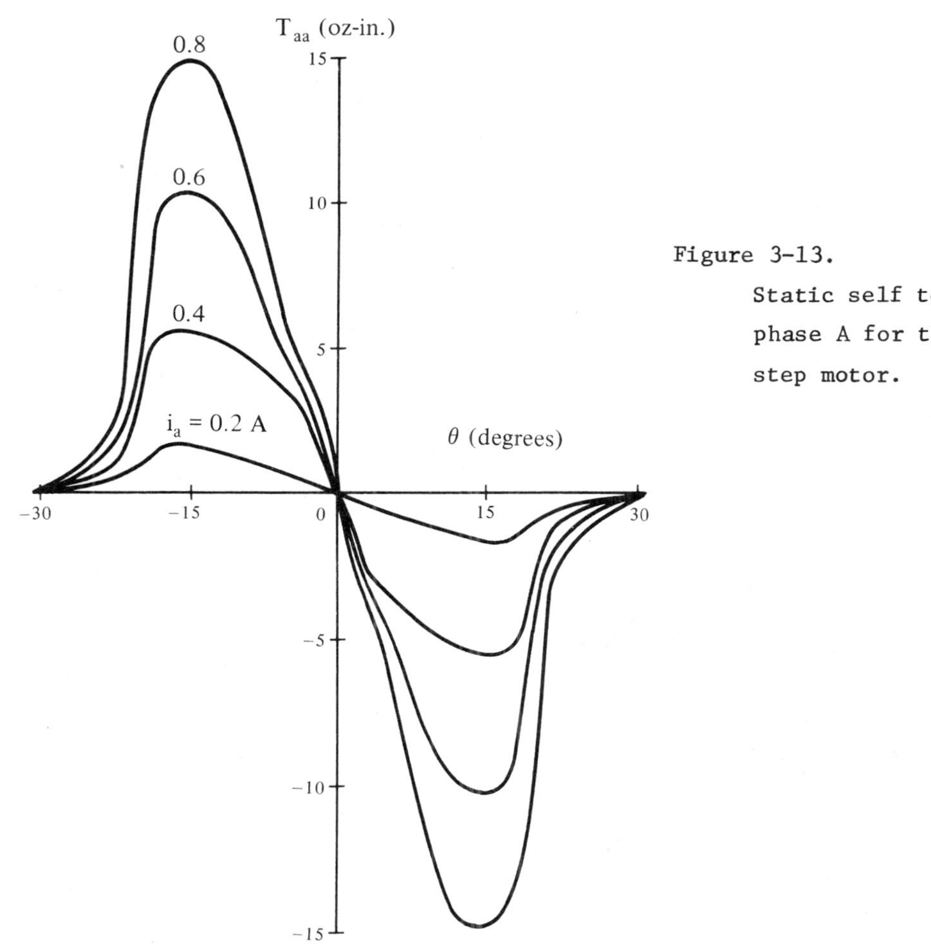

Figure 3-13. Static self torque curves of phase A for the SM024-0018-FE step motor.

The variation of self torque with respect to θ is now approximated by the fundamental component of the torque curve in Figure 3-13. This approximation is valid if the torque curve does not possess high-frequency components. Again, greater accuracy can be obtained, if necessary, by use of higher-order terms. The self torque is written as

$$T_{jj} = -K_{jj}(i_j)\sin(6\theta - \alpha_j) \tag{3-29}$$

$$j = a, b, c, d,$$

where α_j is the reference phase displacement for phase j.

In order to incorporate the effect of i_j on K_{jj} in Eq. (3-29), a linear dependence is assumed. This provides an adequate first approximation. Higher-order terms may be used to provide greater accuracy, if needed. Figure 3-14 shows the variation of K_{aa} with i_a, as will the linearization.

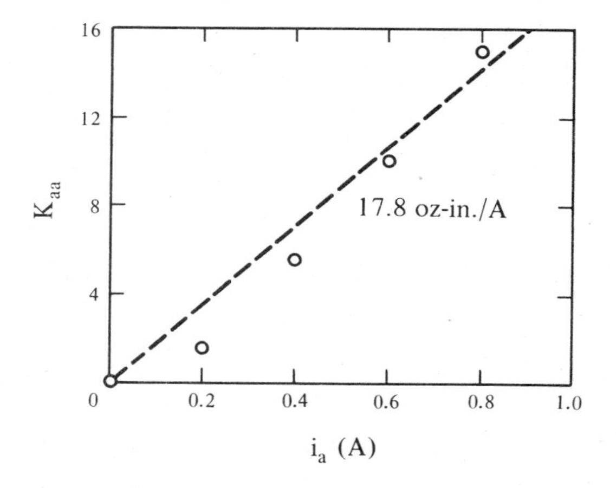

Figure 3-14. Static self torque coefficient, K_{aa}, and its linearization for the SM024-0018-FE step motor.

The self torques then become

$$T_{aa} = -17.8\,|i_a|\,\sin(6\theta)$$

$$T_{bb} = -17.8\,|i_b|\,\sin(6\theta + \tfrac{\pi}{2})$$

$$T_{cc} = -17.8\,|i_c|\,\sin(6\theta + \pi)$$

$$T_{dd} = -17.8\,|i_d|\,\sin(6\theta + \tfrac{3\pi}{2}) \tag{3-30}$$

In order to determine the mutual torques, it is necessary to energize two phases of the motor at one time, and measure the developed torque as a function of the rotor position for different values of currents. If $T_{(total)jk}$ is the developed torque when i_j and i_k are energized, j, k = a, b, c, d, j ≠ k, then the mutual torque T_{jk} is given by

$$T_{kj}(\theta, i_j, i_k) = T_{(total)jk}(\theta, i_j, i_k) - T_{jj}(\theta, i_j) - T_{kk}(\theta, i_k) \quad (3\text{-}31)$$

where T_{jj} and T_{kk} are available from previous measurements.

Figures 3-15 through 3-22 show the graphical interpretation of Eq. (3-31) for various values of currents and various j and k. From these and other figures the mutual torques are approximated by

$$T_{ab} = -2.2\sqrt{i_a i_b}\ \sin(6\theta - \frac{3\pi}{4})$$

$$T_{ac} = -2.5\sqrt{i_a i_c}\ \sin(6\theta)$$

$$T_{ad} = -2.2\sqrt{i_a i_d}\ \sin(6\theta + \frac{3\pi}{4})$$

$$T_{bc} = -2.2\sqrt{i_b i_c}\ \sin(6\theta - \frac{\pi}{4})$$

$$T_{bd} = -2.5\sqrt{i_b i_d}\ \sin(6\theta + \frac{\pi}{2})$$

$$T_{cd} = -2.2\sqrt{i_c i_d}\ \sin(6 + \frac{\pi}{4}) \quad (3\text{-}32)$$

These expressions show a marked difference from the expressions in Eq. (3-9).

Equations (3-26), (3-30) and (3-32) are used in place of Eqs. (3-8) and (3-9) to obtain the developed torque of the motor.

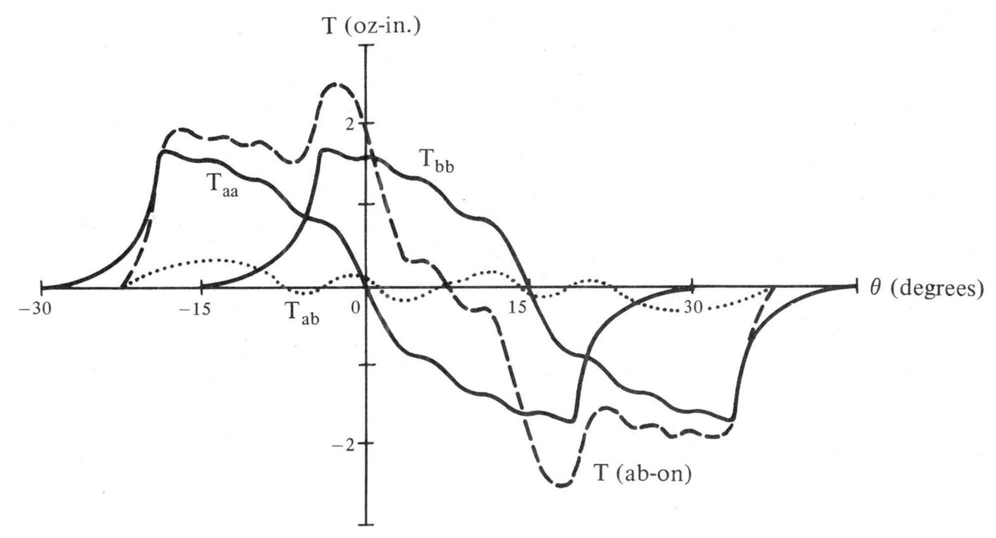

Figure 3-15. Graphical computation of the mutual torque, T_{ab}, for $i_a = i_b = 0.2$ amp. SM024-0018-FE step motor.

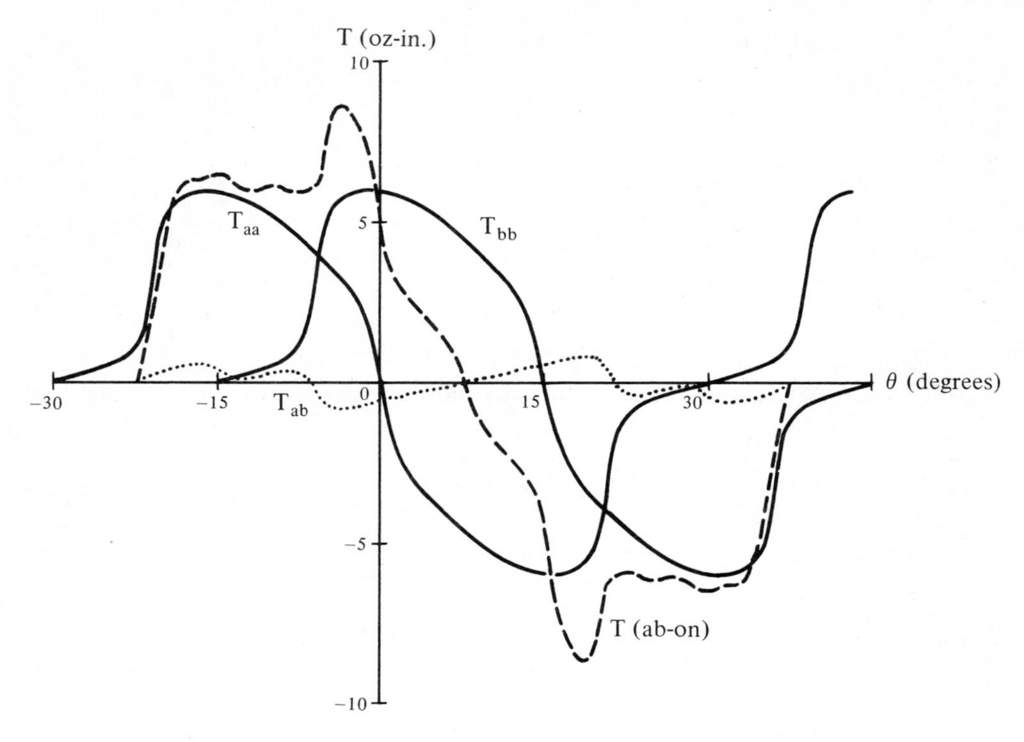

Figure 3-16. Graphical computation of the mutual torque, T_{ab}, for $i_a = i_b = 0.4$ amp. SM024-0018-FE step motor.

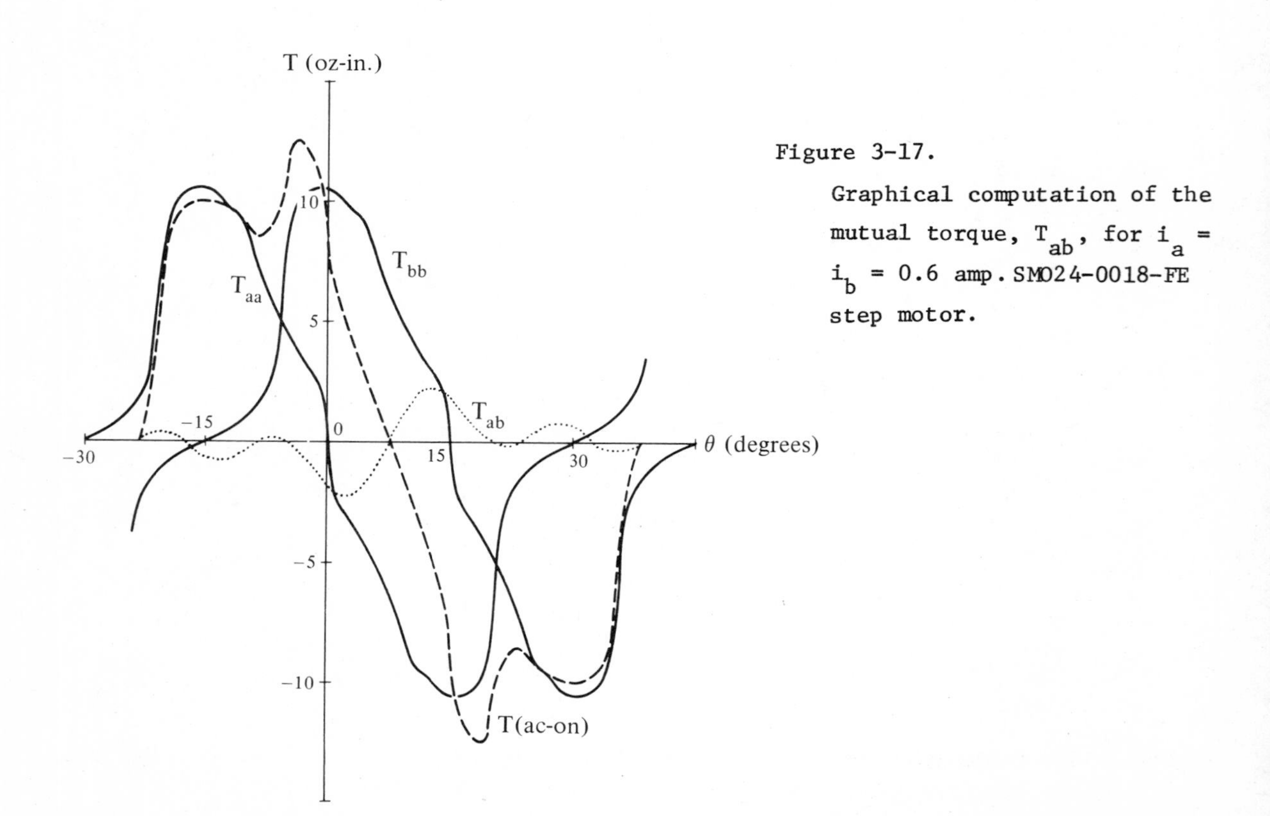

Figure 3-17.

Graphical computation of the mutual torque, T_{ab}, for $i_a = i_b = 0.6$ amp. SM024-0018-FE step motor.

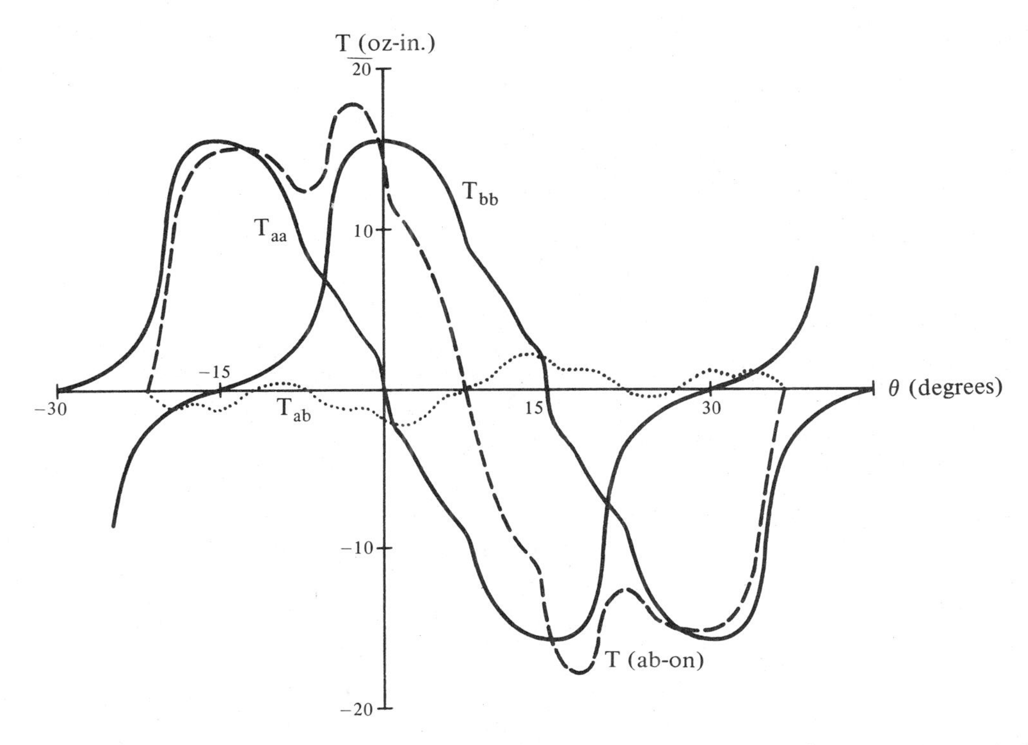

Figure 3-18. Graphical computation of the mutual torque, T_{ab}, for $i_a = i_b = 0.8$ amp. SM024-0018-FE step motor.

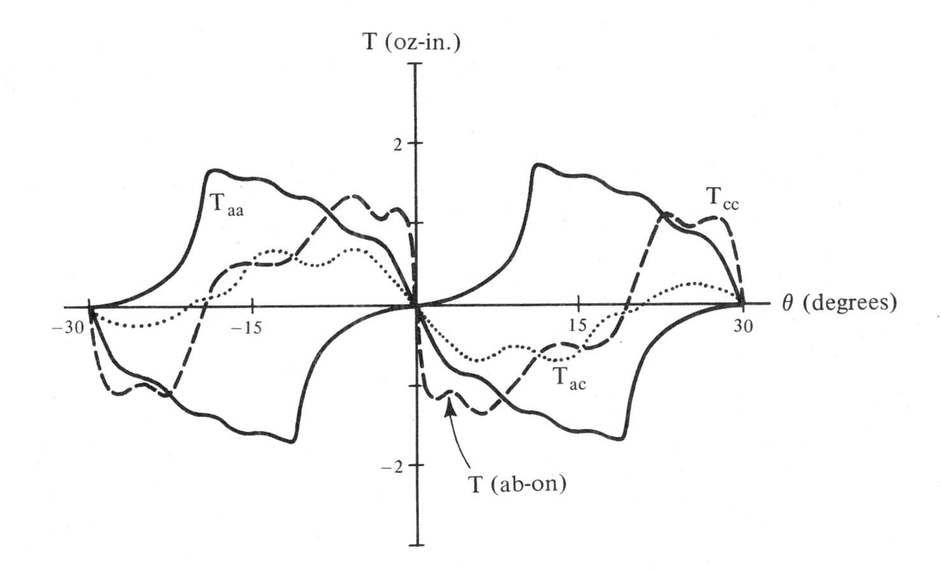

Figure 3-19. Graphical computation of the mutual torque, T_{ac}, for $i_a = i_c = 0.2$ amp. SM024-0018-FE step motor.

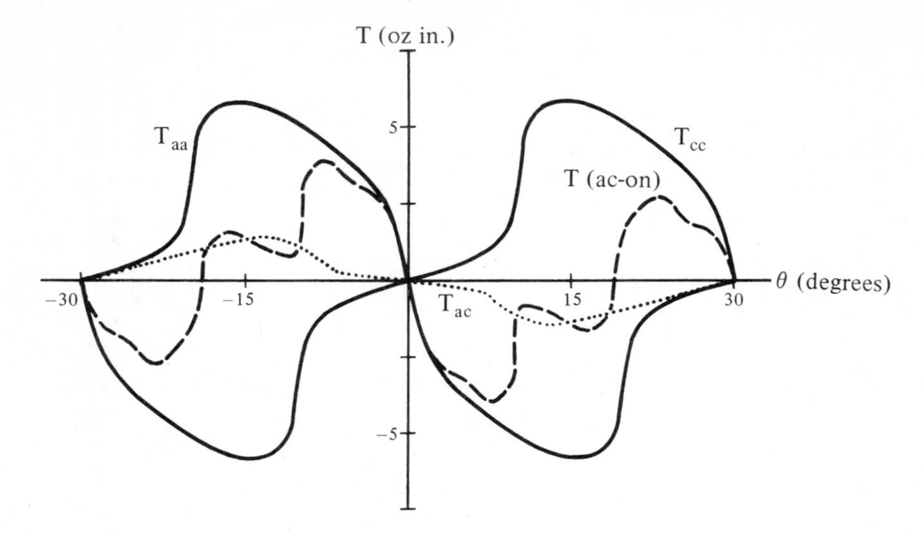

Figure 3-20. Graphical computation of the mutual torque, T_{ac}, for $i_a = i_c = 0.4$ amp. SM024-0018-FE step motor.

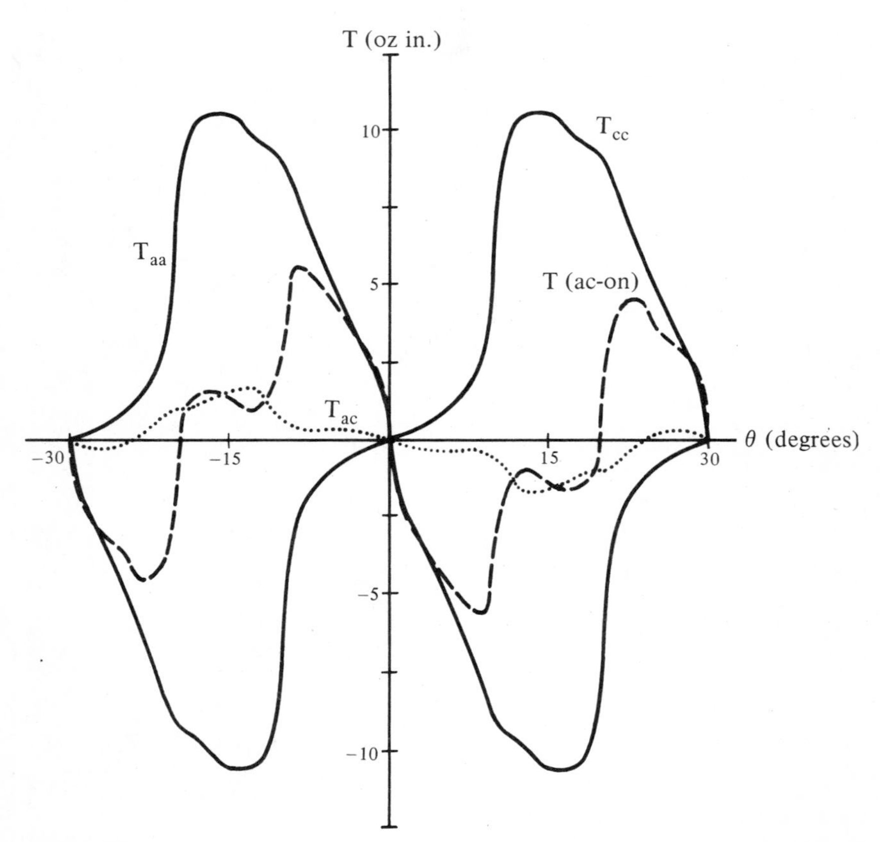

Figure 3-21. Graphical computation of the mutual torque, T_{ac}, for $i_a = i_c = 0.6$ amp. SM024-0018-FE step motor.

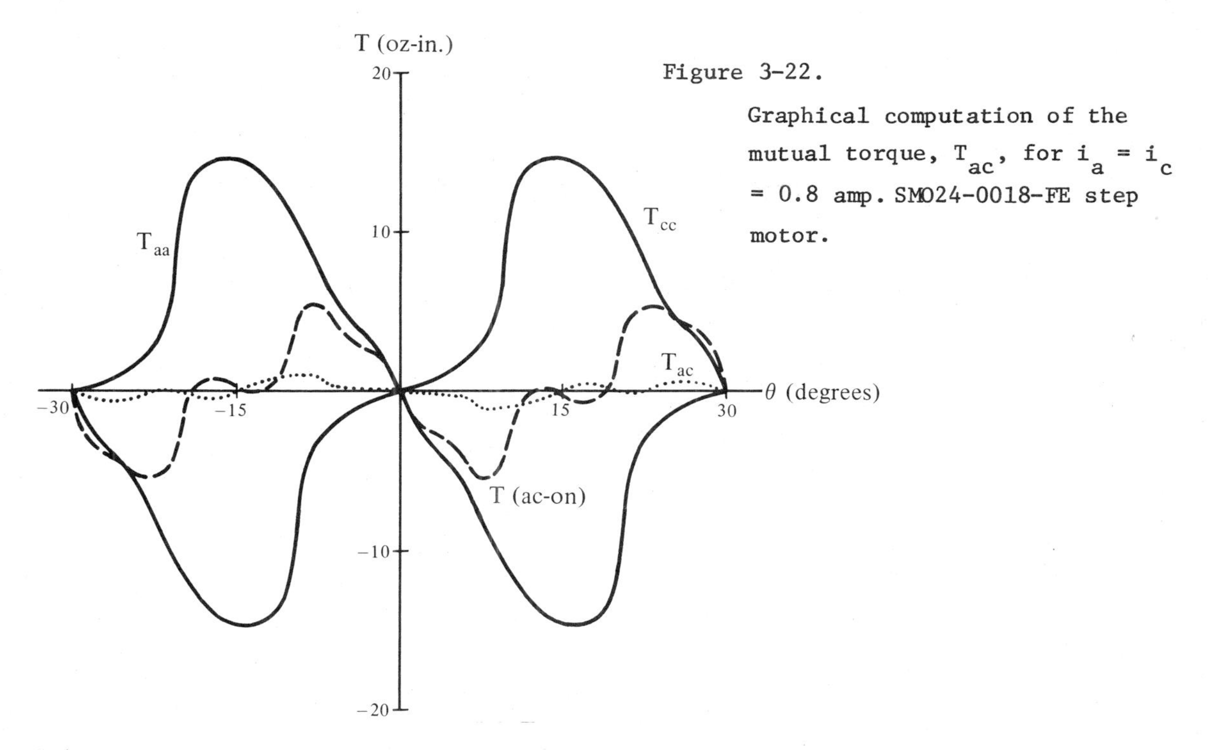

Figure 3-22.

Graphical computation of the mutual torque, T_{ac}, for $i_a = i_c$ = 0.8 amp. SMO24-0018-FE step motor.

3-4 COMPUTER SIMULATION

The mathematical model of the single-stack VR step motor is utilized here to illustrate its use in simulating the dynamic response of the motor. The complete model of the motor consists of

(1) the differential equations for current, Eq. (2-24),

(2) the torque expressions, Eqs. (3-30) and (3-32),

(3) the rotor dynamics, Eq. (3-10).

This model is programmed onto a digital computer to be used with an integrating subroutine. Figures 3-23, 3-24 and 3-25 show the flow diagrams for the logical operations in the computer program. The motor and the drive system have the following parameters:

Applied Voltage: 5.2 volts

Steady-State Current: 0.87 amp.

Type of Drive: one phase on

Rotor Inertia: .00014 oz.in.sec^2

Motor Resistance: 6 ohms

External Resistance: None

The single-step response is simulated with the following inertial loads:

a. load inertia = 0.00038 oz-in-sec^2

b. load inertia = 0.00203 oz-in-sec^2

c. load inertia = 0.0041 oz-in-sec^2.

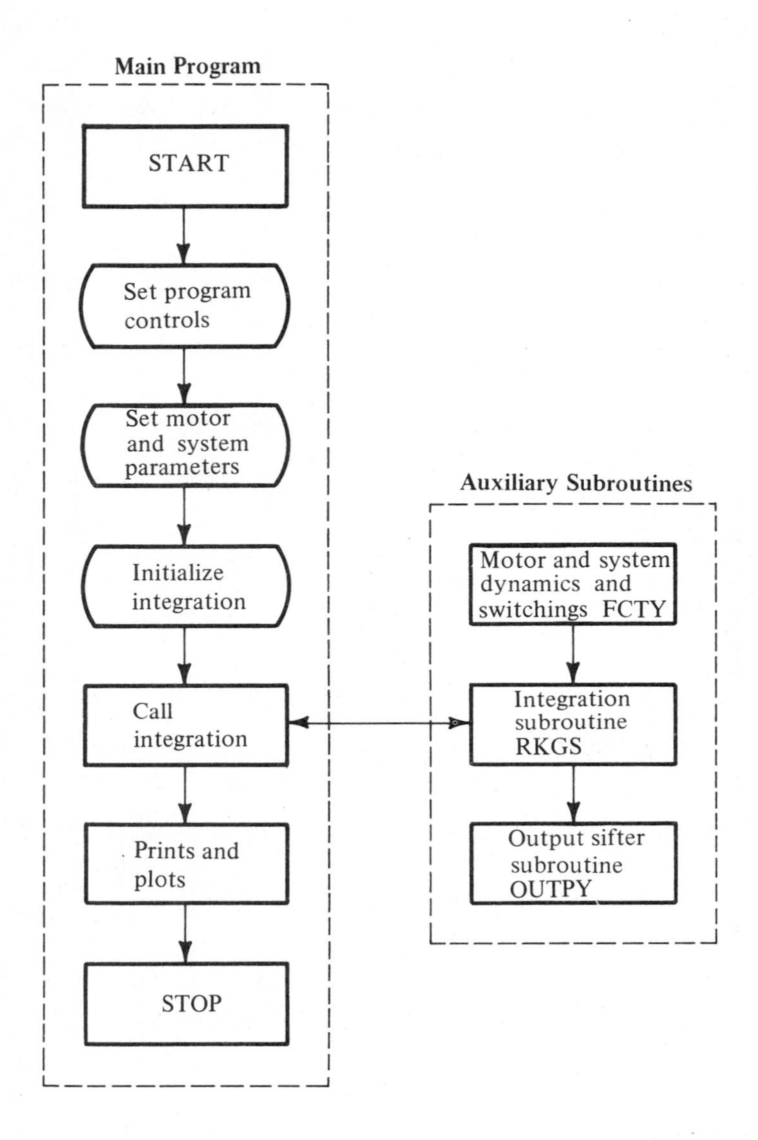

Figure 3-23. Flow Chart illustrating a step motor simulation program.

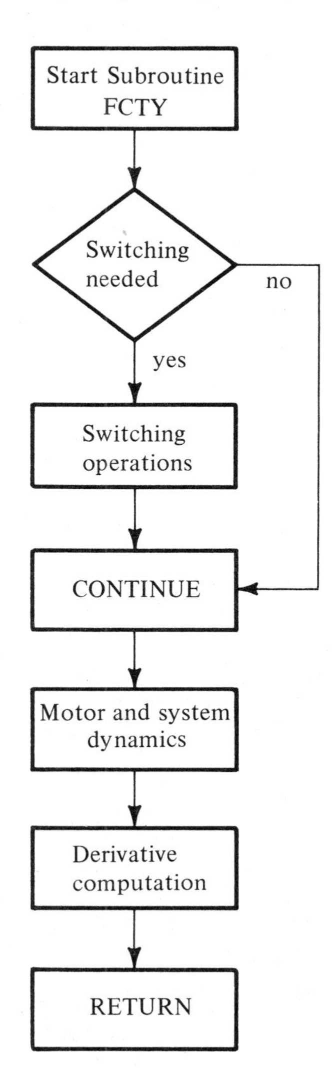

Figure 3-24.
Flow Chart illustrating the operations in the subroutine FCTY.

Figure 3-25.
Flow Chart illustrating the operations in the subroutine OUTPY.

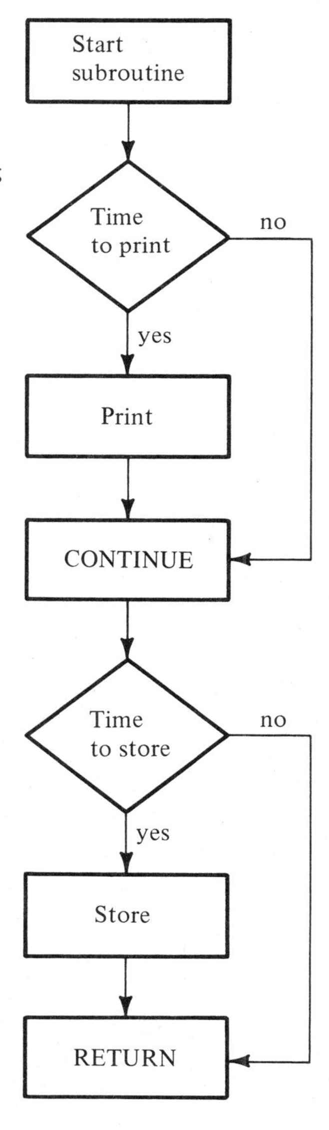

Figures 3-26 through 3-28 show the actual responses of the motor, and Figures 3-29 through 3-31 show the simulated responses obtained with the use of the computer program.

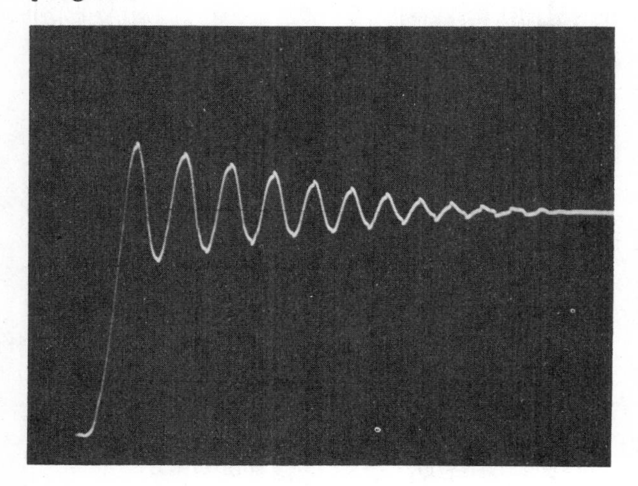

Figure 3-26.
Single-step response of the SM024-0018-FE step motor. Inertial load = 0.00038 oz.in. sec^2 (20 msec/div).

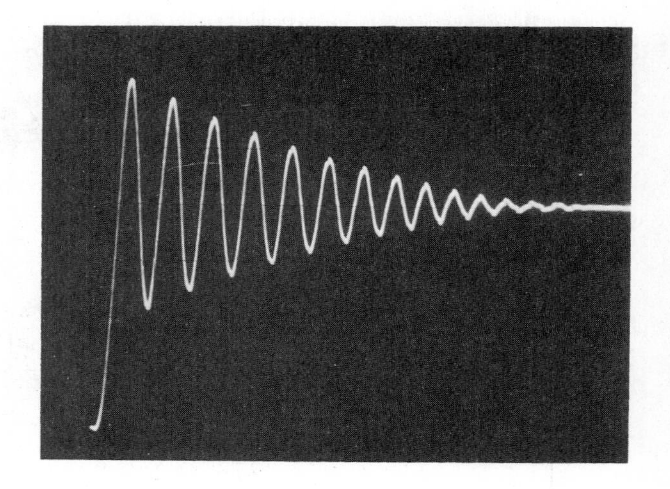

Figure 3-27.
Single-step response of the SM024-0018-FE step motor. Inertial load = 0.00203 oz.in. sec^2 (50 msec/div).

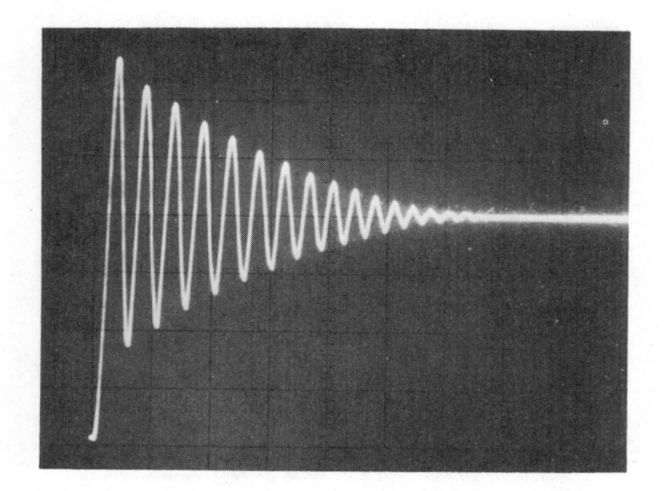

Figure 3-28.
Single-step response of the SM024-0018-FE step motor. Inertial load = 0.0041 oz.in. sec^2 (100 msec/div).

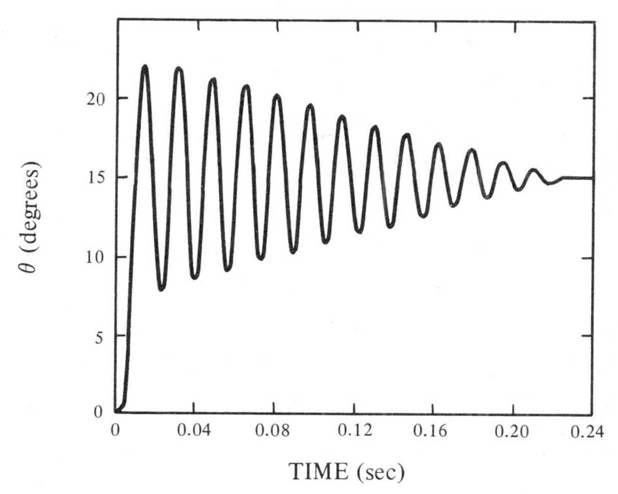

Figure 3-29.

Simulated single-step response of the SM024-0018-FE step motor. Inertial load = 0.00038 oz.in.sec^2.

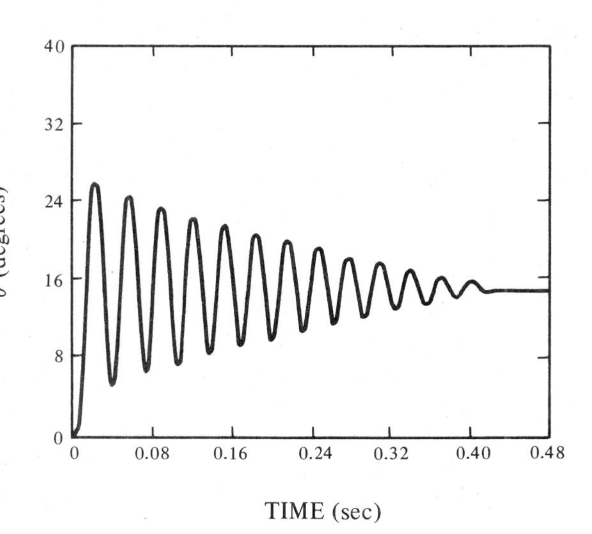

Figure 3-30.

Simulated single-step response of the SM024-0018-FE step motor. Inertial load = 0.00203 oz.in.sec^2.

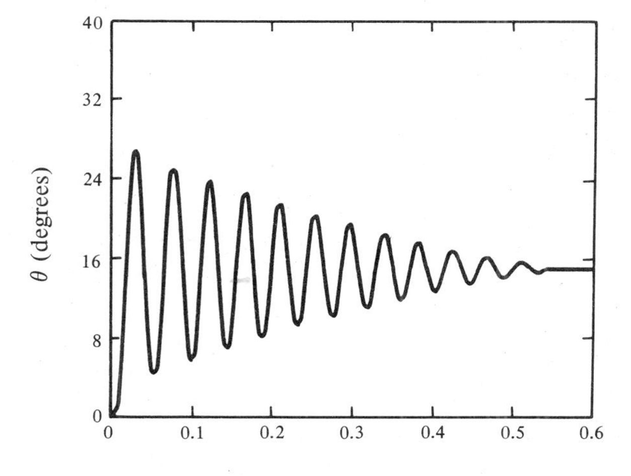

Figure 3-31.

Simulated single-step response of the SM024-0018-FE step motor. Inertial load = 0.0041 oz.in.sec^2.

4 COMPUTER SIMULATION OF STEP MOTOR SYSTEMS

Gurdial Singh
Department of Electrical Engineering
University of Illinois at Urbana-Champaign
Urbana, Illinois

4-1 INTRODUCTION

The dynamic behavior of a step motor is, in general, highly nonlinear. Consequently, it is inadequate to approximate its performance by linear models. One of the methods of obtaining realistic solutions to the analysis and design of step motor systems is computer simulation. Since the mathematical models of various types of step motors have been made available in the preceeding chapters, it becomes a simple matter to program these on computers.

In the past analog computer simulation of step motors has been conducted.[1] However, it has proven to be practical only for single-step performance studies. For multiple stepping, the initial conditions at the beginning of each new step would have to be reset, thus making the procedure extremely tedious. The only advantage of using an analog computer is that the change of the performance of the motor can be investigated when the motor parameters are being changed continuously on line.

On the other hand, digital computer simulation offers a great deal of flexibility. The computer can be used for both analysis and design. Standard FORTRAN or a simulation language like the IBM 360 Continuous System Modeling Program (CSMP)[2] may be used. It has been found that the use of FORTRAN provides greater flexibility for simulating step motor systems than the use of CSMP or any other simulation language. This is because step motor systems are inherently digital in nature and involve a considerable amount of switching operations, which translate to simple logical statements in FORTRAN. Efficiency of the program is also better with FORTRAN, although some extra effort may be required at the programming stage.

In this paper, the use of FORTRAN is emphasized. A programmer who understands a simulation program written in FORTRAN can generally modify it for use with CSMP, but the reverse is usually not true. To illustrate the simulation procedure, the SM048AB step motor manufactured by the Warner Electric Brake and Clutch Company is used. This motor is a three-phase, multiple-stack, variable-reluctance step motor.

It has 16 rotor teeth and a resolution of 48 steps-per-revolution. Therefore, the step size is 7.5 deg. The construction and principle of operation of this type of motor is discussed in detail in Chapter 1.

Before any step motor system can be simulated, some basic information regarding the system and the simulation procedure is needed.

4-2 THE MATHEMATICAL MODEL OF THE SYSTEM

The mathematical model of a step motor consists of the system of differential equations which describe the dynamic behavior of the step motor system. It should be in the form of a system of first-order coupled differential equations. It is important that these differential equations accurately represent the system to be simulated since the simulation results are only as valid as the model which is used to generate them. In the case of the SM048AB step motor with a linear magnetic circuit, connected to a load with inertia and friction, these equations are

$$\dot{\theta} = \omega \tag{4-1}$$

$$\dot{\omega} = \frac{1}{J}[T - B\omega - T_F] \tag{4-2}$$

$$\frac{di_a}{dt} = \frac{1}{L_a}[v_a - i_a r_a + 16L_1 i_a \omega \mathrm{Sin}(16\theta)] \tag{4-3}$$

$$\frac{di_b}{dt} = \frac{1}{L_b}[v_b - i_b r_b + 16L_1 i_b \omega \mathrm{Sin}(16\theta - \frac{2\pi}{3})] \tag{4-4}$$

$$\frac{di_c}{dt} = \frac{1}{L_c}[v_c - i_c r_c + 16L_1 i_c \omega \mathrm{Sin}(16\theta + \frac{2\pi}{3})] \tag{4-5}$$

with

$$T = -K_o\left[|i_a|\mathrm{Sin}(16\theta) + |i_b|\mathrm{Sin}(16\theta - \frac{2\pi}{3}) + |i_c|\mathrm{Sin}(16\theta + \frac{2\pi}{3})\right] \tag{4-6}$$

where

θ = the angular position of the rotor, $\theta = 0$ represents the equilibrium of phase a

ω = the angular velocity of the rotor

i_a = the current in phase a winding

i_b = the current in phase b winding

i_c = the current in phase c winding

T = the torque developed by the motor

B = the viscous frictional coefficient of the motor and load

T_F = the coulomb frictional torque of the rotor and load

v_a = the voltage applied to phase a winding

v_b = the voltage applied to phase b winding

v_c = the voltage applied to phase c winding

r_a = the resistance of phase a winding circuit

r_b = the resistance of phase b winding circuit

r_c = the resistance of phase c winding circuit

$L_a = L_o + L_1 \text{Cos}(16\theta)$ = the inductance of phase a winding

$L_b = L_o + L_1 \text{Cos}(16\theta - 2\pi/3)$ = the inductance of phase b winding

$L_c = L_o + L_1 \text{Cos}(16\theta - 2\pi/3)$ = the inductance of phase c winding

L_o = the average value of inductance of each phase winding

L_1 = the peak magnitude of the variation of the inductance of each phase winding with rotor position

$J = J_M + J_L$ = the total inertia on the motor shaft

J_M = the inertia of the rotor of the motor

J_L = the inertia of the load

K_o = the torque constant of the motor

It should be noted that Eqs. (4-1) through (4-6) are valid only for the motor specified. For a multiple-stack VR step motor with more than three phases and different step size, these equations should be modified accordingly. If there are external dynamics connected to the rotor of the step motor, these must also be properly modeled. For example, if there is a viscous damper connected to the rotor, Eqs. (4-1) and (4-2) are to be replaced by

$$\frac{d\theta}{dt} = \omega \tag{4-7}$$

$$\frac{d\theta_d}{dt} = \omega_d \tag{4-8}$$

$$\frac{d\omega}{dt} = \frac{1}{J}[T - B\omega - (\omega - \omega_d)K_D] \tag{4-9}$$

$$\frac{d\omega_d}{dt} = \frac{K_D}{J_R}(\omega - \omega_d) \tag{4-10}$$

where

θ_d = the angular position of the damper rotor

ω_d = the angular velocity of the damper rotor

$J = J_M + J_L + J_H$ = the total inertia on the motor shaft

J_R = the inertia of the damper housing

K_D = the coefficient of viscous friction of the damper fluid

All other variables have the same meaning as in Eqs. (4-1) through (4-6).

4-3 THE NUMERICAL VALUES OF THE PARAMETERS

It is important that the actual numerical values of all the parameters of the system be accurately obtained since the simulation results will be directly dependent on them. All the numerical values should be consistently represented in one system of units. This will insure proper magnitudes of the variables and output quantaties. Table 4-1 shows the units for all the step motor parameters and variables appearing in Eqs. (4-1) through (4-6) in two different systems; the MKS and the oz-in-sec systems.

TABLE 4-1

Variable	Units - MKS system	Units - oz-in-sec system
θ	radian	radian
ω	radian/sec	radian/sec
i_a, i_b, i_c	ampere	ampere
v_a, v_b, v_c	volt	volt
r_a, r_b, r_c	ohm	ohm
L_a, L_b, L_c, L_1, L_2	henry	henry
J, J_M, J_L	Newton meter sec^2	oz-in-sec^2
T	Newton meter	oz-in
B	Newton meter sec	oz-in-sec
T_F	Newton meter	oz-in

4-4 THE CONTROL SCHEME TO BE SIMULATED

The control scheme consists of the details of the controls which drive the step motor system. Included in the control information are the input voltages to the windings and the drive circuitry; the suppression circuitry; the mode of operation of the step motor (single step, open-loop slewing, delayed last-step damping, etc.); and the length of time of the process. Several control schemes will be discussed in detail later in the paper.

4-5 THE METHOD OF INTEGRATION

One must select the numerical method to be used in integrating the system of differential equations which represent the step motor system. Many methods are available, and the accuracy of the simulation generally depends on the sophistication of the integration method. It has been found that for normal step motor simulations the fourth-order fixed-increment Runge-Kutta method (RKGS), available from the IBM Scientific Subroutine Package,[3] is generally adequate.

4-6 THE OUTPUT FORMAT

The output format merely specifies which variables are to be printed, which are to be plotted, their frequency of printing, etc.

4-7 THE SIMULATION PROGRAM

Having collected all of the information that is necessary to perform the simulation, we now consider the actual simulation program. Figure 4-1 shows the basic structure of a computer program for simulating a step motor system. The complete program consists of a main program and three auxiliary subroutines.

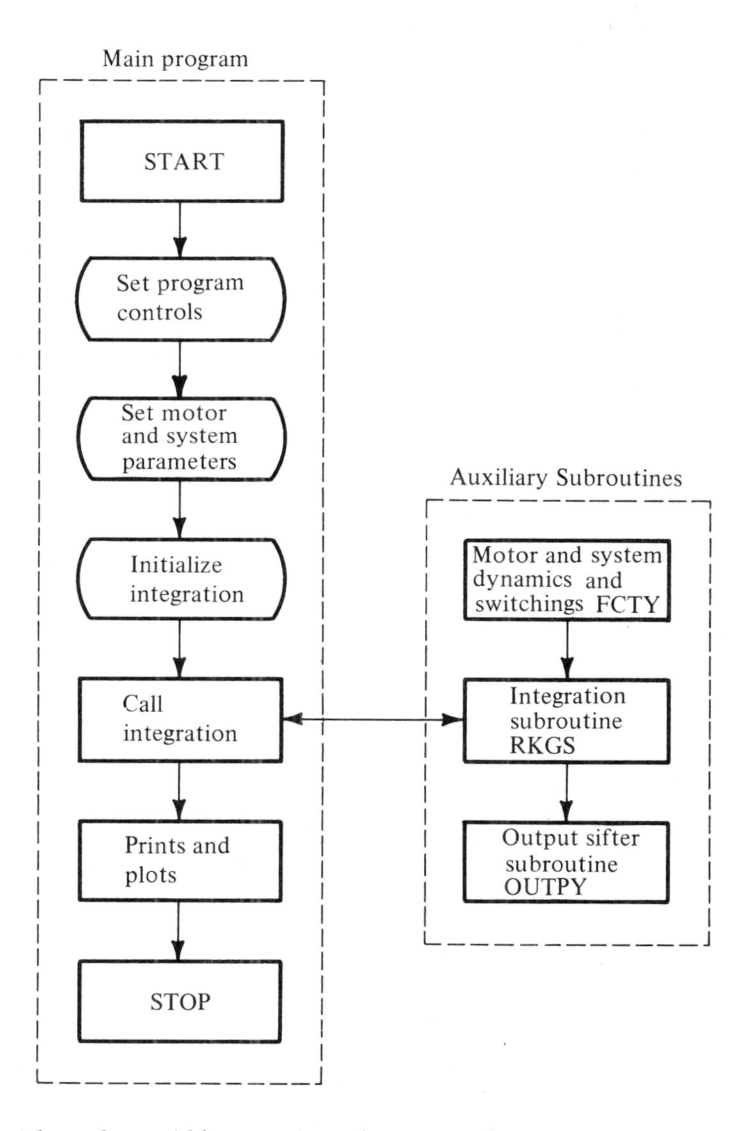

Figure 4-1. Flow chart illustrating the general step motor system simulation program.

The system parameters and the initial conditions are first set in the main program after which the control is transferred to the integration subroutine RKGS (or any other similar integrating subroutine). To integrate the dynamic equations of the system the integration subroutine works between the function subroutine FCTY and the output subroutine OUTPY. The subroutine FCTY calculates the derivatives of the variables at each time instant and passes the information to RKGS. The dynamics of the system and the switching operations are defined in this subroutine. The output subroutine OUTPY receives the information from RKGS after the successful completion of each integration step. This information is sifted, printed or stored for later plotting and/or printing according to the output format requirements. Figures 4-2 and 4-3 show the flow charts illustrating the basic structure of the subroutines FCTY

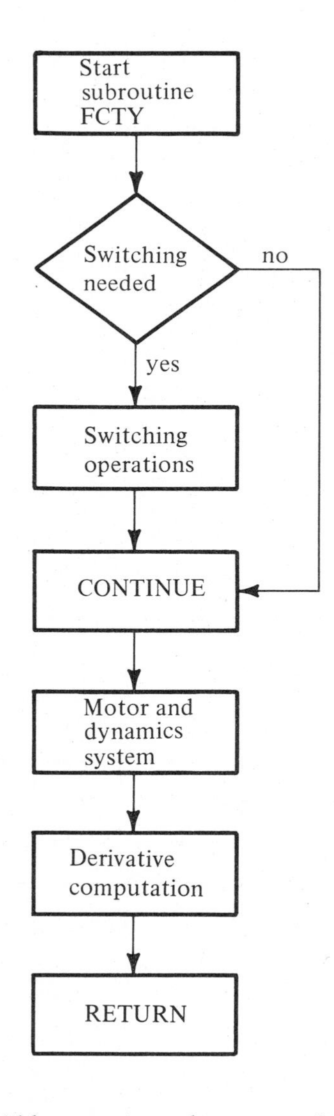

Figure 4-2. Flow chart illustrating the operation inside subroutine FCTY.

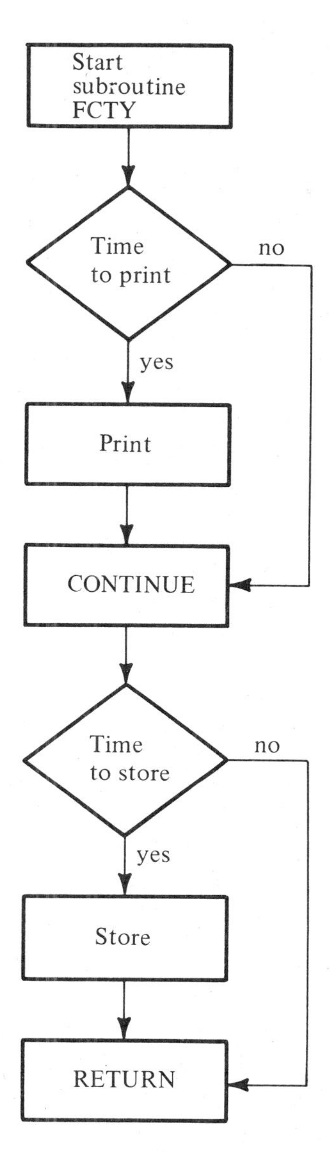

Figure 4-3. Flow chart illustrating the flow of information in subroutine OUTPY.

and OUTPY, respectively. Figure 4-3 is self-explanatory. In Figure 4-2, the motor and the system dynamics, and the derivative computation, represent the differential equations of the system. For the SM048AB step motor with a damper, these would be Eqs. (4-3) through (4-10). The block representing the switching operations in Figure 4-2 contains the details of the control scheme to be used with the appropriate phase switchings. To illustrate the programming of this block, several typical control schemes are now considered.

4-8 EXAMPLES

Figure 4-4 shows the switching scheme for a simple open-loop slewing operation with a fixed distance cutoff for a three-phase step motor. NUM indicates the number of switchings which have taken place and NSW indicates the total number of switchings (or steps) which are to be executed. SW1 indicates the time at which a switch is to

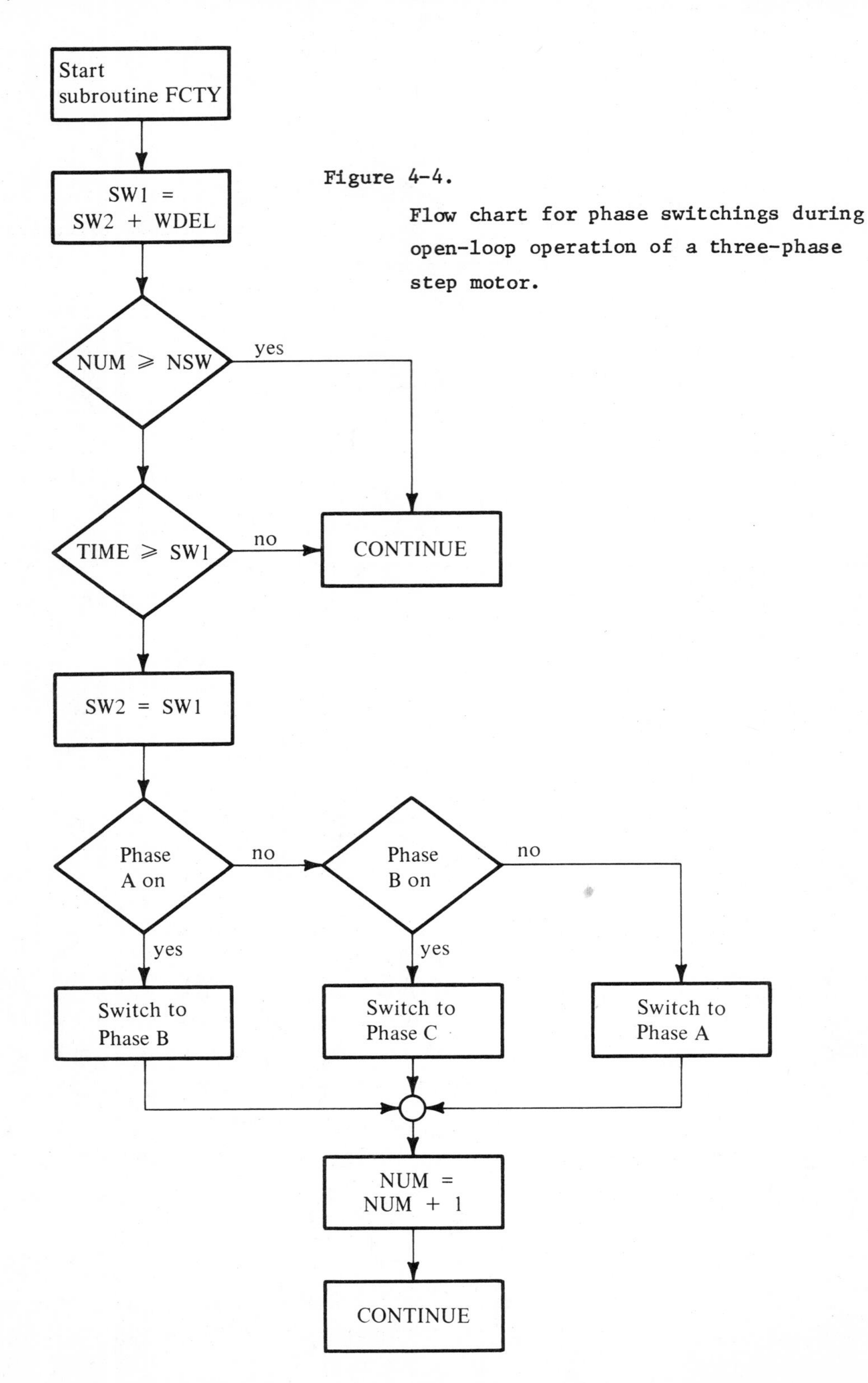

Figure 4-4.
Flow chart for phase switchings during open-loop operation of a three-phase step motor.

take place. SWDEL determines the time increment for the next switch (which can be a constant time increment, a ramp function, an exponential function, or any other

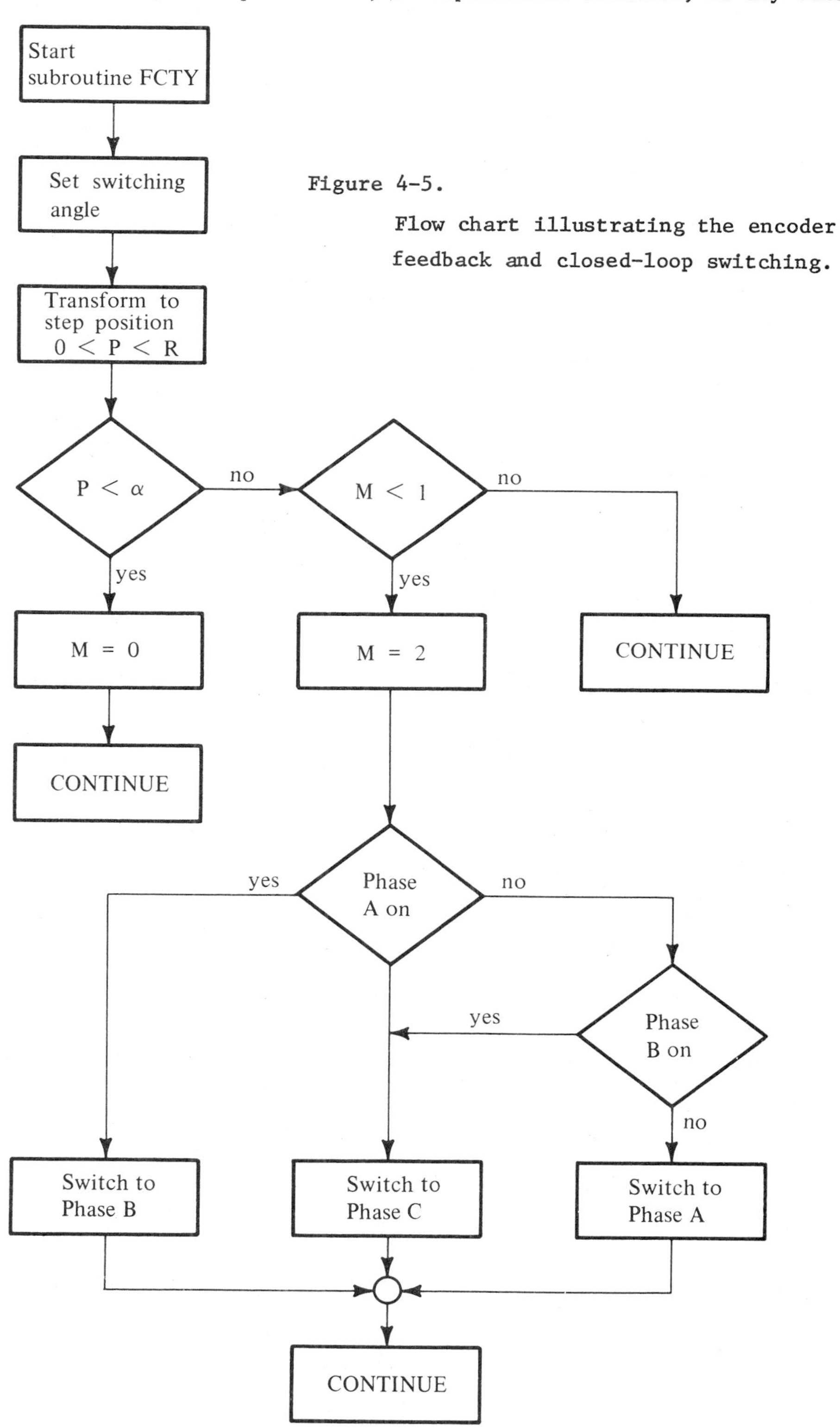

Figure 4-5.

Flow chart illustrating the encoder feedback and closed-loop switching.

variable function, depending on the type of open-loop control scheme which is to be simulated).

The flow chart of Figure 4-5 shows the switching scheme for simple closed-loop operation of a three-phase step motor with encoder feedback. The switching angle α is the position of the rotor beyond each equilibrium position where the phase switching takes place (See Chapter 11). R is the step increment of the motor (7.5 degrees for the SM048AB) and P is the angular position of the step motor normalized to within a step. In equation form P can be written as

$$P = \theta - R.\ \text{Integer}\left\{\frac{\theta}{R}\right\} \tag{4-11}$$

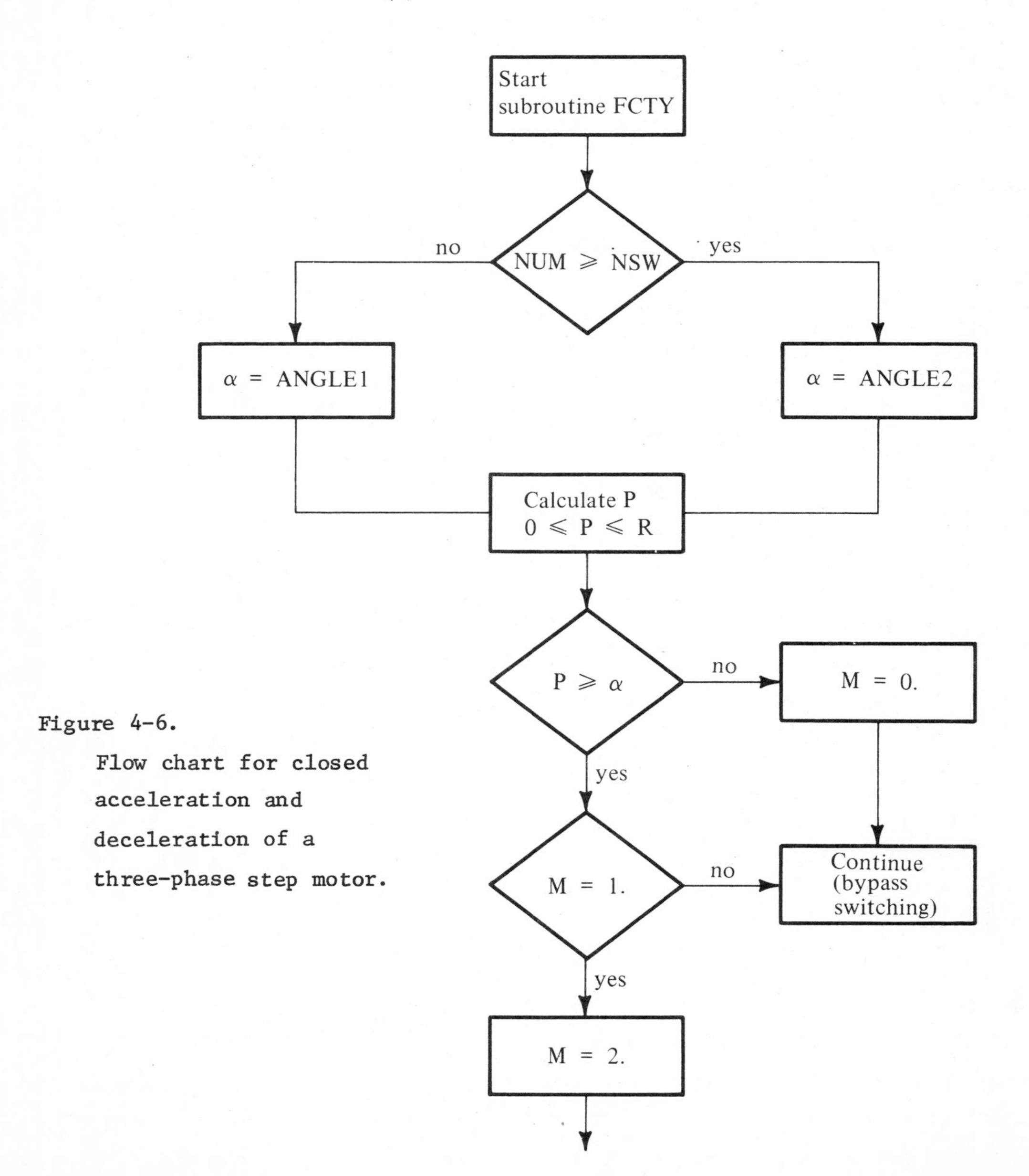

Figure 4-6.
Flow chart for closed acceleration and deceleration of a three-phase step motor.

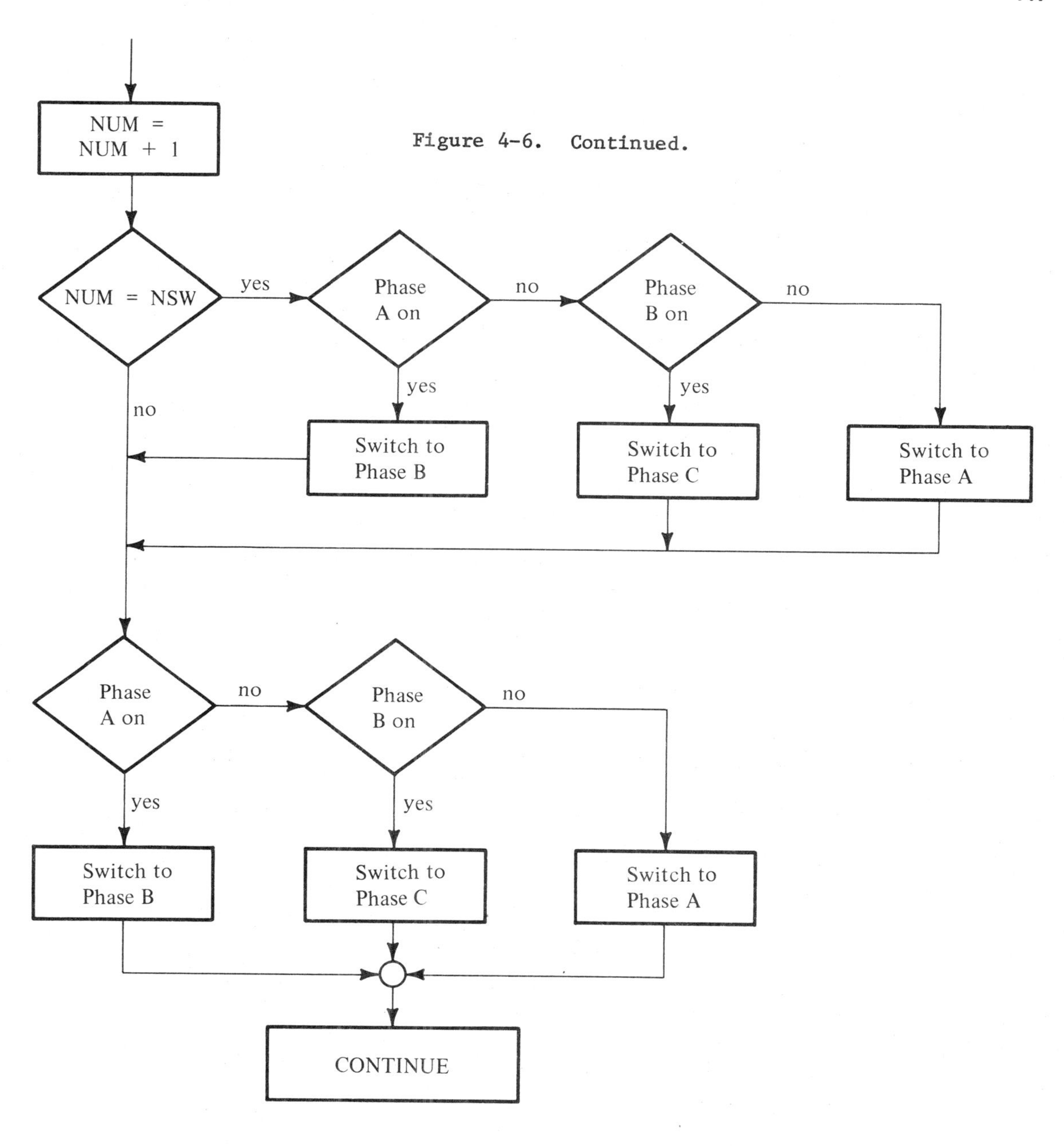

Figure 4-6. Continued.

If deceleration by pulse injection is to be simulated in the closed-loop operation, the flow chart of Figure 4-6 can be programmed. The motor accelerates with the switching angle α = ANGLE1 till NSW switchings have taken place; then, a pulse is injected and the motor decelerates with the switching angle α = ANGLE2.

If dual voltage is to be used, either with open-loop or closed-loop operation, the flow chart of Figure 4-7 can be programmed. This set of logical operations should be located immediately prior to the normal phase switching section.

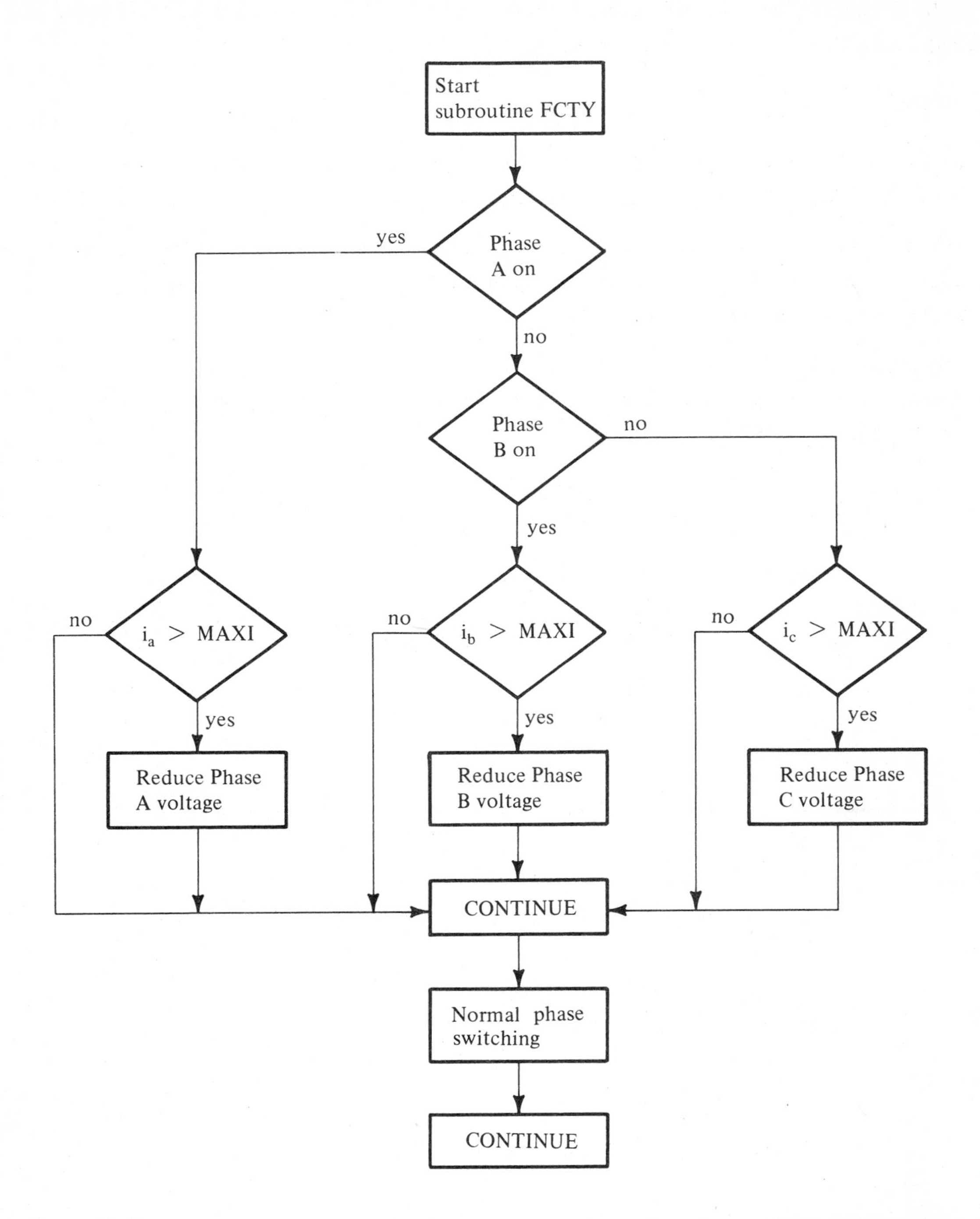

Figure 4-7. Flow chart for dual voltage operation of a three-phase step motor.

In a similar manner, any other control scheme or any other drive circuitry can be programmed for simulation purposes.

REFERENCES

1. G. Singh, "Analog Computer Simulation of Stepping Motor Performance," M.S. Thesis, University of Illinois, Urbana, Illinois, June 1968.

2. IBM Application Program, "System/360 Continuous System Modeling Program (360A-CX-16X)," Users Manual No. H20-036701, International Business Machines Corporation, White Plains, New York, 1966.

3. IBM Application Program, "System/360 Scientific Subroutine Package (360A-CM-03X) Version III," Programmers Manual, International Business Machines Corporation, White Plains, New York, 1968.

5 MAGNETIC CIRCUIT AND FORMULATION OF STATIC TORQUE FOR SINGLE-STACK PERMANENT MAGNET AND VARIABLE RELUCTANCE STEP MOTORS

H. D. Chai
IBM System Products Division
Endicott, New York

5-1 INTRODUCTION

Most of the currently available material on stepping motor analysis is limited to a single-step response of a second-order system. Bailey[1] compares the stepping motor to position and velocity servos, using Laplace transforms with single-step and also periodic-impulse inputs. O'Donohue[2] develops the transfer function for a step motor for a single-step response with zero initial conditions. He assumes the inductance term to be zero and erroneously relates the voltage linearly with displacement instead of velocity. Kieburtz[3] improves on the transfer function by using the linearized velocity term in the voltage equation. However, he also neglects the inductance term.

Departing from the above transfer function approaches which assume a simplified excitation torque in the equation of motion, Snowdon and Madsen[4] develop a torque equation as a function of the angular displacement and time-varying excitation current. They develop the equation by assuming permeance which is sinusoidal in space. With it various properties of the motor are explained. Their results do not explain the existence of a cogging torque for PM motors in the absence of the excitation.

The first comprehensive PM stepping motor analysis was published by Robinson[5] in 1969. A modified version was published in the IEEE Transactions on Industrial Electronics and in Control Instrumentation in 1969[6]. In the report, Robinson presents both single and multi-step response using the constant current and phase plane techniques. From the analysis he concludes: (1) that a step motor can be defined in terms of a natural frequency and a damping ratio; (2) that the motor cannot respond if the applied torque is greater than 0.707 times the stall torque; and (3) that the motor cannot follow a sequential set of step commands if the rotor lags the command position by more than two steps.

In his derivation of the motor torque, Robinson uses a permanent magnet dipole as the rotor. With this model, the resulting torque is sinusoidal, which is correct in form; however, it should be mentioned that actual rotors are far from being permanent magnets. Consequently, the accurate electromagnetic phenomena taking place in the toothed gap cannot be explained with this model.

A second comprehensive work is reported by G. Singh in Chapter 2. He develops a general mathematical model to study dynamic characteristics of both permanent magnet (PM) and variable reluctance (VR) motors of single- and multi-stack types. He begins with the development of a mathematical model for multi-stack motors (Warner Electric type) and extends the resulting formulation to the single-stack VR and PM motors. In the single-stack model, he assumes that the winding distribution around the stator is sinusoidal which, in reality, is not true. His formulation does not show the existence of the cogging torque in PM motors.

This paper begins with a critical examination of conventional single-stack PM step motors (Superior Electric type). A four-phase bifilar-wound motor is selected since it is most common. A physically reasonable linear magnetic circuit is developed to describe the magnetic phenomena of the motor. Due to the existence of redundancy in the circuit, it turns out that only a section of the circuit needs to be examined to formulate the static motor torque.

With the simplified magnetic circuit and appropriate permeance variation around the gap, this paper develops the static torque for both PM and VR motors. (Although the model is developed from a PM motor, it can be applied easily to VR motors by eliminating the PM component of the circuit.) In addition, it establishes the existence of cogging torque.

This paper extends the torque formulation to dynamic conditions by replacing the static current term in the torque equation with a time-varying term. First, the behavior of a PM motor is examined with a-c excitation. Finally, a formulation of multi-step response for a bifilar-wound, four-phase PM motor with commutating capacitors is given.

5-2 MAGNETIC CIRCUIT

Figure 5-1 shows cut-away views of a 1.8° four-phase PM step motor. An axially magnetized cylindrical permanent magnet which is bound by a pair of soft-iron rotor caps is mounted on a shaft. With the polarity of the magnet shown, the PM flux leaves the left-side rotor and enters the right-side rotor. The rotor has 50 teeth. The left- and right-side rotor teeth are offset by 1/2 tooth pitch in order to provide an additive torque generation on the shaft. This will become clear as we examine the magnetic circuit (Figure 5-2).

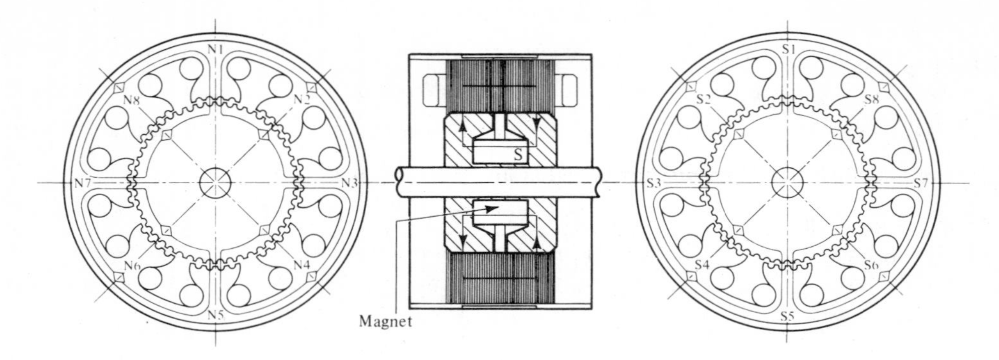

Figure 5-1. Cut-away view of a 1.8° PM step motor.

Teeth on both the stator poles and rotors have the same pitch. For proper motor operation, the pole pitch must be related to the tooth pitch by:

$$P_p = \left(m \pm \frac{1}{n}\right) P_t \quad , \tag{5-1}$$

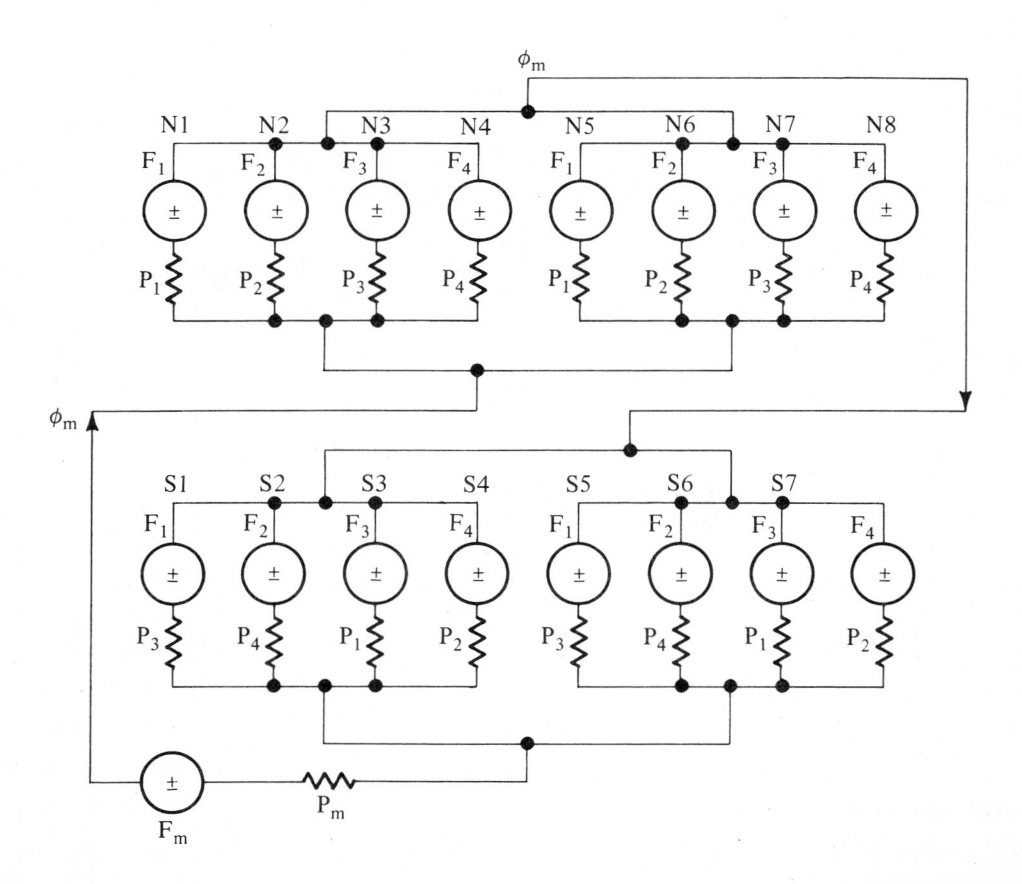

Figure 5-2. Magnetic equivalent circuit of the step motor in Figure 5-1.

where:

P_p is a pole pitch

P_t is a tooth pitch

m is a positive integer suitably chosen from packaging considerations

n is a number of steps used for one tooth pitch rotation or a number of phases.

For the motor shown in Fig. 5-1, $P_t = 7.2^\circ$, n = 4, m = 6, and the positive sign is used. This gives the pole pitch of 45°. Actually a value for m other than 6 could be used. However, m = 6 gives the most balanced pole arrangement with the opposite poles having the same angular displacement relationship with the rotor teeth.

Now, referring to Figure 5-3 for the actual bifilar winding arrangement, we see that a pair of coils are wound around poles* N1 (S1), N3 (S3), N5 (S5), and N7 (S7) unbroken. With one winding excited, the direction of the winding is such that when the fluxes at N1 (S1) and N5 (S5) enter, toward the rotor, they leave the rotor at N3 (S3) and N7 (S7). The same is true for poles N2 (S2)/N6 (S6) and N4 (S4)/N8 (S8).

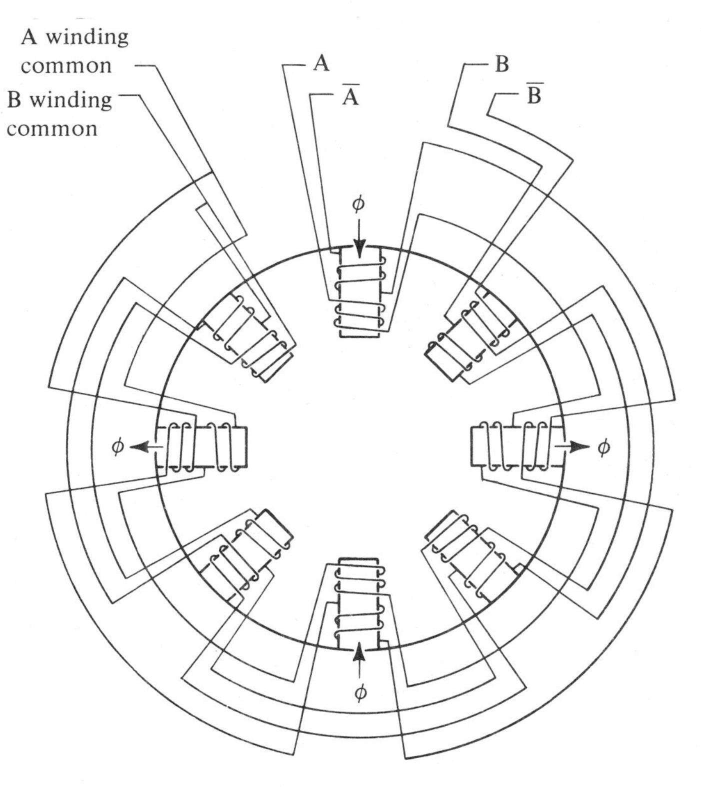

Figure 5-3. Bifilar winding configuration.

Examination of Figures 5-1 and 5-3 reveals that the relationships between the rotor and opposite poles are magnetically identical. Thus we see that there are four identical magnetic circuits in the motor (Figure 5-2).

In Figure 5-2, F_s represent mmf sources and P_s, permeances. As mentioned previously, the flux due to the PM leaves the left rotor and enters the right rotor. Also coils are wound such that $F_1 = -F_3$ and $F_2 = -F_4$. Suppose we set $F_1 = NI$. Then $F_3 = -NI$. At the left rotor, the flux due to NI is additive to the flux due to PM at the poles N1 and N5, and subtractive at the poles N3 and N7. On the other hand, at the right rotor, the flux due to NI is subtractive to the PM force at the poles S1 and S5 and additive at the poles S3 and S7. Now, by offsetting the teeth between the right and left rotors by 1/2 tooth pitch, the four additive poles (N1, N5, S3 and S7) see the same permeance condition P_1. The same is true for the subtractive poles.

Thus we see that the magnetic analysis of the single-stack bifilar-wound PM motors, which are very common in the market, reduces to the analysis of one of the four identical circuits shown in Figure 5-2. When PM is absent ($F_m = P_m = 0$), one of the circuits defines the single-stack VR motors.

5-3 FORMULATION OF STATIC TORQUE

Figure 5-5 shows a magnetic circuit which is equivalent to one of the four identical circuits shown in Figure 5-2. A permanent magnet path is added to simulate PM motors. This circuit simulates the five-tooth rotor, four-pole stator PM motor shown in Figure 5-4. This is the simplest four-phase arrangement whose analytical results can be easily extended to practical PM and VR motors.

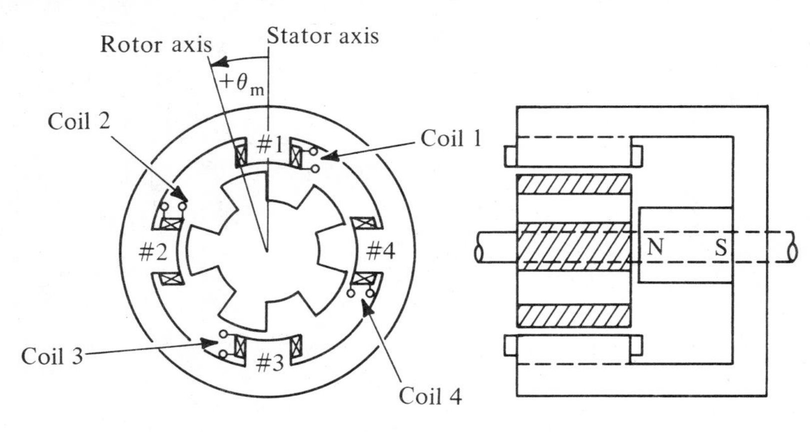

Figure 5-4. Schematic diagram of a five-tooth rotor, four-pole stator step motor.

* The poles N1 and S1 (N2 and S2, N3 and S3...) are physically the same. However, they are distinguished here because the left- and right-rotors are physically separate and offset by 1/2 tooth pitch.

By using the equivalent circuit, we will develop expressions for a static torque on the rotor and for fluxes through the poles. But first it is necessary to

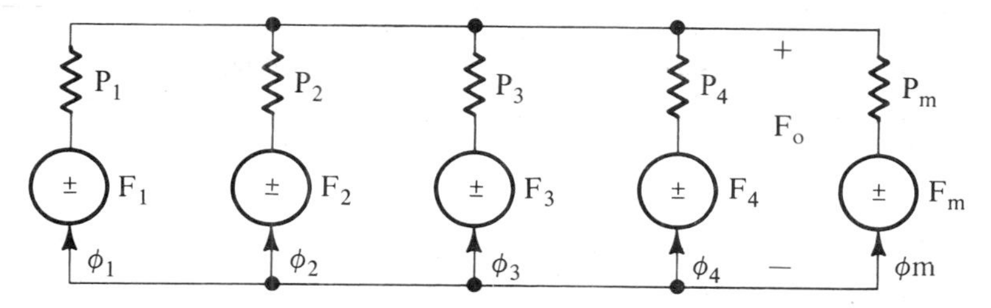

$P_1, \ldots, P_4$, and P_m are the air gap permeances at the stator poles #1, . . . , #4 and the permanent magnet (henrys).

$\phi_1, \ldots, \phi_4$, and ϕ_m are the fluxes in the stator poles #1, . . . , #4 and the permanent magnet (webers).

$F_1, \ldots, F_4$, and F_m are the excitation mmf's in the stator poles #1, . . . , #4 and the permanent magnet (amp-turns).

F_0 is the mmf drop across the stator-rotor gap including the excitation source.

Figure 5-5. Magnetic equivalent circuit of Figure 5-4.

obtain the expression of mmf drop F_o. Since the equivalent circuit in Figure 5-5 has only two nodal points, the Norton equivalent circuit provides a solution for F_o very quickly (Figure 5-6). From the figure, we see that

$$F_o = \frac{\sum_{i=1}^{4} P_i F_i + P_m F_m}{\sum_{i=1}^{4} P_i + P_m} \tag{5-2}$$

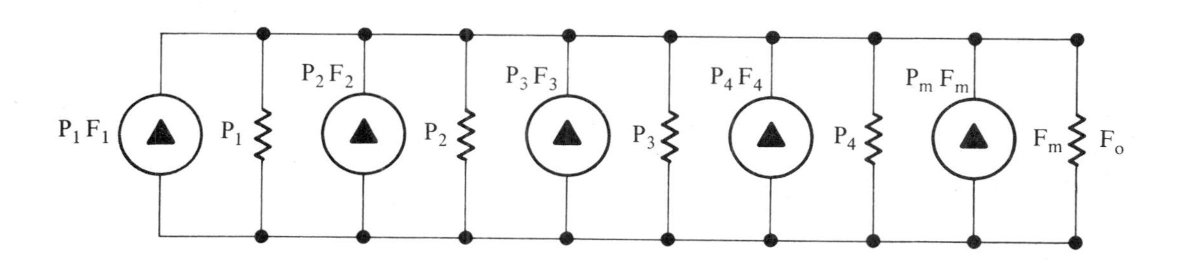

Figure 5-6. Norton's equivalent circuit of Figure 5-5.

and

$$\phi_i = (F_i - F_o)P_i \qquad i = 1, 2, 3, 4. \tag{5-3}$$

The torque on the rotor is:

$$T = -\frac{1}{2}\sum_{i=1}^{4} \phi_1^2 \frac{\partial R_i}{\partial \theta_m}$$

$$= \frac{1}{2}\sum_{i=1}^{4} \frac{\phi_i^2}{P_i^2} \frac{\partial P_i}{\partial \theta_m}$$

$$T = \frac{1}{2}\sum_{i=1}^{4} (F_i - F_o)^2 \frac{\partial P_i}{\partial \theta_m}$$

$$= \frac{N_R}{2} \sum_{i=1}^{4}(F_i - F_o)^2 \frac{\partial P_i}{\partial \theta_e} \quad , \qquad (5\text{-}4)$$

where:

R_i is reluctance $(1/P_i)$

θ_m is a mechanical angle related to an electrical angel θ_e by

$$\theta_m = \frac{\theta_e}{\text{No. of rotor teeth}} = \frac{\theta_e}{N_R} \qquad (5\text{-}5)$$

for the five-tooth rotor (Figure 5-4), N_R is 5.

Equations (5-3) and (5-4) determine the fluxes and the rotor torque once the necessary parameters are known. F_i is a given quantity (e.g., $F_i = NI$). P_m and F_m are parameters related to the permanent magnet. These values can be approximately determined for a given magnet. P_i is the gap permeance with properties that

1. It is periodic with respect to θ_e
2. P_1, P_2, P_3 and P_4 differ only in angular displacements
3. It should provide a rotor torque due to the permanent magnet in the absence of the excitation mmf.

Thus, using the reference rotor and stator axes as shown in Figure 5-4, we assume that the P_i's can be represented by

$$P_1 = P_o + P\cos\theta_e + P_4\cos 4\theta_e$$

$$P_2 = P_o + P\cos(\theta_e - 90^o) + P_4\cos 4(\theta_e - 90^o)$$

$$P_3 = P_o + P\cos(\theta_e - 180^o) + P_4\cos 4(\theta_e - 180^o)$$

$$P_4 = P_o + P\cos(\theta_e - 270^o) + P_4\cos 4(\theta_e - 270^0)$$

or

$$P_1 = P_o + P\cos\theta_e + P_4\cos 4\theta_e$$

$$P_2 = P_o + P\sin\theta_e + P_4\cos 4\theta_e$$

$$P_3 = P_o - P\cos\theta_e + P_4\cos 4\theta_e$$

$$P_4 = P_o - P\sin\theta_e + P_4\cos 4\theta_e \tag{5-6}$$

Then

$$\frac{\partial P_1}{\partial \theta_e} = -P\sin\theta_e - 4P_4\sin 4\theta_e$$

$$\frac{\partial P_2}{\partial \theta_e} = P\cos\theta_e - 4P_4\sin 4\theta_e$$

$$\frac{\partial P_3}{\partial \theta_e} = P\sin\theta_e - 4P_4\sin 4\theta_e$$

$$\frac{\partial P_4}{\partial \theta_e} = -P\cos\theta_e - 4P_4\sin 4\theta_e \tag{5-7}$$

From (5-6) and (5-7)

$$\sum_{i=1}^{4} P_i = 4P_o + 4P_4\cos 4\theta_e \approx 4P_o \qquad P_o \gg P_4$$

$$\sum_{i=1}^{4} \frac{\partial P_i}{\partial \theta_e} = -16P_4\sin 4\theta_e$$

$$\sum_{i=1}^{4} P_i + P_m \approx 4P_o \qquad \text{since } P_m \ll P_o$$

5-4 COGGING TORQUE

PM motors possess detenting torque even in the absence of driving currents. This is due to flux from the permanent magnet. To study the cogging torque characteristics, we set $F_i = 0$, $i = 1, 2, 3, 4$. From (5-2)

$$F_o \approx \frac{1}{4}\left(\frac{P_m}{P_o}\right)F_m \quad ,$$

and

$$\sum_{i=1}^{4} \frac{\partial P_i}{\partial \theta_e} = -16P_4\sin 4\theta_e.$$

Thus

$$T = -\frac{N_R}{2}\left(\frac{P_m}{P_o}\right)^2 F_m^2 P_4 \sin 4\theta_e \qquad (5-8)$$

Equation (5-8) tells that there is a four-cycle variation of the torque due to the permanent magnet alone for one rotor tooth pitch rotation, i.e., 360° electrical degree rotation (Figure 5-7). The stable points are at θ_e = 0, 90°, 180°, and 270°; the unstable points are at θ_e = 45°, 135°, 225°, and 315°. The positive torque

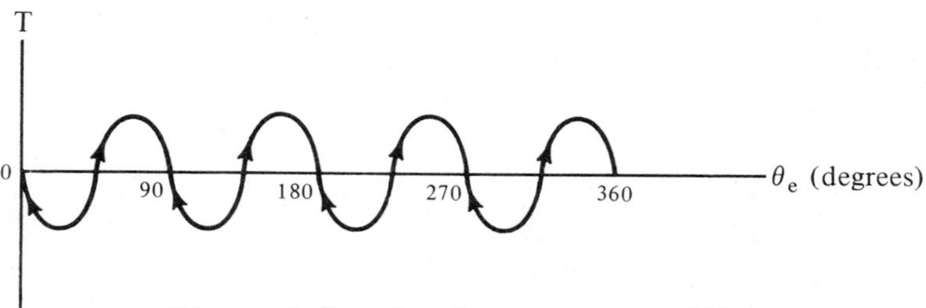

Figure 5-7. Cogging torque profile.

corresponds to the counterclockwise motion. For VR motors, F_m is zero. Hence, the cogging torque is zero.

One interesting observation can be made here. The cogging torque is present due to the existence of the 4th harmonic permeance component. If we can design rotor and stator pole teeth such that the 4th harmonic component of permeance is absent, the cogging torque will be eliminated while maintaining the fast rate of torque buildup in the initial stage of excitation. The absence of the cogging torque is important in some applications from a human factors viewpoint.

5-5 STATIC TORQUE OF PM MOTORS

A. Single-Phase Energization

In practical motors the coils providing F_1 and F_3 are connected in series such that when $F_1 = -F = -NI$, then $F_3 = F = NI$.

The same is true for F_2 and F_4. For single-phase energization analysis, we set F_2 and F_4 to zero; let's determine the expression for torque.

$$\sum_{i=1}^{4} P_i F_i = F(-P_1 + P_3) = -2FP\cos\theta_e$$

$$F_o = \frac{1}{4P_o}(F_m P_m - 2FP\cos\theta_e)$$

$$T = -\frac{N_R}{2}\left(\frac{FF_m P_m P}{P_o}\right)\left[\sin\theta_e - \frac{FP}{F_m P_m}\sin 2\theta_e\right] + \Delta T \quad , \qquad (5-9)$$

where ΔT is a torque containing 4th or higher harmonic terms.

Now $F_m P_m$ can be expressed as

$$F_m P_m = \left(\frac{B_o \ell_m}{\mu_m}\right)\left(\frac{\mu_m A_m}{\ell_m}\right) = B_o A_m \quad ,$$

where B_o and μ_m are, respectively, the magnetic flux density and incremental permeability defined in Figure 5-8. A_m is an effective cross-sectional area of the magnet. Equation (5-9) may now be written as

$$T = -\frac{N_R}{2}(NI)(B_o A_m)\left(\frac{P}{P_o}\right)[\sin\theta_e - \frac{NI}{B_o A_m} P\sin2\theta_e] + \Delta T. \tag{5-10}$$

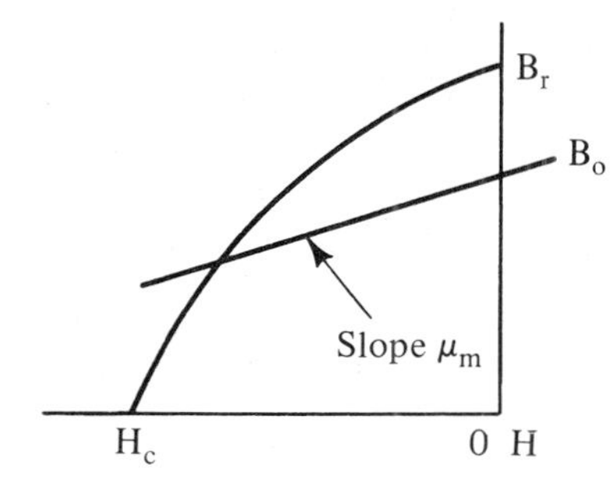

Figure 5-8. Demagnetization curve of permanent magnet.

For the case when $NIP << B_o A_m$, the second harmonic term in Eq. (5-10) can be neglected, and we have

$$T \cong -\frac{N_R}{2}(NI)(B_o A_m)\left(\frac{P}{P_o}\right)\sin\theta_e . \tag{5-11}$$

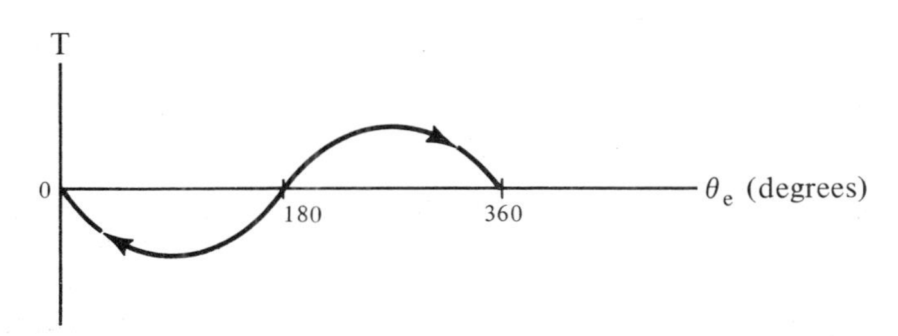

Figure 5-9. PM motor torque profile.

As shown in Figure 5-9, for each rotor tooth pitch rotation the torque goes through one cycle variation. $\theta_e = 0$ and 360° are stable positions and $\theta_e = 180^o$ is an unstable position.

From Eq. (5-11) it is seen that $T \sim NI$ and $B_o A_m$.

For a more general case where the second harmonic term is not negligible, Eq. (5-10) can be put into a normalized form as

$$T_n = -(\sin\theta_e - K_a \sin2\theta_e) \quad , \tag{5-12}$$

where

$$T_n = \frac{T}{\frac{N_R}{2}(NI)(B_o A_m)\left(\frac{P}{P_o}\right)} ,$$

and

$$K_a = \frac{NIP}{B_o A_m} .$$

From Eq. (5-12) a series of curves can be obtained with K_a as a parameter (Figure 5-10).

A few interesting observations can be made from Figure 5-10. For K_a less than 0.5, the stable and unstable points are the same as those for $K_a = 0$. For K_a greater than 0.5, the stable points shift from the previous ones and approach 90° and 270° as

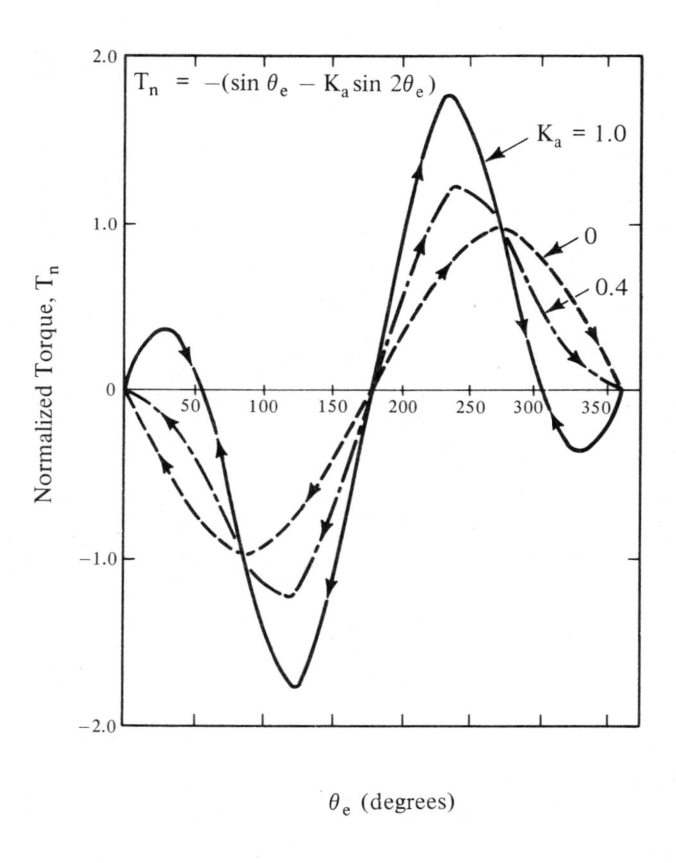

Figure 5-10. PM motor torque profile when second harmonic term is not neglected.

K_a increases. The greater the value of K_a, the greater the maximum torque. The maximum torque points shift to 135° and 225° as K_a increases.

B. Double-Phase Energization

In order to increase the torque output, it is common practice to energize two sets of windings. This is equivalent to letting

$$F_1 = -F_3 = -F = -NI$$

$$F_2 = -F_4 = -F = -NI$$

We wish to find the resulting torque.

Now

$$\sum P_i F_i = F(-P_1 - P_2 + P_3 + P_4)$$

$$= -2\sqrt{2}FP\sin(\theta_e + 45^o)$$

$$F_o = \frac{1}{4P_o}[P_m F_m - 2\sqrt{2}FP\cos(\theta_e - 45^o)]$$

$$T = -\sqrt{2}\,\frac{N_R}{2}\left(\frac{FF_m P_m P}{P_o}\right)[\sin(\theta_e - 45^o) - \frac{FP}{F_m P_m}\sqrt{2}\sin 2(\theta_e - 45^o)] + \Delta T \qquad (5\text{-}13)$$

ΔT is a torque containing 4th or higher harmonic terms. Compare Eq. (5-9) with Eq. (5-13). Equation (5-13) results if we replace F and θ_e in Eq. (5-9) by

$$F \rightarrow \sqrt{2}F$$

and

$$\theta_e \rightarrow \theta_e - 45^o.$$

Thus the excitation of two phases with NI results in the shift in the detent position and the increase in holding torque by a factor of $\sqrt{2}$ for the first harmonic term and by a factor of 2 for the second harmonic term.

5-6 STATIC TORQUE OF VR MOTOR

With the permanent magnet absent ($B_o = 0$), Eq. (5-10) reduces to

$$T = \frac{N_R}{2}(NI)^2\,\frac{P^2}{P_o}\,\sin 2\theta_e \quad . \qquad (5\text{-}14)$$

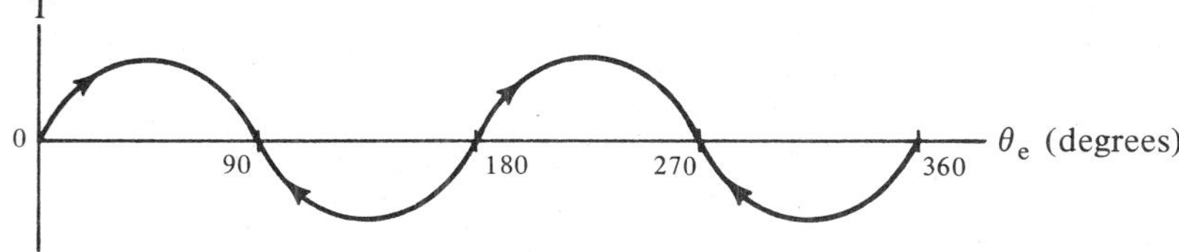

Figure 5-11. VR motor torque profile with $F_1 = -NI$ and $F_3 = NI$.

As shown in Figure 5-11, for each rotor tooth rotation the torque goes through two cycle variations. $\theta_e = 0$ is unstable while 90° is stable. Thus, if the rotor is moved by $\theta_e = 0^+$ it will rotate in the CCW (positive) direction and oscillate around $\theta_e = 90^\circ$ until it settles down at the 90° position.

From Eq. (5-14) it is seen that

$$T \sim (NI)^2$$

A problem with this type of energization is that when we turn off F_1 and F_3 and energize $F_2 = -F_4 = NI$, the rotor finds itself in a stable position. Hence it will not move. Therefore, in practice, coils producing F_1, F_2, F_3 and F_4 are separately excited for dynamic excitation. Unlike PM motors, the F_1 and F_3 (also F_2 and F_4) coils are not connected in series, but they are separate. To find an expression of torque, let

$$F_1 = NI \quad ,$$

and

$$F_2 = F_3 = F_4 = F_m = 0.$$

Then, from Eq. (5-2),

$$F_o = \frac{P_1(NI)}{4P_o} \cong \frac{NI}{4P_o}(P_o + P\cos\theta_e) \quad ,$$

and

$$T = \frac{N_R}{32} P(NI)^2\{[-9 + \frac{1}{4}\left(\frac{P}{P_o}\right)^2]\sin\theta_e + 3\left(\frac{P}{P_o}\right)\sin2\theta_e - \frac{1}{4}\left(\frac{P}{P_o}\right)^2\sin3\theta_e\}$$

$$\approx -\frac{9N_R}{32} P(NI)^2\sin\theta_e \qquad (5\text{-}15)$$

Figure 5-12 shows that the static torque variation is sinusoidal as in the single-phase energization of a PM motor. The stable positions are at $\theta_e = 0$ and 360°; the unstable position is at 180°.

An increase of torque by $\sqrt{2}$ results in two-phase excitation (e.g. $F_1 = F_2 = NI$, and $F_3 = F_4 = 0$). Similar to PM motors, the stable points shift by 45 electrical degrees.

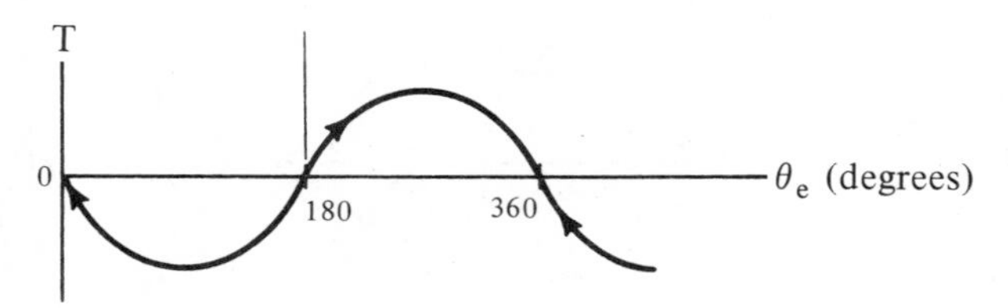

Figure 5-12. VR motor torque profile with $F_1 = NI$.

5-7 TORQUE CHARACTERISTICS WITH A TWO-PHASE A-C EXCITATION

In this section, we extend the static torque formulation to a two-phase alternating-current excitation. The windings are excited in the following way:

$$F_1 = -F_3 = -NI_1\cos\omega t = -F_1\cos\omega t$$

$$F_2 = -F_4 = -NI_2\sin(\omega t + \xi) = -F_2\sin(\omega t + \xi) \tag{5-16}$$

The general formula for torque is:

$$T = \frac{N_R}{2}\sum_{i=1}^{4}(F_i - F_o)^2 \frac{\partial P_i}{\partial \theta_e} \; . \tag{5-17}$$

Now

$$\sum P_i + P_m \cong 4P_o$$

$$\sum P_i F_i = 2P(F_1\cos\theta_e + F_2\sin\theta_e)$$

$$F_o = \frac{2P(F_1\cos\theta_e + F_2\sin\theta_e) + P_m F_m}{4P_o} \tag{5-18}$$

$$T = 2N_R\, PF_o\{F_2\cos\theta_e\sin(\omega t + \xi) - F_1\sin\theta_e\cos\omega t\} - 2P_4\sin 4\theta_e[F_1^2\cos^2\omega t + F_2^2\sin^2(\omega t + \xi)] - 4P_4F_o^2\sin 4\theta \tag{5-19}$$

Let's look at a special case when $F_1 = F_2 = F = NI$, and $\xi = 0$. Then

$$F_o = -\frac{PF}{2P_o}\cos(\omega t - \theta_e) + \frac{P_m F_m}{4P_o}$$

$$T = \frac{N_R}{2}\left[\frac{(P_m F_m)(PF)}{P_o}\sin(\omega t - \theta_e) - \frac{(PF)^2}{P_o}\sin 2(\omega t - \theta_e)\right] + \Delta T \;\; , \tag{5-20}$$

where ΔT contains terms with $\sin 4\theta_e$, i.e.,

$$\Delta T = -\frac{8}{5}N_R P_4 \sin 4\theta_e \left\{\left[F^2 + 2\left(\frac{P_m F_m}{4P}\right)^2\right] + 2\left(\frac{PF}{2P_o}\right)^2\cos^2(\omega t - \theta_e) - 4\left(\frac{PF}{2P_o}\right)\left(\frac{P_m F_m}{4P_o}\right)\cos(\omega t - \theta_e)\right\} \; .$$

Now we can set

$$\theta_e = N_R\theta_m = N_R(\omega_m t - \delta_m) = N_R\omega_m t - \delta_e \tag{5-21}$$

where

ω_m = Angular velocity of rotor

δ_m = Initial angle of rotor.

Then

$$\omega t - \theta_e = \omega t - N_R\omega_m t + N_R\delta_m \quad . \tag{5-22}$$

We see here that the average torque supplied by the motor is zero unless

$$\omega = N_R\omega_m \quad ,$$

or

$$\omega_m = \frac{\omega}{N_R} \quad . \tag{5-23}$$

Thus we find that the step motor behaves like a synchronous motor with many poles. For instance with a 60 Hz input, the motor with 50 rotor teeth will rotate at 72 rpm. This low speed has an important application in machine tool controls.

Now Eq. (20) reduces to

$$T = \frac{N_R}{2}[\frac{(P_mF_m)(PF)}{P_o}\sin\delta_e - \frac{(PF)^2}{P_o}\sin 2\delta_e] + \Delta T \quad . \tag{5-24}$$

The effect of inductance can be understood from the voltage equation

$$e = Ri + L\frac{di}{dt} \tag{5-25}$$

Let's assume that the excitation is sinusoidal with frequency ω. Then, using complex notation, Eq. (5-25) can be written in the form

$$\overline{E} = R\overline{I} + j\omega L\overline{I}$$

or

$$\overline{I} = \frac{\overline{E}}{R + j\omega L} \tag{5-26}$$

The magnitude of the current is

$$|\overline{I}| = \frac{|\overline{E}|}{\sqrt{R^2 + (\omega L)^2}} \tag{5-27}$$

The substitution of Eq. (5-27) into Eq. (5-24) gives

$$T = \frac{N_R}{2}[(P_mF_m)\left(\frac{P}{P_o}\right)N\frac{|\overline{E}|\sin\delta_e}{\sqrt{R^2 + (\omega L)^2}} - \frac{P^2}{P_o}N^2\frac{|\overline{E}|^2}{[R^2 + (\omega L)^2]}\sin 2\delta_e] + \Delta T \quad . \tag{5-28}$$

For the case $P_m F_m >> PF$,

$$T = \frac{N_R}{2}(P_m F_m)\left(\frac{P}{P_o}\right)\frac{N|\overline{E}|}{\sqrt{R^2 + (\omega L)^2}} \sin\delta_e$$

$$= T_{max}\sin\delta_e \quad . \tag{5-29}$$

The torque characteristic of Eq. (5-29) as a function of ω is given in Figure 5-13.

Equation (5-29) and Figure 5-14 explain why the rotor stalls when the pulse rate is increased to a higher value. Referring to Figure 5-14, suppose that the load was T_L and the rotor was running at an excitation frequency of ω_o with the power angle at δ_o. If we now increase the frequency to ω_1 with the same load, the electromagnetic torque becomes less than the load torque, and the rotor stalls.

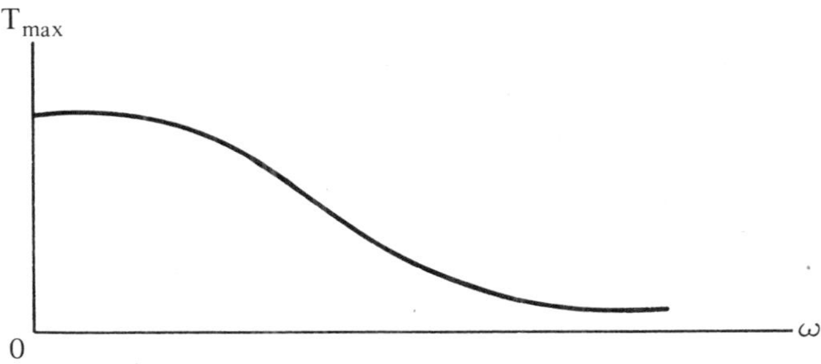

Figure 5-13. Torque profile with a.c. excitation as a function of excitation frequency ω.

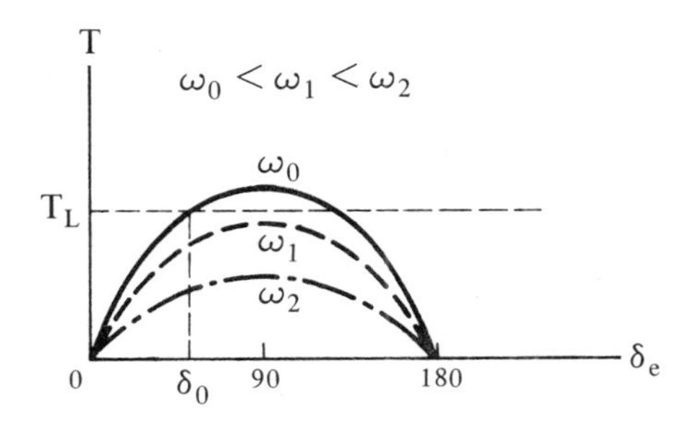

Figure 5-14. Torque profile with a.c. excitation as a function of δ_e.

5-8 FORMULATION FOR MULTI-STEP RESPONSE

In this section, we will use the bifilar winding with a commutating capacitor as a drive circuit (Figure 5-15). In the bifilar winding motors the mmf sources in Figure 5-5 have the relations that

$$F_1 = -F_3$$

$$F_2 = -F_4 \quad .$$

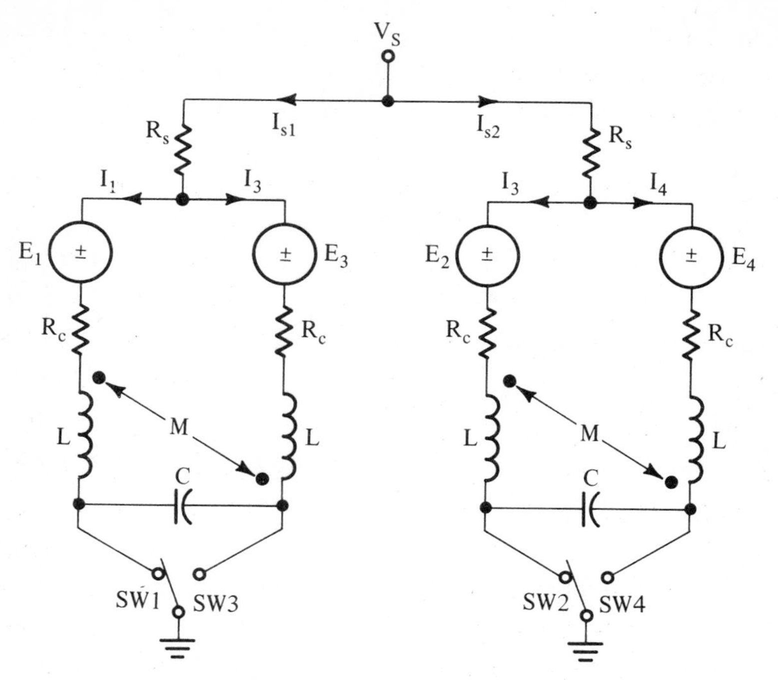

V_s is supply voltage.

R_s is series resistor to limit current and to reduce rise time.

I_1, I_2, I_3 and I_4 are phase currents.

I_{s1} and I_{s2} are currents from the supply voltage.

E_1, E_2, E_3 and E_4 are back emf voltages.

R_c is winding resistance.

L is winding inductance.

M is mutual inductance.

C is commutating capacitor.

Figure 5-15. Bifilar winding.

Neglecting the 4th harmonic terms of the permeances, we obtain the torque expression from Eq. (5-4) as

$$T_e = -2N_R P F_o (F_1 \sin\theta_e - F_2 \cos\theta_e) \quad , \tag{5-30}$$

where

$$F_o = \frac{1}{4P_o}[2PF_1\cos\theta_e + 2PF_2\sin\theta_e + P_m F_m] \quad .$$

We extend the static torque in Eq. (5-30) to the dynamic case by assuming that the mmf's F_1 and F_2 vary with time. This extension is valid if eddy current and hysteresis effects are small.

Referring to Figure 5-15, we observe that at any instant two switches are on. A total of four switching modes exist (Table 1) - SW1 and 2 (SW3 and 4) are complementary. They cannot be simultaneously on.

Table 5-1

Switching Modes

Mode	SW1	SW2	SW3	SW4
1	ON	ON		
2	ON			ON
3			ON	ON
4		ON	ON	

The mechanical equation (equation of motion) is

$$J\ddot{\theta}_m + K_d\dot{\theta}_m + T_f \frac{\dot{\theta}_m}{|\dot{\theta}_m|} + T_L(t) = T_e \quad , \tag{5-31}$$

where

K_d is damping constant

T_f is friction torque

T_L is load torque

J is total inertia

In terms of the electrical angle θ_e,

$$\frac{J}{N_R}\ddot{\theta}_e + \frac{K_d}{N_R}\dot{\theta}_e + F_f \frac{\dot{\theta}_e}{|\dot{\theta}_e|} + T_L(t) = T_e \quad . \tag{5-32}$$

By setting

$F_1 = N(I_1 - I_3)$

$F_2 = N(I_2 - I_4)$

Equation (5-30) becomes

$$T_e = 2N_R P F_o[-(I_1 - I_3)\sin\theta_e + (I_2 - I_4)\cos\theta_e] \tag{5-33}$$

where

$$F_o = \frac{P}{2P_o}[(I_1 - I_3)\cos\theta_e + (I_2 - I_4)\sin\theta_e + \frac{P_m F_m}{2P}] \tag{5-34}$$

When we neglect the first two harmonic terms in F_o, we can express the torque in a simpler form.

$$T_e = K_t[-(I_1 - I_3)\sin\theta_e + (I_2 - I_4)\cos\theta_e] \quad , \tag{5-35}$$

where

$$K_t = \frac{N_R}{2} NB_oA_m \frac{P}{P_o} \quad , \text{ the torque constant.}$$

Now the electrical equations (Kirchhoff's voltage equations) are:

$$V_s = R_s(I_m + I_n) + R_cI_m + L\frac{dI_m}{dt} - M\frac{dI_n}{dt} + E_m \tag{5-36}$$

$$V_s = R_s(I_m + I_n) + R_cI_n - M\frac{dI_m}{dt} + L\frac{dI_n}{dt} + E_n + \frac{1}{C}\int I_n dt \quad , \tag{5-37}$$

where the subscript m signifies current with switches on while n signifies those with switches off. There are two sets of equations corresponding to Eqs. (5-36) and (5-37). For example, for the mode 1 condition:

$$I_m = I_1 \text{ and } I_2$$
$$I_n = I_3 \text{ and } I_4 \quad .$$

Because the coils are bifilar wound (Figure 5-3), $M \approx L$ and $E_m \approx -E_n$. Using the approximations in Eqs. (5-36) and (5-37), we obtain after some manipulation:

$$\frac{dI_m}{dt} = \frac{1}{2}\{\frac{V_s - E_m}{L} - \frac{R_s + R_c}{L} I_m - [\frac{R_s}{L} + \frac{1}{(2R_s + R_c)C}]I_n\} \tag{5-38}$$

$$\frac{dI_n}{dt} = -\frac{1}{2}\{\frac{V_s - E_m}{L} - \frac{R_s + R_c}{L} I_m - [\frac{R_s}{L} - \frac{1}{(2R_s + R_c)C}]I_n\} \quad . \tag{5-39}$$

The back emf terms (E_m and E_n) are approximately given by:

$$E_1 = -\frac{K_g}{N_R}\dot{\theta}_e\sin\theta_e$$

$$E_2 = +\frac{K_g}{N_R}\dot{\theta}_e\cos\theta_e$$

$$E_3 = +\frac{K_g}{N_R}\dot{\theta}_e\sin\theta_e$$

$$E_4 = -\frac{K_g}{N_R}\dot{\theta}_e\cos\theta_e \tag{5-40}$$

where K_g represents a back emf parameter.

Equations (5-32), (5-38) and (5-39) form a complete set of differential equations that describe the multi-step operation. By using the torque expression given in (5-35), and either measured values or those supplied by motor manufacturers for R_c, L, K_t, K_g, etc., we can obtain an approximate solution. For a more accurate simulation the spatial variation of inductance, Eq. (5-34) in place K_t, and current-dependent back-emf terms should be used.

5-9 COMMENTS

The permeance model given in the paper uses only the d-c, first, and fourth-harmonic components. The model satisfactorily explains the motor characteristics observed in practice. However, for a rigorous treatment, the air-gap permeance model at the pole faces should be

$$P_i = P_o + \sum_{m=1}^{\infty} P_{im} \cos m\left[\theta_e - \frac{\pi}{2}(i - 1)\right] + \sum_{n=1}^{\infty} P_{in} \sin n\left[\theta_e - \frac{\pi}{2}(i - 1)\right] \quad ,$$

where i = 1, 2, 3 and 4.

This expression takes care of any type of tooth geometry. Sine terms are zero when the tooth geometry is symmetric. However, when it is desired to have the counterclockwise torque different from the clockwise torque, the tooth geometry has to be non-symmetric. In this case, sine terms will be present.

In this paper, a four-phase motor is used to formulate the static torque. However, the similar approach can be applied to n-phase motors. As an example, for a three-phase motor, Figure 5-5 will have four branches with three gap permeances, three mmf sources, a permanent magnet permeance and mmf due to the permanent magnet. The gap permeances will be offset by 120 electrical degrees with each other. With these modifications, the similar approach given in this paper can be applied to the study of the three-phase motors.

The formulation given in Section 5-3 can be easily applied to determine motor parameters such as self- and mutual-inductances and back emf terms.

5-10 SUMMARY

A critical examination of a very commonly used four-phase, bifilar-wound PM step motor resulted in a simple magnetic circuit. Based on this circuit, static torque formulas were obtained. For PM motors, the torque is proportional to NI; for VR motors, the torque is proportional to $(NI)^2$. It was also shown that the existence of the cogging torque was attributable to the 4th harmonic permeance component.

The static torque formulation was extended to dynamic analysis. With the a-c excitation, it was shown that step motors behave like synchronous motors with many poles. Finally, the formulation for a multi-step response was given for the bifilar winding drive circuit with commutating capacitors.

REFERENCES

1. Bailey, S. J., "Incremental Servos, Part II: Operation and Analysis," Control Engineering, Vol. 7, pp. 97-102, Dec. 1960.

2. O'Donohue, J. P., "Transfer Function for a Stepper Motor," Control Engineering, Vol. 8, pp. 103-104, Nov. 1961.

3. Kieburtz, R. B., "The Step Motor - The Next Advance in Control Systems," IEEE Transactions on Automatic Control, pp. 98-104, Jan. 1964.

4. Snowdon, A. L. and Madsen, E. W., "Characteristics of a Synchronous Inductor Motor," AIEE Transactions, Applications and Industry, Vol. 81, pp. 1-5, March 1962.

5. Robinson, D. J., "Dynamic Analysis of Permanent Magnet Stepping Motors," NASA TN D-5094, Lewis Research Center, Cleveland, Ohio, March 1969.

6. Robinson, D. J. and Taft, C. K., "A Dynamic Analysis of Magnetic Stepping Motors," IEEE Transactions on Ind. Elect. and Control Instr., Vol. IECI 16, No. 2, pp. 111-125, Sept. 1969.

6 PERMEANCE MODEL AND RELUCTANCE FORCE BETWEEN TOOTHED STRUCTURES

H. D. Chai
IBM System Products Division
Endicott, New York

6-1 INTRODUCTION

With the advancements in the digital computer equipment field and the rapid growth of high-power, low-cost solid-state devices, step motors are finding applications in such diverse fields as paper transports, print-head carriers, and disk drives in computer systems. They are also used in radar tracking and machine-tool operations.

With this increased usage of the motors, there is a need for a simple mathematical model that can be used by engineers in the design and application of step

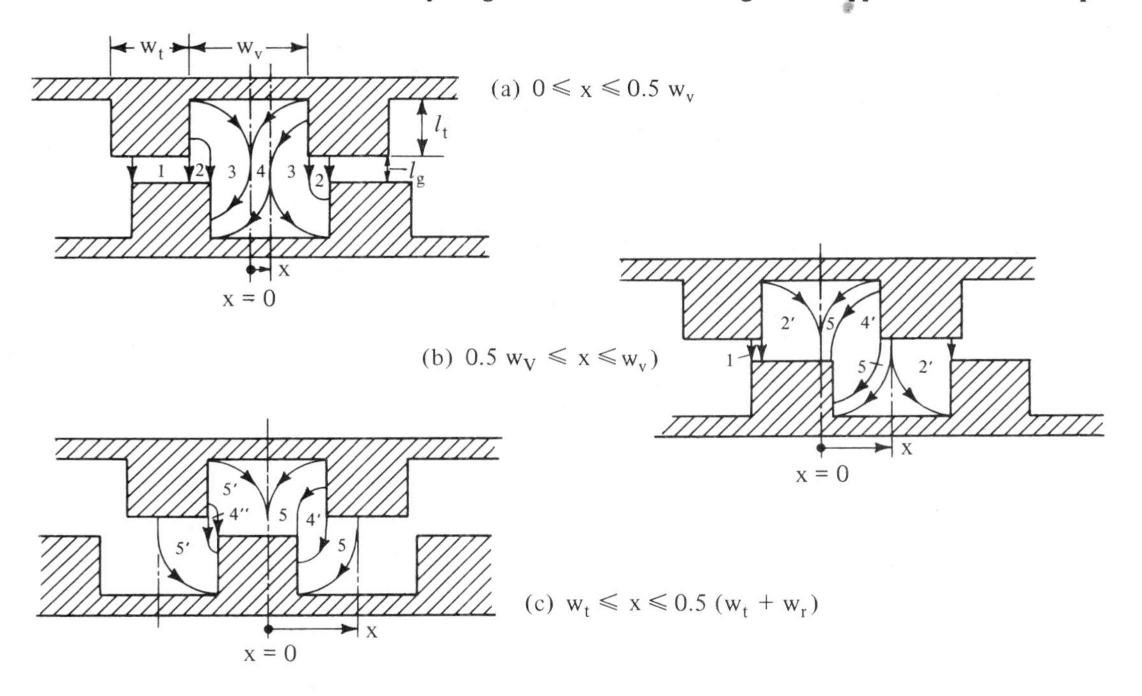

Figure 6-1. Assumed flux pattern between toothed poles $0.5W_v \geq W_t \geq W_v$.

motors. With this objective, this paper develops permeance formulas using the field patterns assumed in Figure 6-1. The permeances and their spatial derivatives are normalized for compactness and generality. Normalized values are used to develop a series of graphs for handy reference, and formulas for the reluctance force are developed using the normalized values. A number of examples are given to show how the formulas and graphs may be applied in practice.

The formulation given in the paper is not to be confined to step motors alone. It can be applied to any design which makes use of reluctance force. Also the permeance formulas developed here may be used as a basis for comprehensive analytical work. For example, the permeance data can be put into Fourier series from which expressions for motor constants (torque constant, generator or back emf constant, and inductances) can be obtained. With the derived motor constants, dynamic equations can then be solved.

6-2 PERMEANCE

The permeance formula is developed using the field patterns given in Fig. 6-1. Field lines are assumed to consist of straight line segments and circular arcs. These lines represent satisfactorily those observed in actual situations. In Figure 6-1 the tooth width w_t is equal to or greater than one-half the valley width w_v. This is the condition usually found in step motors. When the tooth width is greater than the valley width, the holding torque decreases. The same holds true when the tooth width is much smaller than one-half the valley width.

Using the permeance formula

$$P = \mu_0 \int \frac{dA}{\ell} \quad \text{(in Henries)}, \tag{6-1}$$

where:

dA = the differential cross section of the flux tube
ℓ = the length of the flux tube,
μ_0 = permeability of air ($4\pi \times 10^{-7}$ H/m)

we obtain for the case

$$0.5w_v \leq w_t \leq w_v$$

$$P_1 = \frac{\mu_0 (w_t - x) t}{\ell_g} \qquad 0 \leq x \leq w_t$$

$$P_2 = \frac{2}{\pi} \mu_0 t \ln\left(1 + \frac{0.5\pi x}{\ell_g}\right) \qquad 0 \leq x \leq 0.5w_v$$

$$P_{2'} = \frac{2}{\pi} \mu_0 t \ln\left(1 + \frac{0.25\pi w_v}{\ell_g}\right) \qquad 0.5w_v \leq x \leq w_t$$

$$P_3 = \frac{\mu_0 t}{\pi} \ln\left[1 + \frac{\pi(0.5w_v - x)}{\ell_g + 0.5\pi x}\right] \qquad 0 \le x \le 0.5w_v$$

$$P_4 = \frac{\mu_0 t x}{\ell_g + 0.5\pi(w_v - x)} \qquad 0 \le x \le 0.5w_v$$

$$P_{4'} = \frac{\mu_0 t(w_v - x)}{\ell_g + 0.5\pi(w_v - x)} \qquad 0.5w_v \le x \le w_v$$

$$P_{4''} = \frac{\mu_0 t(x - w_t)}{\ell_g + 0.5\pi(x - w_t)} \qquad w_t \le x \le w_v$$

$$P_5 = \frac{2}{\pi} \mu_0 t \ln\left[\frac{\ell_g + 0.25\pi w_v}{\ell_g + 0.5\pi(w_v - x)}\right] \qquad 0.5w_v \le x \le w_v$$

$$P_{5'} = \frac{2}{\pi} \mu_0 t \ln\left[\frac{\ell_g + 0.25\pi w_v}{\ell_g + 0.5\pi(x - w_t)}\right] \qquad w_t \le x \le w_v \tag{6-2}$$

where:

w_t = tooth width

w_v = valley width

ℓ_g = gap length

t = thickness

x = relative displacement between opposing teeth.

6-3 NORMALIZATION

The permeance expressions given in (6-2) are greatly simplified using the normalization

$$\alpha = \frac{P}{\mu_0 t} \tag{6-3}$$

We also define normalized derivatives

$$\beta = \frac{\ell_g}{\mu_0 t}\frac{dP}{dx} = \frac{d\alpha}{dX} \tag{6-4}$$

The resulting expressions are for permeance:

$$\alpha_1 = W_t - X \qquad 0 \le X \le W_t$$

$$\alpha_2 = \frac{2}{\pi} \ln(1 + 0.5\pi X) \qquad 0 \le X \le 0.5W_v$$

$$\alpha_{2'} = \frac{2}{\pi} \ln(1 + 0.25\pi W_v) \qquad 0.5W_v \le X \le W_t$$

$$\alpha_3 = \frac{1}{\pi}\,\ell n\left(1 + \pi\,\frac{0.5W_v - X}{1 + 0.5\pi X}\right) \qquad 0 \le X \le 0.5W_v$$

$$\alpha_4 = \frac{X}{1 + 0.5\pi(W_v - X)} \qquad 0 \le X \le 0.5W_v$$

$$\alpha_{4'} = \frac{W_v - X}{1 + 0.5\pi(W_v - X)} \qquad 0.5W_v \le X \le W_v$$

$$\alpha_{4''} = \frac{X - W_t}{1 + 0.5\pi(X - W_t)} \qquad W_t \le X \le W_v$$

$$\alpha_5 = \frac{2}{\pi}\,\ell n\left(\frac{1 + 0.25\pi W_v}{1 + 0.5\pi(W_v - X)}\right) \qquad 0.5W_v \le X \le W_v$$

$$\alpha_{5'} = \frac{2}{\pi}\,\ell n\left(\frac{1 + 0.25\pi W_v}{1 + 0.5\pi(X - W_t)}\right) \qquad W_t \le X \le W_v \qquad (6\text{-}5)$$

Normalized derivatives are:

$$\beta_1 = -1 \qquad 0 \le X \le W_t$$

$$\beta_2 = \frac{1}{1 + 0.5\pi X} \qquad 0 \le X \le 0.5W_v$$

$$\beta_{2'} = 0 \qquad 0.5W_v \le X \le W_t$$

$$\beta_3 = -\frac{1}{2}\left(\frac{1}{1 + 0.5\pi(W_v - X)} + \frac{1}{1 + 0.5\pi X}\right) \qquad 0 \le X \le 0.5W_v$$

$$\beta_4 = \frac{1 + 0.5\pi W_v}{[1 + 0.5\pi(W_v - X)]^2} \qquad 0 \le X \le 0.5W_v$$

$$\beta_{4'} = -\frac{1}{[1 + 0.5\pi(W_v - X)]^2} \qquad 0.5W_v \le X \le W_v$$

$$\beta_{4''} = \frac{1}{[1 + 0.5\pi(X - W_t)]^2} \qquad W_t \le X \le W_v$$

$$\beta_5 = \frac{1}{1 + 0.5\pi(W_v - X)} \qquad 0.5W_v \le X \le W_v$$

$$\beta_{5'} = \frac{-1}{1 + 0.5\pi(X - W_t)} \qquad W_t \le X \le W_v \qquad (6\text{-}6)$$

where:

$$X = x/\ell_g$$

$$W_t = w_t/\ell_g$$

$$W_v = w_v/\ell_g$$

$$T = t/\ell_g$$

Thus, for

$$0 \leq X \leq 0.5W_v \qquad \text{(Figure 6-1a)}$$

$$\alpha = \alpha_1 + \alpha_4 + 2(\alpha_2 + \alpha_3)$$

$$\beta = \beta_1 + \beta_4 + 2(\beta_2 + \beta_3) \tag{6-7}$$

For

$$0.5W_v \leq X \leq W_t \qquad \text{(Figure 6-1b)}$$

$$\alpha = \alpha_1 + \alpha_{4'} + 2(\alpha_{2'} + \alpha_5)$$

$$\beta = \beta_1 + \beta_{4'} + 2(\beta_{2'} + \beta_5) \tag{6-8}$$

For

$$W_t \leq X \leq 0.5(W_t + W_v) \qquad \text{(Figure 6-1c)}$$

$$\alpha = \alpha_{4'} + \alpha_{4''} + 2(\alpha_5 + \alpha_{5'})$$

$$\beta = \beta_{4'} + \beta_{4''} + 2(\alpha_5 + \alpha_{5'}) \tag{6-9}$$

Note that the α's and β's are dimensionless quantities, which give permeances and their derivatives from Eqs. (6-3) and (6-4). Figure 6-2 gives a series of curves for the normalized permeance and derivative.

In agreement with physical situations, the spatial derivative of permeance is zero at both $x = 0$ and $0.5(w_t + w_v)$. Also the permeance and its derivative formulas match at the boundaries ($x = 0.5w_v$ and w_t) between three regions (see Figure 6-1).

6-4 SINUSOIDAL APPROXIMATION

Figure 6-2 reveals that the permeance variation with respect to displacement is fairly sinusoidal. Hence, for some applications it is convenient to approximate the permeance in a sinusoidal form. This gives:

$$\alpha(x) = 0.5\left\{(\alpha_{max} + \alpha_{min}) + (\alpha_{max} - \alpha_{min})\cos\frac{2\pi}{W_t + W_v}X\right\} \tag{6-10}$$

and

$$\beta(x) = \frac{d\alpha}{dx} = -\frac{\pi}{W_t + W_v}(\alpha_{max} - \alpha_{min})\sin\frac{2\pi}{W_t + W_v}X \quad , \tag{6-11}$$

where

$$\alpha_{max} = W_t + \frac{2}{\pi}\,\ell n\,(1 + 0.5\pi W_v)$$

$$\alpha_{min} = \frac{W_v - W_t}{1 + 0.25\pi(W_v - W_t)} + \frac{8}{\pi}\,\ell n\,\frac{1 + 0.25\pi W_v}{1 + 0.25\pi(W_v - W_t)}$$

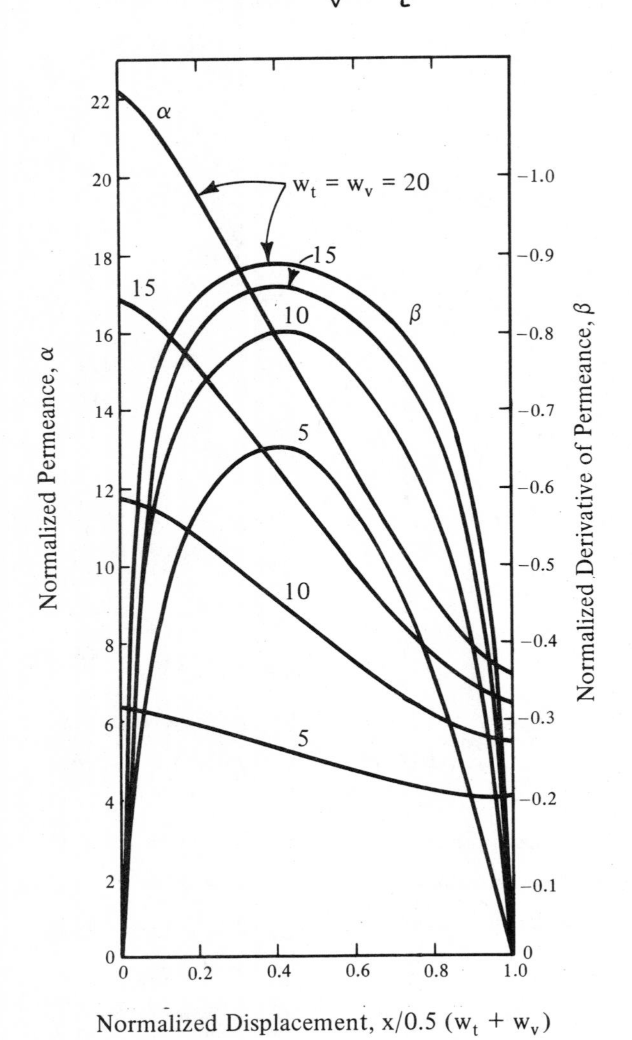

Figure 6-2. Normalized permeance and derivative versus displacements.

6-5 RELUCTANCE FORCE

A very important application of the permeance model is in calculating the reluctance force between opposing toothed structures such as step motors.

In evaluating the reluctance force per pole, let us assume that the pole has n teeth. Then the reluctance per pole is:

$$R = \frac{1}{nP} ,$$

where P is permeance per tooth.

The force per pole is

$$F = \frac{1}{2}\phi^2 \frac{dR}{dx}$$

$$= -\frac{1}{2n}\phi^2\left(\frac{1}{P^2}\frac{dP}{dx}\right) ,$$

where ϕ = flux through the pole.

In terms of the normalized permeance and its derivative

$$F = -1.39 \times 10^4 \frac{\phi^2}{nt\ell_g}\left(\frac{\beta}{\alpha^2}\right) \quad \text{lb/pole} \tag{6-12}$$

or

$$F = -5.78 \times 10^{-7} w_t^2 B^2 \left(\frac{nt}{\ell_g}\right)\left(\frac{\beta}{\alpha^2}\right) \quad \text{lb/pole} \tag{6-13}$$

or

$$F = -1.41 \times 10^{-7} n\left(\frac{t}{\ell_g}\right) F^2 \beta \quad \text{lb/pole} \tag{6-14}$$

where

ϕ = net flux at the pole (10^3 lines or maxwells).

n = number of teeth per pole.

t, ℓ_g, w_t = thickness, gap length, and tooth width, respectively, (10^{-3} inches).

$B = \frac{\phi}{nw_t t}$, flux density in the pole (10^3 gauss).

F = mmf drop across the gap (amp-turns).

Equation (6-12) is useful when the flux is known. In design work, it is more convenient to use Eq. (6-13) because we can put an upper limit on the value of the flux density. For ordinary soft iron materials, the saturation flux density is 20,000 gauss. When a magnetic circuit is unsaturated so that we can safely neglect the mmf drop in the iron, Eq. (6-14) can be used with F equal to an applied amp-turns.

6-6 EXAMPLES

The following examples, using Figure 6-3 as a hardware model, clarify the preceding equations and graphs.

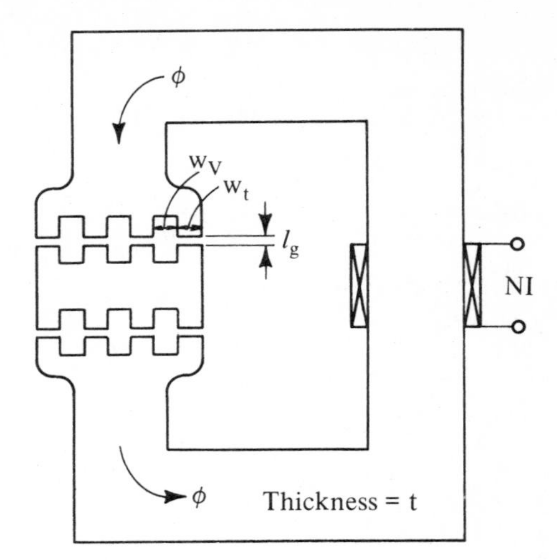

Figure 6-3. Hardware model.

Example 6-1

Given:

$$w_t = 50 \times 10^{-3} \text{ inch}$$

$$w_v = 50 \times 10^{-3} \text{ inch}$$

$$\ell_g = 5 \times 10^{-3} \text{ inch}$$

$$t = 500 \times 10^{-3} \text{ inch}$$

Find: Maximum and minimum permeance per tooth.

Solution: First, normalize the given parameters.

$$W_t = \frac{w_t}{\ell_g} = 10$$

$$W_v = \frac{w_v}{\ell_g} = 10$$

From Figure 6-2,

$$\alpha_{max} = 11.8 \text{ at } x = 0$$

$$\alpha_{min} = 5.5 \text{ at } x = 50 \times 10^{-3} \text{ inch}$$

From Eq. (6-3),

$$P_{max} = 11.8\mu_0 t/\text{tooth}$$

$$P_{min} = 5.5\mu_0 t/\text{tooth}$$

We see that the permeance is directly proportional to thickness, and the ratio of maximum to minimum permeance is a little more than 2.

Example 6-2

Using the values given in Example 6-1, find the location and value of maximum derivative of the permeance per tooth.

Solution: From the curve for $\beta(W_t = W_v = 10)$ the location is

$$\frac{X}{0.5(W_t + W_v)} \cong 0.4$$

or

$$X \cong 0.2(10 + 10) = 4$$

$$x_{max} = 4\ell_g = 20 \times 10^{-3} \text{ inch}$$

$$\beta_{max} = 0.8$$

From Eq. (6-4),

$$\left.\frac{dP}{dx}\right|_{max} = 80\mu_0/\text{tooth}$$

Example 6-3

Assuming the negligible reluctance of the iron in Figure 6-3, find the maximum reluctance of the magnetic circuit.

Solution: There are 4 teeth (n = 4) on each pole, and two poles are in series.

The reluctance per pole gap

$$R_{max} = \frac{1}{nP_{min}} = \frac{1}{4(5.5)\mu_0 t} = \frac{0.0455}{\mu_0 t} \text{ per pole.}$$

Total reluctance

$$R_t = 2R_{max} = \frac{0.091}{\mu_0 t}$$

Example 6-4

Let us apply 200 amp-turns to the device (Figure 6-3). Again neglecting the mmf drop in the iron, find the maximum reluctance force on the armature.

Solution: Since we neglect the mmf drop in the iron all the mmf drop appears across the pole gaps. Therefore

$$F = \frac{NI}{2} = 100 \text{ amp-turns.}$$

The maximum force occurs at

$$x_{max} = 20 \times 10^{-3} \text{ inch (from Example 6-2)}$$

The corresponding β is 0.8. Using Eq. (6-14)

$$F_{max} = +1.413 \times 10^{-7}(4)(10^2)(10^4)(0.8) \quad \text{lb/pole}$$

$$= 4.522 \times 10^{-1} \quad \text{lb/pole}$$

or

$$F_{max} = 0.9044 \quad \text{lb total}$$

Example 6-5

Let us suppose that the flux density at the teeth is 10 Kgauss.

Find the corresponding reluctance force on the armature at $x = 20 \times 10^{-3}$ inch.

Solution: First, from Fig. 6-2,

$$\left.\frac{\beta}{\alpha^2}\right|_{X = 0.4} = \frac{-0.8}{9.1^2} = -0.965 \times 10^{-2}$$

Then, using Eq. (6-13),

$$F = (5.78 \times 10^{-7})(2.5 \times 10^3)(10^2)(4)(10^2)(0.965 \times 10^{-2}) \quad \text{lb/pole}$$

$$= 0.558 \text{ lb/pole}$$

or

$$F_{total} = 1.116 \text{ lb}$$

6-7 DISCUSSION

Figures 6-4 through 6-7 give normalized permeance and derivatives with varying tooth to valley ratios. They are provided as convenient graphs for those who are involved in design work that utilizes the reluctance force. The range of pitches $(w_t + w_v)$ should cover most cases in practical design work.

The figures show that the permeance decreases as the tooth-to-valley ratio (w_t/w_v) decreases. This is obvious from Figure 6-1. As the tooth becomes thinner for a given pitch $(w_t + w_v)$, there is more air space resulting in a higher reluctance or lower permeance.

β curves reveal that for a given pitch and a fixed mmf drop across the gap there is an optimum tooth-to-valley ratio at which the holding force is maximum. A one-to-one tooth-to-valley ratio does not give an optimum condition as many design engineers may already know from experience. As the ratio is decreased, the maximum force increases. A further decrease in the ratio results in a decrease in the maximum force. In design work, it is suggested that one use a tooth-to-valley ratio around 2/3. Even though the graph shows the greater maximum force with the ratio less than

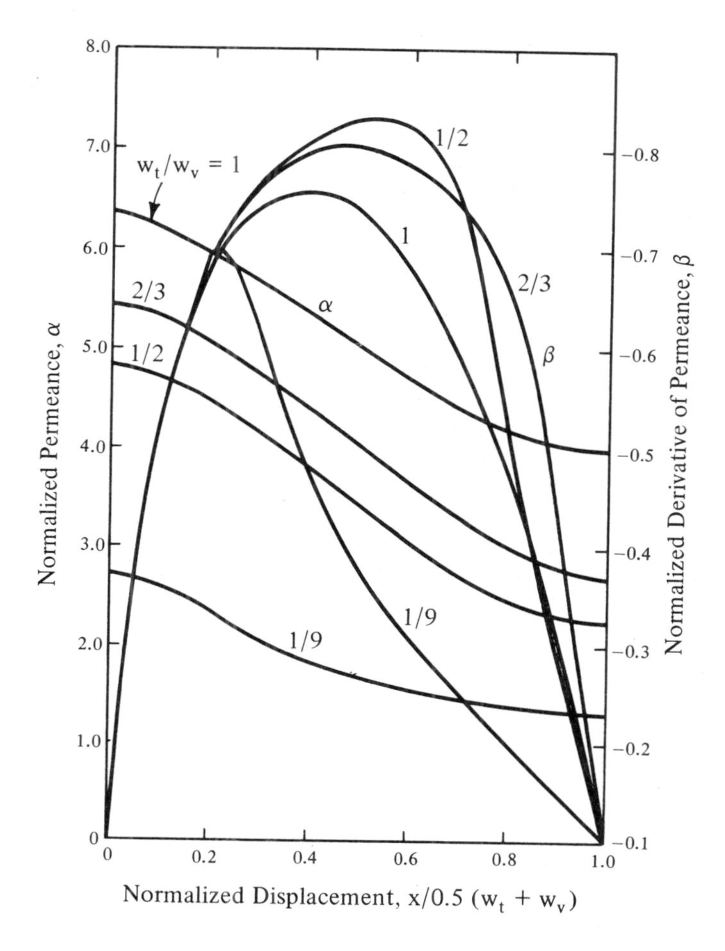

Figure 6-4. Normalized permeance and derivative versus displacements ($W_t + W_v = 10$).

2/3, the decrease in tooth width will saturate the teeth quicker. This quicker saturation of teeth will result in a decrease in the mmf drop across the gap for a given ampere-turns at the coil. This decrease will result in the decreased maximum force [see Eq. (6-14)]. Another point to note is that the location of the maximum force is not fixed at one point but is a function of the tooth-to-valley ratio.

In practical design work, our objective is to develop a maximum force in a minimum volume. With fixed physical dimensions, we assign a reasonable amount of flux at the gap. (The saturation flux density for commonly used magnetic materials is around 20 Kgauss.) Then we calculate the necessary ampere-turns required to develop the desired flux. In this situation, the mmf drop in the iron cannot be neglected. The mmf drop across the gap is no longer fixed as the opposing teeth are displaced. Hence Eq. (6-14) cannot be simply applied. Equations (6-12) and (6-13) are more convenient to use.

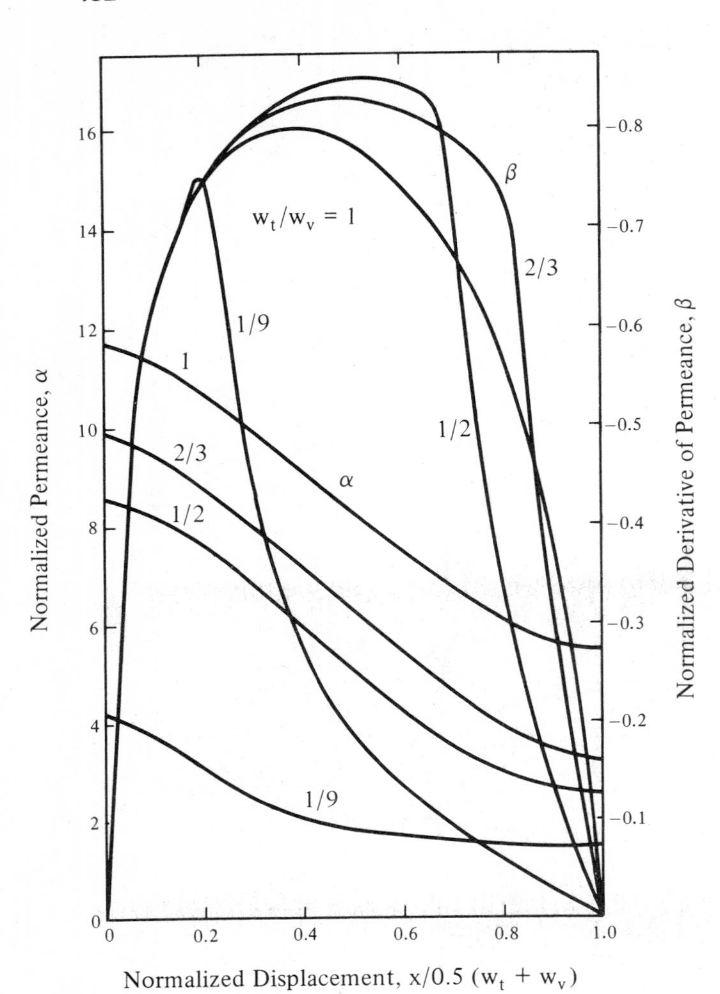

Figure 6-5. Normalized permeance and derivative versus displacements ($W_t + W_v = 20$).

Figure 6-6. Normalized permeance and derivative versus displacements ($W_t + W_v = 30$).

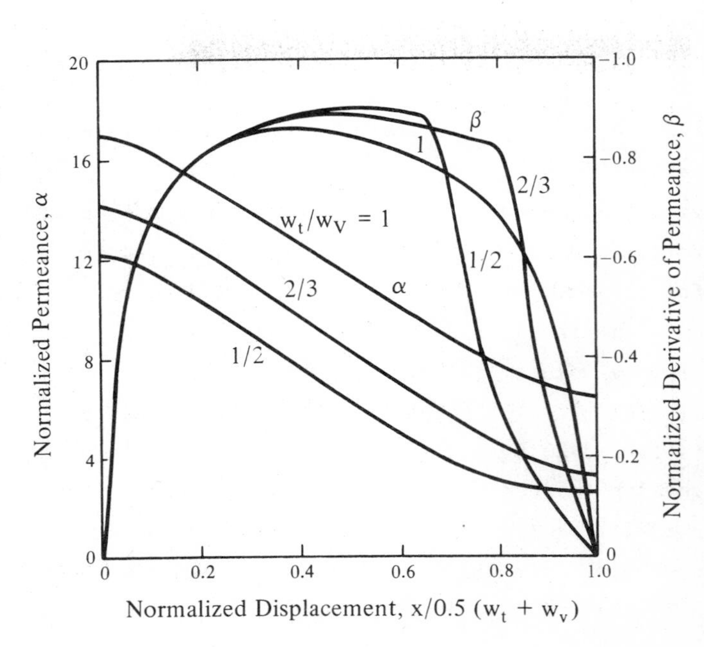

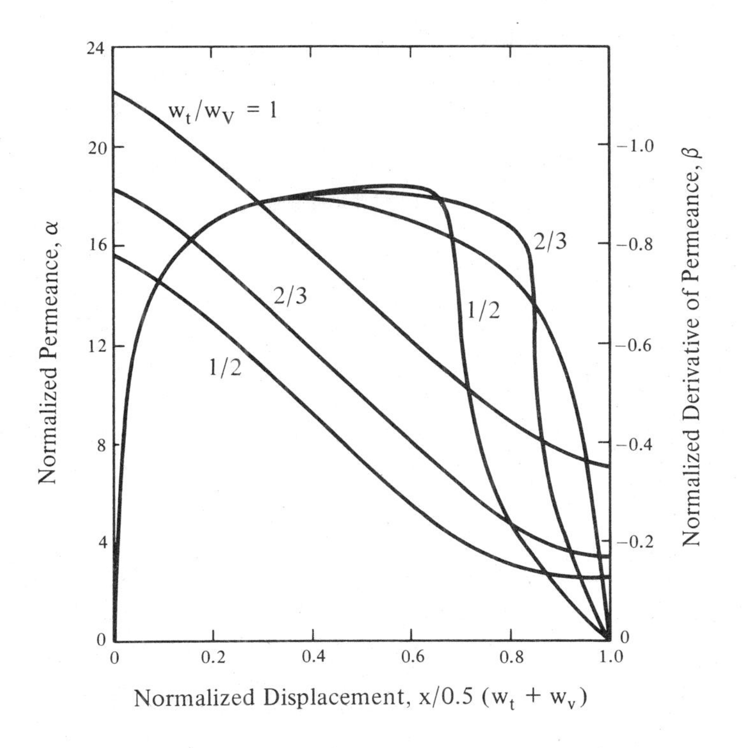

Figure 6-7. Normalized permeance and derivative versus displacements ($W_t + W_v = 40$).

The assumption that the field lines are perpendicular to the magnetic surface (Figure 6-1) is valid for nonsaturated conditions. This can be seen from the relationship,

$$\frac{\tan\theta_{air}}{\tan\theta_{iron}} = \frac{1}{\mu_{iron}}$$

where θ_{air} and θ_{iron} are, respectively, the angles of incidence of field in air and in iron, and where μ_{iron} is the relative permeability of the iron. For a μ_{iron} of 2000 and θ_{iron} of 89 degrees (incident field on the side of the teeth), we find that θ_{air} is about 1.5 degrees.

As the teeth begin to saturate, θ_{air} will increase. This implies that the effective length of the flux path in air decreases resulting in a higher permeance (see Eq. (6-1)) for the flux emanating from the sides of the teeth. Also the constant magnetic potential surface will deviate from the physical boundary surface of the teeth.

7 PRACTICAL DESIGN CONSIDERATIONS OF STEP MOTORS

K. S. Kordik
Warner Electric Brake & Clutch Co.
Beloit, Wisconsin

7-1 INTRODUCTION

The design of step motors requires many considerations. The user may have space, heat rejection, and input power limitations. The step angle and step accuracy, as well as power output, are usually rigid requirements. All of these factors must be considered before a motor is selected. If a special motor is designed these factors become even more important. If the application involves several motors, cost becomes a very important consideration to the user, but this is always an important factor for the designer. Since a step motor is almost always used in incremental motion applications, this requires minimum rotor inertia with good torsional shaft stiffness and the ability to withstand radial loads.

The temperature rise and the effects of temperature on materials used for insulating the motor, and mechanical structures, must be taken into consideration. Temperature can affect the size of the airgap.

When designing a step motor, it must be thought of as a sub-system which will become part of a more complex system. The motor can be broken down into additional sub-systems as indicated in Figure 7-1.

The output element is usually a shaft made from stainless steel for corrosion resistance or for its non-magnetic properties. The output element could also be a lead screw.

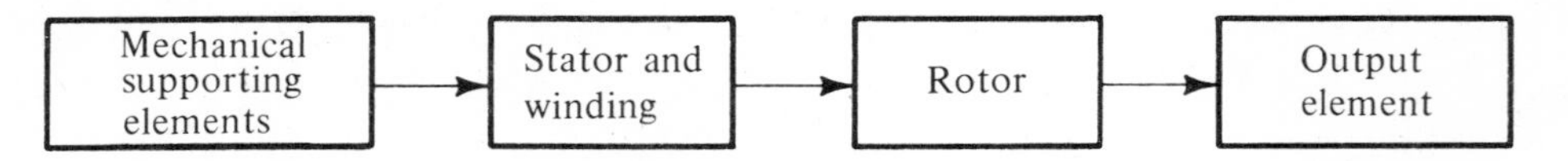

Figure 7-1. Block diagram of a step motor.

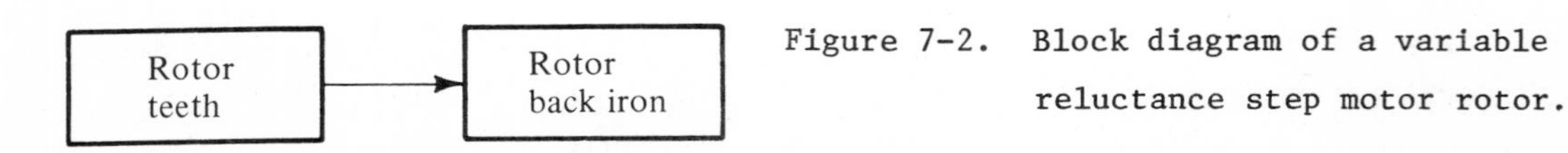

Figure 7-2. Block diagram of a variable reluctance step motor rotor.

The important rotor elements are shown in Figure 7-2.

The rotor back iron supports the rotor teeth, transmits torque between the rotor teeth and the output element and provides a magnetic flux path between rotor teeth of opposite polarity.

Magnet flux in the airgap reacts with the rotor teeth to produce torque, The teeth also serve as a flux path to the back iron. The rotor is usually constructed from soft magnetic iron, either laminated or solid. It usually contains silicon to decrease eddy currents. Laminated construction reduces eddy currents and hysteresis affects even further.

The important elements of the stator are shown in Figure 7-3.

The stator teeth form the stationary reaction member for the torque produced by the rotor and provide a flux path between the airgap and poles. The stator poles support the stator windings which transform electrical energy into magnetic energy. Some motors have only teeth to support the windings. These motors would have one tooth per pole. The stator back iron supports the teeth and poles and provides a flux path between the poles of the motor. The stator is made up of silicon iron laminations. Motors have utilized solid iron rings for the back iron.

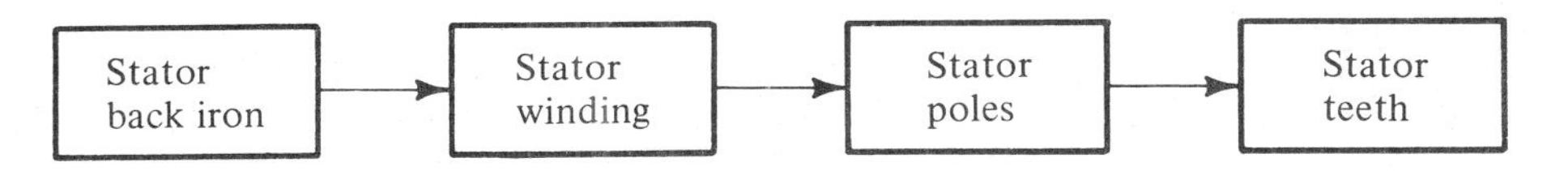

Figure 7-3. Block diagram of a step motor stator.

The mechanical supporting elements consist of bearings, supports for the bearings, and a stator supporting member which ties the bearing supports together. Precision step motors use sealed or shielded pre-lubricated ball bearings. Ball bearings are used because low friction is essential, particularly in small motors. The airgap in a variable reluctance motor is usually very small, 2-5 thousands. To maintain this airgap requires a precision bearing with low wear and minimum radial play. Some motors must be able to operate with a high thrust or axial load. The supporting elements can be stainless or low carbon steel, aluminum or zinc. Aluminum is frequently used in small motors because of its light weight, corrosion resistance and good thermal conductivity.

After searching through existing literature on step motors, there seems to be almost a complete lack of design information. Most articles deal with control application and control of step motors.

7-2 DESIGN CONSIDERATIONS

Torque

Torque is always used when rating a step motor. This can be static holding torque or torque developed at some speed. There is no industry wide standard for specifying torque. The designer must know which torque is referred to when designing a motor for a specific torque rating. If a motor is designed to perform a specific function, the average torque required can be calculated. To move a load incrementally requires an accelerating torque to start it and a decelerating torque to stop it. The inertia of a load, distance it must move and time allowed for the motion to occur affect the torque required.

$$T = J_L \alpha \tag{7-1}$$

$$\alpha = \frac{d\omega}{dt} \tag{7-2}$$

$$\omega = \frac{d\theta}{dt} \tag{7-3}$$

Conbining Equations (7-1) and (7-3), we have

$$\alpha = \frac{\frac{d(d\theta)}{dt}}{dt} = \frac{d^2\theta}{dt^2} \tag{7-4}$$

Assuming a constant acceleration, resulting in a constant torque, Equation (7-4) becomes

$$\alpha = \frac{2\theta}{t^2} \tag{7-5}$$

The torque required if acceleration occurs for the entire distance is

$$T = J_L \frac{2\theta}{t^2} \tag{7-6}$$

Since normal practice for short motions is to accelerate for half the distance and decelerate for the other half, the required acceleration torque will be double that given in Equation (7-6). Equation (7-6) now becomes

$$T = 2J_L \frac{2\theta}{t^2} \tag{7-7}$$

The angle of motion is normally given in degrees. The formula will be more useful if degrees are used directly.
Therefore

$$T = \frac{4\pi J_L \theta}{180t^2} \tag{7-8}$$

where θ is in degrees.

The motor rotor inertia J_m should not be ignored. Adding J_m results in

$$T = \frac{4\pi(J_L+J_m)\theta}{180t^2} \tag{7-9}$$

The torque formula can also be written as

$$T = \frac{4\pi(J_L+J_m)\theta\times10^6}{180t^2} \qquad (7\text{-}10)$$

with t in milliseconds.

All step motors have some inductance in the stator windings; therefore, torque is not developed instantaneously. The torque will be developed as current builds up, neglecting eddy current effects. If the average inductance of the winding is known or the inductance for a similar motor is known, this time can be calculated.

$$E = L\frac{di}{dt} \qquad (7\text{-}11)$$

Assuming a constant change in current with initial current equal to zero

$$E = L\frac{I}{t} \qquad (7\text{-}12)$$

Let t = tau or current buildup time

$$tau = \frac{LI}{E} = \frac{L}{R} \qquad (7\text{-}13)$$

If L is in millihenries, tau will be in milliseconds.

The formula for average torque in Equation (7-10) can now be corrected to include current and torque buildup time.

$$T = \frac{4\pi(J_L+J_m)\theta\times10^6}{180(t+tau)^2} \qquad (7\text{-}14)$$

If the inertias are given in oz-in^2, the torque will be in oz-in. Initial motor designs are usually made for static holding torque. The average torque is multiplied by 1.3 to arrive at the peak holding torque. The accuracy of the torque calculated depends on the accuracy of tau.

Factors Affecting Static Holding Torque

A. Rotor and stator diameter - torque varies directly with rotor diameter.

B. Airgap length - torque increases with decreasing airgap, decreasing with increasing airgap. In small and medium size variable reluctance motors, the airgap ranges from two to five thousandths of an inch.

C. Flux density - torque varies as the square of the magnetic flux density in the airgap. Increasing flux density increases torque.

D. Tooth width - for one phase "ON" operation holding torque increases as tooth width decreases. Figure 7-4 illustrates the effects of tooth width variations.

E. Rotor and stator length. Torque varies directly with stack length.

F. Rotor tooth and stator bore surface finish. The smoother the finish, the higher the holding torque. The smoothness of the surface finish determines the effective airgap. A rough surface produces a mechanical airgap which is smaller than the effective or magnetic airgap.

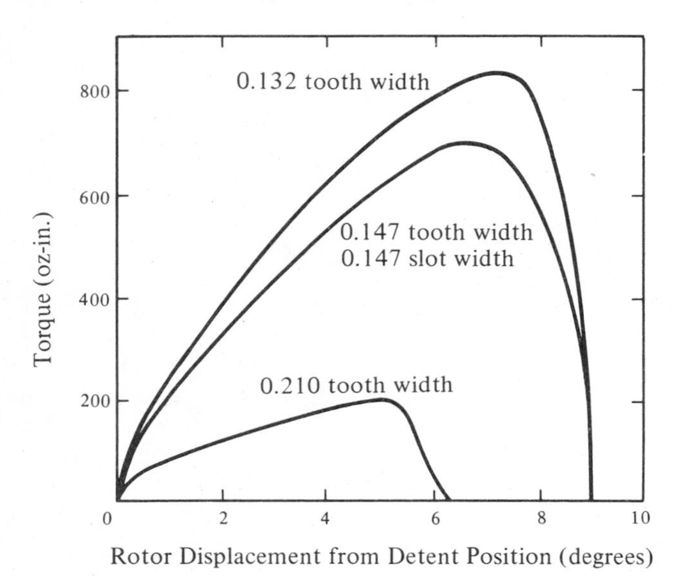

Figure 7-4. Holding torque variations with changing tooth width. Constant pole width and constant excitation.

Airgap Length

The minimum permissible airgap length can be determined from the tolerances of mechanical parts and dimensional changes due to thermal expansion of various parts.

The mechanical parts which affect airgap are stator, rotor, shaft, bearings, bearing brackets and housing, or nearly every part in the motor, plus external radial loads and shaft deflection due to unbalanced magnetic pull.

If the stator, housing and bearing brackets are bored as an assembly, runout and clearance between these parts may be ignored when calculating the minimum allowable airgap. If the shaft and rotor are machined as an assembly, rotor to shaft runout can be neglected.

This leaves bearing to bearing bracket clearance, external ring runout, internal ring runout, bearing to shaft runout and rotor runout. For calculation purposes, one-half the T.I.R. values are used. Thermal expansion of structural parts must also be considered, keeping in mind the temperature differential between various parts.

Unbalanced magnetic pull should be calculated using the runouts as calculated previously for a starting point. The amount of deflection which occurs due to the unbalanced magnetic pull depends on the bearing spread as well as shaft and rotor stiffness. A nominal airgap may be assumed and if this is inadequate it can be increased until calculations show that the rotor clears under the worst conditions.

The unbalanced magnetic pull may be calculated from the following formula:

$$\text{Force} = \frac{B^2 S}{72} \quad \text{LB} \tag{7-15}$$

B = kilo-maxwells per sq. in.

S = gap area

From Equation (7-15) we have

Pull = Flux density^2 × combined area of energized poles/72

To determine the unbalanced magnetic pull the rotor displacement from its true centered position must be divided by the airgap length. The formula now becomes

$$\text{Unbalanced magnet pull} = \frac{B^2 S d}{72 g} \tag{7-16}$$

d = the rotor displacement from true position

g = airgap length.

The shaft deflection can be calculated using the double integration method or a graphical integration method.

A simplified approach may be taken which assumes that the load is concentrated at the center of the shaft and that the shaft is a simple beam. This would be a worst case condition.

$$Y = \frac{Pl^3}{48\ El} \tag{7-17}$$

Y = the shaft deflection

P = unbalanced magnetic pull

l = the distance between bearings

I = the moment of inertia of the shaft

E = modulus of elasticity of rotor shaft

After the airgap is selected the design of the magnetic circuit can commence.

Winding Selection

Normally the operating current and voltage will have been selected before the design of the motor is started. These may be changed in order to meet the time constant requirements or slew speed requirements.

Once the rotor size and tooth size, stack length, and airgap flux density have been determined, the winding ampere turns is calculated.

Normally this is calculated using a digital computer to compute the airgap flux density with the rotor displaced from detent position in small increments. From these figures the static holding torque curve is computed. If the results do not agree with the desired results, adjustments are made in tooth width, windings or both. Sometimes other adjustments must be made.

After the number of turns are determined, a winding resistance must be determined.

For fastest response and maximum output power, tau should be low. Therefore, according to Equation (7-13), R should be as large as possible. From an efficiency standpoint, R should be as small as possible since the power dissipated is

$$P = I^2R \tag{7-18}$$

and

$$1KW = 0.9478 \text{ BTU/Sec.} \tag{7-19}$$

During the time the motor is providing holding torque only, not rotating, all the power must be dissipated by the motor. Step motors are totally enclosed and seldom fan cooled. The heat caused by the power dissipated in the motor must be radiated to the surrounding air from the motor housing or conducted out through heat sinking.

There is always a voltage limitation which limits the value of winding resistance. If the motor is to be used in a small enclosure with little or no ventilation, both current and resistance may have to be a minimum. Equations (7-18) and (7-19) show the relative effects of current and winding resistance on the power to be dissipated by the motor.

A low resistance winding may be used if a fast response constant current supply or a dual voltage source is used to supply the power to the motor.

The minimum resistance possible in a given motor is determined by the winding space available. Power dissipated in a motor is usually a compromise between size and torque.

For a given step angle and frame size, a four-phase motor has more winding room than a three-phase motor. This can result in a lower winding resistance, hence less heat dissipation.

REFERENCES

1. Electromagnetic Devices, Herbert C. Roters, John Wiley & Sons, Inc., New York. 1941.

2. Machine Design, Joseph Edward Shipley, McGraw-Hill. New York. 1956

3. Marks' Mechanical Engineers Handbook, Theodore Boumeister, McGraw-Hill. New York. 1958.

8 CONTROL ASPECTS OF STEP MOTORS

B. C. Kuo
Department of Electrical Engineering
University of Illinois at Urbana-Champaign
Urbana, Illinois

8-1 INTRODUCTION

One of the basic features of a step motor is that when the motor is energized it will rotate and come to rest at one of its distinct detent positions. Incremental rotary motion can be obtained with a step motor by energizing and de-energizing successive phases of the motor in turn. Another distinctive characteristic of the step motor is the versatility of its control. A step motor, depending upon the application requirements, can be controlled in a great variety of ways. For instance, in performing the single-step operation, there are many ways of damping out the oscillations at the end of the step motion. The control for driving the motor in a continuous slewing operation can be classified as open loop and closed loop; and even within these two categories there are many different control schemes. In general, given a step motor, one can obtain different levels of performance by the motor, depending upon the ingenuity and complexity of controls applied to the motor.

Although the control schemes and drive circuits discussed in this chapter are devised for the variable-reluctance-type step motors, the principle of control may be applied to any electrically driven step motor.

8-2 STEP MOTOR DRIVER CIRCUITS

The block diagram of Figure 8-1 shows the basic elements of a step motor control system which consists of sequence logic, power switching circuitry, power supply, current or voltage limiting circuitry, and a feedback encoder when necessary. The difference between the open-loop control and the closed-loop control is, essentially, the latter has the encoder for sensing the position and/or the speed of the motor and the feedback connection.

The function of the power driver in Figure 8-1 is to accept low-level input logic signals in the form of a digital pulse train and control the sequence of the

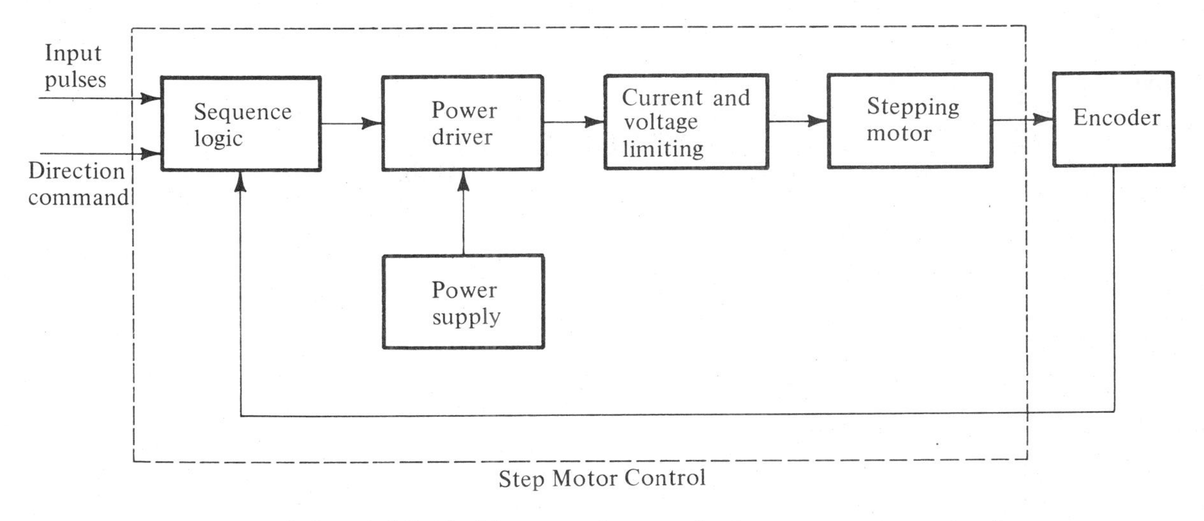

Figure 8-1. A block diagram of a typical step motor control.

high currents to the motor phases to produce the discrete angular motion. The sequence logic section accepts the input pulses along with the direction command, and supplies a low-level signal to each of the power switching circuits. This logic section contains an n-stage ring counter, where n equals the number of phases of the motor. The input pulses to the ring counter cause it to shift either right or left depending on the direction control and, consequently, to sequence the voltage of the motor phases clockwise or counterclockwise. The ring counter may be initially set so that either one or more motor phases are always on at a given time. For example, a four-phase step motor with one phase on and having a specific direction of rotation would have the switching sequence: A-B-C-D-A-B .... The same motor with two phases on at a time would have the sequence: AB-BC-CD-DA-AB .... Timing diagrams for these two sequences are shown in Figure 8-2. The sequence logic may be arranged to switch the motor phases in many different ways depending upon the type of step motor used.

A typical power stage for the control of a three-phase step motor is shown in Figure 8-3 along with its voltage suppression circuitry. To turn a motor phase on, a low-voltage signal from the sequence logic section, usually the ring circuit, turns on the driver transistor Q_1 which then turns on the power transistor Q_2. This closes the circuit from the supply voltage V through the motor phase, the series resistor R, and the transistor Q_2. A current now builds up through this closed circuit. The differential equation for the activated phase is

$$V = (R + R_m)i + L\frac{di}{dt} + i\frac{dL_m}{d\theta}\frac{d\theta}{dt} \tag{8-1}$$

where R_m is the resistance of the motor windings, θ is the motor displacement and L_m is the winding inductance. The equivalent circuit for the charging of a motor phase is shown in Figure 8-4. The last term in Equation (8-1) is known as the back emf of the motor. If the back emf is neglected, the solution of Equation (8-1) with a constant applied voltage V is

$$i(t) = \frac{V}{R + R_m} - \left(\frac{V}{R + R_m} - I_0\right) e^{-t/\tau} \tag{8-2}$$

where I_0 is the initial current in the motor phase when the phase is turned on, and

$$\tau = \frac{L_m}{R + R_m} \tag{8-3}$$

Equation (8-2) is obtained also with the assumption that the inductance L_m does not vary with the motor displacement θ. In a multiple-stack motor, the inductance of one phase of a step motor can be approximated by

$$L_m = L_1 + L_2 \cos n\theta \tag{8-4}$$

where n is the number of steps per revolution divided by the number of motor phases.

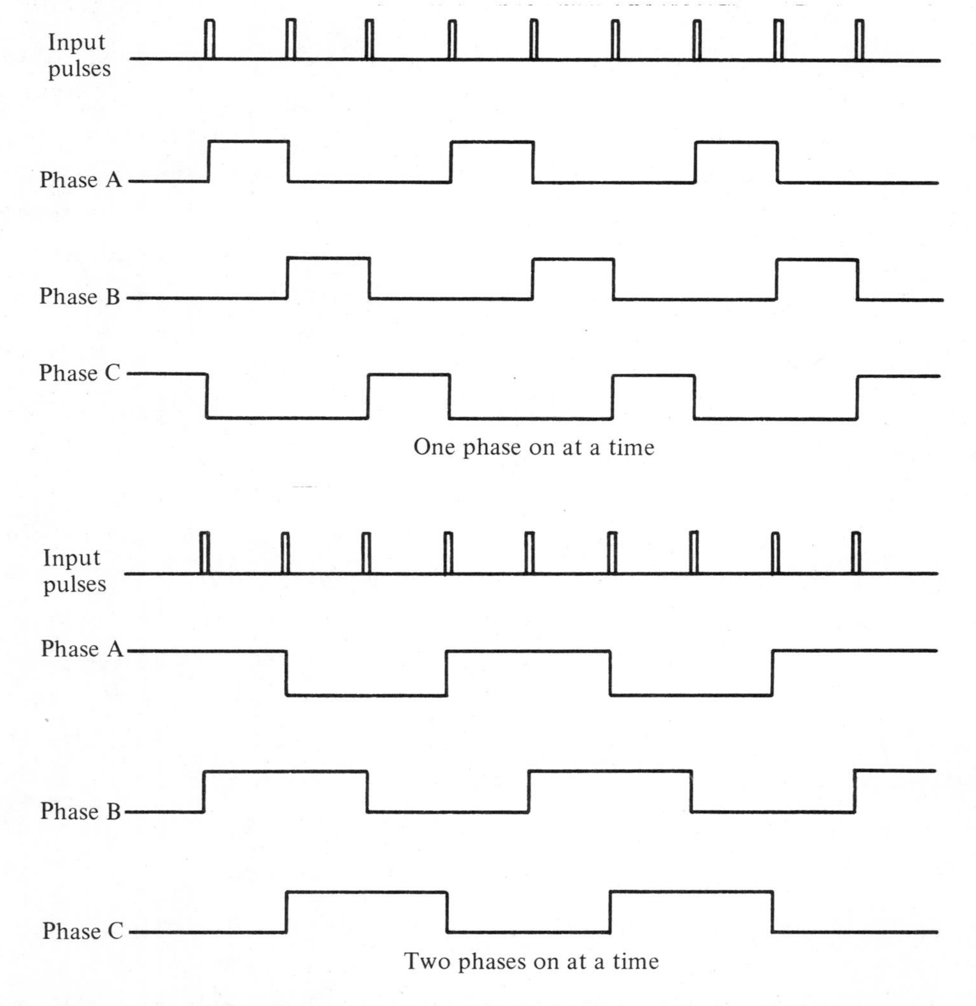

Figure 8-2. Timing diagram for phase sequencing of a VR step motor.

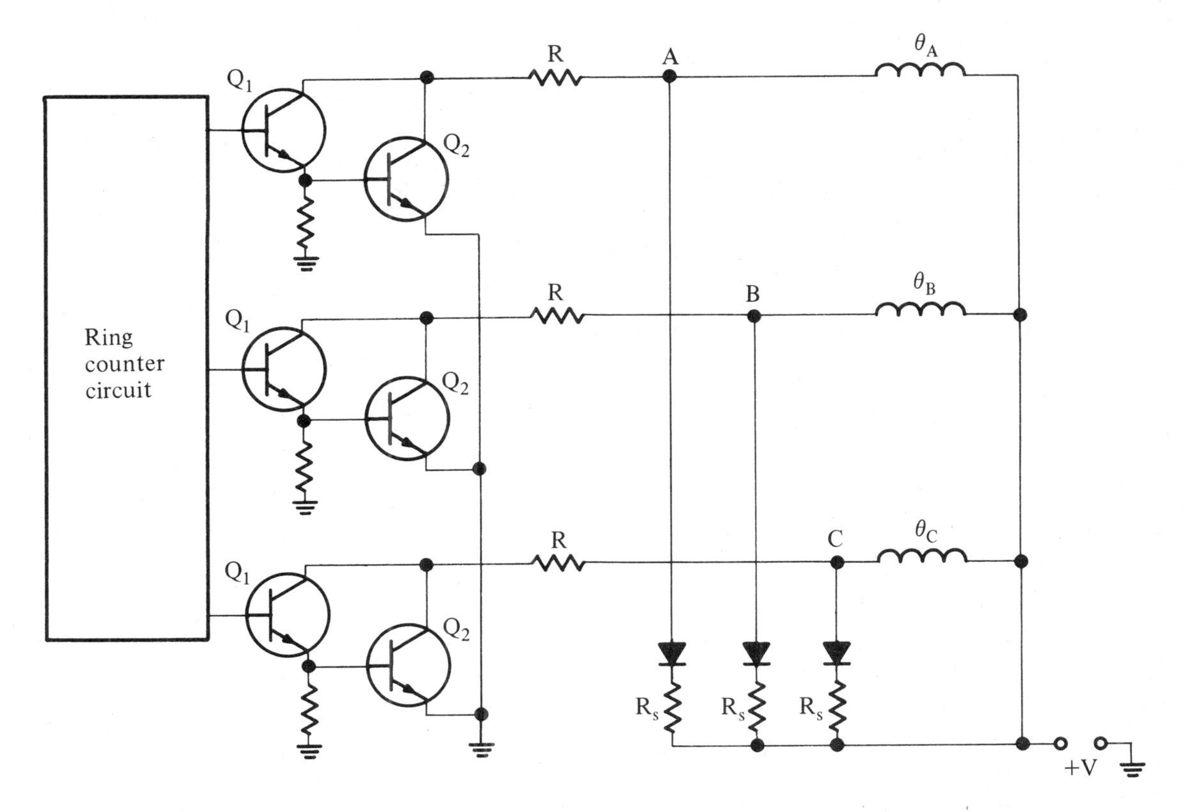

Figure 8-3. A typical power stage for the control of a three-phase step motor.

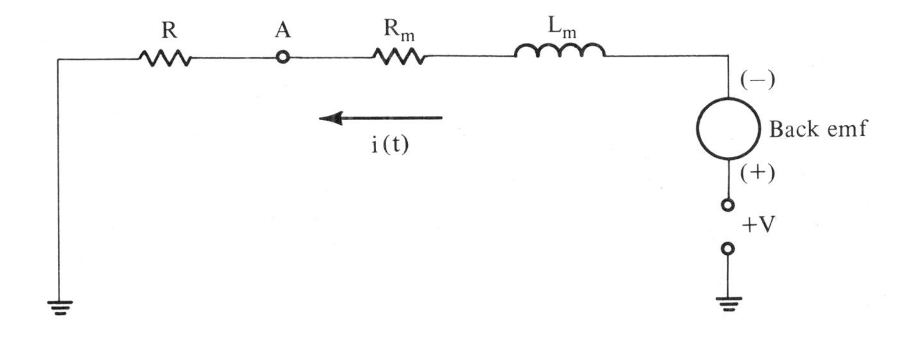

Figure 8-4. Equivalent circuit of the charging of a motor phase.

The waveform of L is shown in Figure 8-5a, where the maximum inductance occurs at the stable equilibrium position, and the minimum inductance occurs at the unstable equilibrium position. Substituting Eq. (8-4) into Eq. (8-1) and simplifying, we have

$$V = \left(R + R_m - nL_2 \frac{d\theta}{dt} \sin n\theta\right) i + \left(L_1 + L_2 \cos n\theta\right) \frac{di}{dt} \tag{8-5}$$

Since both the angular velocity, $d\theta/dt$, and the sine factor are positive in the back emf term, the effect of this term on the charging current is that the initial time constant is increased. Figure 8-6 shows typical charging current waveforms with and without the back emf effect. The dip in the curve is due to the typically underdamped case where the motor velocity rises to a peak initially.

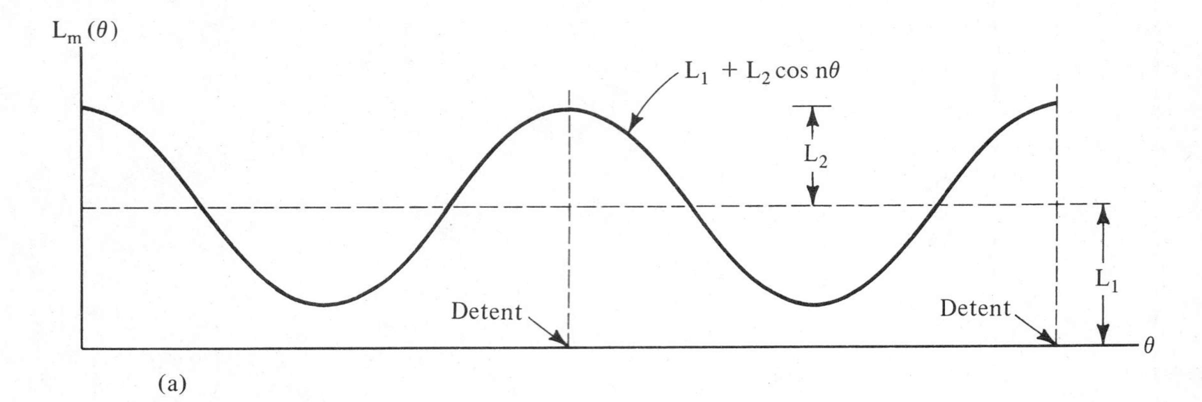

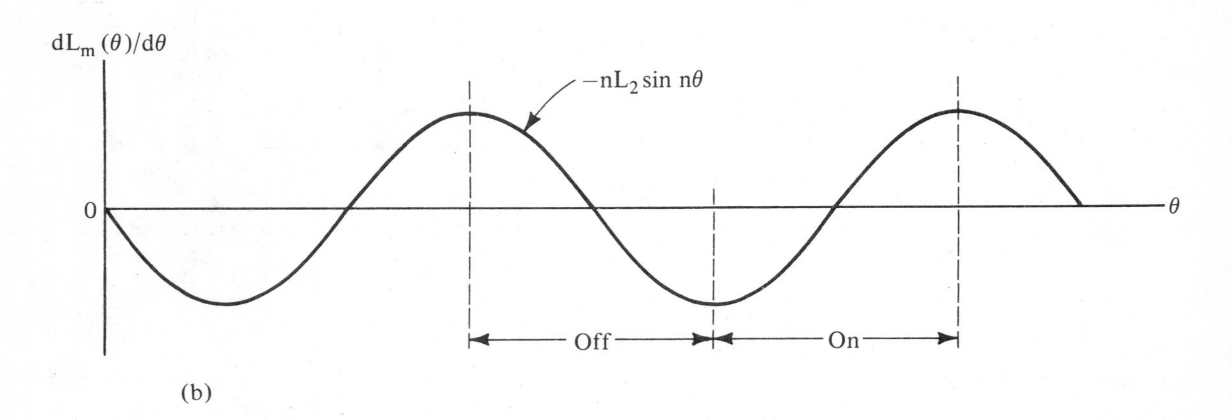

Figure 8-5. (a) Motor inductance as function of displacement.

(b) Derivative of motor inductance with respect to displacement.

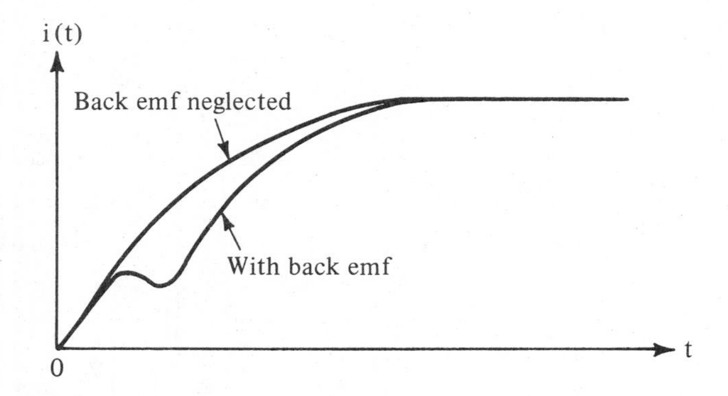

Figure 8-6. Typical charging currents of step motor with and without back emf effect.

Suppression Circuits

When a motor phase is ready to be turned off, a low-level voltage signal is applied to the base of the transistor Q_1 which results in turning off Q_1 and Q_2. The current in the motor phase then decays through the fly-back diode and the discharge resistance R_s. The equivalent circuit for the discharge of a motor phase is shown in Figure 8-7. The differential equation for the discharge current in Figure 8-7 is written

$$\left(L_1 + L_2 \cos n\theta\right) \frac{di}{dt} + \left(R_m + R_s - nL_2 \frac{d\theta}{dt} \sin n\theta\right) i = 0 \qquad (8\text{-}6)$$

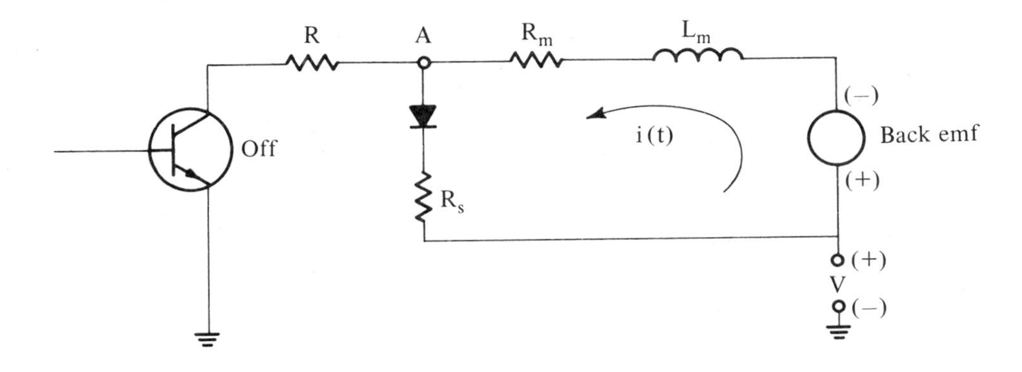

Figure 8-7. Equivalent circuit of the discharge of a motor phase.

Figure 8-5b illustrates the variation of $dL(\theta)/d\theta$ which has a direct effect on the motor back emf. For a typical stepping operation the approximate angular position over which the phase is turned on and off is shown in Figure 8-5b. It is shown that during the current decay the back emf is opposite that during the charging or on cycle. Thus, as shown in Figure 8-8, the initial decay in the current is faster due to the back emf.

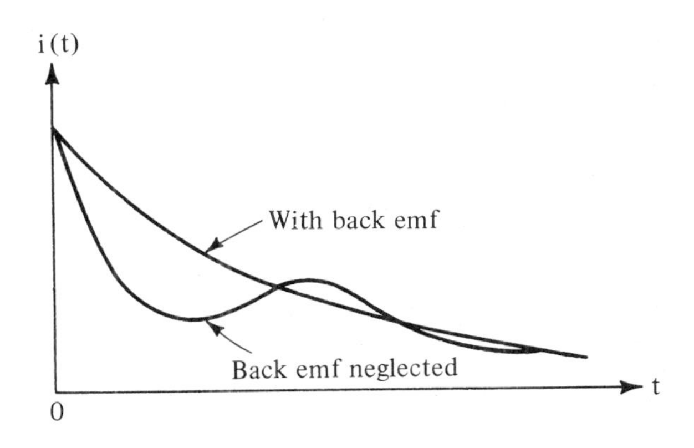

Figure 8-8. Typical current decay of step motor.

At low speeds the fly-back diode may be used without the suppression resistor, R_s, since the motor resistance will sufficiently decay the current during the period in which this phase is off. At higher speeds, larger values of R_s must be used to decay the current more rapidly in order to maintain a high average torque. Large values of R_s, however, cause a high reverse voltage across the output transistor Q_2.

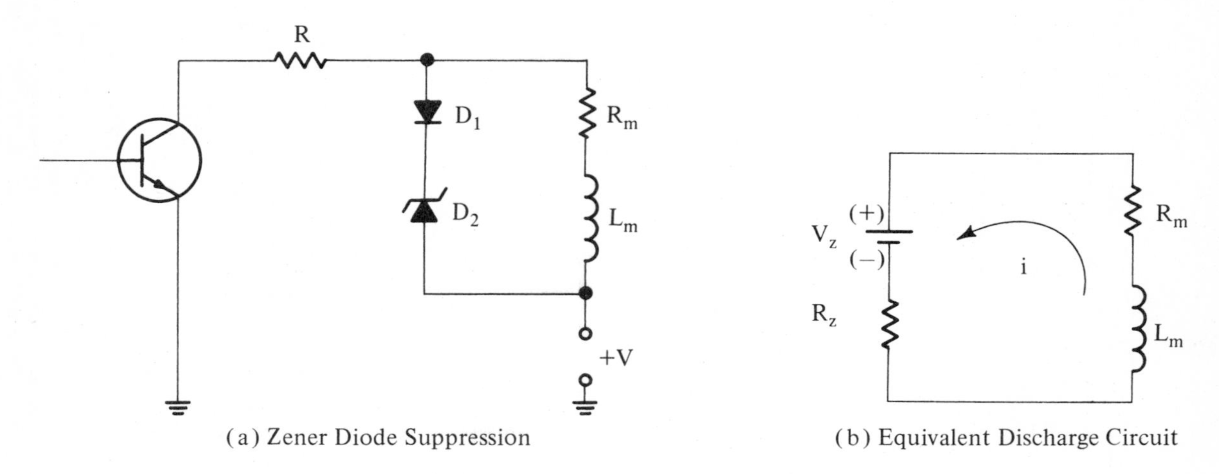

(a) Zener Diode Suppression (b) Equivalent Discharge Circuit

Figure 8-9. Zener diode suppression circuit.

The zener diode suppression circuit in Figure 8-9a is more effective at decaying currents at very high stepping rates. During the charging period, diode D_1 prevents the zener diode from conducting. During the discharge period the voltage across the inductance rises until the zener breakdown occurs and current begins to conduct. An equivalent circuit for this suppression circuit is shown in Figure 8-9b, where V_z is the zener voltage and R_z the zener resistance plus the resistance due to the diode D_1. The current satisfies the following equation during the period of zener breakdown.

$$0 = L\frac{di(t)}{dt} + (R_m + R_z)i + V_z \tag{8-7}$$

Given an initial current I_0 and solving for i(t) from the last equation gives

$$i(t) = I_0 e^{-t/\tau} + \frac{V_z}{R_m + R_z}(1 - e^{-t/\tau}) \tag{8-8}$$

where

$$\tau = \frac{L}{R_m + R_z}$$

This equation is valid for $0 < i(t) \leq I_0$. When $i(t) = 0$ the zener comes out of conduction and the current remains at zero. Figure 8-10 shows the current waveforms for two different zener voltages. The time at which the current reaches zero, t_0, can be found by setting i = 0 in Eq. (8-8).

$$t_0 = \tau \, \ln\left[\frac{I_0(R_m + R_z) + V_z}{V_z}\right] \qquad (8\text{-}9)$$

For $V_z = 0$ the current decays as in a diode suppression circuit and $t_0 = \infty$. As V_z is increased, t_0 approaches zero. Figure 8-11 shows the relationship.

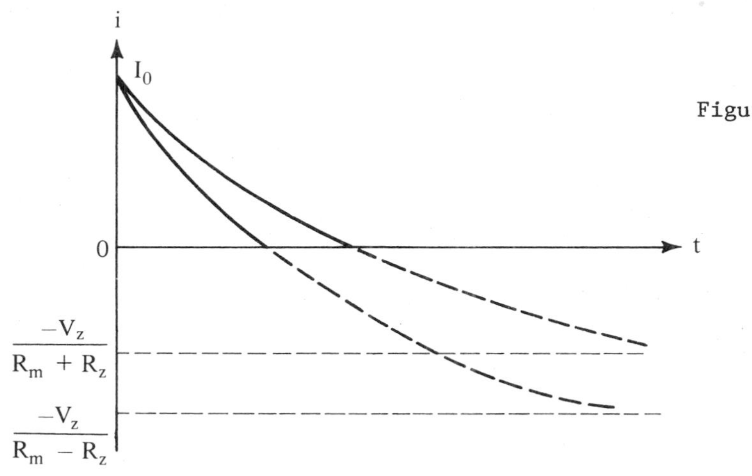

Figure 8-10. Current decay in the zener diode suppression circuit.

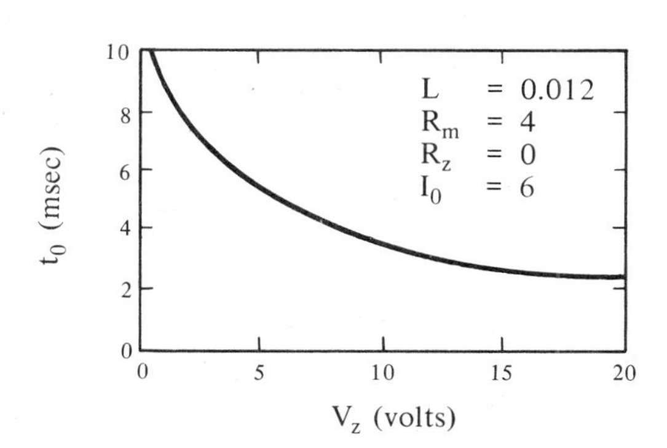

Figure 8-11. Relationship between time at which motor current reaches zero and zener voltage.

For small motors the current may be suppressed by allowing the output transistor to go into second breakdown as shown in Figure 8-12. However, this usually raises the cost of the output transistor and results in extra heating of this transistor.

A capacitor may be shunted across the motor windings for current suppression. However, this is usually not practical because of the size of the capacitor needed. However, it is often helpful to use a small capacitor with a diode suppression circuit or second transistor to eliminate voltage spikes. (Figure 8-13)

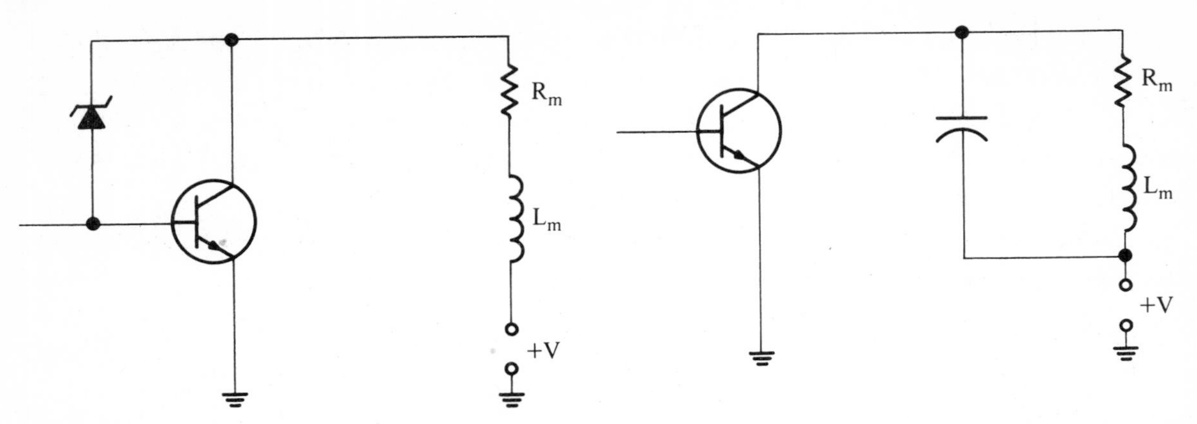

Figure 8-12. Current suppression by second breakdown transistor.

Figure 8-13. Current suppression by capacitive damping.

In single-stack motors there is significant mutual magnetic coupling between the phases of the motor. Thus, the rate of decay of the current in the de-energized phase affects the current rise in the energized phase, and vice versa. Consider the two-phase driver circuit shown in Figure 8-14. Assume that at time $t = 0$, Q_1 is turned on and Q_2 is turned off, and that there is no initial current in the suppression loop for Q_1 and a current $I_0 = V/(R_m + R)$ in the inductance L_{m2}. The equations for i_1 and i_2 are then

$$V = L_m \frac{di_1}{dt} + M \frac{di_2}{dt} + (R + R_m)i_1 \tag{8-10}$$

$$0 = L_m \frac{di_2}{dt} + M \frac{di_1}{dt} + (R_m + R_s)i_2 \tag{8-11}$$

Solving these equations for di_1/dt and di_2/dt yields

$$\frac{di_1(t)}{dt} = \frac{1}{\Delta} [VL - (R + R_m)Li_1 + M(R_m + R_s)i_2] \tag{8-12}$$

$$\frac{di_2(t)}{dt} = \frac{1}{\Delta} [-VM + M(R + R_m)i_1 - L(R_m + R_s)i_2] \tag{8-13}$$

where $\Delta = L^2 - M^2$.

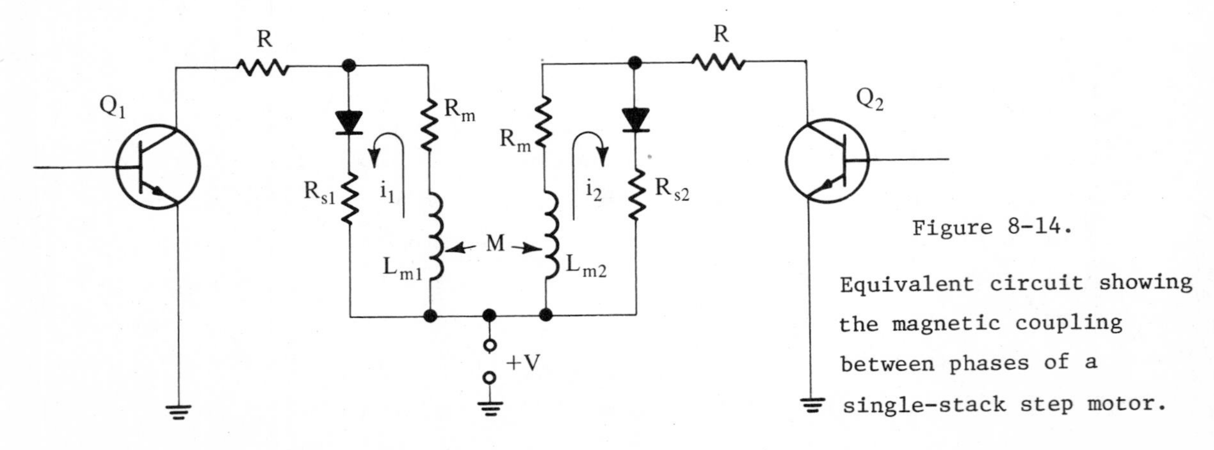

Figure 8-14. Equivalent circuit showing the magnetic coupling between phases of a single-stack step motor.

Let the solutions to these two differential equations be $i_1(t)$ and $i_2(t)$, when the initial currents are I_{10} and I_{20}, respectively. Then, it can be shown from Eqs. (8-12) and (8-13) that if the initial currents are changed to I_{10} and $-I_{20}$, and the mutual inductance is negative, $-M$, the solutions of these equations will be $i_1(t)$ and $-i_2(t)$, respectively.

The responses for $i_1(t)$ and $i_2(t)$ for M = -8, -4, 0, 4, and 8 mh, and the following parameters are plotted in Figure 8-15:

$V = 24$ v, $L_m = 12$ mh, $R = 1$ ohm, $R_m = 5$ ohms,

$R_s = 15$ ohms, $i_1(0) = 0$, $i_2(0) = 4$ amp.

From Figure 8-15a it is noted that for $i_2(0) = 4$ amp which is positive, and for positive values of the mutual inductance, M, the rising current in coil 1 induces a

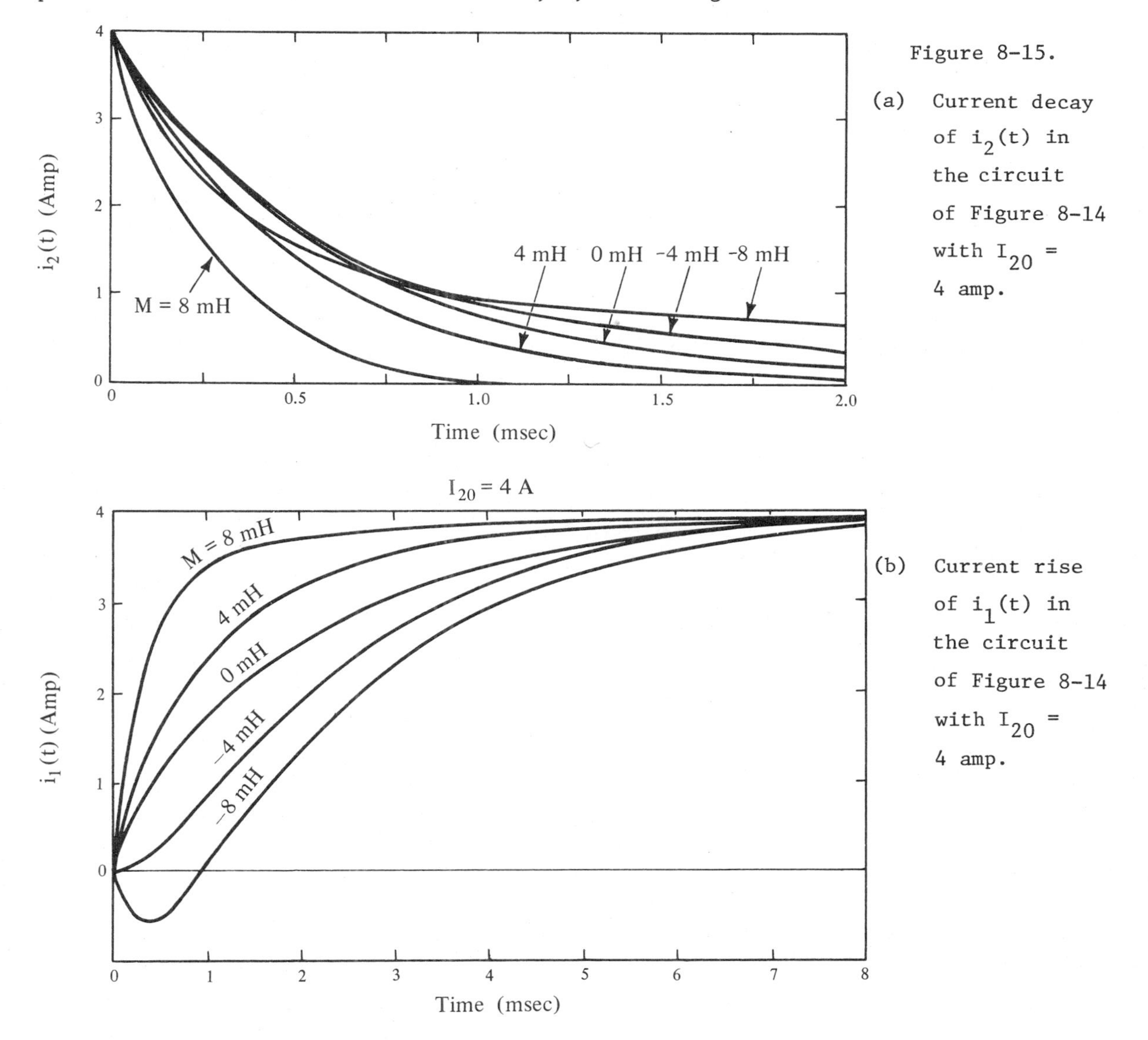

Figure 8-15.

(a) Current decay of $i_2(t)$ in the circuit of Figure 8-14 with I_{20} = 4 amp.

(b) Current rise of $i_1(t)$ in the circuit of Figure 8-14 with I_{20} = 4 amp.

voltage in the suppression circuit which aids the decay of current $i_2(t)$. Similarly, Figure 8-15b shows that the decaying current $i_2(t)$ induces a voltage which aids the rise of $i_1(t)$. These same statements can be made if the initial current $i_2(0)$ and M are negative.

For positive initial current $i_2(0)$ but negative M, the rising current in coil 1 induces a voltage in the suppression circuit which aids the decay of $i_2(t)$ initially but eventually opposes the decay of $i_2(t)$. The decaying $i_2(t)$ induces a voltage which opposes the rise of $i_1(t)$. These same statements are true if the initial current $i_2(0)$ is negative but M is positive.

The effect of the mutual inductance on the initial rise or decay of current can be seen be setting t to zero, $i_1(0) = 0$, and $i_2(0) = I_{20}$ in Eqs. (8-12) and (8-13),

$$\frac{di_1(0)}{dt} = \frac{L_m V - M(R + R_s)I_{20}}{L_m^2 - M^2} \tag{8-14}$$

$$\frac{di_2(0)}{dt} = \frac{MV - L_m(R + R_s)I_{20}}{L_m^2 - M^2} \tag{8-15}$$

With the parameters used in this analysis, these derivatives are tabulated below as a function of the mutual inductance M. Note that for large absolute values of M, the initial decay of $di_2(t)/dt$ is large.

M (mh)	-8	-4	0	4	8
$\frac{di_1(0)}{dt}$	-4400	-250	2000	4750	11600
$\frac{di_2(0)}{dt}$	-9600	-6750	-6666	-8250	-14400

Dual-Voltage Power Supply

For low motor speeds a single voltage power supply is usually sufficient. However, for high speeds and fast acceleration a dual-voltage power supply may be used to increase the rate of current build up in the energized phase. Figure 8-16 shows a typical method of switching the motor phase from a high voltage to a low voltage. When this motor phase is energized, transistors Q_1 and Q_2 are both turned on. The high voltage is then applied to this phase. After the current has reached a desired level or after a fixed time has elapsed, Q_2 is turned off and the low voltage maintains the current at its desired level. When this phase is de-energized, Q_1 is turned off and the current in L_m flows back into the high-voltage power supply.

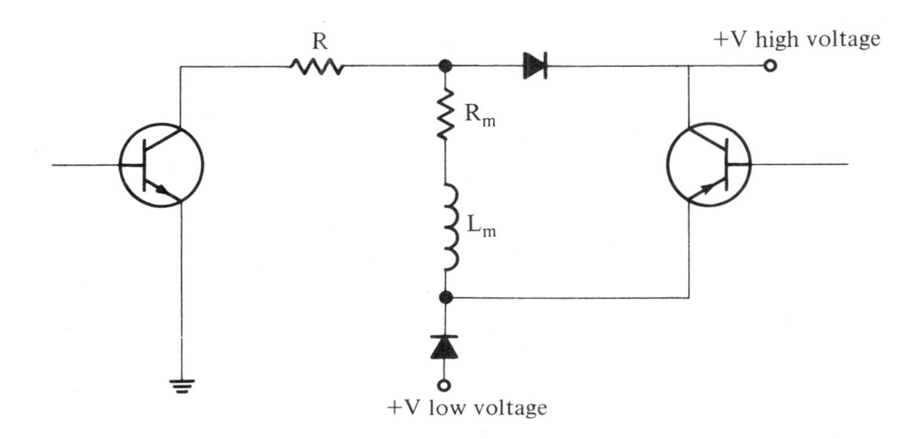

Figure 8-16. Dual voltage power supply of step motor.

8-3 CHOPPED-INPUT CONTROL FOR STEP MOTORS

The conventional method of exciting a step motor with a single power supply involves the application of voltage to a particular phase through series resistors. The controller switches this voltage from phase to phase in a sequence determined by the motion that is desired.

In this section we consider the modification of the controller in such a way that the I^2R loss in the series resistors is eliminated.

The method relies on a chopping scheme incorporated in the control circuit which will supply the motor with a chopped input while maintaining a desirable current level. The series resistors are removed and the input voltage is directly applied to a particular phase. Since the steady-state current level is now much higher than the rated value for the winding, the voltage is switched off when the current reaches a certain predetermined value. The current now decays and when it falls below its rated value, the winding is reexcited. Thus by chopping the input voltage, the current can be maintained within a specific band about the rated value. The scheme for switching the voltage from one phase to the next is not altered.

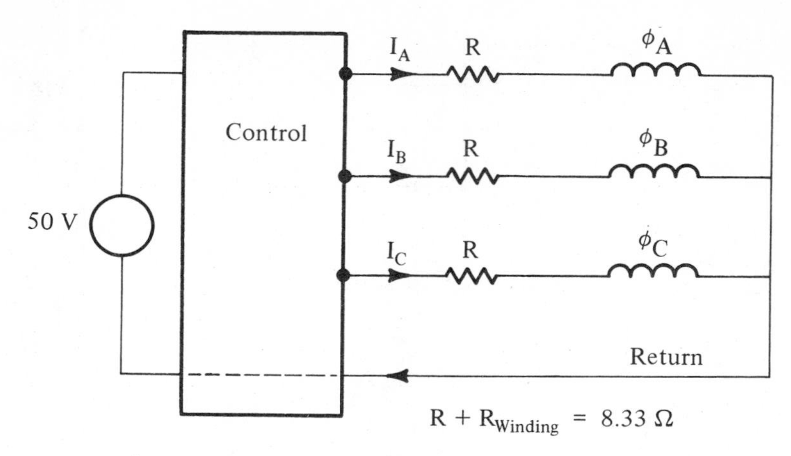

Assume A is excited at t = 0, V_A = 50 V, VB = VC = 0.
At t = T power is switched to ϕB. V_B = 50 V, VA = VC = 0.

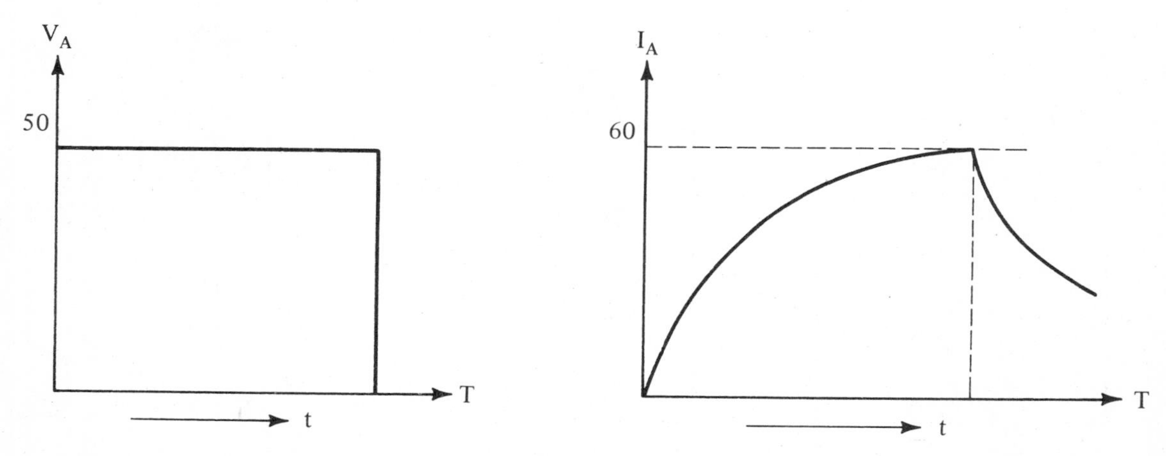

Figure 8-17. Schematic diagram of conventional step motor control and voltage and current waveforms.

A 50-V controller and the Warner Electric SM048AB 6-amp motor have been chosen for the purpose of analysis. Figure 8-17 shows the schematic diagram for exciting the motor using the usual method. The voltage and current waveforms are also indicated in the same figure. Figure 8-18 shows the arrangement and voltage-current waveforms for the chopped scheme.

For the purposes of analysis, the following parameters have been chosen for the motor and controller:

Power supply voltage = 50.0 volts

Rated current of motor = 6.0 amps

Resistance of one winding and the control circuit = 2.5 ohms

Discharge circuit resistance = 10 ohms

Total resistance in each phase when unchopped scheme is used = 50/8 = 8.33 ohms

Maximum and minimum current levels in the chopped scheme = 6.5 and 5.5 amps.

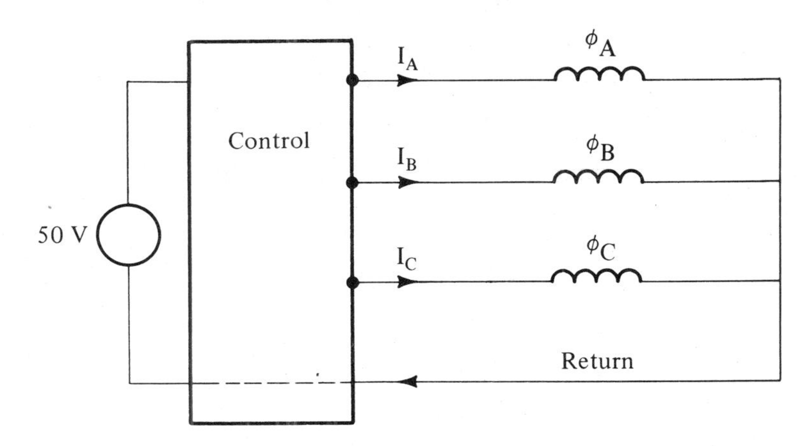

Assume same switching as in Figure 8-17.

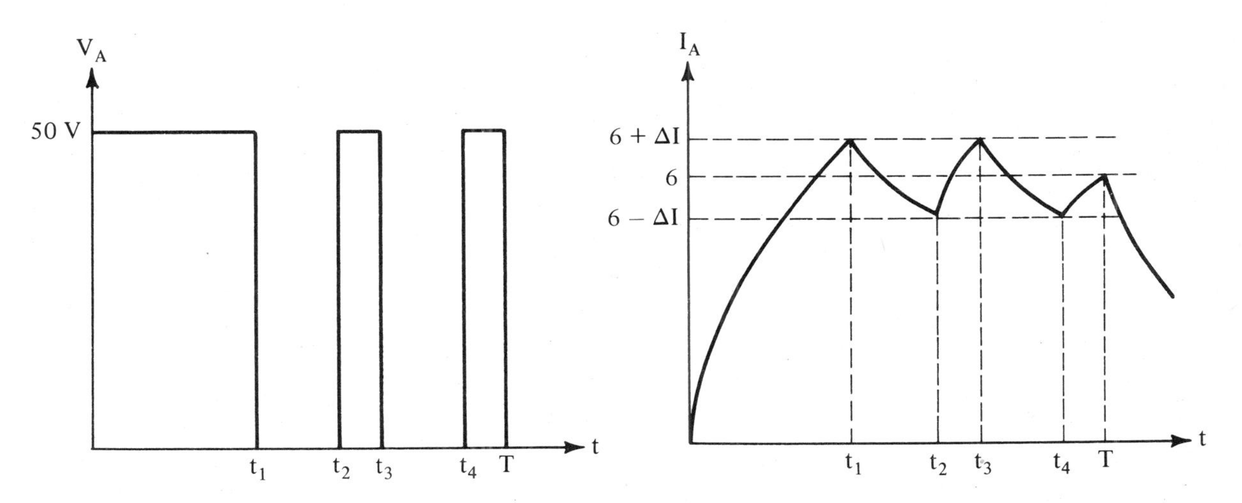

Figure 8-18. Schematic diagram of a chopping step motor controller and typical voltage and current waveforms.

The complete analysis was carried out by computer simulation. The motor was driven with a closed-loop control with an acceleration angle of 2.5 degrees.

The motor is first run without load using both schemes. In subsequent runs increasing loads are applied while the motor stalls. The last case (stalled motor) is equivalent to no forward motion and the motor just maintains the equilibrium position.

The output quantities are: final speed, total power lost in the circuit, distance travelled, phase currents, and amount of chopping required (for chopped scheme only).

The simulation results are shown in Figure 8-19 through 8-27.

Figure 8-19 shows the torque-speed curves obtained for both schemes while Figure 8-20 shows the power outputs (torque speed) as functions of speed. The total power lost in the circuits is shown in Figure 8-21 as a function of speed and in

Figure 8-22 as a function of power output. Since the power output is zero for no load as well as the case of stalled motor, the curves in Figure 8-22 loop back towards the vertical axis. The direction of arrows indicates increased loading and reduced

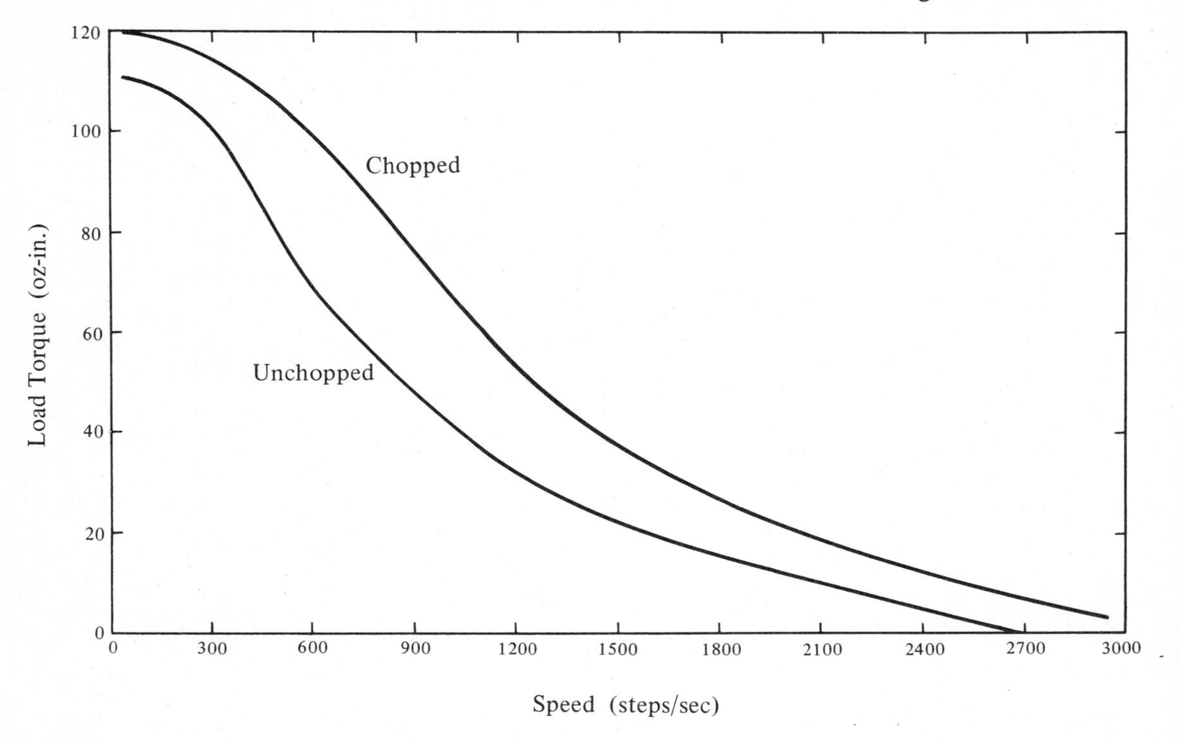

Figure 8-19. Torque-speed curves of step motor with chopped and unchopped controllers.

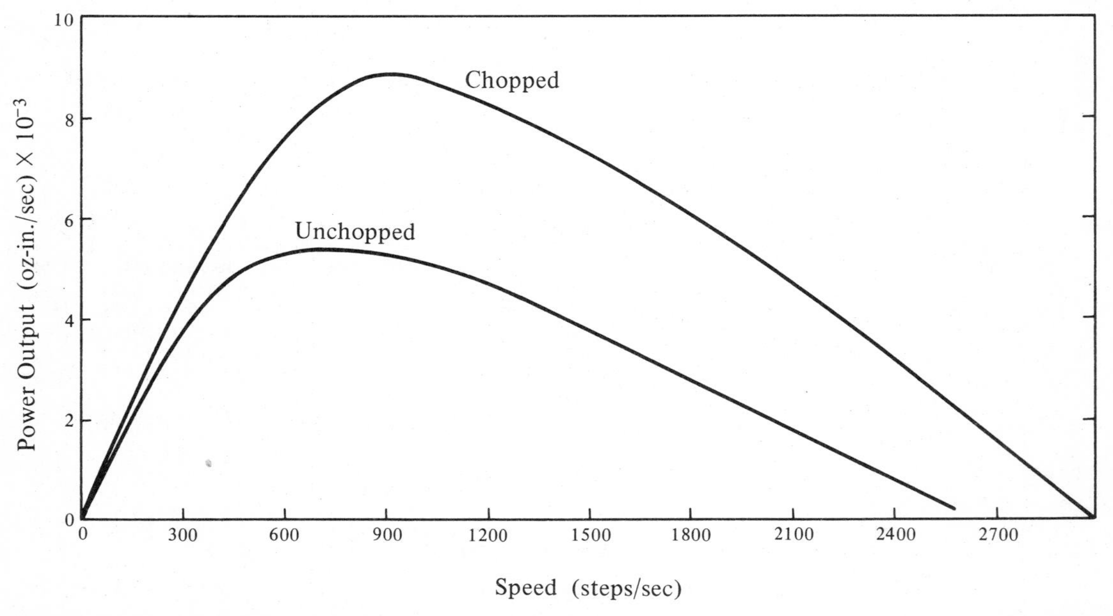

Figure 8-20. Power outputs versus speed of step motor with chopped and unchopped controllers.

speed. The points of maximum load correspond to the points of zero speed. The power losses indicated here are those that would occur if the motor stayed motionless and maintained holding torque.

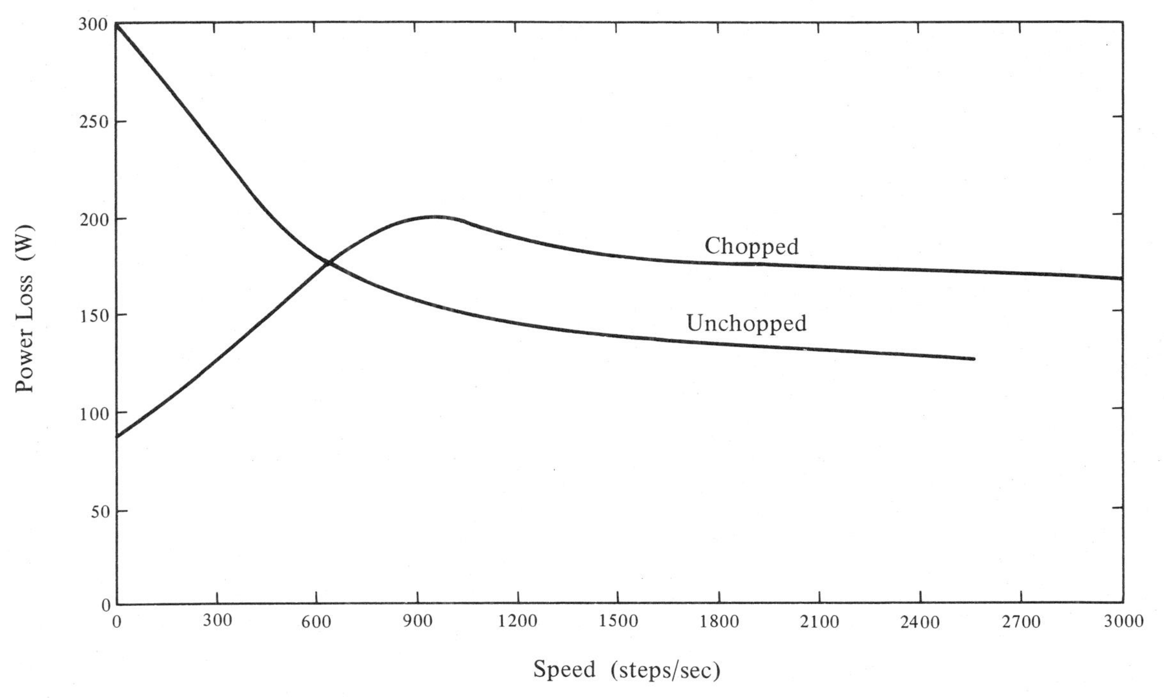

Figure 8-21. Power loss versus speed of step motor with chopped and unchopped controllers.

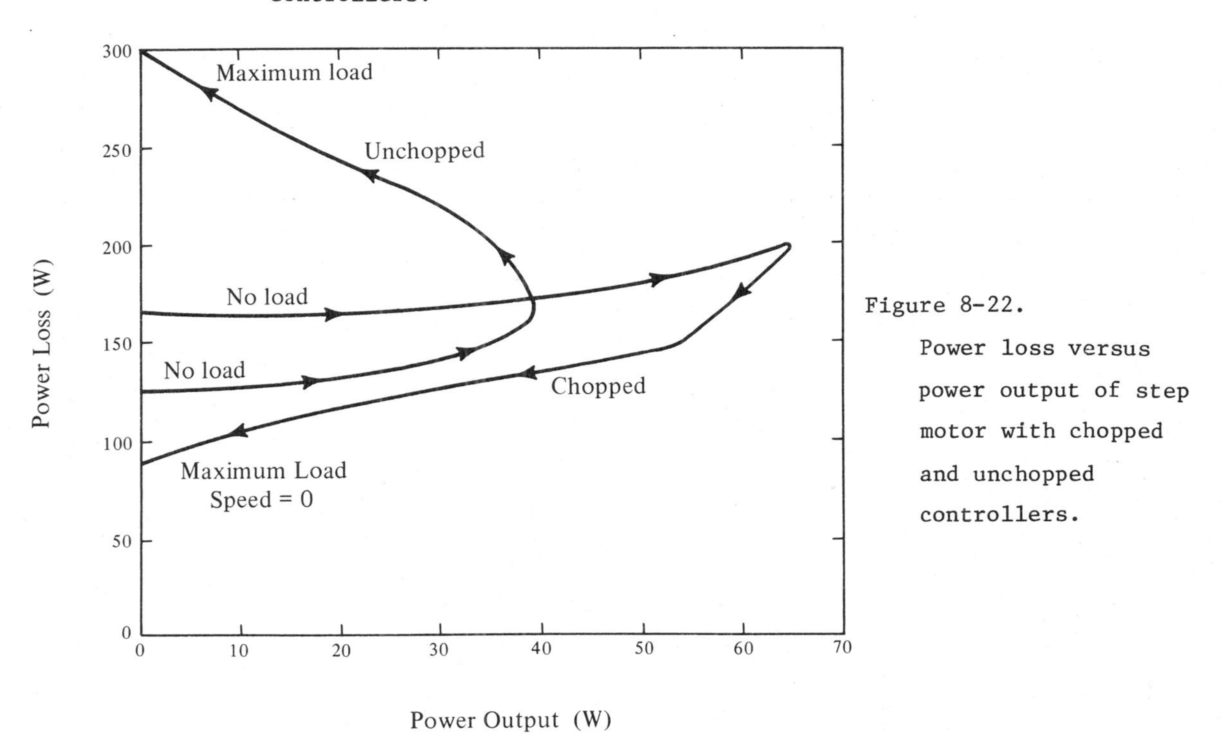

Figure 8-22. Power loss versus power output of step motor with chopped and unchopped controllers.

The amount of chopping necessary to maintain the desired current level is indicated in Figure 8-23 as a function of time. Since the ordinates represent the total number of choppings, the slope of the curve indicates chopping rates of 2 KHz,

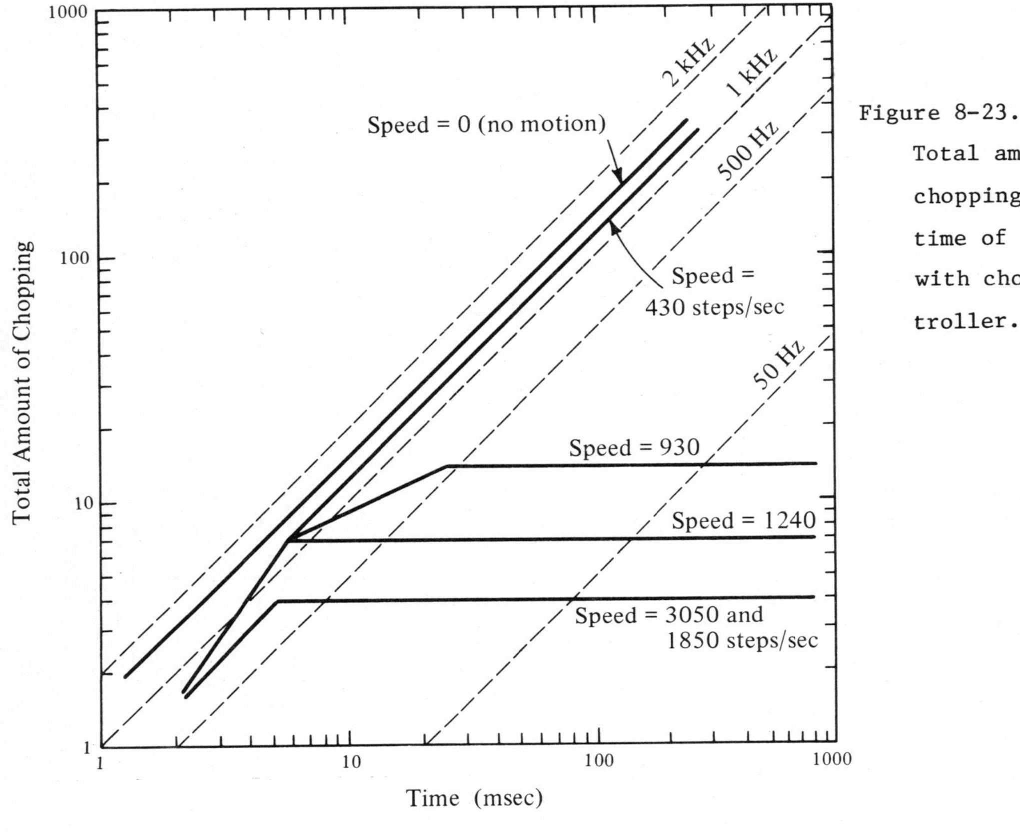

Figure 8-23.
Total amount of chopping versus time of step motor with chopped controller.

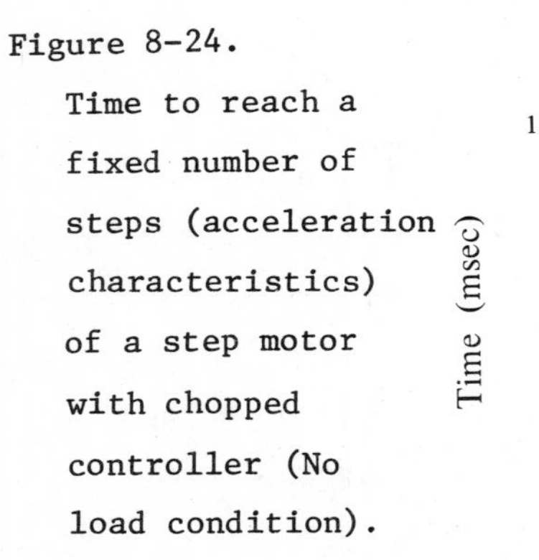

Figure 8-24.
Time to reach a fixed number of steps (acceleration characteristics) of a step motor with chopped controller (No load condition).

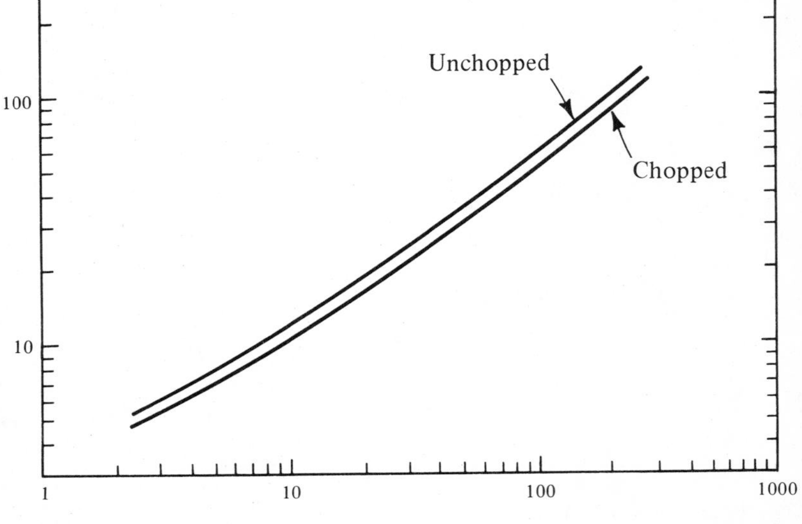

1 KHz, 0.5 KHz, and 50 Hz. It is clear that for speeds higher than about 750 steps/sec, chopping is necessary only during the initial part of the motion (until the motor picks up speed). The maximum chopping rate of 1.5 KHz is obtained when the motor is at a standstill. Figures 8-24 through 8-25 show the acceleration rates for various loadings. The plots are for the distance travelled as a function of time.

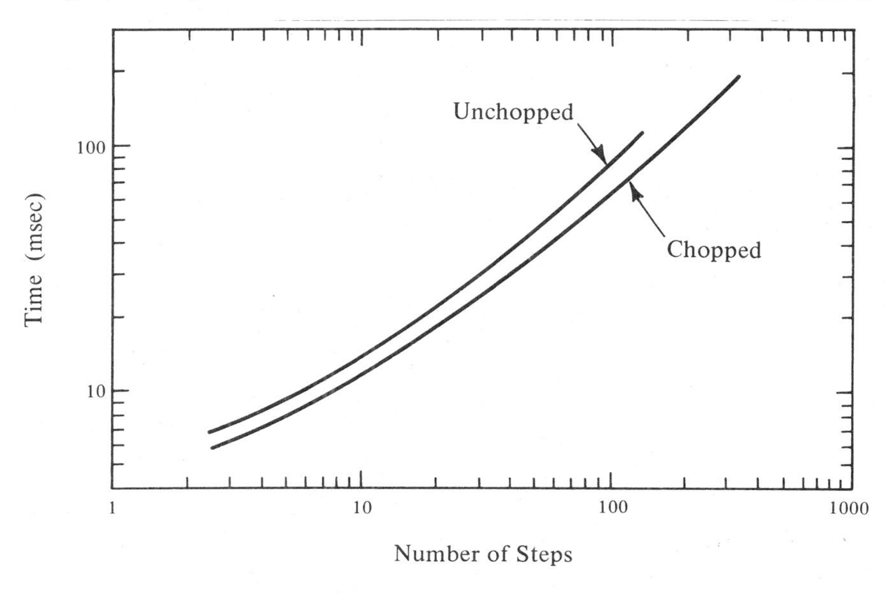

Figure 8-25. Time to reach a fixed number of steps (acceleration characteristics) of a step motor with chopped controller (load = 25 oz-in.).

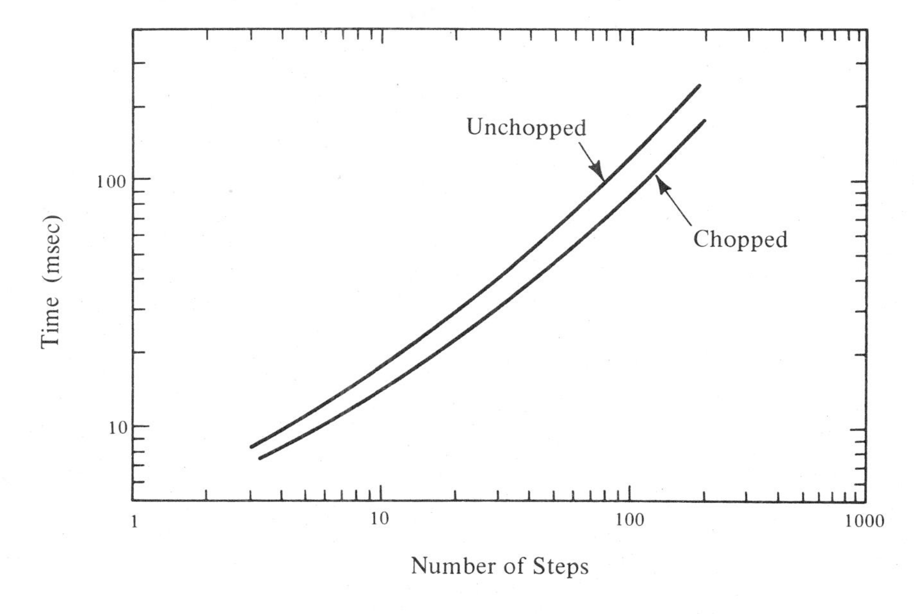

Figure 8-26. Time to reach a fixed number of steps (acceleration characteristics) of a step motor with chopped controller (load = 50 oz-in.).

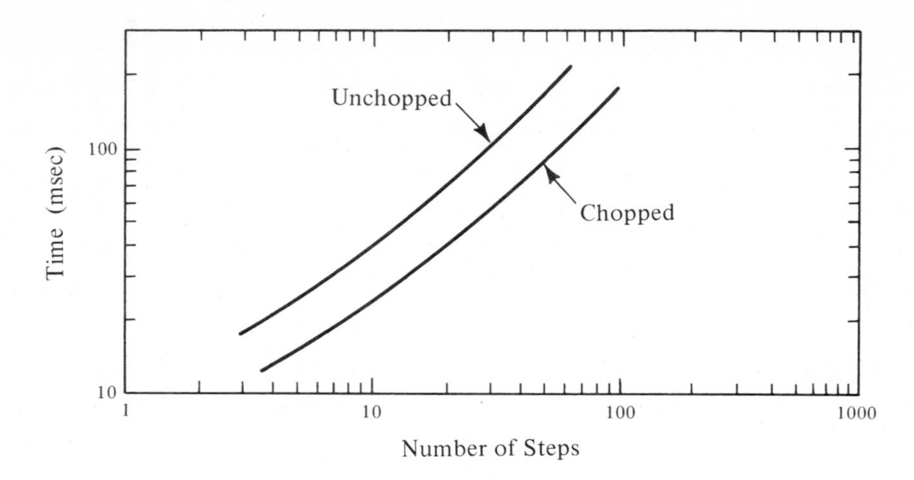

Figure 8-27. Time to reach a fixed number of steps (acceleration characteristics) of a step motor with chopped controller (load = 100 oz-in.).

The results indicate that the performance of the motor is better in terms of torque, output, and acceleration rate when the new scheme is used. It should be mentioned, however, that with the speed of 750 steps/sec there is no need for any chopping. The motor has to switch to the next phase before current can build up to a level where chopping is necessary. But since the charging resistance is much lower (2.5 vs. 8.33 ohms) than in the unchopped case, a higher level of current is maintained. This accounts for the improved performance and also for the increased power loss at higher speeds. Below 750 steps/sec, chopping is effective and power loss is considerably less for the case of chopped input. When there is no motion, the average losses are 88 watts for the chopped scheme and 300 watts for the unchopped scheme.

In conclusion, we can say that at higher speeds the proposed scheme improves the performance of the motor at a cost of increased heat dissipation and the chopping circuit is effectively not in use.

At lower speeds, and especially at standstill, there is an improvement in power loss by chopping, without much sacrifice to performance.

8-4 OPEN-LOOP CONTROL OF THE STEP MOTOR

A step motor when operated in an open-loop manner has several operating modes, depending upon the stepping rate and load conditions underwhich it is operated. These modes are illustrated in Figures 8-28, 8-29, and 8-30. In these figures the angular position of the rotor and the total motor torque are plotted as functions of time. In addition, the three restoring torque curves for the three phases of the motor are superimposed on each other and plotted along the displacement axis to the left of each figure. This enables us to show the state of the excitation of the three phases and to identify how the phase torques combine to produce the total motor torque waveforms at the bottom. The numbers along the waveform at the bottom correspond to the numbers along the curves at the left and are provided to assist in relating the two.

If the step motor is pulsed at a sufficiently slow rate such that it comes to rest at the end of each step, it will move in discrete steps as shown in Figure 8-28. In applications where the distance to be travelled is small and the response time is not critical, this mode of operation provides the simplest solution. In other applications where rapid start and stop motion is desired such as in a paper advance mechanism for a printer, this mode of operation sometimes combined with a viscous damper provides a good solution.

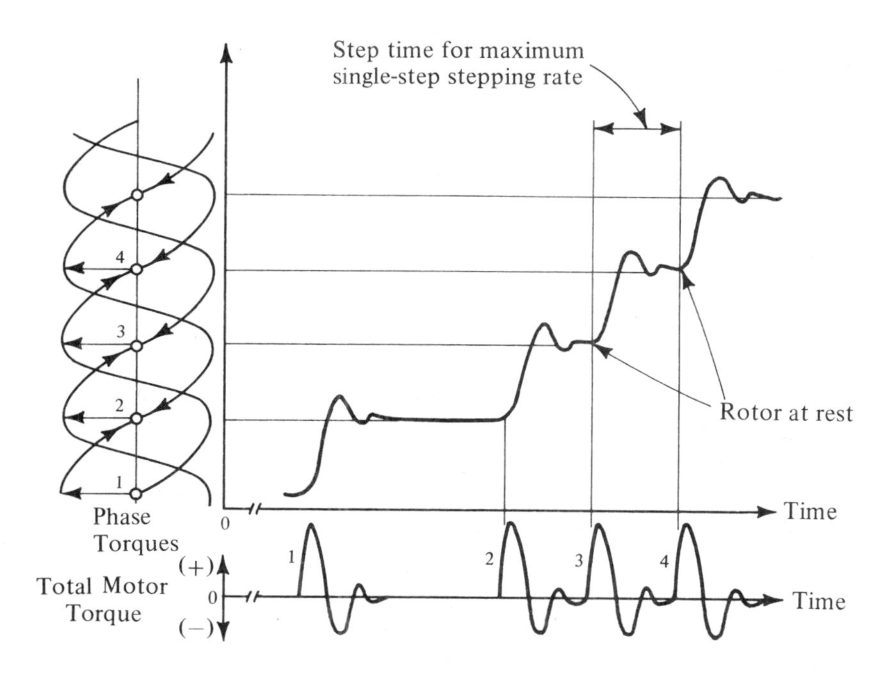

Figure 8-28. Torque and displacement versus time for single-step operation.

If the input pulse rate to the motor is increased, the angular motion of the motor will change from that of discrete steps to a continuous forward motion referred to as "slewing." Figure 8-29 shows this transition from single-stepping between steps 1, 2 and 3 to a "transitional" mode of operation between 3, 4 and 5 to a continuous slewing mode beyond 5. In the transitional mode of operation the motor does not come to rest between steps, thus subsequent switching pulses may come when the motor has either a positive or negative velocity causing the motor to behave in a somewhat unpredictable and erratic way.

Figure 8-29.
Torque and displacement versus time for continuous acceleration.

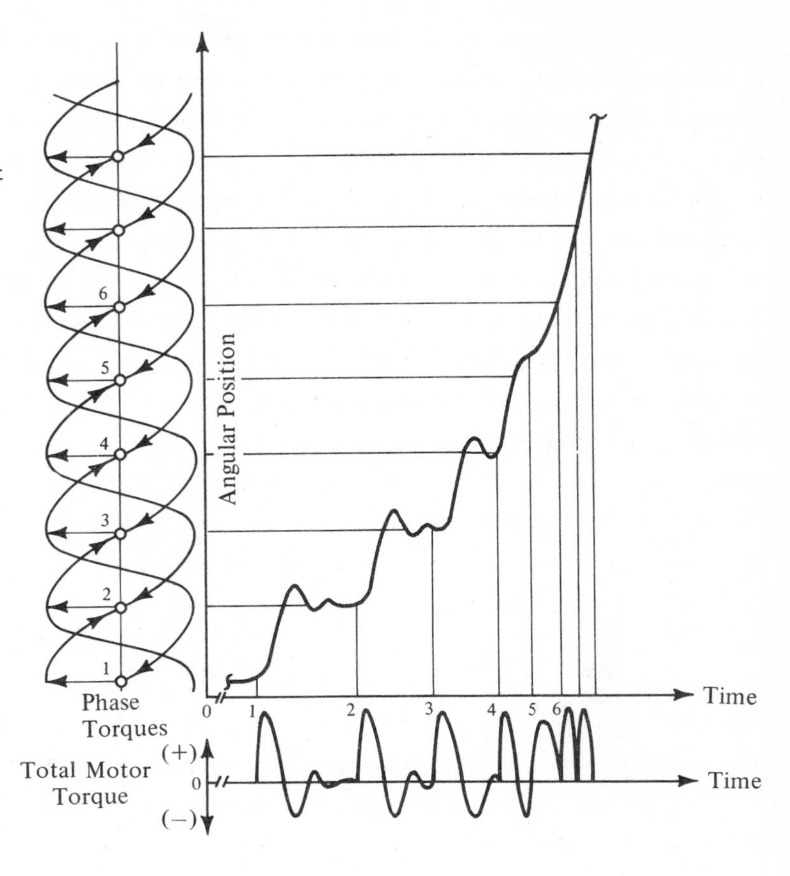

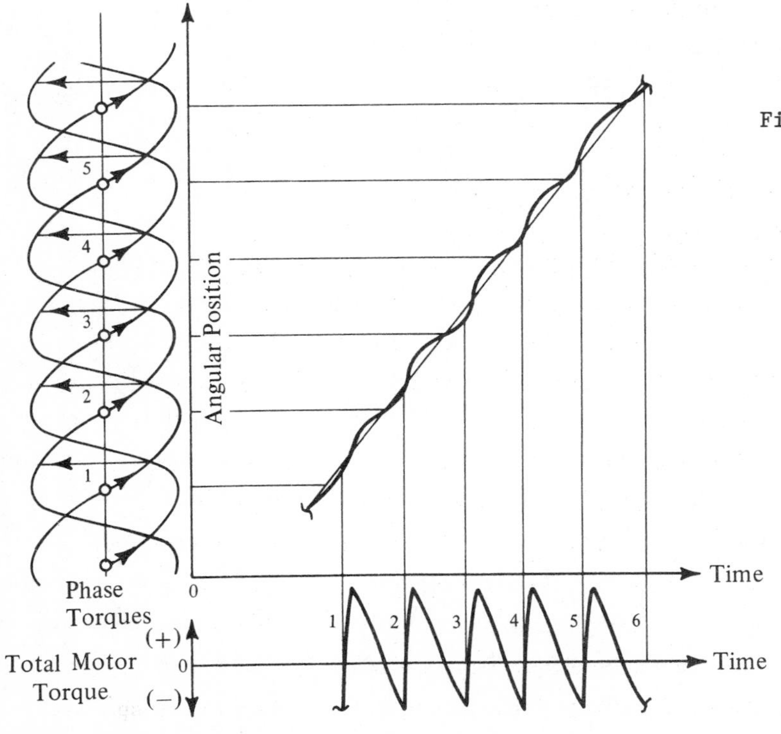

Figure 8-30.
Torque and displacement versus time for steady-state slewing.

Once the motor is in the slewing mode of operation it behaves very similarly to a synchronous motor. Figure 8-30 shows the motor in its steady-state slewing mode. Although it is difficult to see in this figure, the velocity of the motor generally goes through large oscillations about a nominal value. This phenomenon is similar to the hunting characteristic of a synchronous motor.

In the following we shall now illustrate some typical open-loop multiple-stepping performances of the SM036AB motor manufactured by the Warner Electric Brake and Clutch Company. The results show the effect on the motor speed from such factors as load, control voltage, pulse interval, and the rate of decay of pulse intervals.

It has been found both experimentally and by computer simulation that in the transitional mode of operation it is best to bring the motor gradually up to a desired running speed by applying a pulse train with exponentially decaying intervals. It will be seen later that such a pulse train closely approximates the pulse train that would be generated by a closed-loop system.

If we let the time interval between two adjacent pulses by denoted by SW, where

$$SW = SW1 + SW2e^{-t/\tau} \tag{8-16}$$

the pulse interval will decrease with a time constant τ from SW1 + SW2 to SW1. If the motor follows this pulse train its speed will be 1/SW.

The responses illustrated in Figures 8-29 through 8-32 show the behavior of the motor under prescribed open-loop controls.

In Figure 8-29, the motor is driven by a 100-volt controller with current switchings; that is, the control voltage is switched to a low voltage when the motor current reaches the rated value. The motor is not loaded. The input pulse train has pulse intervals described by

$$SW = 0.003e^{-t/\tau} \tag{8-17}$$

Therefore, as time progresses the pulse interval should approach zero. As shown in Figure 8-31, when $\tau = 0.05$ second, the motor stalls after reaching a speed of almost 3500 steps/sec. With the pulse intervals decaying at a lower rate, $\tau = 0.1$, the motor reached the speed of 3750 steps/sec before stalling. Of course, the acceleration is slower in this case.

In the cases of Figures 8-32 and 8-33, the motor is driven by a constant 30-volt controller. Three different load inertias are used.

The pulse rate used in the case of Figure 8-32 is slower than that of Figure 8-33. The values of τ indicated in the figures are somewhat insignificant, since the pulse rate is truly governed by SW1. For Figure 8-32, SW1 = 0, whereas for Figure 8-33, SW1 = 0.005; this indicates that the final desired speed is 2000 steps/sec. Again, the results show that slower decaying pulse rates give higher final speeds.

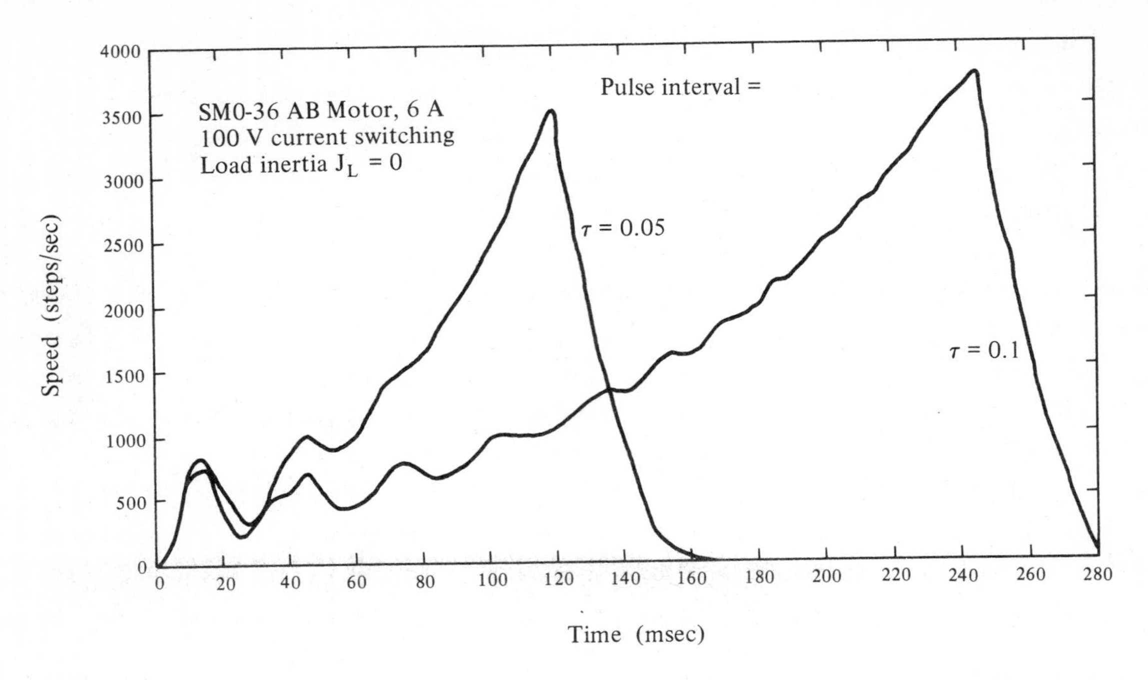

Figure 8-31. Open-loop multiple-stepping response of step motor.

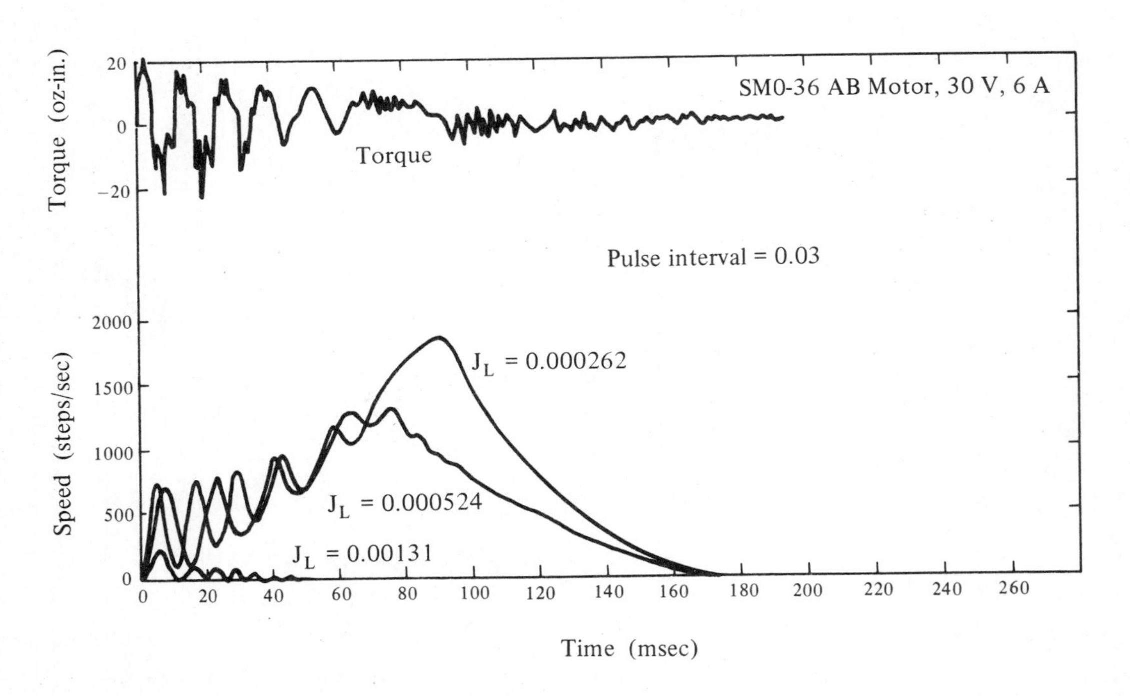

Figure 8-32. Open-loop multiple-stepping response of step motor.

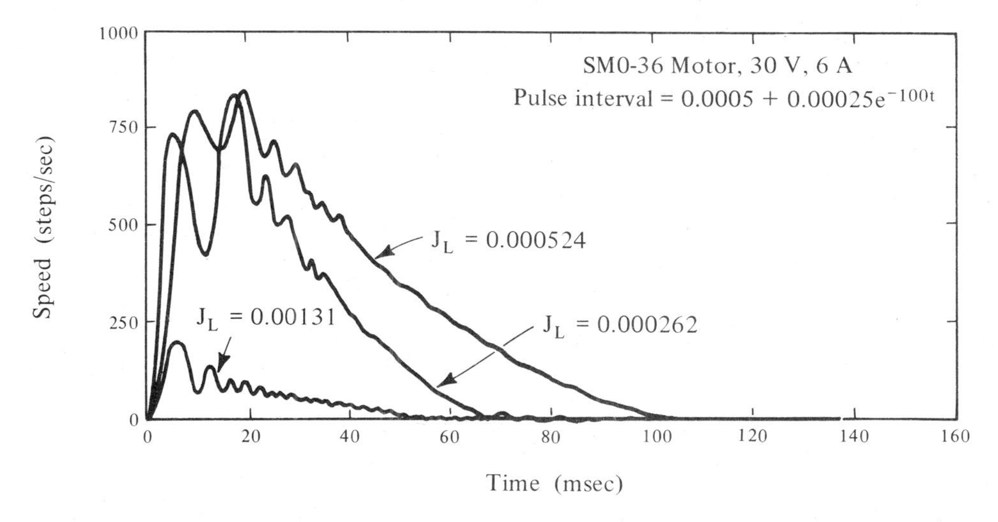

Figure 8-33. Open-loop multiple-stepping response of step motor.

In the cases shown in Figure 8-34, the controller is 30-volt 6-amp. There is no load. Final speed responses of 500 and 600 steps/sec are shown. At these low speeds the motor speed fluctuates after the steady state is reached. The frequency of oscillation is approximately 500 rad/sec, which is close to the natural frequency of the motor. Notice that these speed fluctuations are typical in the open-loop slewing operations of step motors, especially if the load friction is small. These oscillations tend to decrease as the final speed becomes higher. At 1000 steps/sec and 2000 steps/sec these final speeds are quite stable.

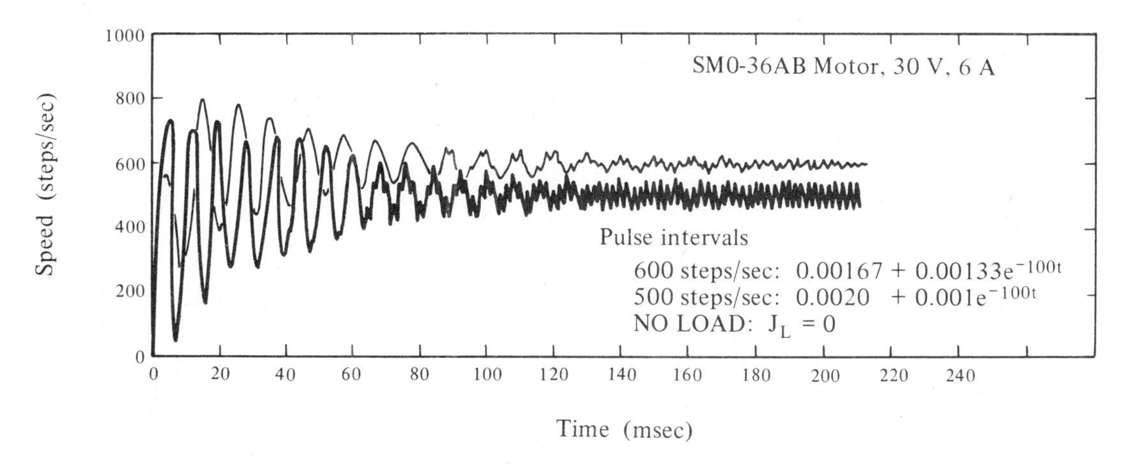

Figure 8-34. Open-loop multiple-stepping response of step motor.

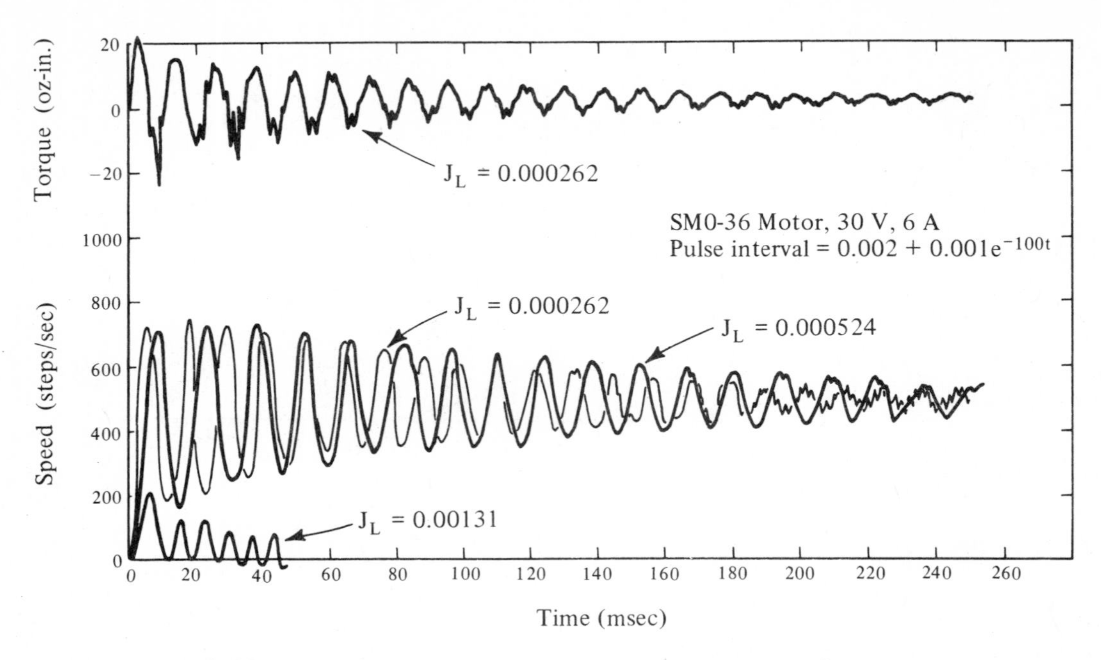

Figure 8-35. Open-loop multiple-stepping response of step motor.

Figure 8-35 illustrates the speed responses of the motor when it is driven by the same controller as in the above case except that an inertial load is added.

One significant finding from these results is that a stable slow speed operation can be achieved by reducing the control voltage (without changing the resistance). For instance, referring to the 500-steps/sec response shown in Figure 8-34, the final speed fluctuation is between 460 to 544 steps/sec with V = 30 volts. With V = 20 volts, the final speed fluctuation is between 477 to 527 steps/sec. The important point is that there is no significant decrease in the average acceleration by decreasing the input voltage, as long as the motor can still reach the desired speed.

9 DAMPING METHODS OF STEP MOTORS

B. C. Kuo and G. Singh
Department of Electrical Engineering
University of Illinois at Urbana-Champaign
Urbana, Illinois

9-1 INTRODUCTION

It is well known that the transient response of most step motors is generally oscillatory. This characteristic in the step motor performance is particularly evident when the motor is performing a single-step type of operation, or when it attempts to stop after a slewing operation. The presence of this oscillation is generally unacceptable in present-day systems, and a damping scheme is generally included as part of the control strategy. Although several damping methods are available, no particular one is applicable to all practical situations. Depending on the required performance of a particular system, a specific scheme may or may not be satisfactory. With the present state of the art most incremental control systems have to be compromised to some extent in performance due to inadequate damping.

In this paper some of the well-known damping methods, as well as some new methods, are presented. These may be classified into three categories:

1. Inertia Dampers: This includes viscous inertia dampers, hysteresis dampers, and eddy current dampers. The viscous inertia damper, which is the most commonly used in this category, will be described in the next section.
2. Damping by Electronic Switching Schemes: There are a great variety of damping schemes available, based on the control of the switchings of the motor. Of the five schemes in this category presented in this paper, two are new and three have been well established in principle.
3. Damping by Modification of Motor Parameters: This category of damping schemes includes the use of additional stator windings and additional rotor windings, either of which could be excited or unexcited. Also included are effects such as modification of tooth shape for improved damping.

A total of nine damping methods are presented in the following sections. The advantages and disadvantages of each scheme are discussed and some typical results are shown.

9-2 DAMPING BY USE OF VISCOUS INERTIA DAMPERS

The viscous inertia damper is a commonly used mechanical damping device for motion control. In such a damper, a large inertia, typically several times larger than the inertia of the rotor of the motor plus load is loosely coupled to the motor shaft by a viscous-type coupling. Such couplings can be formed by using silicones, or other viscous fluids, and certain types of solid friction surfaces.

Figure 1 shows the schematic diagram of a stepping motor coupled to a viscous-inertia damper. The following system parameters are used:

T_m = motor torque; B = total frictional coefficient of motor and load; J_L = load inertia; J_m = motor inertia; J_H = inertia of damper housing; J_R = inertia of damper rotor; K_D = viscous frictional coefficient of damper. From Figure 9-1, the state equations are written for the system:

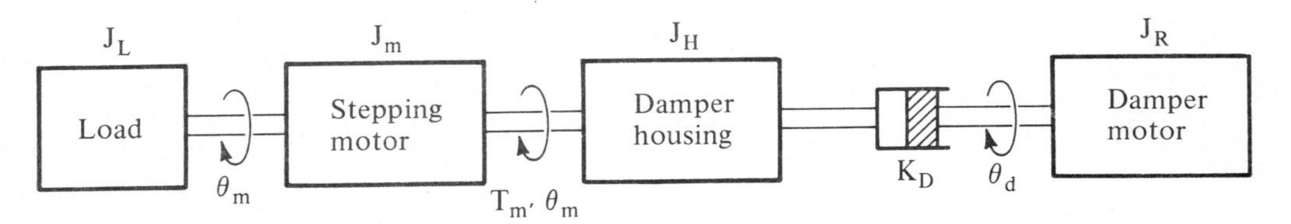

Figure 9-1. Schematic diagram of a step motor coupled to a viscous-inertia damper.

$$\frac{d\omega_m}{dt} = \frac{1}{J} T_m - \frac{B}{J} \omega_m - (\omega_m - \omega_d) \frac{K_D}{J} \tag{9-1}$$

$$\frac{d\omega_d}{dt} = \frac{K_D}{J_R} (\omega_m - \omega_d) \tag{9-2}$$

$$\frac{d\theta_m}{dt} = \omega_m \tag{9-3}$$

$$\frac{d\theta_d}{dt} = \omega_d \tag{9-4}$$

where

$$J = J_L + J_m + J_H \tag{9-5}$$

Figure 9-2 illustrates typical step responses of a stepping motor which is coupled to a viscous-inertia damper with various damper parameters. The step responses with and without a load are shown in the figure. The load inertia in

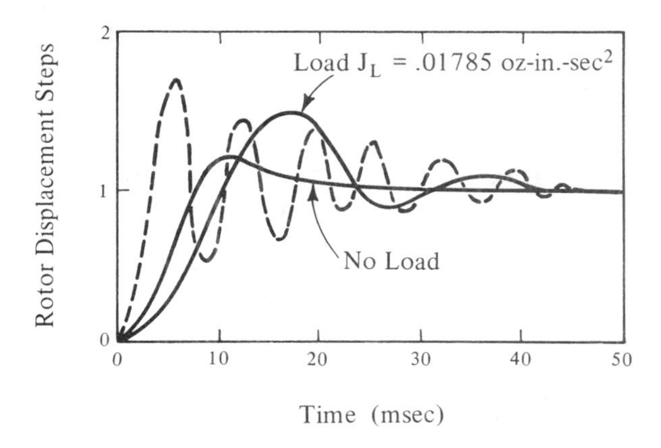

Figure 9-2. Typical single-step responses of a variable-reluctance step motor with viscous-inertia damping.

this case is considered to be relatively heavy for the particular motor. The undamped response, without load, is shown by the dotted curve. The damper apparently reduces the oscillation considerably. It should be noted that the viscous inertia damper always adds inertia to the system; therefore, the step response of the damped system will be slower than the undamped system.

By using different viscous fluids in the damper, the viscous frictional coefficient, K_D, can be changed by several orders of magnitude to give different damping effects.

The viscous inertia damper is relatively inexpensive and can be easily attached to any system if space permits. Its disadvantage is that it slows down the system considerably.

9-3 DAMPING BY USE OF ELECTRONIC SWITCHING SCHEMES

Electronic damping is probably the most popular approach towards improving the transient response of step motors. The convenience of modern electronic hardware makes easy the implementation of any switching scheme which appears plausible. As a result, a great number of these schemes have been attempted. In this section, five such schemes are presented and discussed. Two of these are new while three have been well established, and quite frequently implemented in practice.

Damping by Multiple-Phase Excitation

It has been found that single-stack variable-reluctance step motors provide improved damping characteristics when more than one phase is energized. Consequently, a very common damping scheme involves the use of two phases on at a time for driving the motor. The combined effects of mutual inductances, saturation and eddy currents, are responsible for this improvement. Figures 9-3 through 9-6 show the improvement obtained with this scheme as compared to one-phase-on excitation, for a typical

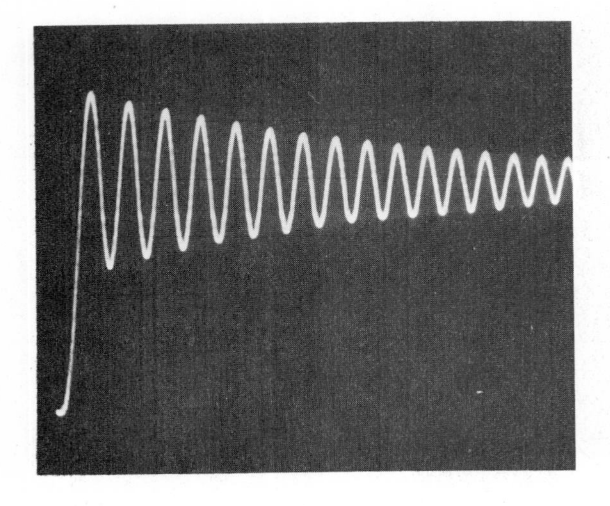

Figure 9-3. Single-step response with one-phase-on excitation; no load (10 msec/div.).

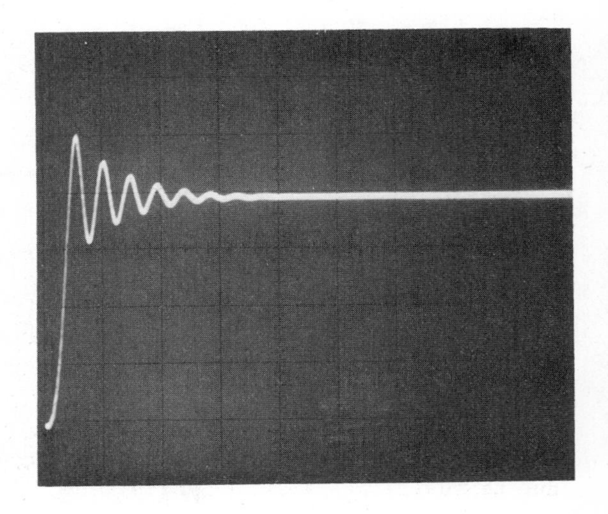

Figure 9-4. Single-step response with two-phase-on excitation; no load (10 msec/div.).

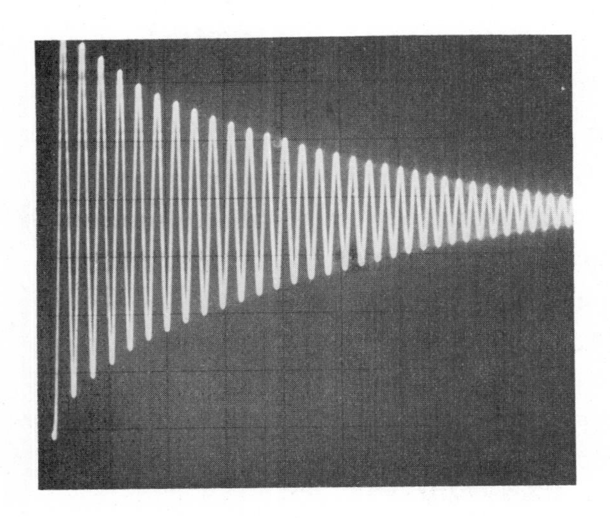

Figure 9-5. Single-step response with one-phase-on excitation; with inertial load (100 msec/div.).

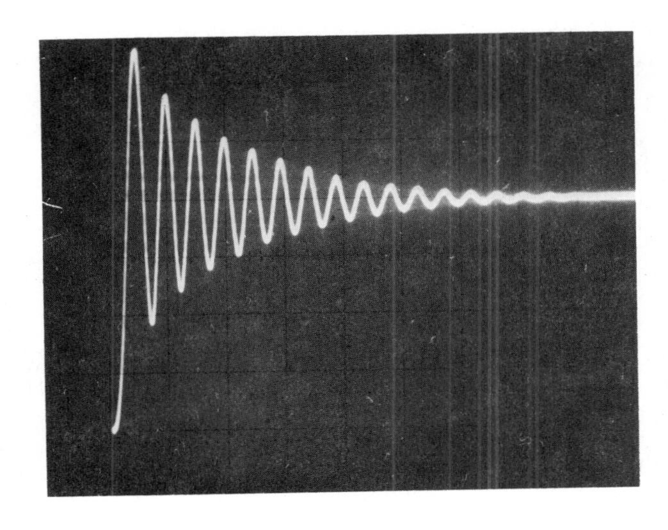

Figure 9-6. Single-step response with two-phase-on excitation; with inertial load (50 msec/div.).

single-stack motor (SM024-0045-HS of Warner Electric Brake and Clutch Company). In all of these figures, the motor is excited by 28 volts, and there is no additional suppression or series resistance in the circuit. Figure 9-3 shows the response with one-phase-on excitation and Figure 9-4 with two-phases-on excitation. There is no mechanical loading on the motor. Figures 9-5 and 9-6 show the responses under the same excitations, except the motor has an inertial load of 0.00177 oz-in-sec^2 coupled to its shaft. The rotor inertia is 0.00018 oz-in-sec^2.

Similar or greater improvement is possible by use of other forms of multiple-phase excitation. In one such scheme, a three-phase motor is driven by the usual one-phase-on method, but just before a step is completed, the remaining two phases are turned on with partial voltage (generally half voltage).

The use of multiple-phase excitations provides a simple and convenient method for obtaining improvement in the damping characteristics. Its disadvantage is that, due to additional phase currents, the motor will have to dissipate a greater heat loss, or be appropriately derated. Also, these schemes are effective with single-stack motors only, and do not provide much improvement when used with multiple-stack motors.

The Bang-Bang Damping Scheme

This well-known electronic damping scheme uses back phasing during a certain period of the step motion to achieve a deadbeat response. It is applicable with the one-phase-on or two-phases-on type of excitation. The scheme will be described for the one-phase-on excitation by use of the static torque curve shown in Figure 9-7. Assume that phase C of the motor is already on and a step is to be made by shifting the excitation from phase C to phase A. Since the initial torque developed by the motor is high, oscillation in the rotor displacement will result if phase A is left on. In order to improve on the oscillatory response, a "braking" torque may be

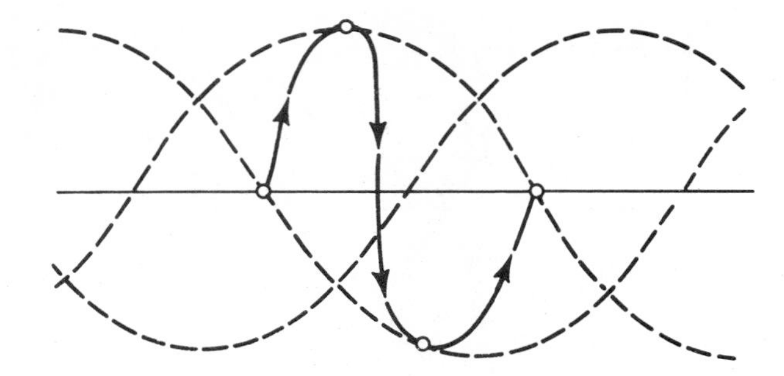

Figure 9-7. Static torque curves showing the principle behind bang-bang damping of the single-step response.

applied by shifting the excitation from phase A back to phase C at some appropriate instant of time so that the rotor will just reach the one-step position without overshoot. Finally, at the instant the step is completed, phase C is deenergized, and phase A is energized to hold the rotor in the proper position. Therefore, the proper switching sequence for this type of electronic damping is C-A-C-A if phase C is initially excited. Figure 9-8 illustrates a typical step response obtained when this scheme is carried out under the ideal situation.

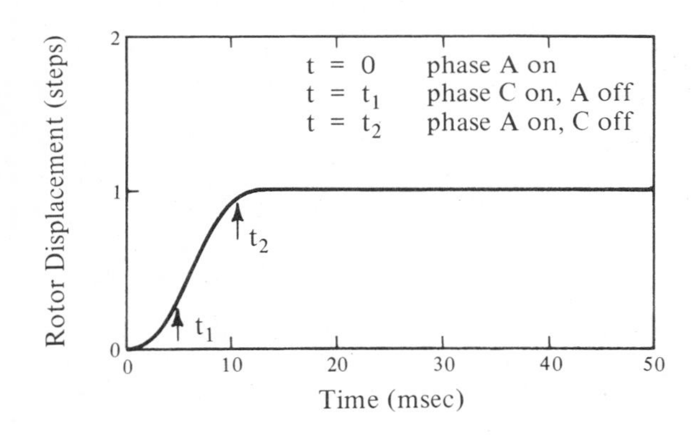

Figure 9-8. Single-step response of a VR step motor with bang-bang damping.

The bang-bang scheme is simple to implement and, theoretically, it should always work. However, in practice, the response of the motor is very sensitive to the timing of the back-phase pulsing, as well as to the slightest load changes. Due to the high sensitivity the bang-bang scheme rarely provides satisfactory results in practice.

Delayed-Last-Step Damping

This scheme involves a delay in time in the turning on of the last pulse to the motor, and is applicable when the step motor is required to perform two or more steps. The method actually utilizes the highly oscillatory nature of the step response to its advantage. In fact, if the overshoot of the single-step response could reach 100 percent, it is possible to obtain a multiple-step response without any overshoot.

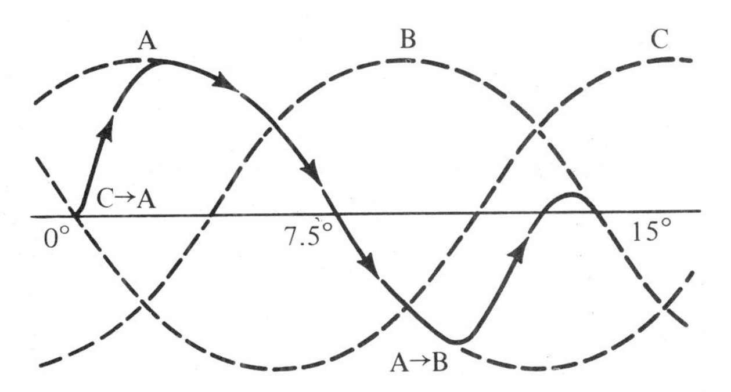

Figure 9-9. Static torque curves showing the principle behind delayed-last-step electronic damping.

The scheme is now described for a two-step motion by use of the static torque curves of Figure 9-9. If phase C of the motor is initially excited, and it is desired to make two consecutive steps, we use the sequence, C-A-B. Since the single-step response has a large overshoot, we let phase A stay on for an extended duration (usually around the time at which peak overshoot occurs), and then switch the excitation from A to B. As shown in Figure 9-9, the torque which corresponds to phase B at the peak of oscillation of the rotor position due to the excitation on A is rather small. Therefore, if phase B is switched on at the rotor position which corresponds to this small torque magnitude, and if the rotor is already near the final desired position due to the overshoot from the previous step, the motor will stop at the desired step position with little or no overshoot. Figure 9-10 shows a typical time response for a two-step motion with this scheme.

The main difference between the delayed-last-step scheme and the bang-bang scheme is that in the present case there is no back phasing of the pulse sequence.

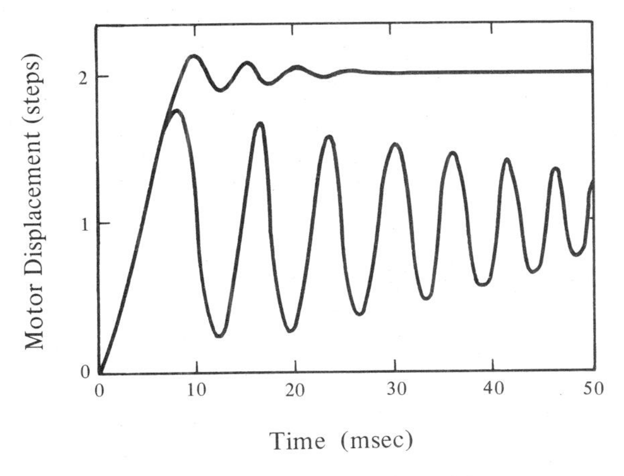

Figure 9-10. Typical step responses of a variable-reluctance step motor with delayed-last-step damping.

The delayed-last-step damping is less sensitive to load variations than the bang-bang scheme, but oscillations still will occur for certain types of load variances. Drag friction may be of assistance with this method up to a certain extent. However, too much drag friction will slow down the step response. The method can be used with open or closed-loop control; however, if more than a few steps are involved, the scheme is more effective with closed-loop control. This is because the ultimate damping is dependent on the consistency of the previous steps.

Damping by Use of Variable Voltage in the Two-Phases-On-Scheme

A new damping scheme is presented here which energizes two phases of the motor at a time, with different voltages on each phase. The higher voltage is applied for stepping purposes, and the lower voltage for damping purposes. Thus, the phase which has the higher voltage is referred to as the stepping phase and the phase with the lower voltage is referred to as the damping phase. The following two cases are considered:

Case 1. In this case, the phase which is discharging during normal stepping is excited partially with a lower voltage. The switching scheme can be illustrated by the following table:

	Time Instants				
	t_0	t_1	t_2	t_3	t_4
Stepping Phase (HV)	A	B	C	A	B
Damping Phase (LV)	C	A	B	C	A

As can be noticed from the above table, this scheme will be directionally dependent. If the step motor is to travel in the opposite direction, the scheme becomes

	Time Instants				
	t_0	t_1	t_2	t_3	t_4
Stepping Phase (HV)	A	C	B	A	C
Damping Phase (LV)	B	A	C	B	A

which does not yield the same equilibrium positions as the previous scheme. If, however, the two voltages are equal, then the scheme corresponds to the normal two-phases-on operation, and there is no directional dependence.

The single-step response is tested with the SM024-0045-HS step motor (made by Warner Electric Brake and Clutch Company). Figures 9-11 and 9-12 show the step responses in the case when the motor is not loaded, and Figure 9-13 shows the step response when an inertial load of 0.00177 oz-in-sec.^{2} is attached to the rotor shaft.

Case 2. In this case the previous phase is not energized but the next forward phase is partially energized for damping purposes. The switching scheme is

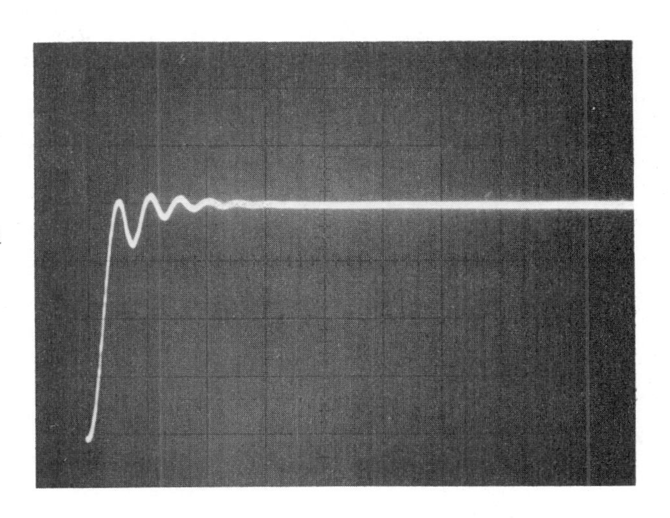

Figure 9-11. Single-step response, two-phase-on variable-voltage damping, HV = 28 v, LV = 14 v; no load (10 msec/div.).

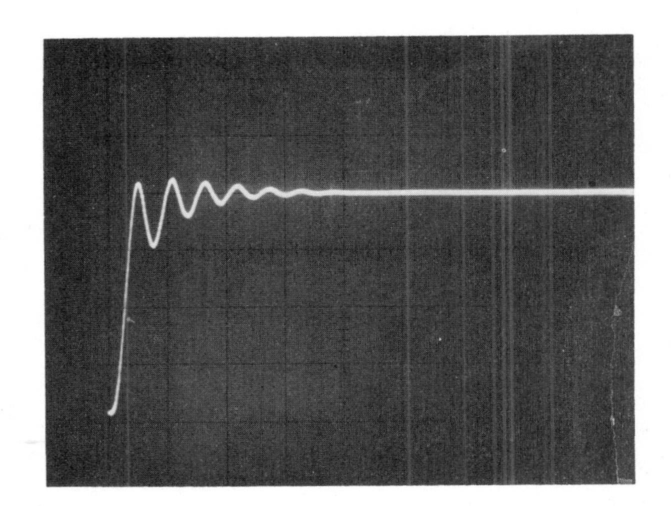

Figure 9-12. Single-step response, two-phase-on variable-voltage damping, HV = 28 v, LV = 16.5 v; no load (10 msec/div.).

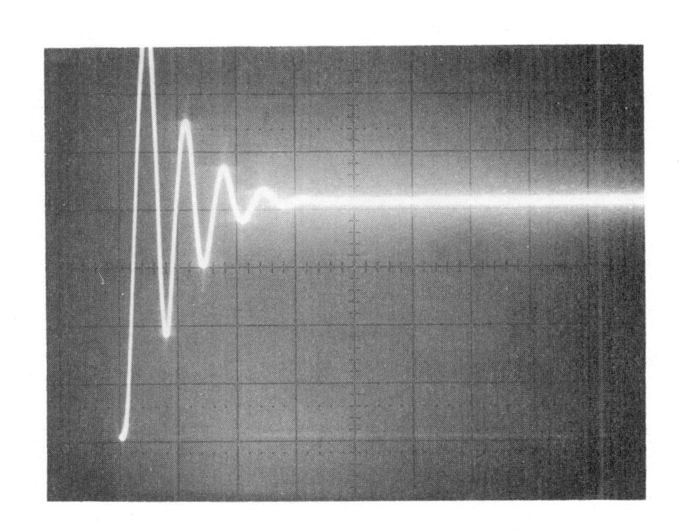

Figure 9-13. Single-step response, two-phase-on variable-voltage damping, HV = 28 v, LV = 14 v; with inertial load (50 msec/div.).

	Time Instants				
	t_0	t_1	t_2	t_3	t_4
Stepping Phase (HV)	A	B	C	A	B
Damping Phase (LV)	B	C	A	B	C

This scheme will also be directionally dependent, unless the two voltages are equal, in which case the scheme corresponds to the normal two-phases-on operation.

The step responses obtained with this scheme are shown in Figures 9-14 and 9-15. The step motor used is again the SM024-0045-HS.

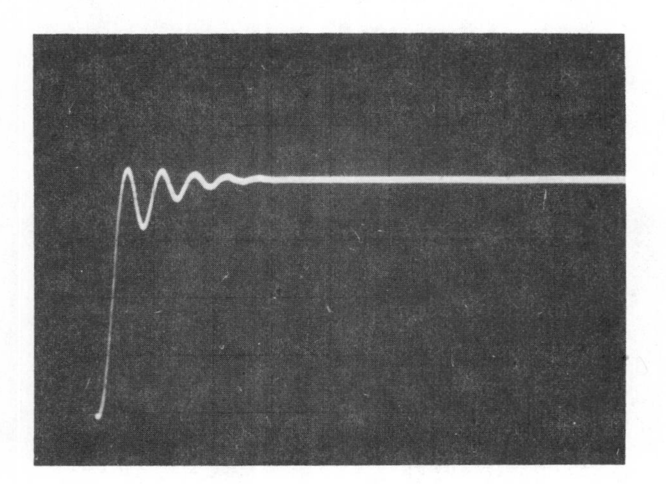

Figure 9-14. Single-step response, two-phase-on variable-voltage damping, HV = 28 v, LV = 14 v; no load (10 msec/div.).

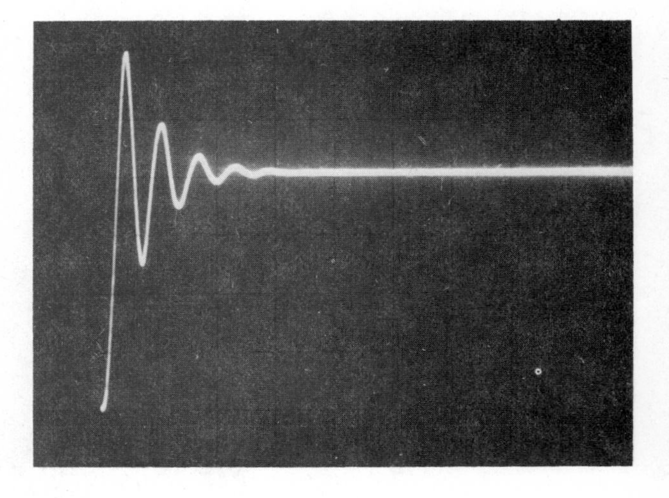

Figure 9-15. Single-step response, two-phase-on variable-voltage damping, HV = 28 v, LV = 14 v; with inertial load (50 msec/div.).

It is clear from Figures 9-11 through 9-15 that this method of damping does provide a marked improvement in the single-step response. Due to its directional dependence, this scheme would be useful for applications where accurate stepping is required only in one direction. The requirement of two different voltages may be a disadvantage in certain situations.

Damping by Use of Variable Turn-Off in the One-Phase-On Scheme

The standard one-phase-on stepping scheme is modified here such that the phase which is turned off has a slow current decay. This results in a significant improvement in the damping of the single step response. Although increased winding inductance could yield slower current decay, it would not be desirable since the current buildup would also be delayed. In this scheme, rapid current buildup is desired to maintain a fast rise time.

The single-step response is tested with the SM024-0035-AA step motor (made by the Warner Electric Brake and Clutch Company). The motor is operated with the standard one-phase-on scheme, standard two-phases-on scheme and the modified one-phase-on with variable turn-off. The motor is driven with 28 volts in each case, and there are no additional series and suppression resistors. Figures 9-16 through 9-18 show the single-step responses of the unloaded motor with these three different schemes. Figures 9-19, 9-20 and 9-21 show the step response of the motor with a small inertial load, again with the same three schemes. Figures 9-22 and 9-23 show the decay of the phase voltage as a result of slow turn-off, and Figure 9-24 shows the normal turn-off of voltage. To study the effect of repetitive stepping, the unloaded motor is driven at 20 steps per second with one-phase-on, two-phases-on, and modified one-phase-on with variable turn-off schemes. Figures 9-25, 9-26 and 9-27 show the responses for these three cases.

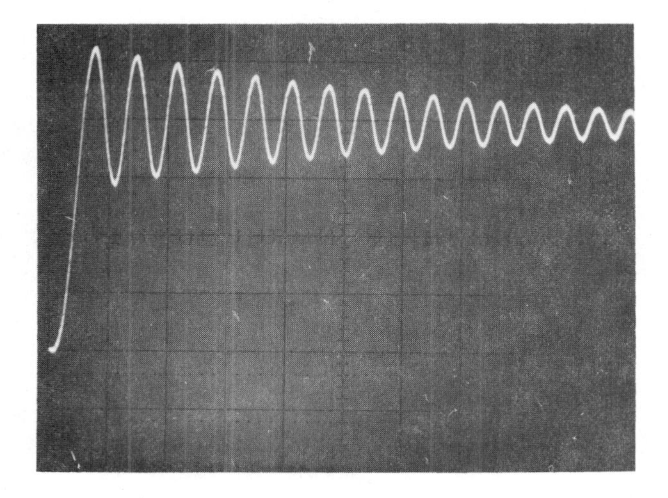

Figure 9-16. Single-step response, one-phase-on; no load (10 msec/div.).

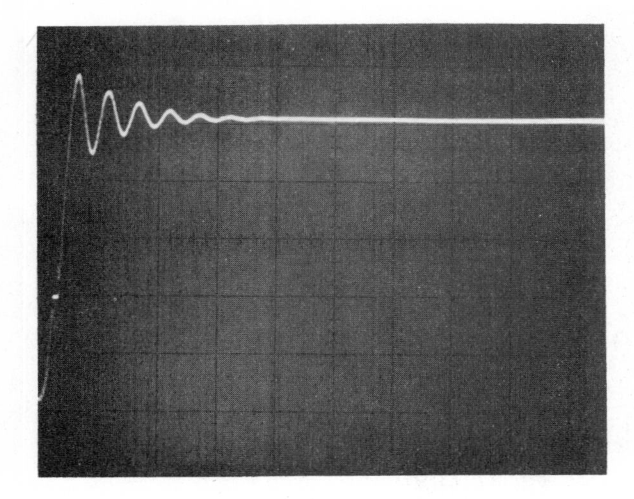

Figure 9-17. Single-step response, two-phase-on; no load (10 msec/div.).

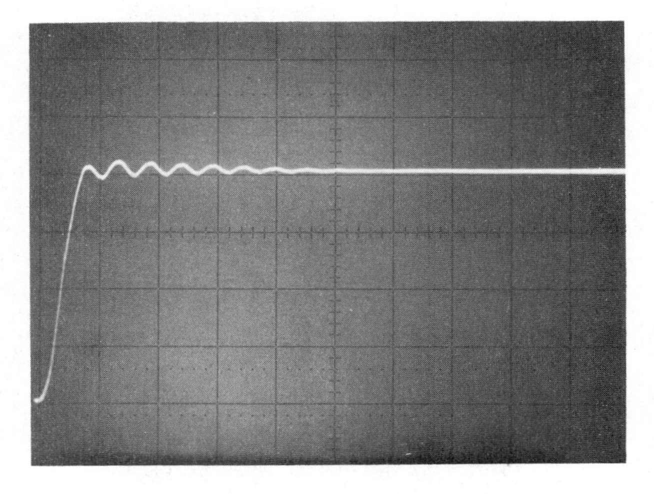

Figure 9-18. Single-step response, modified one-phase-on with delayed-turn-off damping; no load (10 msec/div.).

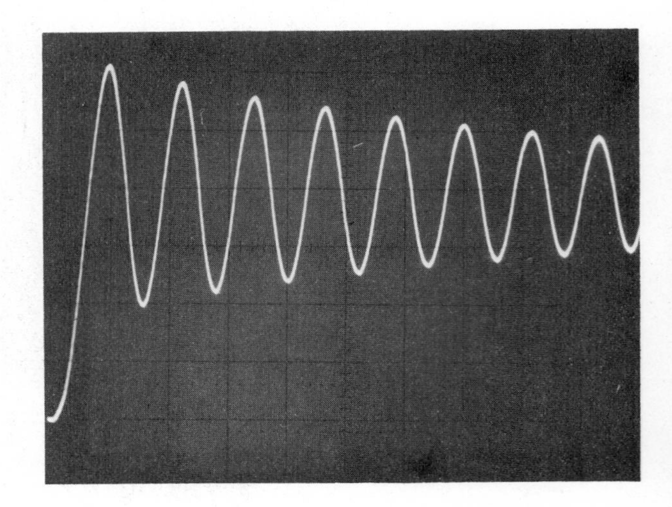

Figure 9-19. Single-step response, one-phase-on; light inertial load (10 msec/div.).

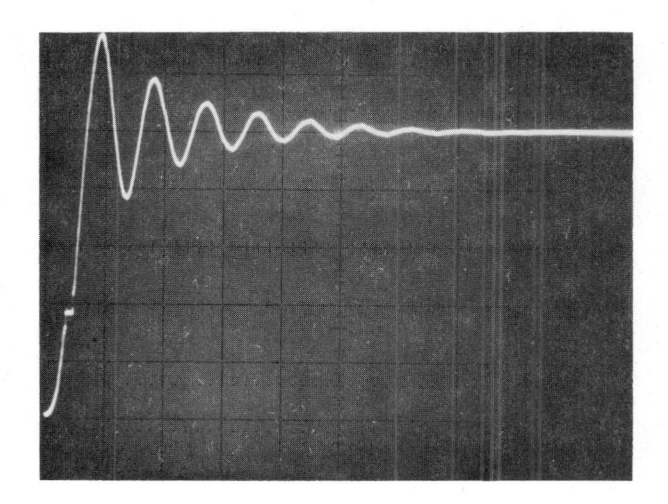

Figure 9-20. Single-step response, two-phase-on; light inertial load (10 msec/div.).

Figure 9-21. Single-step response, one-phase-on with delayed-turn-off damping; light inertial load (10 msec/div.).

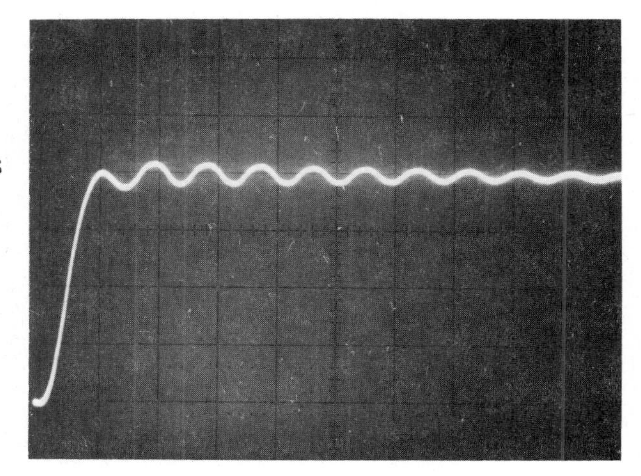

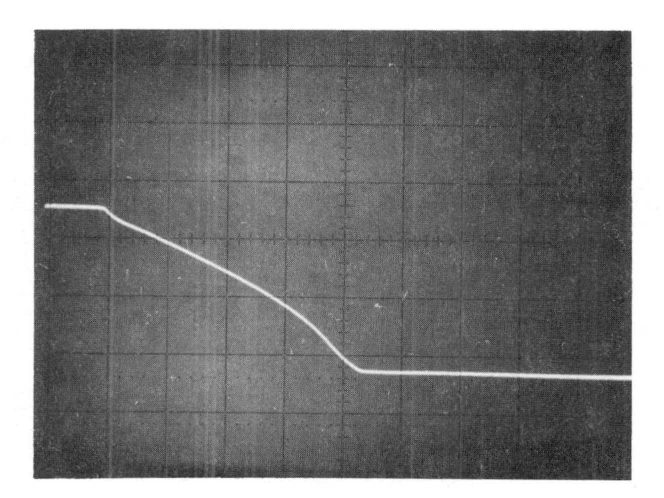

Figure 9-22. Phase voltage decay during single-step operation. One-phase-on with delayed turn-off damping; no load (1 msec/div.).

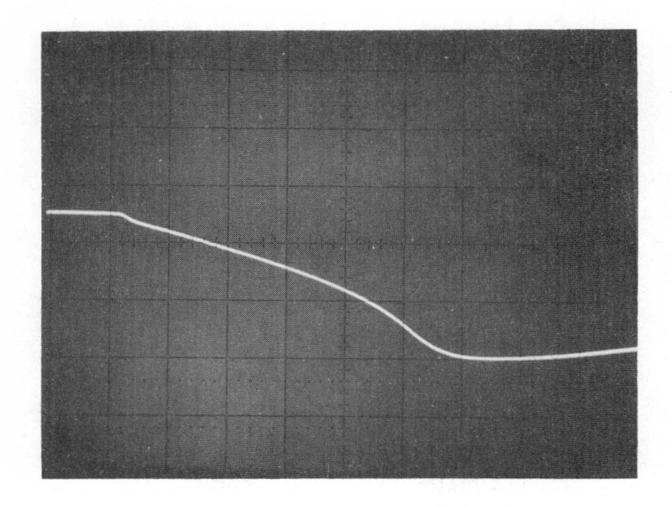

Figure 9-23. Phase voltage decay during single-step operation, one-phase-on with delayed-turn-off damping; light inertial load (1 msec/div.).

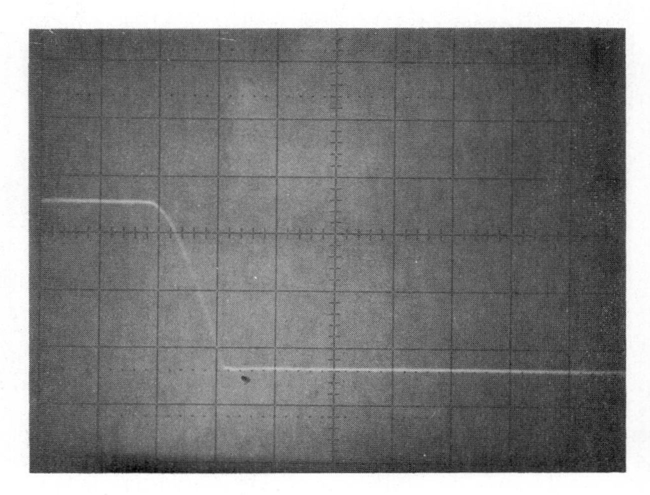

Figure 9-24. Phase voltage decay during single-step operation, one-phase-on; no load (2 msec/div.).

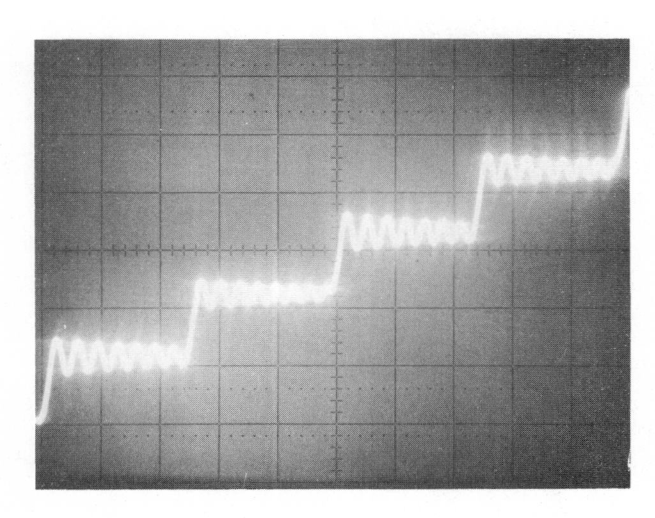

Figure 9-25. Slewing response (20 steps/sec) with one-phase-on excitation; no load (20 msec/div.).

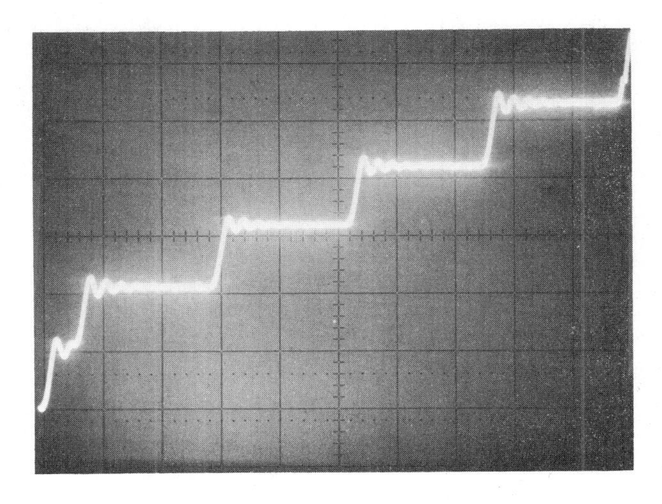

Figure 9-26. Slewing response (20 steps/sec) with two-phase-on excitation; no load (20 msec/div.).

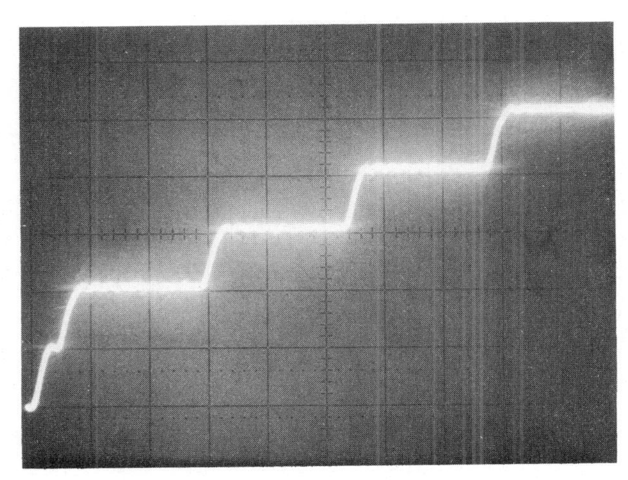

Figure 9-27. Slewing response (20 steps/sec) with one-phase-on delayed-turn-off damping; no load (20 msec/div.).

Figures 9-16 through 9-17 show that there is a marked improvement in the single-step response by use of the delayed turn-off scheme. This scheme is generally advantageous to the variable-voltage two-phase-on scheme. This is due to the fact that it is not directionally dependent, does not need an extra power supply, and requires a minor change of the ordinary drive circuit.

If high speed slewing is required by the motor, this scheme would pose a hinderance, and means should be made to switch it out of the drive circuit.

9.4 DAMPING BY MODIFICATION OF MOTOR PARAMETERS

In this section several approaches to damping are discussed which utilize additional windings or modified slot-tooth shapes. The advantage with these approaches is that the damping schemes are essentially contained in the motor itself, and nothing has to be done externally. A disadvantage is that by modifying the motor to improve its damping, other performance characteristics would have to be generally compromised to some extent.

Damping by Use of Auxiliary Stator Windings

To study the effect of damping by using auxiliary windings on a standard single-stack VR step motor, an SM024-0035-AA motor (made by the Warner Electric Brake and Clutch Company) was modified to contain two similar sets of stator windings. Each tooth of the motor has both windings wound on it. Figure 9-28 shows the schematic diagram of the two windings.

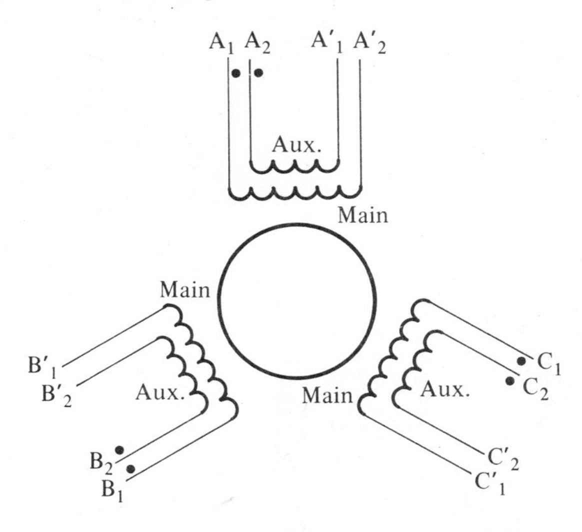

Figure 9-28. Schematic diagram of step motor with auxiliary stator windings.

For test purposes, the main winding is energized with 16V d.c., and is stepped using the one-phase-on scheme. The auxiliary winding is energized differently in each test. No additional series or suppression resistances are used. Figures 9-29 through 9-33 show the various single-step responses. Figures 9-29 shows the step response when the auxiliary winding is open circuited, and Figure 9-30 shows the same when the auxiliary windings are shorted. Figures 9-31 through 9-33 show the single-step responses when the auxiliary windings are energized at different current levels.

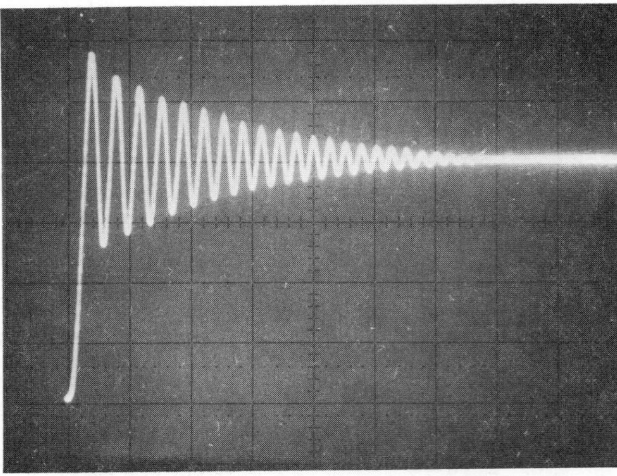

Figure 9-29. Single-step response of step motor with damping windings, auxiliary windings open; one-phase-on excitation (20 msec/div.).

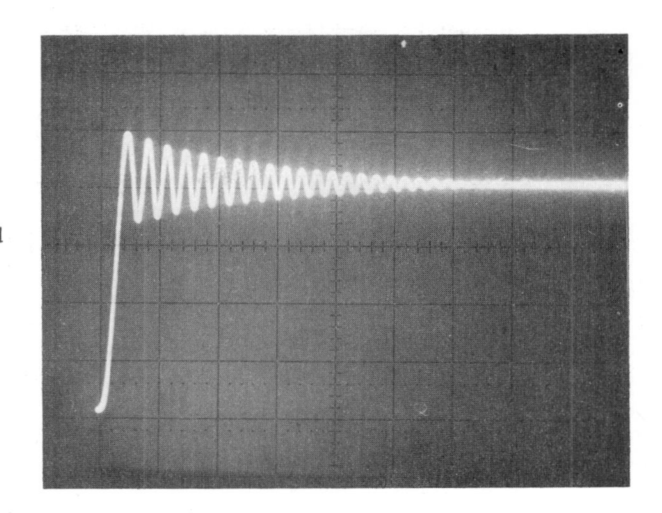

Figure 9-30. Single-step response of step motor with damping windings, auxiliary windings short circuited (20 msec/div.).

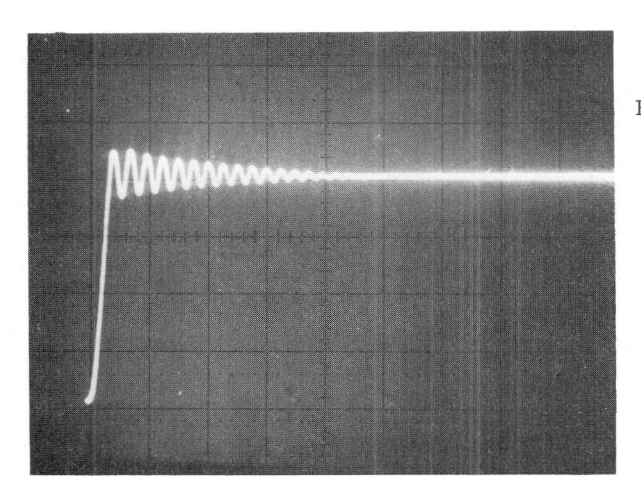

Figure 9-31. Single-step response of step motor with damping windings, auxiliary windings energized all at once; 1/6 amp/phase (20 msec/div.).

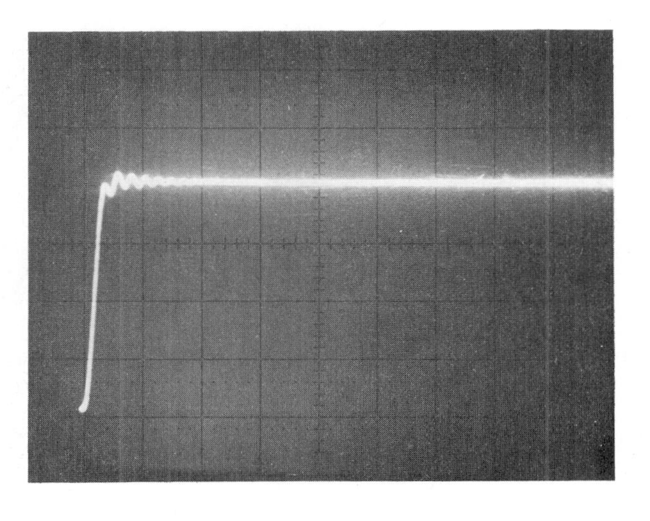

Figure 9-32. Single-step response of step motor with damping windings, auxiliary windings energized all at once; 1/2 amp/phase (20 msec/div.).

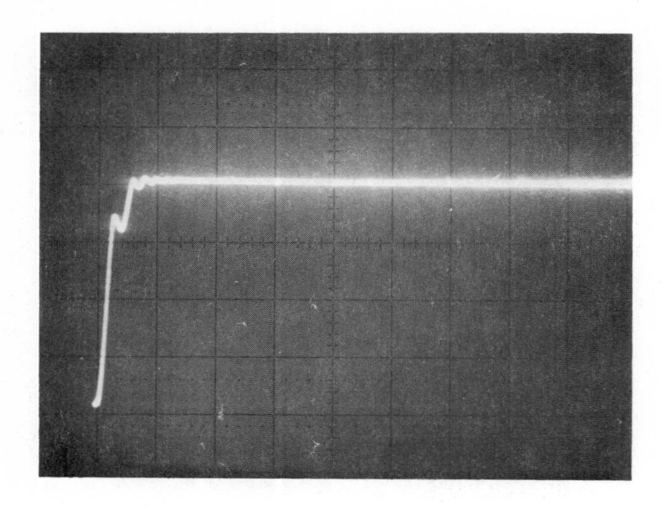

Figure 9-33. Single-step response of step motor with damping windings, auxiliary windings energized all at once; 1 amp/phase (20 msec/div.).

The test results show that the use of auxiliary stator windings can improve the damping of the step response. The rise time is increased but not significantly. The improvement in step response is not without limitations. The motor torque and output capacity may be reduced, since only part of the windings are actively engaged in driving the rotor.

Damping by Use of Additional Rotor Windings

A common method of damping transient oscillations in synchronous machines involves the use of damper bars or shorted coils on the rotor poles. To investigate the usefulness of this approach for damping step motors, shorted coils were placed in the teeth of a multiple-stack VR step motor.

Figures 9-34 and 9-35 show the single-step response with the rotor coils open and shorted, respectively.

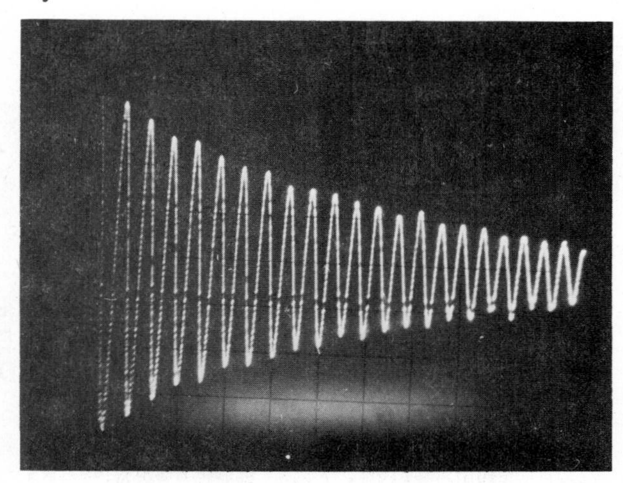

Figure 9-34. Single-step response of step motor with rotor windings for damping, rotor windings open circuited; no load (20 msec/div.).

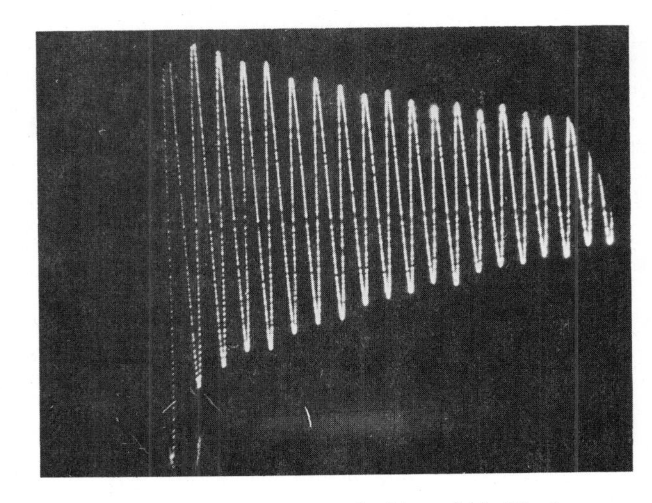

Figure 9-35. Single-step response of step motor with rotor windings for damping, rotor windings short circuited; no load (20 msec/div.).

Although extensive test results are not available, Figures 9-34 and 9-35 do show that this method of damping does not produce any marked improvement in the response of these types of motors.

Damping By Modifying Motor Dimensions

This method of damping involves the modification of motor dimensions such as slot-tooth proportions, pole width and air gap. Very little is yet known on how these parameters actually affect the damping characteristics. However, it is worth mentioning that this approach could yield the most fruitful results due to its fundamental nature.

The size and shape of slots and teeth play a major role in the shape of the static torque curve and consequently on the damping characteristics of the motor. Figure 9-36 illustrates the effect of changing tooth width on static torque.

Similarly, the pole width, air gap and other motor dimensions will have an effect on the damping behavior of the motor. It is not known exactly how these parameters affect damping. Also, these parameters will affect the other dynamic characteristics of the motor, and an appropriate compromise between different overall characteristics will be necessary in each case.

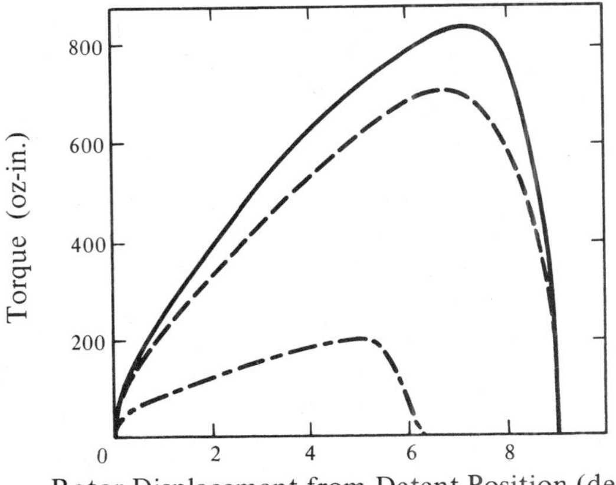

Figure 9-36. Holding torque variations with changing tooth width. Constant pole width and constant excitation.

10 PERMANENT MAGNET STEP MOTORS

T. E. Beling
Sigma Instruments, Inc.
Braintree, Massachusetts

10-1 INTRODUCTION

This chapter is concerned with the description and the control of permanent magnet step motors.

Permanent magnet (PM) motors have several features similar to VR motors. The stator of a PM motor has a number of wound poles which may have a number of teeth as part of its flux distributing member. Also, the rotor is cylindrical and toothed.

The key difference is in the incorporation of a permanent magnet in the magnetic circuit. Most PM motors add the permanent magnet in the rotor assembly. This magnet can be axially charged as in the large size Sigma 18 Series and all 20 Series motors, or radially charged as in the small size Sigma 18 and all 19 Series.

The PM stepping motor operates by means of the interactions between the rotor magnet biasing flux and the magnetomotive forces generated by applied current in the stator windings. If the pattern of winding energization is fixed, there is a series of stable equilibrium points generated around the motor. The rotor will move to the nearest of these, as in the VR motor, and remain there. If then the windings are excited in sequence, the rotor will follow the changing point of equilibrium and rotate in response to the changing pattern.

By virtue of the permanent magnet, there is a "detent" torque developed in the motor even when stator windings are not excited. A restoring torque is generated on the rotor whenever the rotor is moved from the position which has minimum reluctance for the permanent magnet flux. This torque is much weaker than the energized torque, typically a few percent of the maximum torque.

The Series 9, a permanent magnet motor, has a completely different construction. The stator has a permanent magnet, and the method of operation is such that there is no inherent reversibility, so that a reversible motor consists essentially of two separate devices on a single shaft. This motor features very high static detent torque (nearly equal to holding torque) and extreme simplicity of drive.

10-2 CHARACTERISTICS OF PM STEP MOTORS

In presenting the characteristics of step motors, it must be emphasized that the motor is, in a sense, a transfer device between the electrical information presented by the driver and the mechanical motion delivered to the load. The actual functioning of the stepping motor system is heavily dependent upon the load and the driver, and both of these must be clearly specified if performance with a given stepping motor is to be predicted successfully. Obviously, it is not practical to generate performance data for every possible driver and load combination. Further, many systems defy characterization as a load that can be analyzed readily. The most practical approach under these circumstances is for the designer to acquaint himself with the general parameters and problems involved, make a reasonable choice of driver and motor, and then thoroughly test the combination in the actual system. This approach, used in conjunction with application engineering support from the stepping motor manufacturer, has proven considerably more useful than attempts to thoroughly quantify and analyze dynamic systems that rapidly become highly complex.

Dynamics of a Single Step

The motion of the stepping motor rotor in the single-step mode resembles that of a torsional pendulum. Excitation is supplied by the stator flux, and the current rate of rise determines the maximum kinetic energy input to the rotor. Load friction is manifested as damping, while the system inertia consists of the sum of the rotor and load inertias. The effects of variation of friction are shown in Figure 10-1, in which the friction load is varied. Note that the rotor position as a function of time

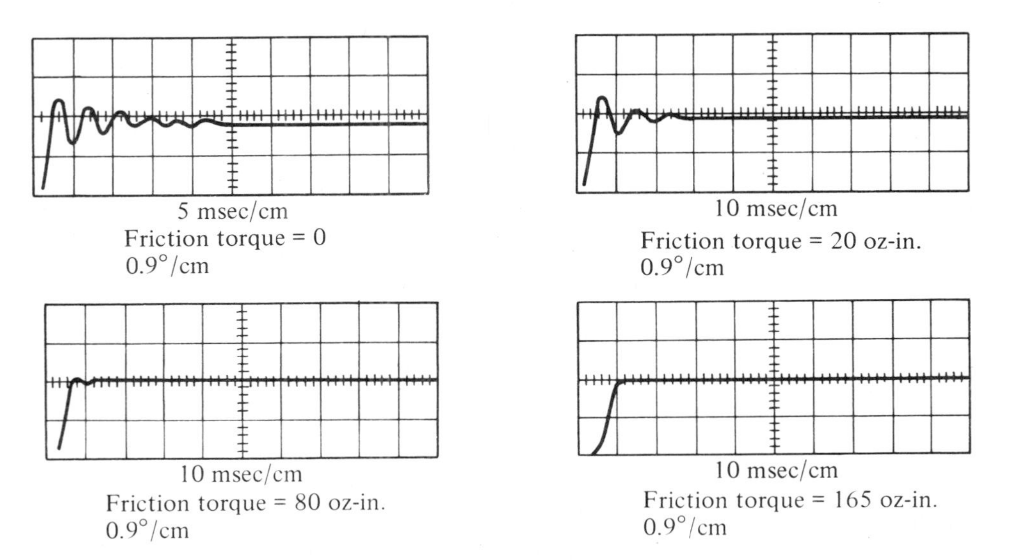

Figure 10-1. Effect of friction on single-step response with unipolar R/L drive, no external inertia (R_S = 3 ohms).

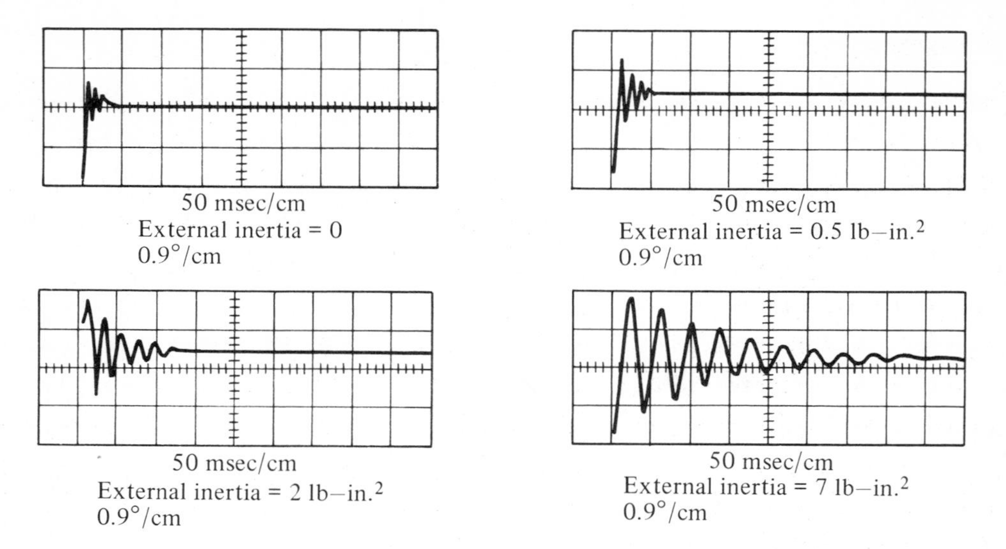

Figure 10-2. Effect of inertia on single-step response with unipolar R/L drive, (R_S = 3 ohms).

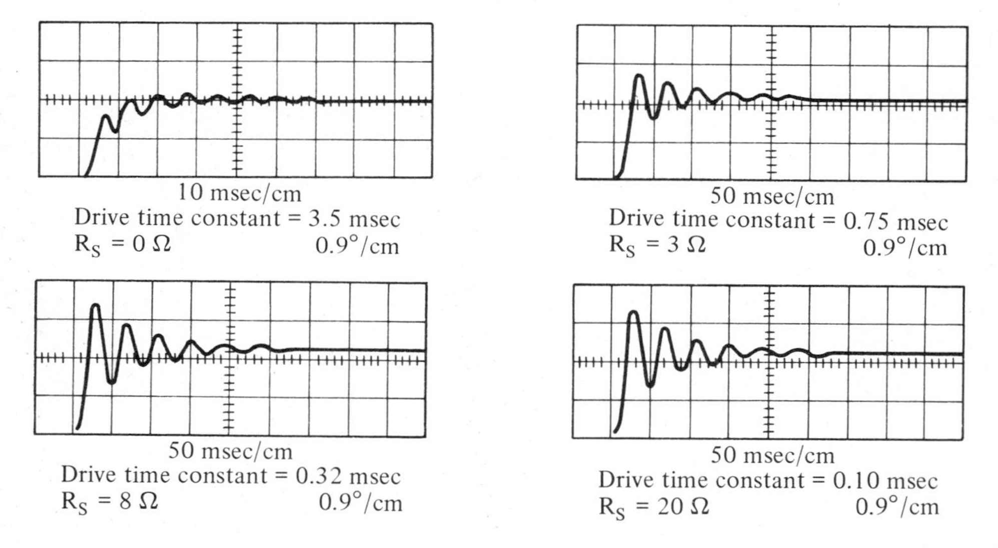

Figure 10-3. Effect of current rise time on single-step response with unipolar R/L drive.

varies from underdamped to critically damped to overdamped as the friction load increases. Increasing inertia lowers the frequency of rotor oscillation (Figure 10-2), while increased current rate of rise increases the magnitude of oscillation (Figure 10-3). Clearly, both the magnitude of oscillation and the settling time are extremely important in most applications in that they also determine the ability of a system to restart or reverse after a stop has occurred. Various means of reducing these effects are discussed later.

Angular Accuracy

The accuracy of a typical step motor is rated at ±3% to ±5% noncumulative. For instance, in a 200 step-per-revolution motor the step angle is 1.8° or 108'; if a step is taken from a reference position, the final angle of the rotor will be 108' ±3.24' for a ±3% error. If 1000 steps are taken, the angular motion of the shaft will be 108,000' ±3.24' - corresponding to five complete revolutions of the output shaft. The point is, of course, that the rated angular error (in this case ±3.24') is the same regardless of the number of steps taken. The actual measured error is apt to be considerably less than the rated value. This high order of accuracy results because of the inherently symmetrical construction and the magnetic averaging over many poles that is characteristic of stepping motors.

Nevertheless, problems do arise in applications in which accuracy is critical. Ordinarily, these are traceable to two basic causes - motor stiffness and winding current balance.

Stiffness refers to the torque-displacement curve that is generated when the rotor is deflected from its energized rest position (Figure 10-4a). The significance of the curve is that the rotor rest position will be such as to balance the load torque; i.e., the rotor will not be at its unloaded position, but at a different point. This effect theoretically would not inherently affect the step-to-step angular motion, but two problems arise. First, load torque may actually vary somewhat from step to step, so that the real position between steps may be in error by corresponding amounts. Second, the system dynamics may change enough so that the step overshoot, and the direction from which the final position is approached, may change, with deleterious effects on measured angular accuracy.

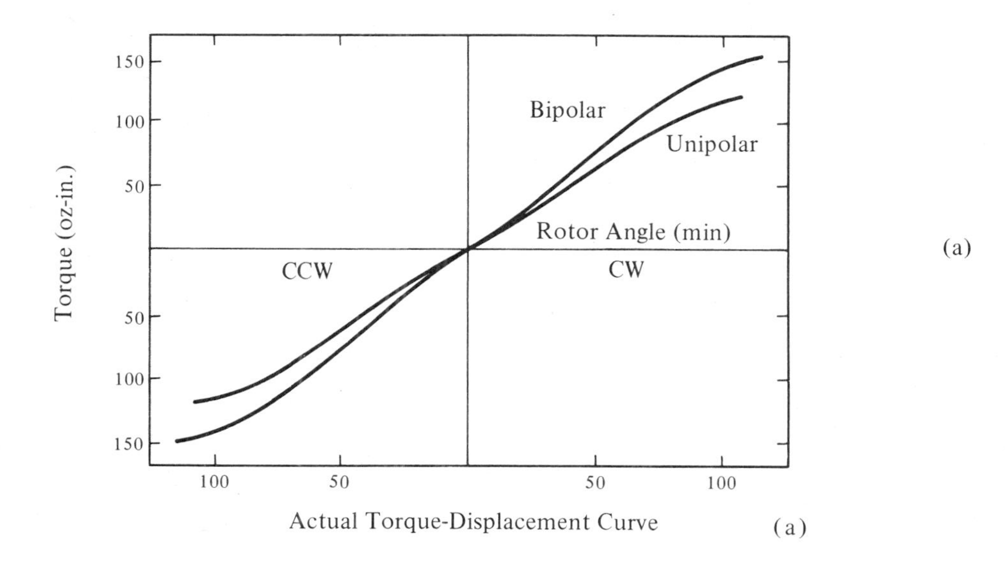

Actual Torque-Displacement Curve (a)

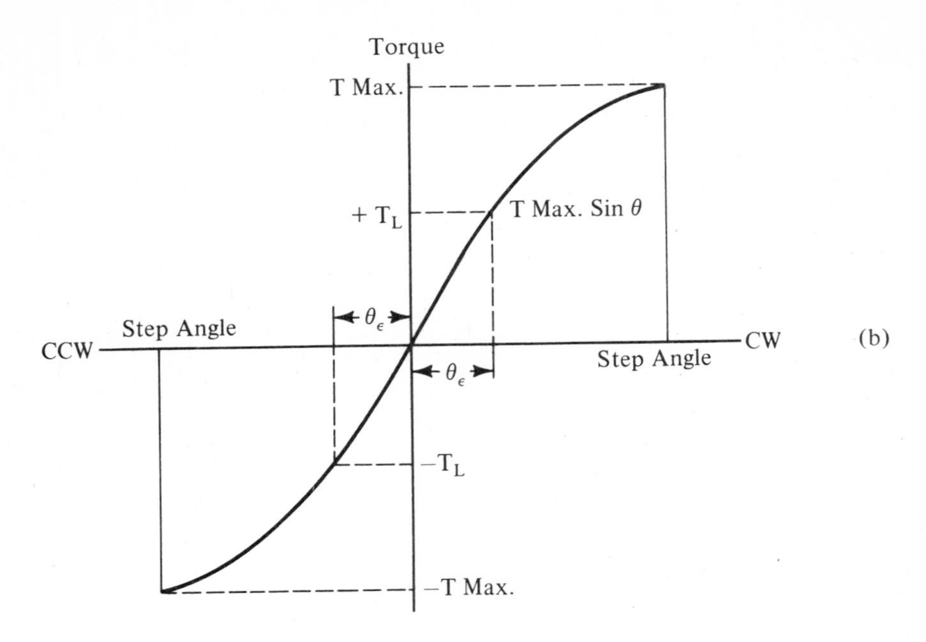

Figure 10-4. Position errors caused by load torque.

For example, if a system has a torque-angle curve as shown in Figure 10-4b and a static friction load of T_L, for clockwise rotation, the static position at the end of a step is not at the origin as drawn but rather at point A. The presence of T_L generates an error θ in the ccw direction, where $\theta_1 \approx \sin^{-1}(T_L/T_{max})$. If the direction of rotation were always clockwise, and T_L were constant, θ_1 would never by observed. However, if the origin were also approached from the counterclockwise direction, the final equilibrium point would be at B, with a clockwise error of θ_2, equal to θ_1. Therefore, the effective error in position observed is equal to $2\sin^{-1}(T_L/T_{max})$.

The effect of system dynamics on this potential error is dramatic. Rotor overshoot (underdamped system response) tends to allow the rotor to approach the origin closer than calculation might indicate, even on a static basis. One cause, of course, if the transfer of rotor kinetic energy to output work ($\theta_1 T_L$); another cause is easier to visualize as "dither" which coaxes the static position closer to the origin.

Another factor in the final rotor rest position is current balance between windings. In many commonly used drive systems, two windings are energized simultaneously. If the current in one of these windings is reduced gradually to zero while the other remains fixed, the shaft will move 1/2 step, with the motion being related linearly to the winding current. Naturally, this factor must be considered in driver design if the highest angular accuracy is required. It should also be noted that the unenergized rotor rest position differs from the rest position with two windings energized by approximately one-half step.

Multiple Step Characteristics

Stepping motor data are available in three main forms:

1. Start-run (slew) curves. The motor is run up to speed gradually and a varying friction load applied until the motor loses synchronism. The resulting data are plotted as a speed-torque curve in Figure 10-5.

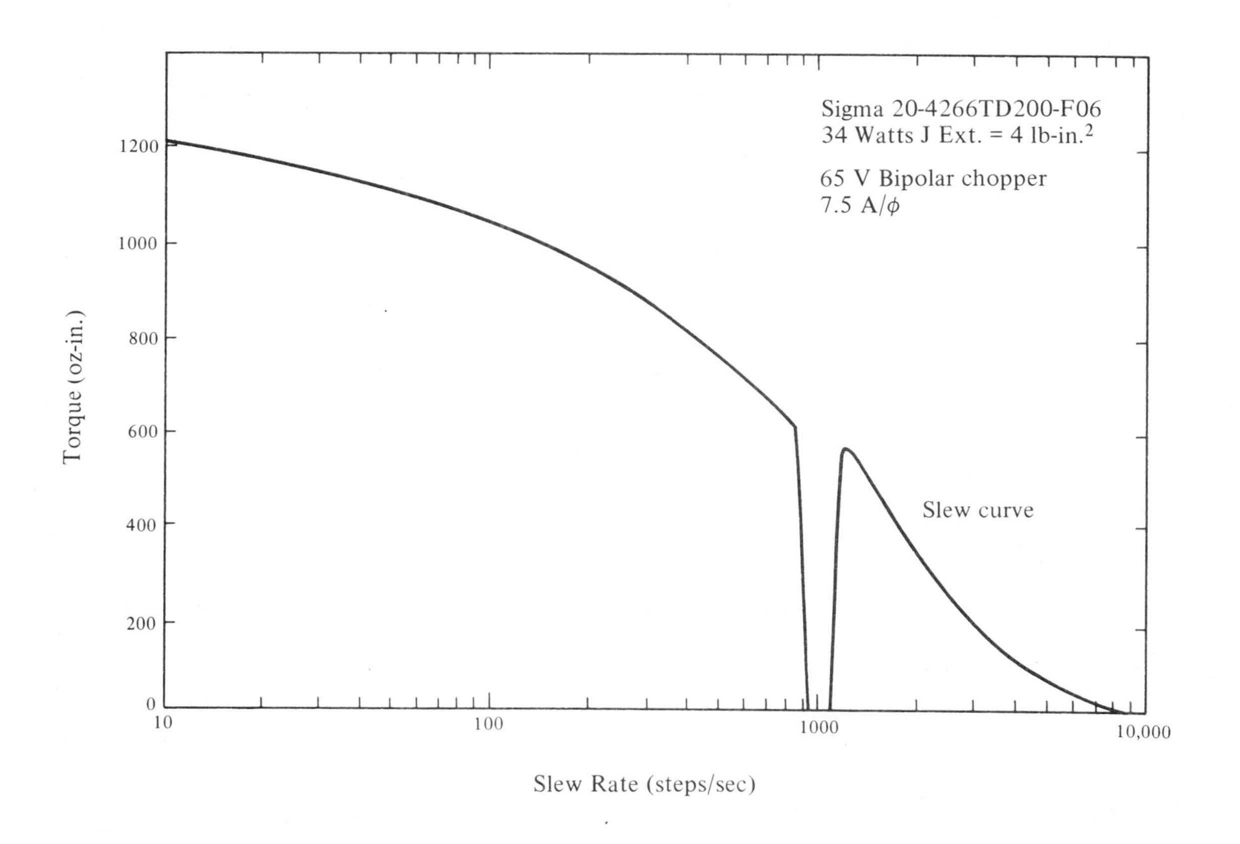

Figure 10-5. Speed-torque curve.

2. Start-stop curves. Friction and inertial loads are applied to the motor at rest. A pulse burst containing a fixed number of pulses (say, 200 pulses for a 200 step/revolution motor) is used to drive the motor. The time period of the pulse train is long compared to the starting transients of the motor dynamic response. The frequency of the pulses is varied to determine the maximum frequency at which the given load can be both started and stopped without error. The three-dimensional curve obtained is usually plotted as a family of inertia curves on a torque-speed plot in Figure 10-6. Generally, these data are combined on one plot.

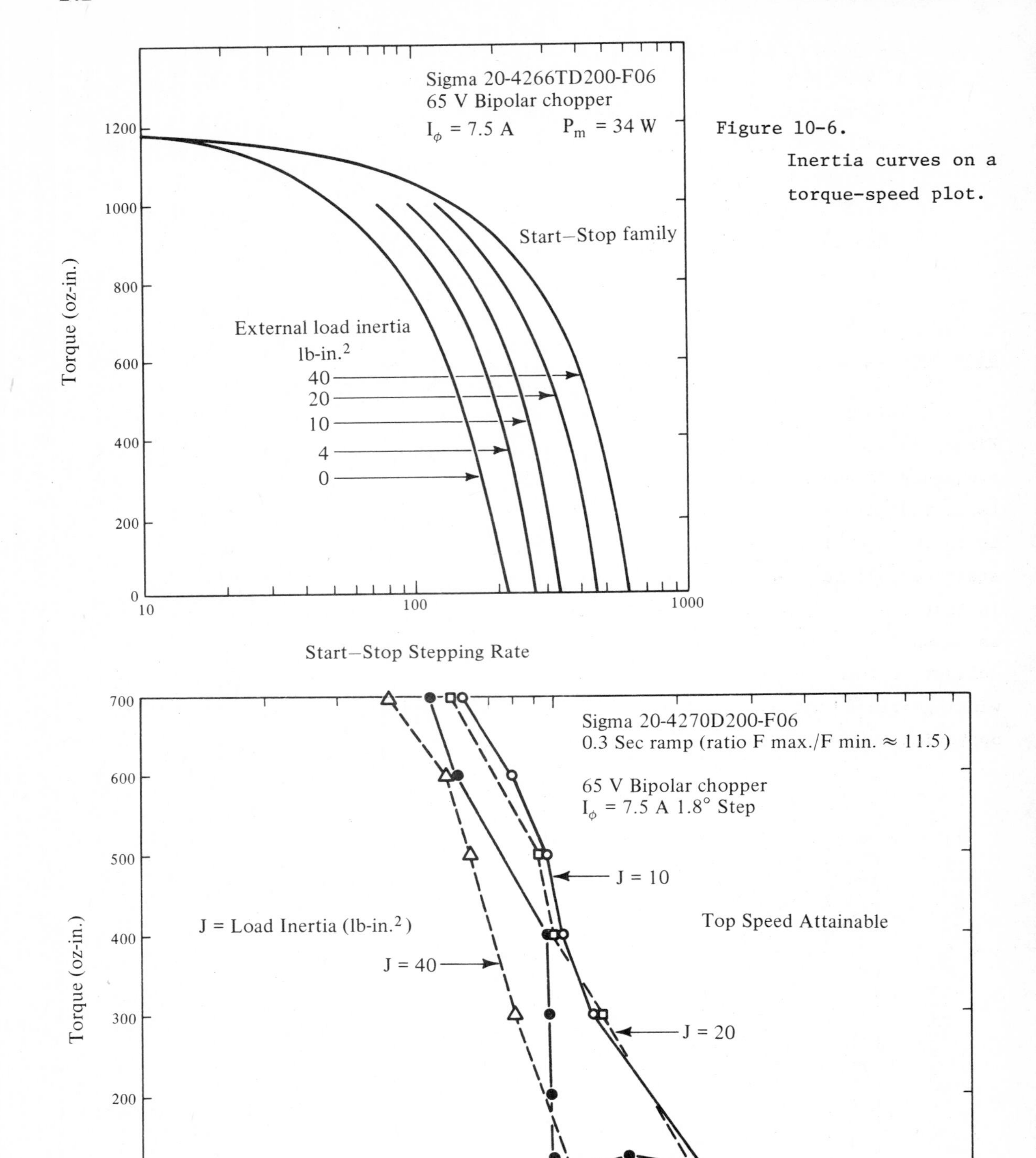

Figure 10-6. Inertia curves on a torque-speed plot.

Figure 10-7. Ramping curves.

3. Finally, for larger motors a ramping curve is generated. Fixed loads are again used on the motor at rest. A pulse train is applied containing a fixed number of pulses, but arranged so that the frequency starts at a low value, accelerates to a top speed, runs at top speed, decelerates to the low speed, and then stops. This type of operation is typical for large loads such as machine tools. The data are taken in much the same way as the start-stop curves, and are plotted as shown in Figure 10-7.

Slew-Mode Data

Examination of slew-mode data on a step motor reveals several interesting points. First, as the rate of rise of the current in the windings is increased, the high frequency torque available increases (Figure 10-8). This fact is of central importance in the design of drive circuits - the goal of most drive circuit design efforts is to achieve high rate of rise of current in the motor windings while limiting steady-state current value to the motor. Another phenomenon that is perhaps less expected is that the low frequency torque increases as the current rate of rise increases, as shown in Figure 10-9. This occurs simply because the rotor is in position to deliver maximum torque at the beginning of the step, and if the current in the windings rises rapidly enough to achieve its final value before the rotor can move appreciably, maximum peak torque will be delivered by the motor.

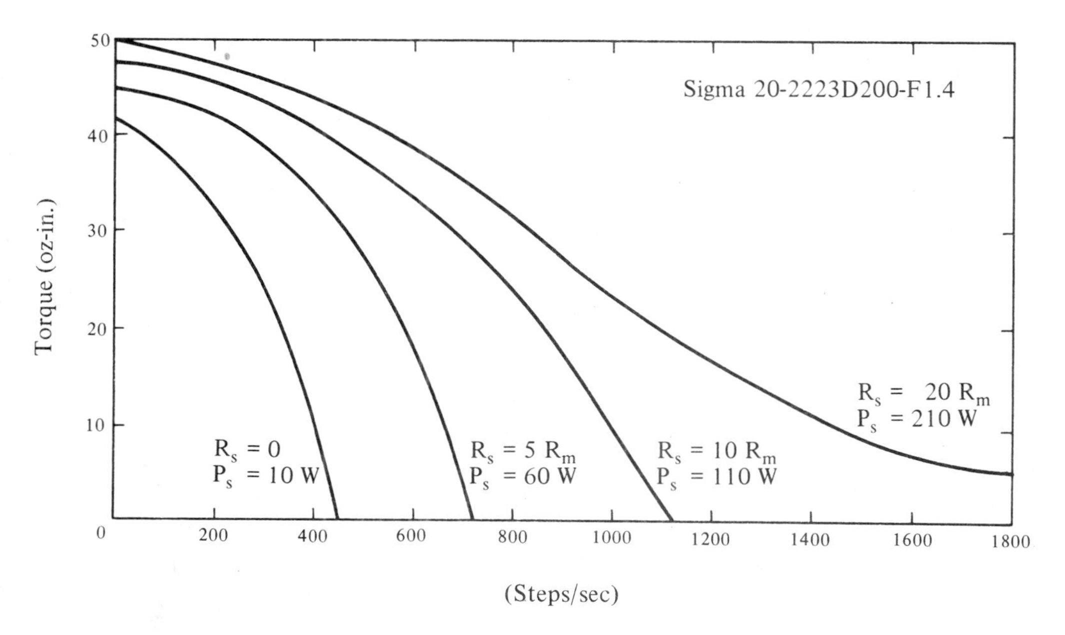

Figure 10-8. Slew-mode data - available torque as function of current.

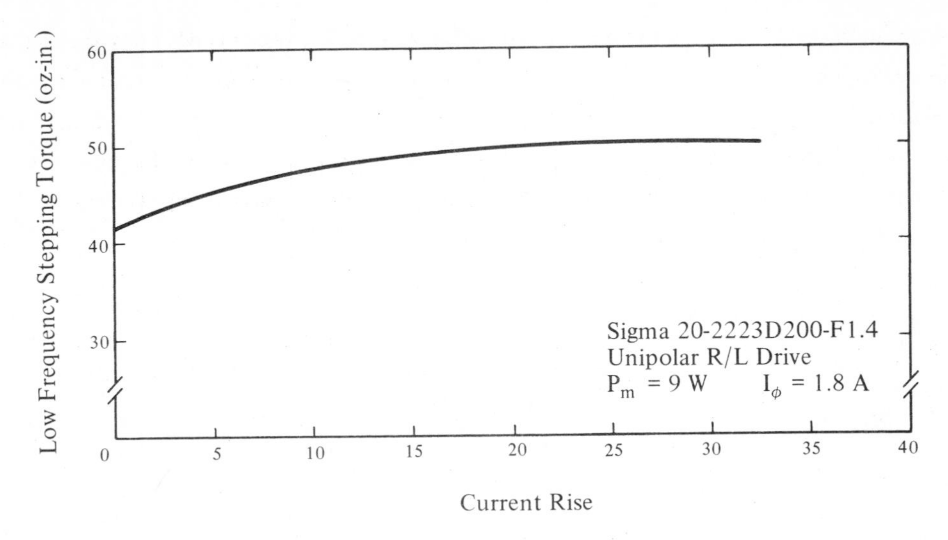

Figure 10-9. Low-frequency torque as a function of current rate of rise.

There is a fixed relationship between the maximum value of energized detent torque (T_{max} of Figure 10-4) or pull-over torque and the maximum load torque that can be stepped. Referring to Figure 10-10, assume the motor is loaded with a torque, T_L. This generates a rotor lag angle of θ_e as previously described. When the motor is commanded to take a step, the coordinates of the torque-displacement curve advance $\pi/2$ electrical degrees or one step (dashed curve). The lag angle θ_e causes a motor restoring torque T_1 (always less than the peak torque T_{max}). With this load, the net excess restoring torque is $T_1 - T_L$, which is positive, and the motor will complete the step.

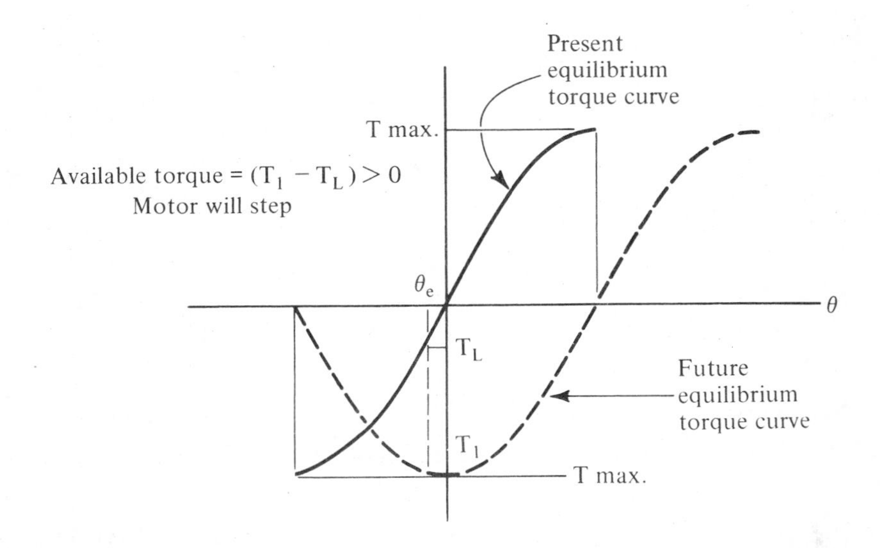

Figure 10-10. Maximum stepping torque.

If the system friction is enough to generate $\pi/4$ electrical degrees of error, the situation becomes that shown in Figure 10-11. In this case T_L is equal to $\sqrt{2}/2T_{max}$ since $T_L = T_{max}\sin\theta_e$. When the motor steps, the origin shifts $\pi/2$ degrees (dashed curve). The motor restoring torque is now $\sqrt{2}/2T_{max}$ which is equal to T_L; hence, in the limit, the motor cannot step since the net restoring torque is zero.

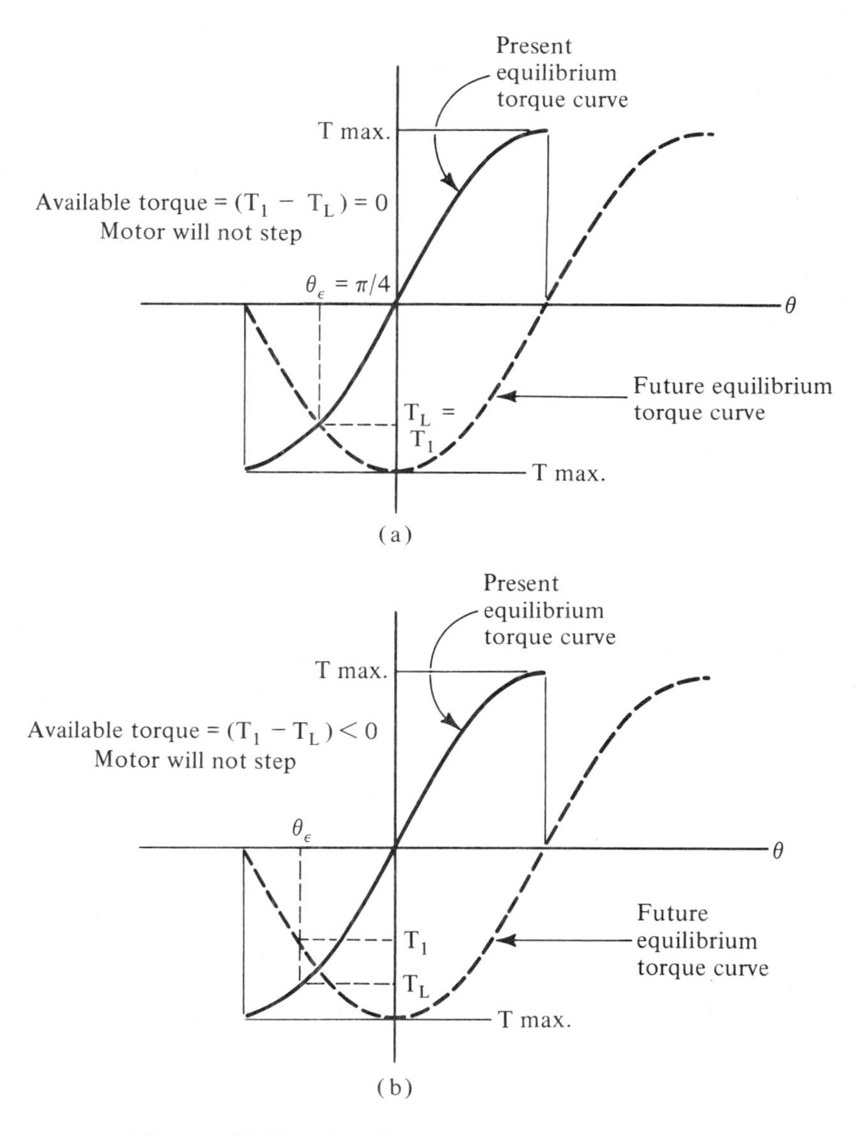

Figure 10-11. Maximum stepping torque.

If the load torque is greater than $\sqrt{2}/2T_{max}$ (i.e. θ_e greater than $\pi/4$), (Figure 10-11a) the motor restoring torque is less than the load torque and the motor cannot step.

Therefore, the maximum frictional load a motor can step is theoretically $\sqrt{2}/2$ times the maximum energized pull-over torque. The precision of this relationship is related to how closely the torque displacement curve can be approximated by the expression

$$T_L = T_{max} \sin\theta_e$$

Under conditions of high current rate of rise, careful control of pulse rate will disclose regions where the motor will not run at all without a torque load (20, 150 Hz region on Figure 10-12). The friction required to achieve stable operation is plotted as the sharply peaked triangles at 50, 30 Hz, etc., on Figure 10-12. These regions are related to the resonant frequency of the rotor. If insufficient damping is present, the periodic excitation of the rotor at its resonant frequence, or a sub-multiple of it, will reinforce the resonance and cause the amplitude of the oscillation to increase. A high current rate of rise further accentuates this oscillation and will carry the rotor to a region such that the next pulse will drive it back instead of forward, and a dithering, unstable operating mode will result. Friction loads damp the rotor oscillation (Figure 10-1) enough so that it will operate in these regions with no difficulty.

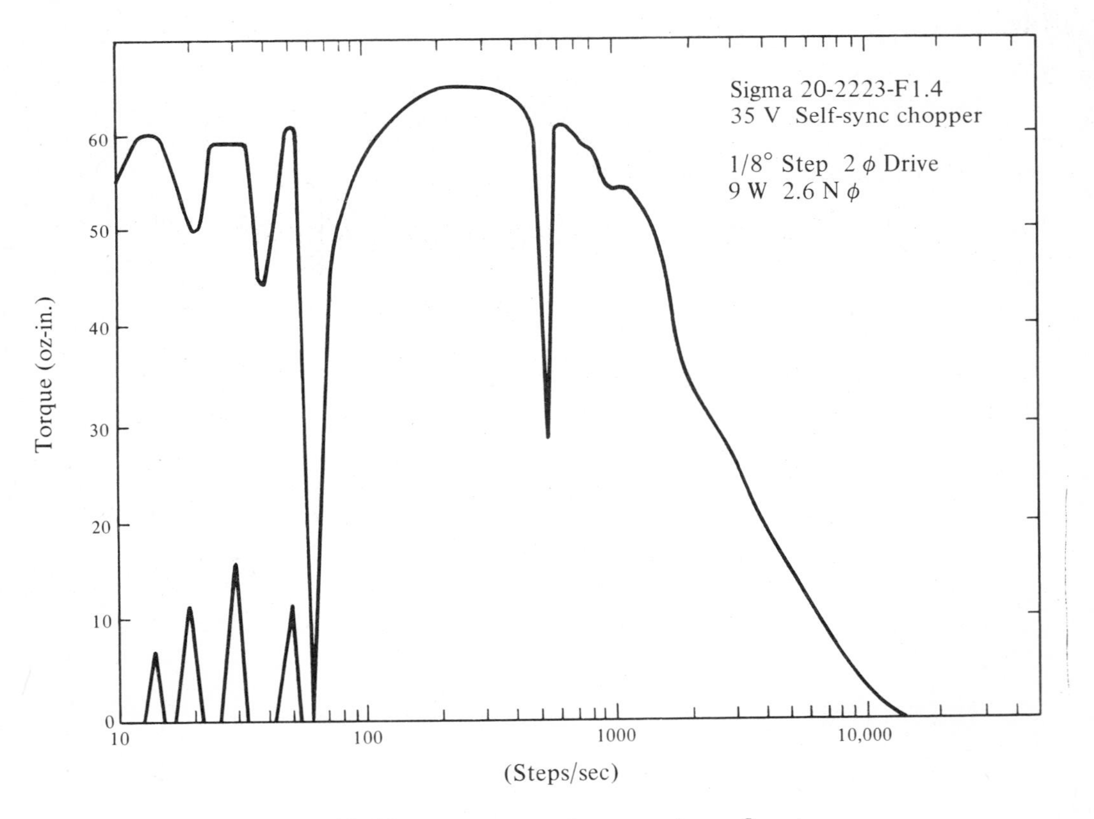

Figure 10-12. Resonant frequencies of rotor.

Another major resonance occurs in the 900-1200 Hz region. Here, a serious loss in torque is suffered. It is basic to all permanent-magnet steppers under conditions of high overdrive, and is related to the phenomena that cause the appearance of "critical speeds" in large multi-horsepower synchronous motors. This resonance is reduced with heavy inertial loading, but at the cost of acceleration characteristics. In general, it is best to avoid operation in this region, or use one of the drive systems that reduce resonance effects.

Start-Stop Mode

In many applications, the motor must start and stop under load, and without acceleration and deceleration times. These requirements severely limit the top speeds available. As a typical example, a Sigma 20-4270 can rotate synchronously in the range of 5000 to 8000 steps/second in the slew mode, but is limited to 300 to 800 steps/second for synchronous start-stops. Start-stop operation with a variable number of steps must be examined critically, since results achievable will depend to a considerable degree upon the number of steps taken, especially when the time period for the pulse train is short with respect to the rotor dynamic response time. This is due to the variation in the rotor-stator lag angle with number of steps, as illustrated in Figure 10-13. Since results will depend upon number of steps up to a total of about 15 steps, it is necessary to examine stepping characteristics carefully under these circumstances.

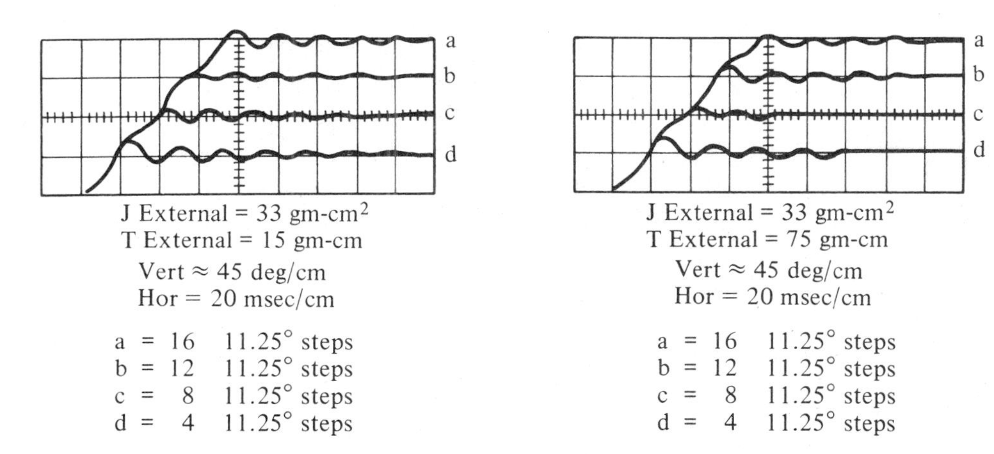

Figure 10-13. Position response of Sigma 18-2013D32-21463, 30 v bipolar R/L, 20 ohms, 0.5 amp/phase.

The amount of time required to settle the load varies with load and drive conditions, but it is frequently desirable to improve settling time characteristics. One method for accomplishing this that works well with fixed load conditions is reverse pulse damping. In this system, a reversing pulse of carefully selected duration is applied to the motor after the last pulse has been initiated. Both the duration and the timing of the damping pulse should be determined experimentally with normal load in place. Actual results of such damping are shown as Figure 10-14. It must be emphasized that the timing values for reverse pulse damping depend upon the load parameters, so that the system should be checked over the full range of variation of load characteristics.

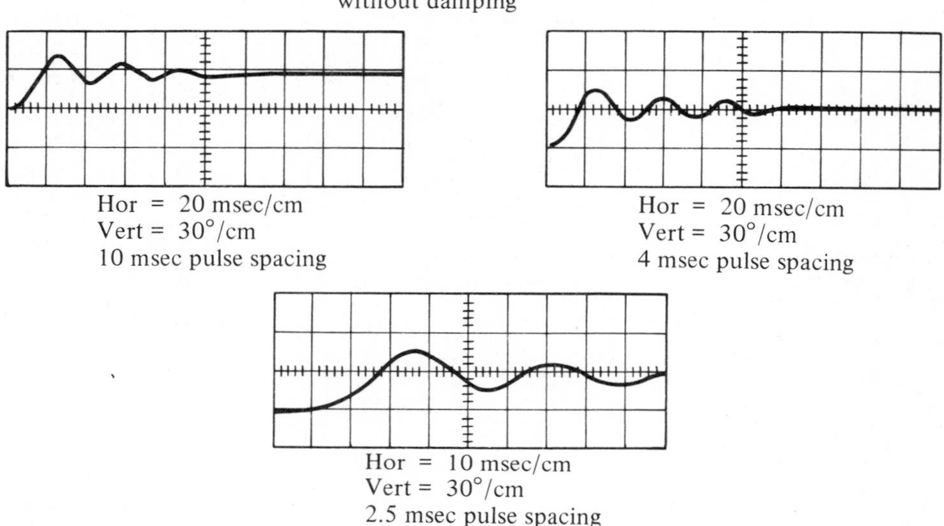

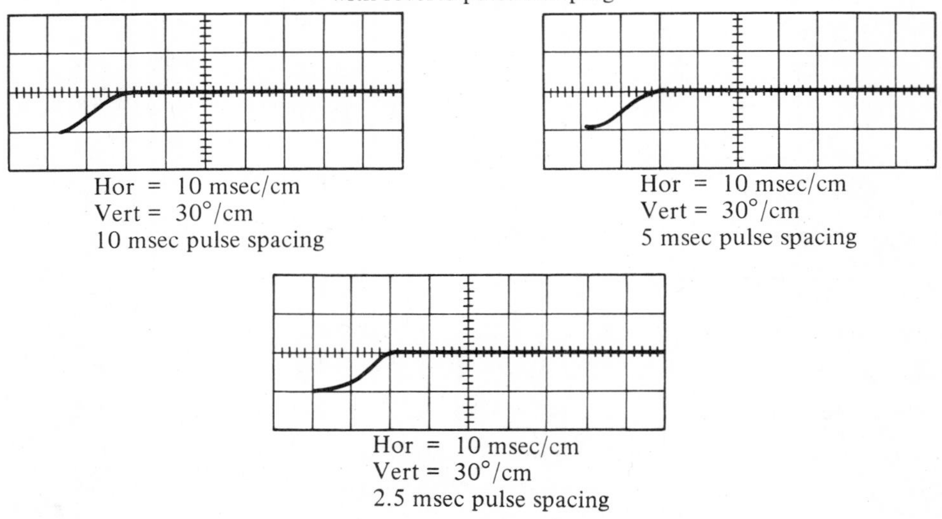

Figure 10-14. Reverse-pulse damping responses.

Another aid in damping is a system such as current feeding where the supply voltage varies with speed. In such a system, the voltage is low at low speeds and less excess kinetic energy is added to the rotor, which improves settling time considerably.

Restarting and Reversing

The ability of a stepping motor and driver system to restart or reverse at the end of a sequence of pulses is related to its settling time under the load in question. Consider Figure 10-15; at the end of a long train of pulses, the rotor will tend to overshoot the final position as indicated. The maximum value of the rotor lag angle can approach 1.8 degrees under conditions of high overdrive, high inertia, or low system friction. It can be seen that there are discrete "windows" in time during which the motor cannot be restarted or reversed because the rotor velocity is too high or in the wrong direction, or both. Likewise, the rotor-stator angle can be improper for the generation of torque when the command for a step arrives even if the rotor velocity is correct. For example, the motor can be reversed easily at point A but it would probably not restart in the same direction because the velocity is a maximum in the wrong direction. Point B is satisfactory for restarting, but not for reversing. At C, the rotor is out of position for either restart or reverse.

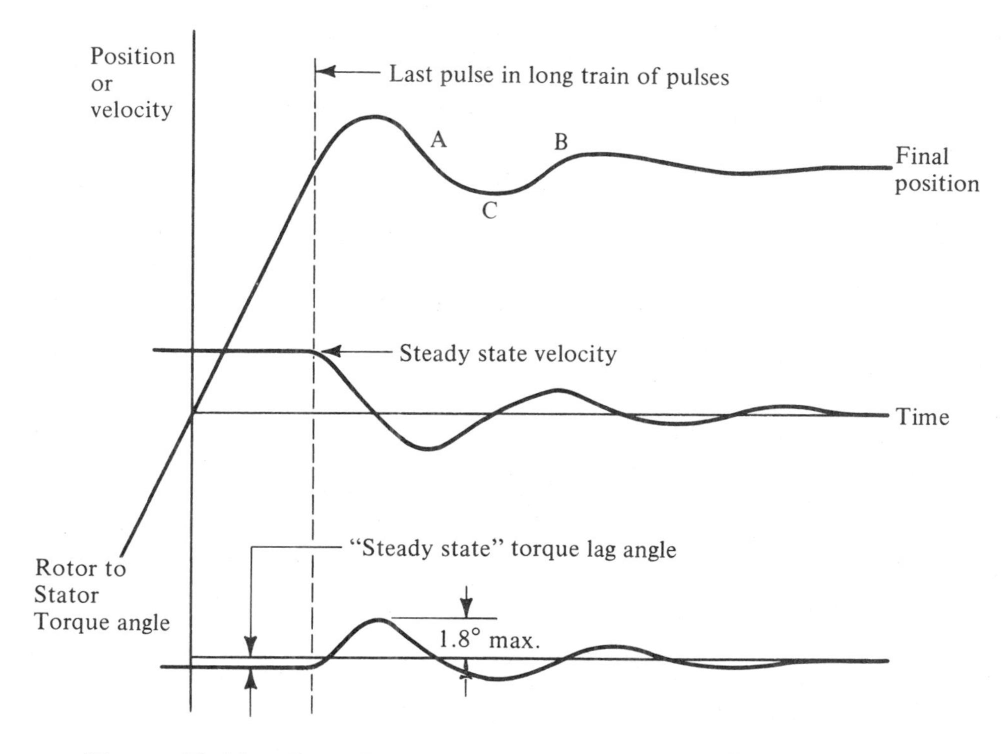

Figure 10-15. Transient responses at the end of a pulse train.

Thus, it may be seen that the restart or reverse frequency is really not so much a question of frequency as it is a matter of time after the last pulse. If the point of restart is chosen optimally, the motor can be restarted at a higher speed than if the motor had been sitting still.

The ideal way to handle this problem is to have a way to measure shaft position or velocity to determine the restart or reverse timing. If this is impractical, the only alternative is to wait until the rotor settles after this last step. This is a question of damping and is discussed in the previous section.

Ramping Characteristics

In the ramping mode, the motor is accelerated and decelerated with fixed load applied. The ideal ramp would be generally exponential for acceleration, and a reverse exponential for deceleration (Figure 10-16). In practice, a simple and widely used compromise is the quasi-linear ramp, 1.5 to 2 time constants of an exponential

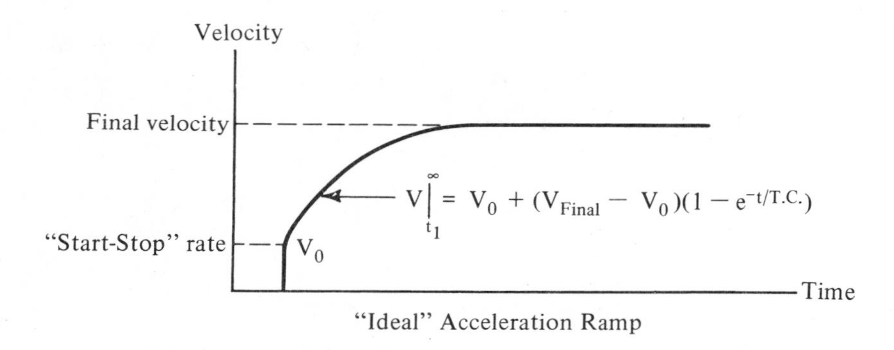

"Ideal" Acceleration Ramp

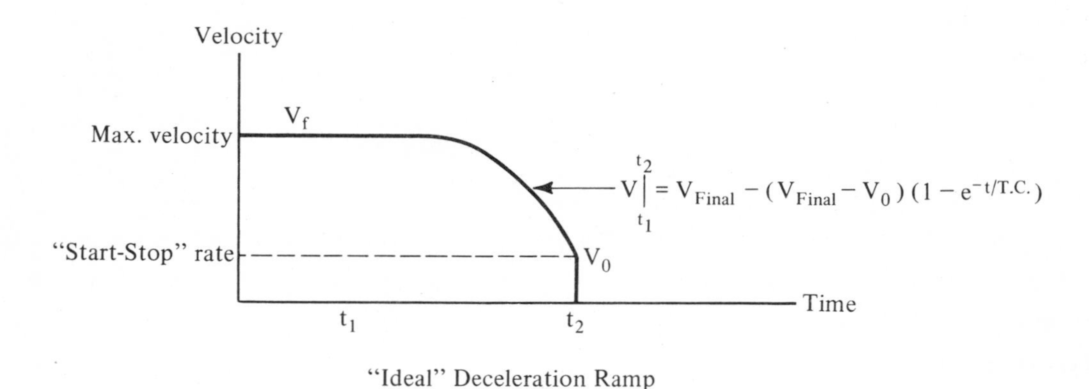

"Ideal" Deceleration Ramp

Figure 10-16. Ideal acceleration ramp and deceleration ramp.

curve, which was used in obtaining data for this manual. Also, a single ramp slope value, generally appropriate to the motor under test, was used. The data obtained in this manner are interesting in several respects (Figure 10-17). First, a certain amount of load inertia aids in running close to the resonance at 900 steps/second. Since the motor must start/stop at a given low frequency (in this case $f_{LO} = f_{HI}/11.5$),

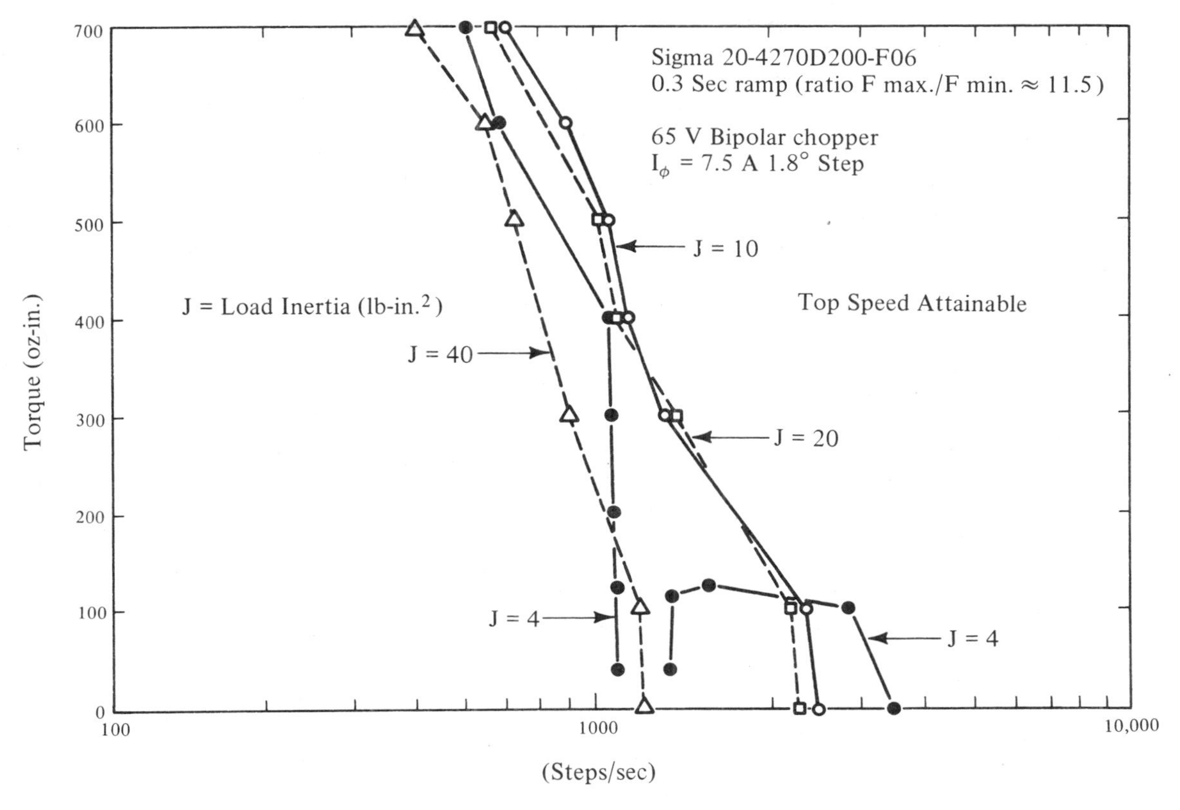

Figure 10-17. Ramping curves.

increasing inertia limits the frequency of operation. The half-step drive (discussed in the following section) aids performance where resonance is a limiting factor, but is somewhat inferior in low frequency torque compared to the full-step drive.

10-3 DRIVE CIRCUITS

Drive circuit design is one of the most important aspects of a stepping motor system. The overall system performance is heavily dependent upon the drive system, not only in available power delivered to the load, but also in such parameters as efficiency, power dissipation, and cost.

A stepping motor drive system, diagrammed in Figure 10-18, accepts a drive signal and converts it to the proper format for driving the motor windings. A power amplifier drives the required current through the windings, and a power return system removes the current from the windings at the termination of the step.

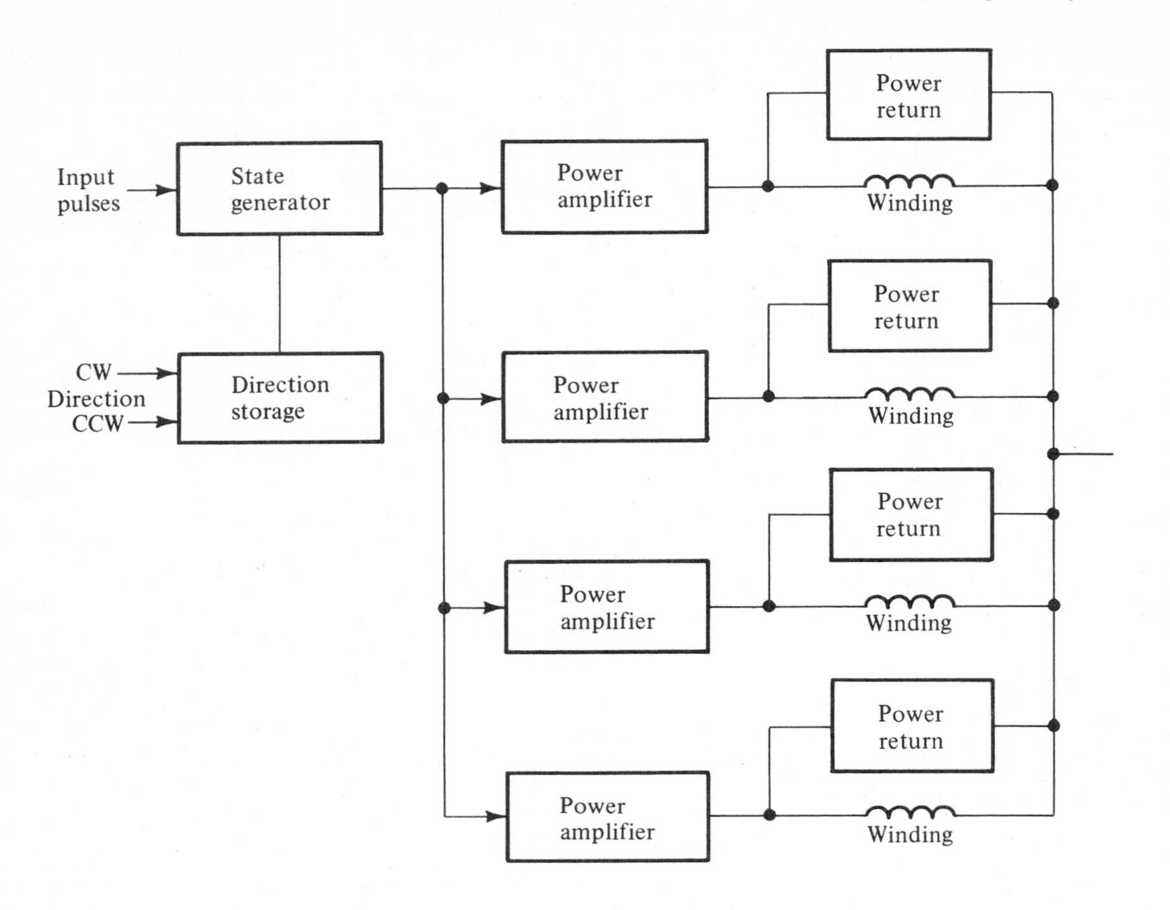

Figure 10-18. Block diagram of step motor driver.

State Generator

As previously discussed, the motor rotates in response to a changing pattern of interactions between the rotor magnetic field and the stator magnetic fields. The function of the State Generator is to create the proper sequence and pattern of states in response to a serial pulse train.

There are two major sequences which will cause the motor to step. One is called "wave" drive, in which only one set of stator poles are energized at a time. The other sequence energizes both poles set simultaneously and is called "two-phase" drive. Either of these will cause an N° per step motor to step by increments of N°, but there is an N°/2 spatial displacement of the stator and rotor between the two sequences. The idealized tooth alignment is shown in Figure 10-19a and b.

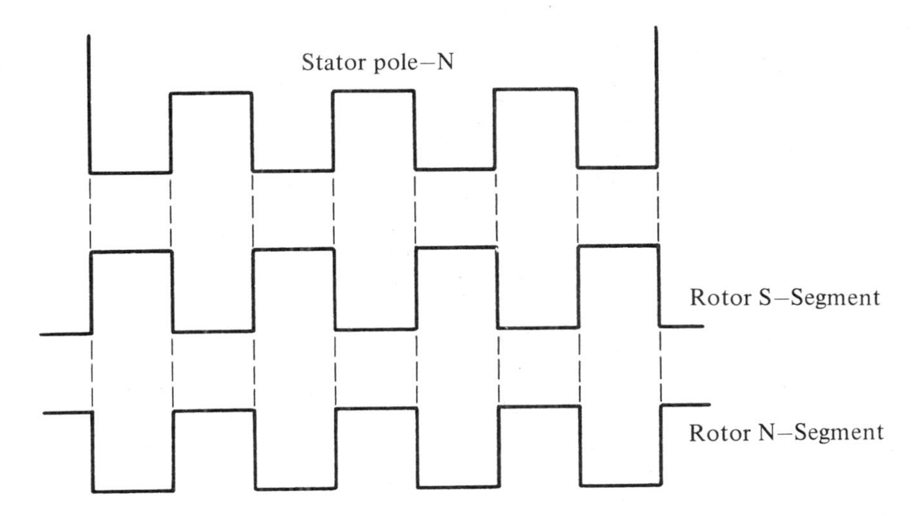

(a) Idealized tooth alignment for wave drive

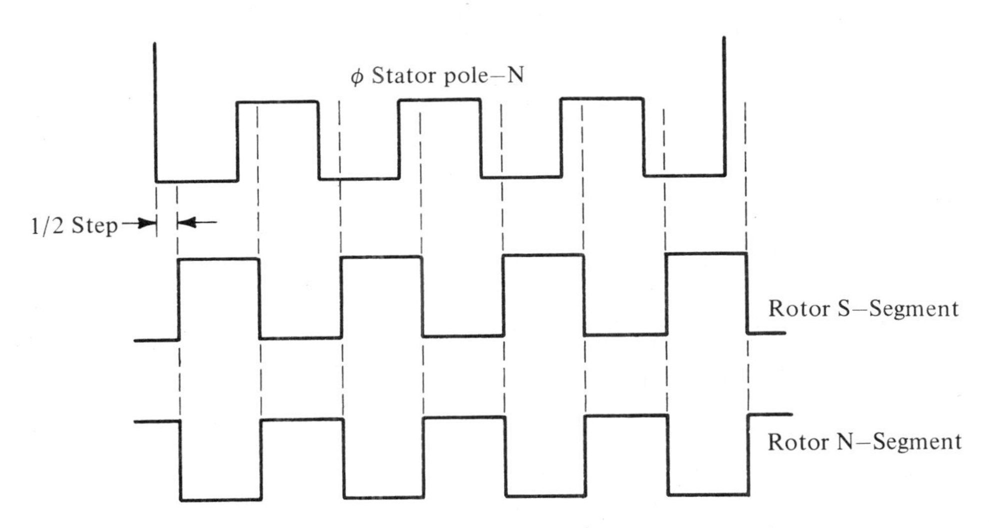

(b) Idealized tooth alignment for two-phase drive

Figure 10-19. Idealized tooth alignment.

Wave Drive

In this sequence, the required currents are shown in Figure 10-20a. The A_1 current energizes all phase A poles to create a North pole at the stator pole teeth. Current B_1 then generates a North pole at the phase B pole teeth. Similarly, the A_2 current generates a South pole at the pole A teeth. Finally, B_2 creates a South pole on the B phase poles. During this sequence, the rotor advances to align the rotor and stator teeth. The logic required to generate these waveforms is shown in Figure 10-20b.

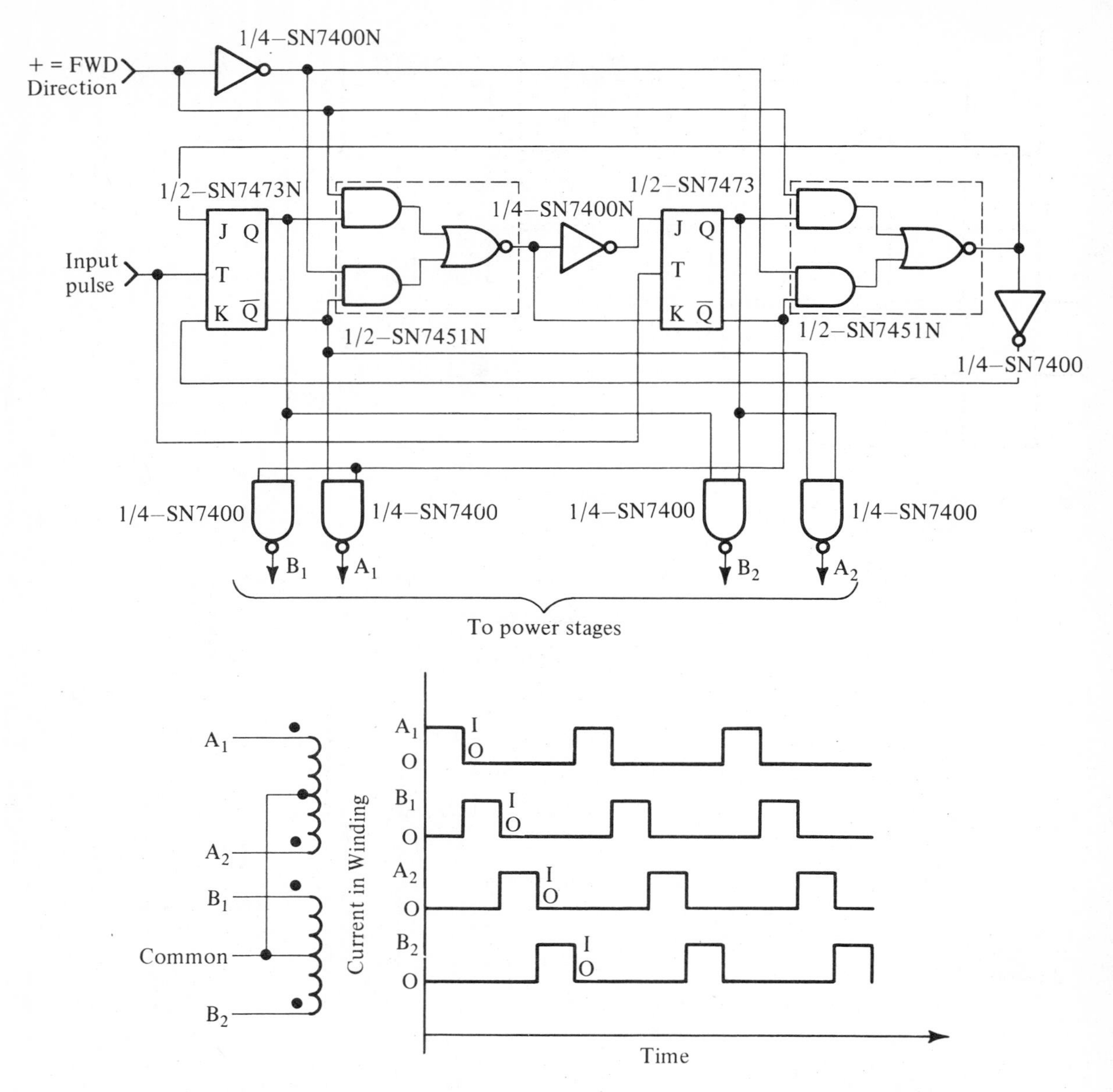

Figure 10-20. (a) Current waveforms of unipolar wave drive.

(b) State generator for wave excitation.

Two-Phase Drive

It can be shown that the ampere-turns on the stator poles per watt of input power are 41% higher if both the ØA and ØB poles are driven. This is done by driving the four windings A_1, A_2, B_1, and B_2 two at a time. Unfortunately, the torque does not increase by 41%. The torque per ampere-turn is higher for the wave drive tooth alignment (Figure 10-21) than for the two-phase alignment. However, there is a net gain in torque per watt of about 20%.

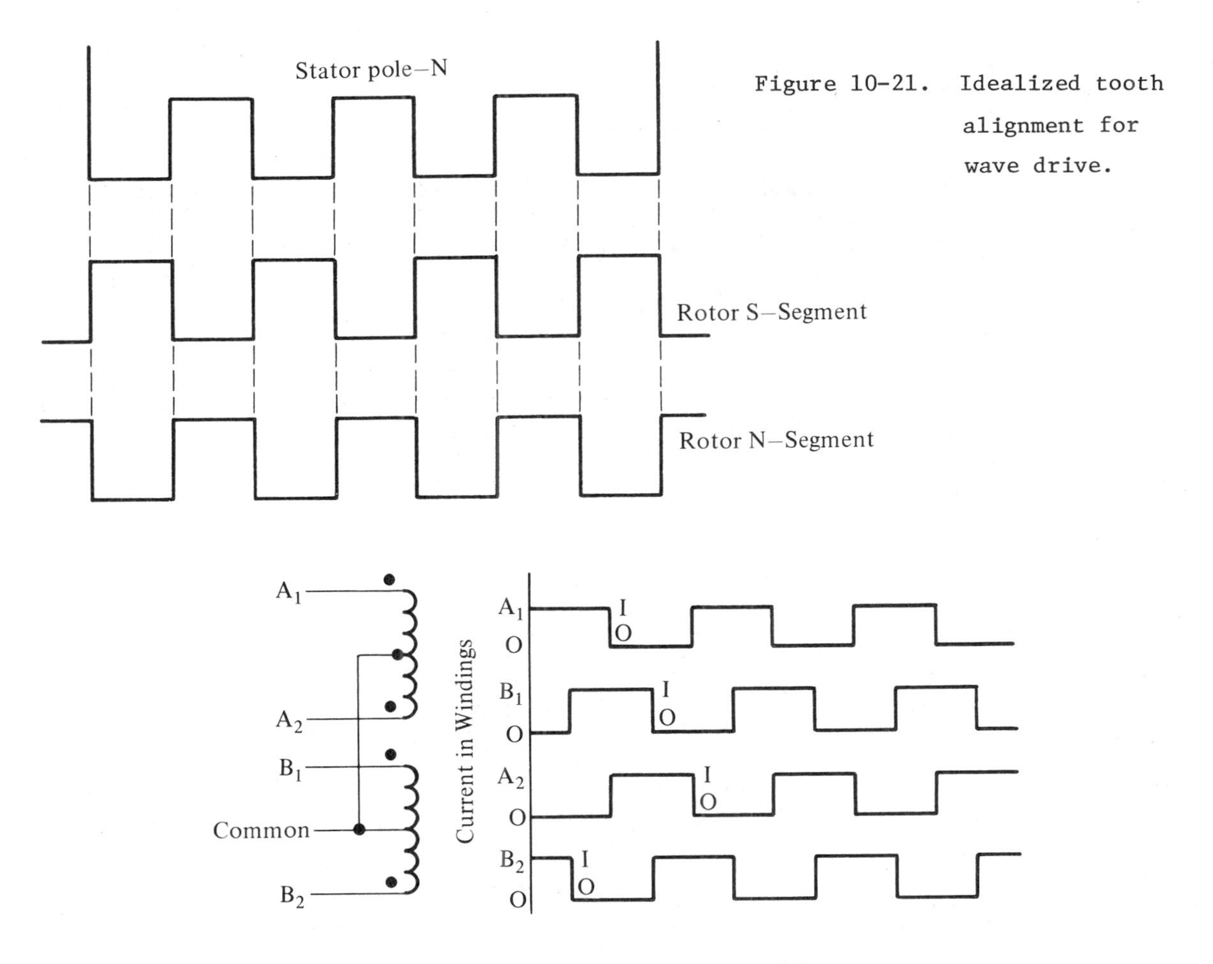

Figure 10-21. Idealized tooth alignment for wave drive.

(a) Unipolar two-phase drive

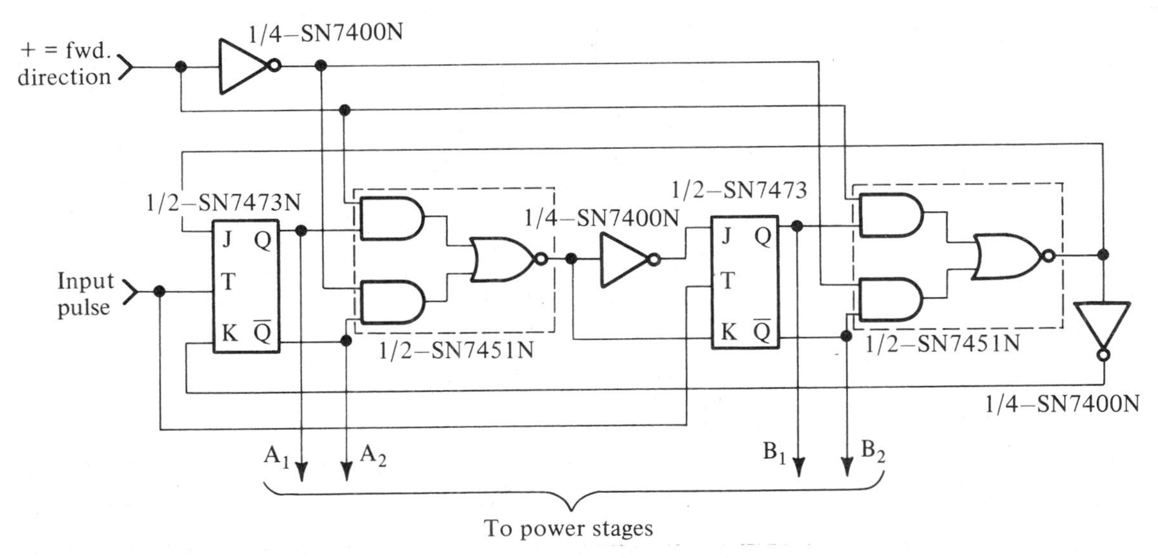

Figure 10-22. (a) Current waveforms for unipolar two-phase drive.

(b) State generator for two-phase excitation.

The sequence of phase currents is shown in Figure 10-22a. Current in A_1 generates a North pole in the A stator poles, while current in A_2 generates a South pole. Likewise B_1 currents create a North pole in the B stator poles and B_2 creates South poles. All four combinations of current in two windings at the time are generated and give rise to four motor steps. The pattern repeats every four steps. The logic is shown in Figure 10-22b. Note that the logic is slightly simpler than the wave drive. For this reason, and the increased performance, two-phase drive is more commonly used than wave drive.

Stepping Motor Operation with Differing Drive Sequences

If four windings of a stepping motor are energized as shown in Figure 10-20a (so-called wave drive), the tooth alignment will be as shown in Figure 10-21, and successive steps will have the normal spacing (1.8° in the case shown). Alternatively, the drive sequence shown in Figure 10-20b (two-phase drive) will result in the tooth alignment as shown, with normal 1.8° steps. If a drive is used that alternates between the wave type and the two-phase drive, the step motor output will be 1/2 the normal step, or 0.9° in the illustration (Figure 10-23). All three of these drive

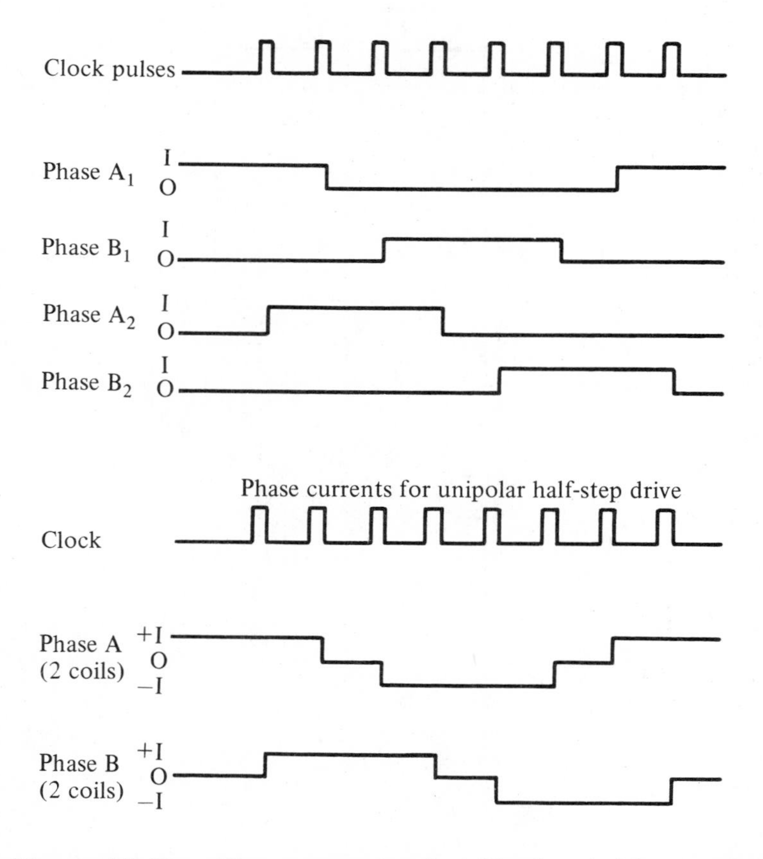

Phase currents for unipolar half-step drive

Figure 10-23. Phase currents for bipolar half-step drive.

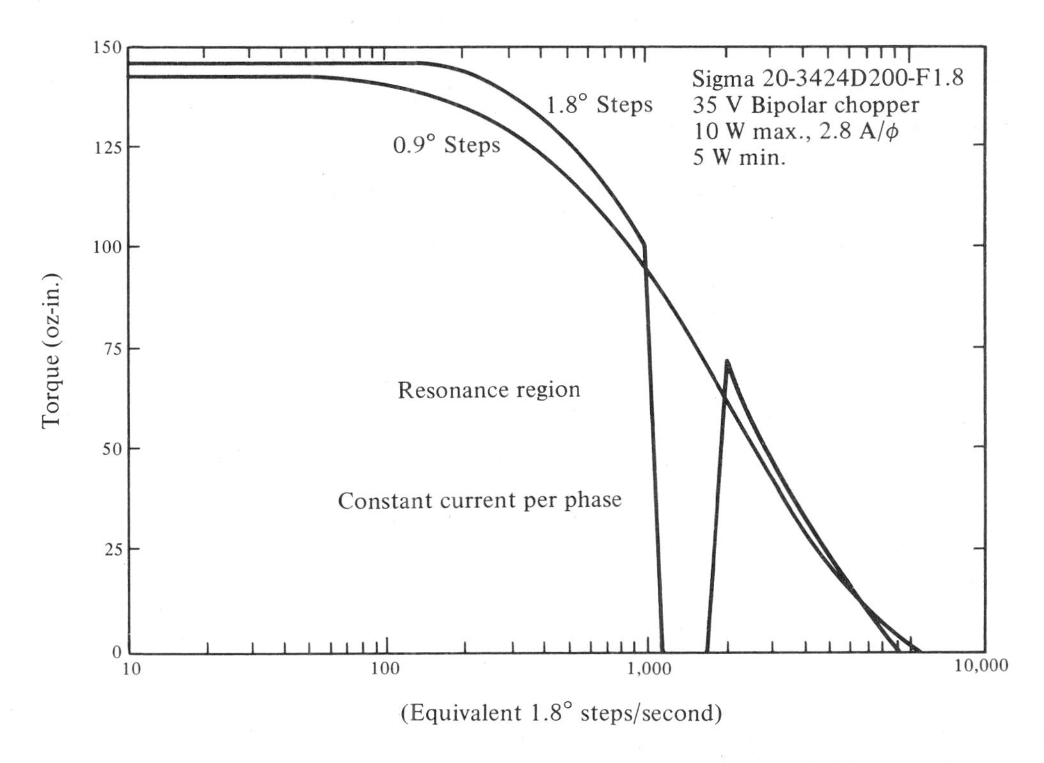

Figure 10-24. Torque-speed curves of half-step drive and full-step drive.

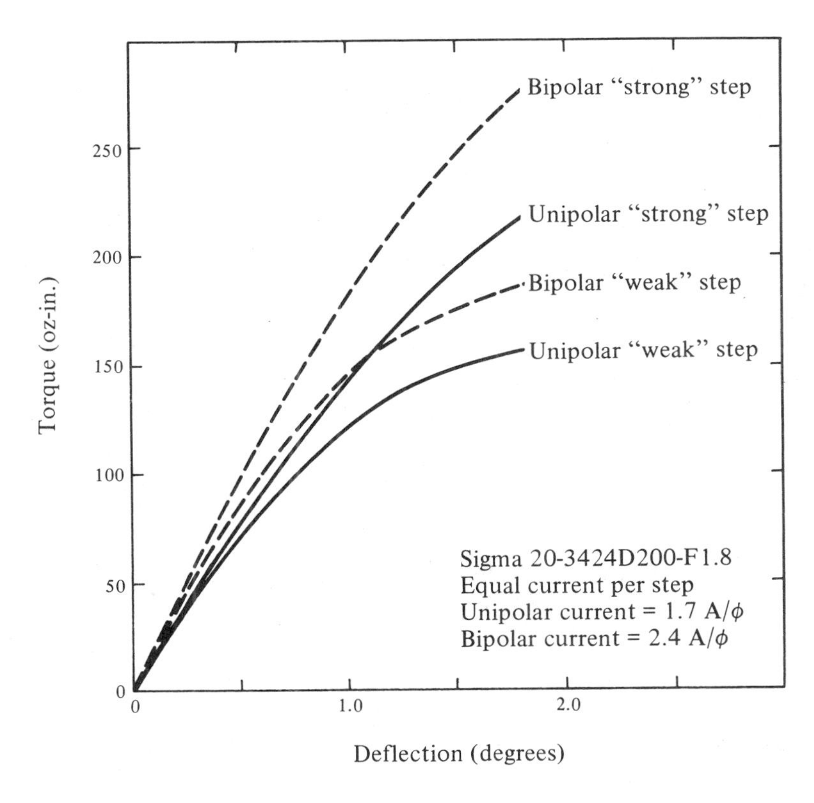

Figure 10-25. Comparison of stiffness on half-step drive.

sequences are useful. Two-phase is the most widely used "normal" drive, since it is somewhat more efficient than wave drive. Half-step drive has some interesting applications in reducing resonance problems. Compare the curves of Figure 10-24 where a motor was run on half-step versus full-step drive. The difficulty with half-step drive is that the two types of steps have somewhat different characteristics (particularly stiffness) due to the differing magnetic alignment (Figure 10-25).

The logic for half-step drive is shown in Figure 10-26 and can be used with either bipolar or unipolar drivers. For unipolar drivers, A_1, A_2, B_1, and B_2 represent individual winding drivers. In the case of bipolar drivers, a positive voltage at A_1 represents the signal that drives a positive current into the A-phase winding pair, and a positive voltage at A_2 drives a negative current into the A-phase winding pair. The same convention holds with B_1 and B_2 with respect to the B-phase winding pair.

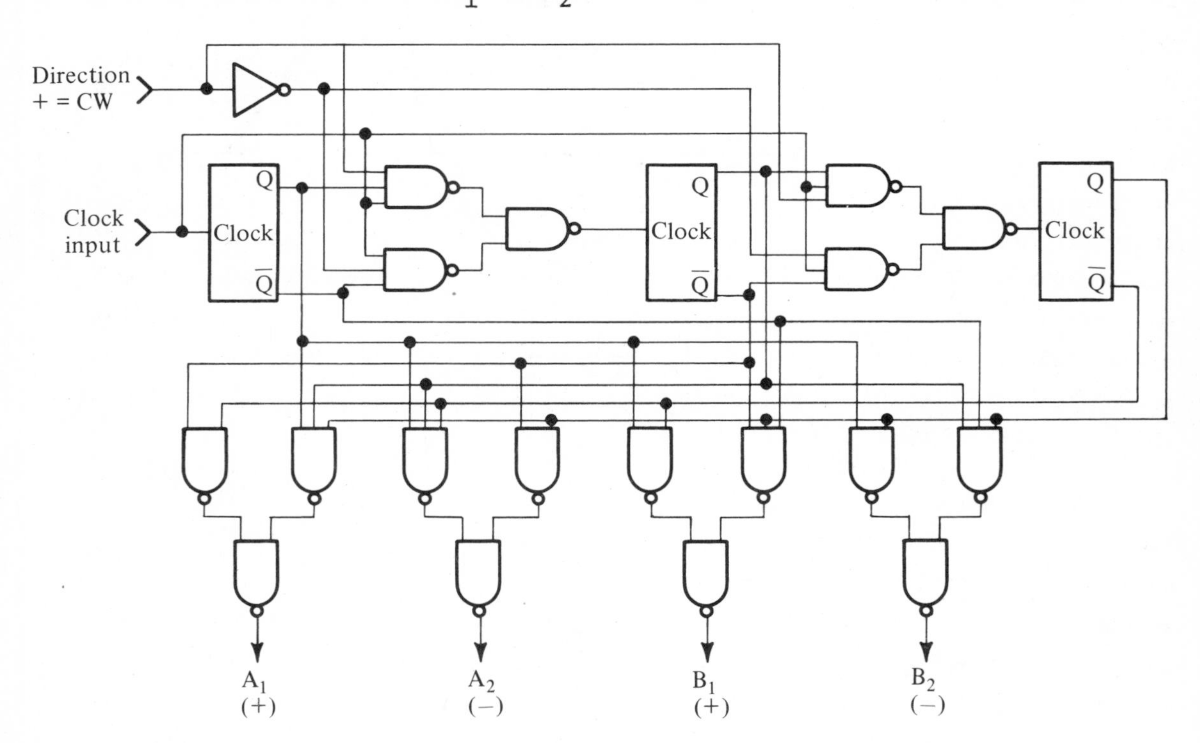

Figure 10-26. Half-step logic diagram, unipolar driver, bipolar current sense.

There is an important point to bear in mind when using a half-step drive system with constant current per winding. There are cases where two windings (or winding sets in bipolar) are energized, and cases when only one winding is energized. When the two windings are energized the motor power is at a maximum and available torque is at a maximum. When only one winding is energized the motor power is reduced by half. This gives rise to a "strong" step and a "weak" step. Figure 10-25 shows the relative magnitude of the steps on a torque displacement curve. In terms of positional accuracy, if the load torque is less than 30-40% of the maximum torque, the stiffness of either step is about the same around the origin. Above this, the weak step is clearly weaker. This difference in stiffness appears as a loss of stepping torque at low frequencies, but as about equal performance at high frequencies. Figure 10-24 shows the speed-torque curves of another motor with full-step and half-step logic using a bipolar drive. The results are similar but lower in torque for unipolar drive.

Power Amplifiers

The step motor winding may be considered as an inductance in series with a resistance. Another effective parallel resistance appears across the winding representing load power, but its effect is noticeable only with high efficiency drivers at very high speeds, and it generally has a negligible effect on driver design. The time constant of the winding, L/R, is typically of the order of 10 milliseconds. Therefore, if a voltage source is impressed across such a winding, 95% of full current is reached in 3 time constants, or 30 milliseconds. The usual drive systems apply current to each winding at 1/4 of the input pulse rate, so that each winding would receive 95% of full current at an input rate of $4/30 \times 10^{-3}$ = 133 steps/second. In most cases, such a severe limitation on motor speed is not acceptable. For instance, a small 200 step/revolution motor which is capable of only 200 steps/second slew speed when driven with a voltage source can deliver 15,000 steps/second slew speed when properly driven.

Stepping motor drive requirements vary from a few volts at perhaps 50 ma to 100 volts or more at 20 amperes. The requirements imposed upon a driver depend not only upon the particular stepping motor involved, but also upon the speed and torque requirements of the system. In general, the windings of stepping motors may be adjusted to differing impedance levels to trade voltage for current. The standard windings reflect what is believed to represent a reasonable match to available semiconductors. In any event, the current rate of rise in a winding is proportional to V/L, so that higher voltages will yield better high speed performance up to the point where resonance problems (or economics!) prevent further improvement. It is not unusual for the source voltage for pulse initiation and termination to be 10, 20, or 30 times the steady state motor voltage. The principal methods for control of this "overdrive" voltage so as to limit steady state current to the correct value will be discussed in this section.

The basic problem of drive circuitry for high speed operation of stepping motors is to provide a high voltage to move current into and out of the winding at pulse transition times, and a low voltage to sustain only the correct current during the steady-state portion of the current pulse. Basically, the required waveform is generated in four principal ways (Figure 10-27):

A. Series resistance is added to the winding or windings.

B. A "chopped" waveform is generated. A high voltage is applied to the windings at the beginning of the step to allow the current to build up, and then the high voltage is time modulated into pulses during the remainder of the step to hold motor current at its correct value.

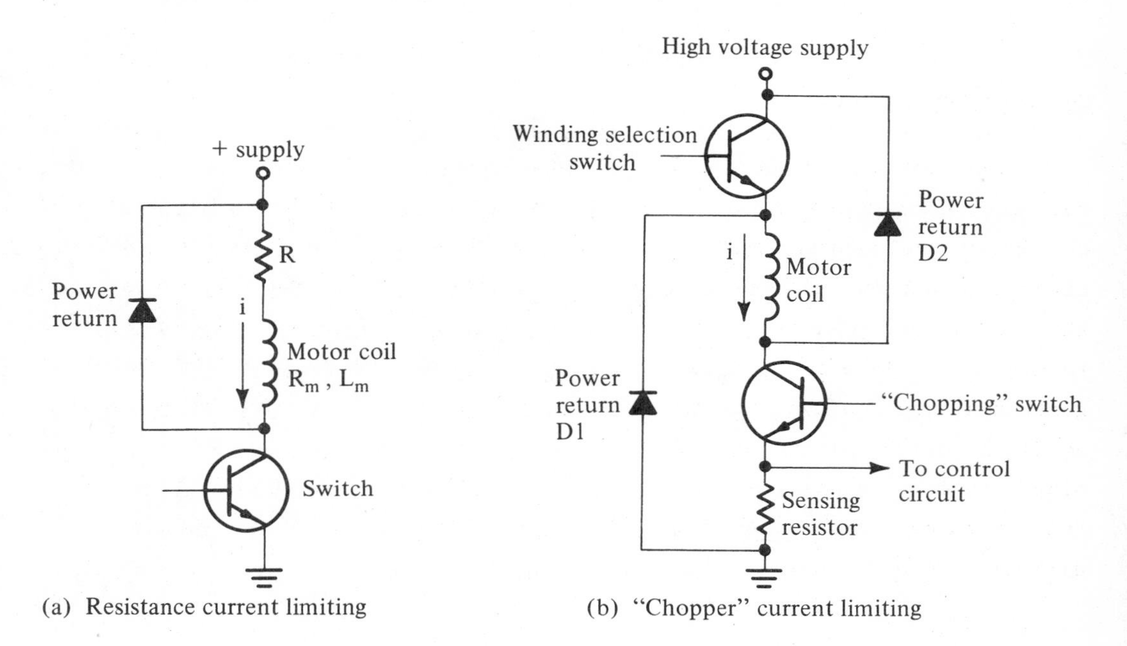

(a) Resistance current limiting

(b) "Chopper" current limiting

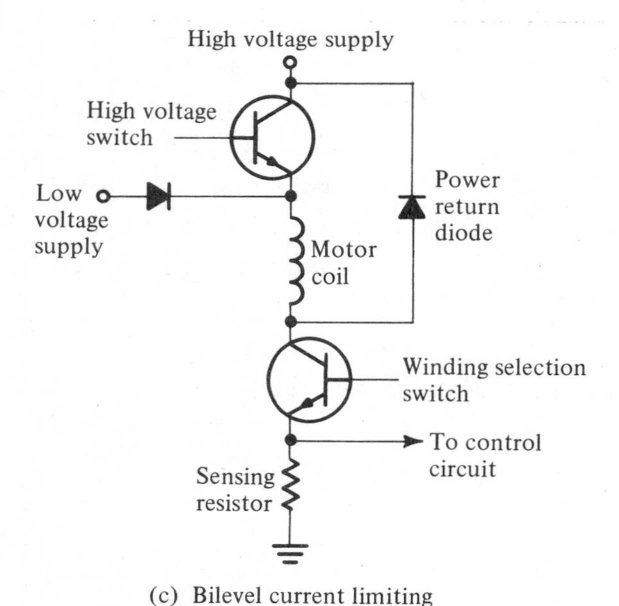

(c) Bilevel current limiting

(d) Variable voltage current limiting

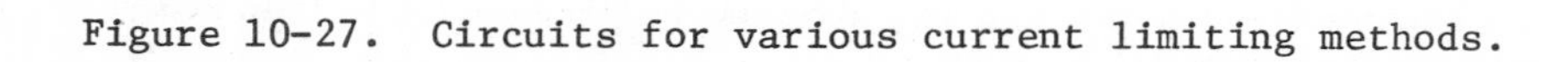

Figure 10-27. Circuits for various current limiting methods.

C. Bilevel drive involves a high voltage source used during pulse initiation and termination, with a low voltage source during the flat top of the pulse.

D. A programmed voltage source is used (phase controlled SCR's, constant current transformers, etc.) to deliver the voltage required to maintain the current constant regardless of pulse repetition rate.

The advantages and limitations of these approaches will be examined in detail in the following sections.

Unipolar Resistance Current Limiting

The most widely used method of driving stepping motors is simply to add one or more resistors in series with the motor winding (or windings), and raise the supply voltage to yield normal motor current under steady state conditions. Some typical circuits are shown as Figure 10-28. For example, in the circuit of Figure 10-28(c), the winding time constant, L/R_M is reduced to $L/R_M + R_S$ on both pulse initiation and termination. However, the actual current rate of rise also depends upon the back EMF generated, and consequently speed and load. In circuits with shared resistors, the available drive voltage at pulse initiation is lower than the power supply voltage, since the other winding currents result in a voltage drop through the common resistor. Pulse termination is aided by this fact, however, since the voltage drop is in the direction to aid removal of current.

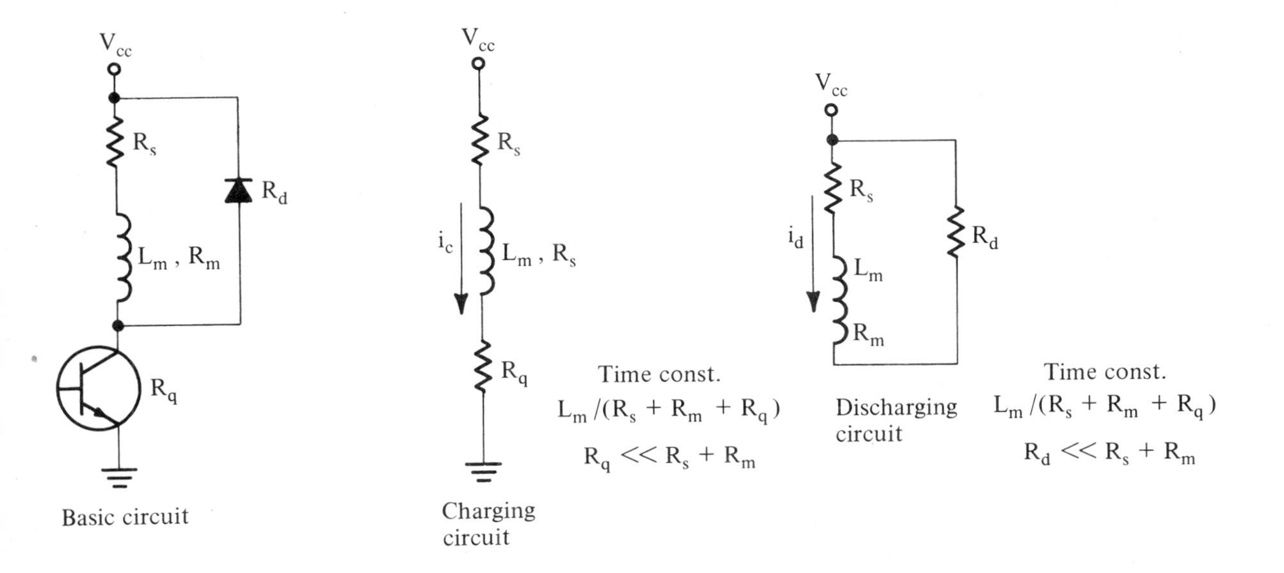

Figure 10-28. Charge and discharge circuit time constants.

Unipolar resistance limited drives are simple and reliable, and are frequently adequate where speed and torque requirements are low. However, there are several inherent problems:

A. If voltage overdrive is raised to improve high speed performance, the system efficiency is very low. For instance, a small (10 watt) motor may use a driver dissipating 200 watts (at 20 times overdrive).

B. The windings in the motor are not completely utilized, since the current duty cycle in the windings is 50%. Complete utilization of the windings (bipolar drive) considerably increases motor output.

C. The voltage applied to the motor windings is a decaying exponential, and the current is a rising exponential. For a given supply voltage, the current rate of rise (and the high speed performance) is considerably less than in a system where the full supply voltage is available until the desired current is reached.

Bipolar Drivers

As previously noted, there is an improvement in performance between wave drive and two-phase drive when two of four windings are used instead of one of four. Similarly, if all four windings are energized all the time, a further theoretical increase of 41% in the ampere-turn per watt factor can be achieved. Again, the full 41% increase in torque is not reached because of the non-linearity of the B-H curve of the iron. However, 25 to 30% improvement in torque per watt is possible.

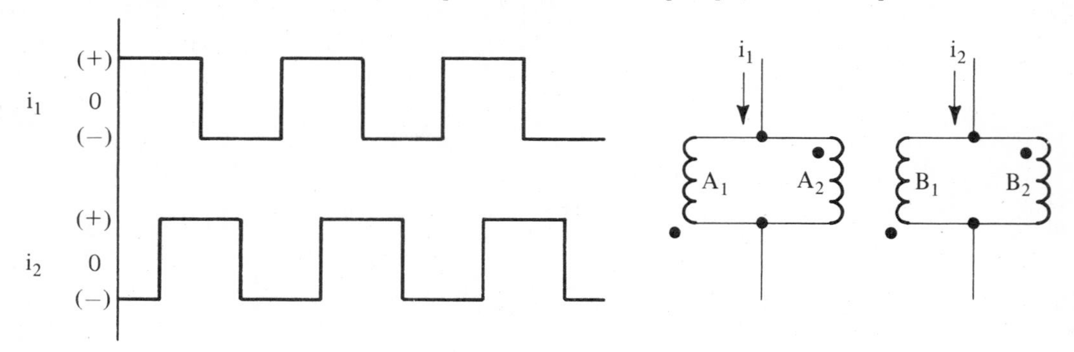

Figure 10-29. Bipolar two-phase drive.

If all 8 wires of the four windings are separate and are cross-connected as in Figure 10-29, currents in A_1 and A_2 will create the same sign of magnetic pole in the A phase stator poles. When the windings are driven with the bidirectional currents of Figure 10-29, alternate magnetic poles are created in the stator as in a unipolar drive, except that the two-phase windings aid each other. The logic required is the same as the unipolar two-phase logic shown in Figure 10-22b. The improved motor efficiency due to bipolar operation may be realized as increased torque and speed for a given motor wattage (20%-40% as compared to unipolar R/L drive), or as a 50% reduction in motor power while maintaining torque and speed characteristics equal to that of the unipolar case.

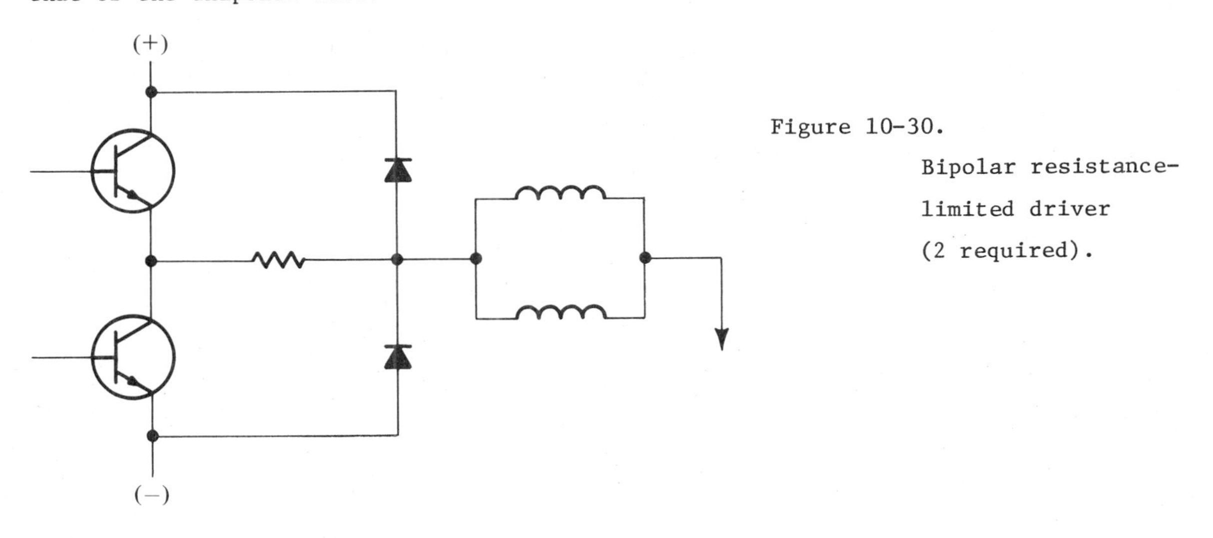

Figure 10-30. Bipolar resistance-limited driver (2 required).

One practical method of obtaining the required drive waveforms is the split supply system shown in Figure 10-30. Here equal value positive and negative supplies are required, but the total number of switches and the overall complexity of the system are approximately the same as for a simple unipolar driver. Because the power switches are essentially in series across the supply voltage, it is necessary to incorporate delay circuitry to prevent one switch from being turned on while another one is conducting. Alternatively the power resistors may be split, as shown in Figure 10-31, so that any slight overlap of switch conduction will not be harmful.

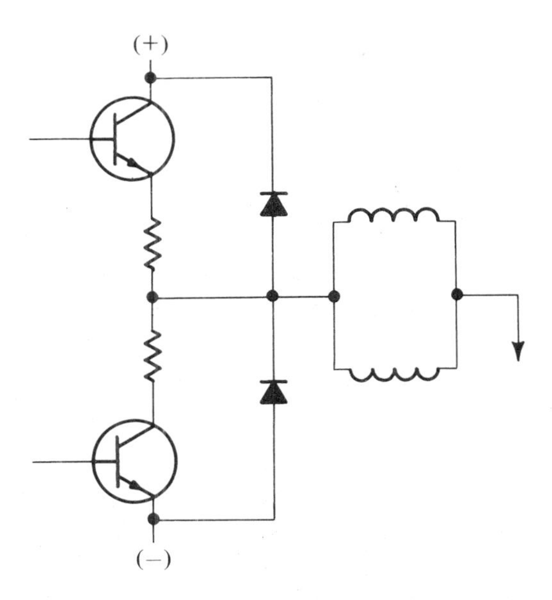

Figure 10-31. Bipolar driver-split resistor.

Another version of bipolar drive uses full bridge operation, and so requires only a single supply voltage. This convenience is obtained at the expense of doubling the required number of power switches to 8 per motor. In this system, the same switch timing considerations apply as in the split supply case, and the use of delay networks can be avoided in the same way, by using extra power resistors.

High Efficiency Systems

Series resistance padding of windings to reduce time constants is inefficient for small stepping motors; for large motors the power dissipated makes resistance padding completely impractical. For instance, a large 200-step motor may require 7.5A/Ø with 0.46 ohm windings. A 30:1 overdrive ratio, necessary for moderately good performance, would result in resistor dissipation of 1500 watts. A high efficiency driver system would dissipate perhaps 50 watts, while providing available motor power output of about 3-4 times as much as the resistive system.

An important consideration in designing high efficiency driving systems is the method used to remove current from the windings at the end of the driving pulse. In low efficiency resistance limited drivers, the current usually is returned through the charging resistance, so that the time constant is the same on discharge as it is on charge. In high efficiency systems, it is desirable to provide a return path to the highest available voltage without merely dissipating the energy stored in the winding, since losses can be substantial at high speeds. For instance, in the case of the 20-4770, the energy stored in the winding, $LI^2/2 = 33 \times 10^{-3}$ watt seconds. At a stepping rate of 3000 pulses/second, this corresponds to a power of $3000 \times 33 \times 10^{-3} = 100$ watts. Further, it is vital to high speed operation to remove the current

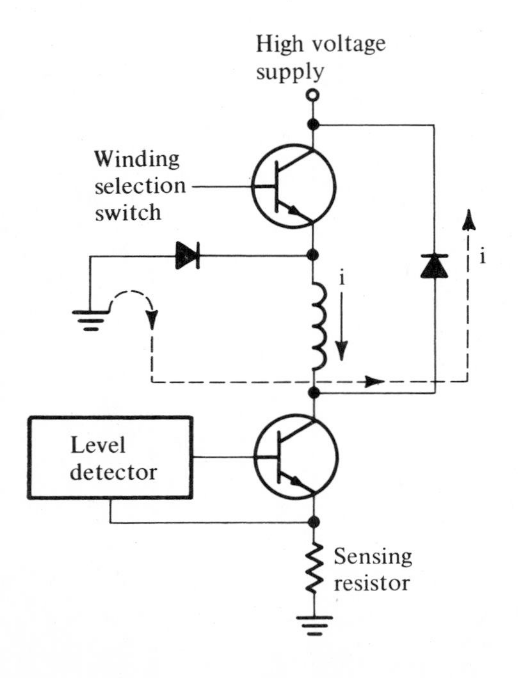

Figure 10-32. Unipolar chopper phase driver (4 required).

from the winding as rapidly as possible. A switching circuit to accomplish rapid discharge of the winding is shown as Figure 10-32. The motor winding is charged with both switches closed. When the switches are opened, current is returned to the supply. This system requires two switches per winding, or a total of eight switches, for unipolar operation. The same system designed as a bridge allows bipolar operation, and a consequent improvement in motor output, with the same total switch count (Figure 10-33). For this reason the basic bridge design is used in several high efficiency systems described in the following section.

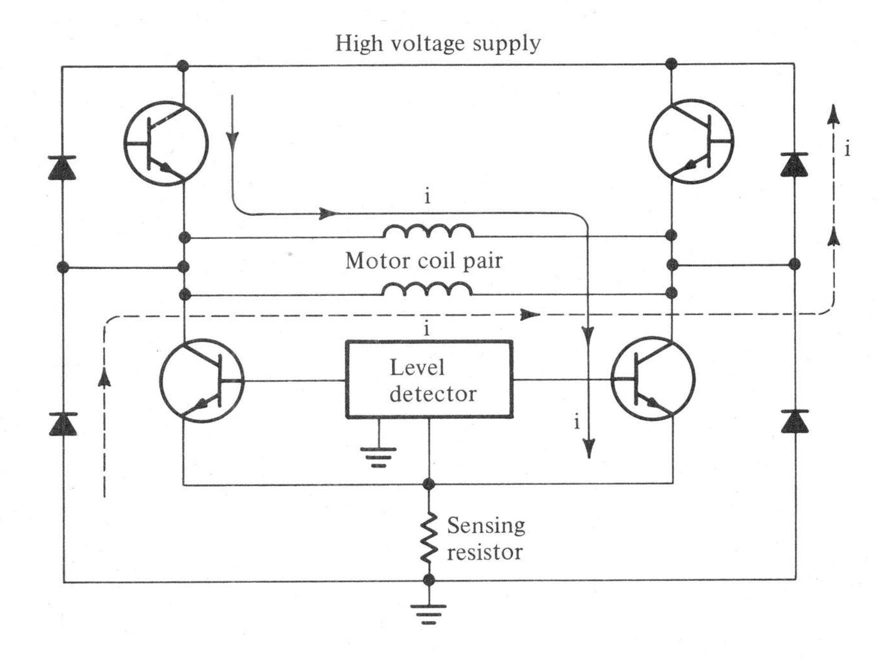

Figure 10-33. Bipolar chopper phase driver (2 required).

Bilevel Drive

One system for obtaining a high current rate of rise efficiently involves the use of two supplies - a high voltage supply for pulse initiation and termination, and a low voltage unit to supply sustaining current during the pulse flat top. One approach to this type of circuit is shown as Figure 10-34. The high voltage supply is turned on until the motor current reaches the operating level, when the high voltage switch is turned off. At the end of the pulse, the lower switch is turned off, and the current in the winding is returned to the high voltage supply as indicated in Figure 10-34. This action results in rapid pulse initiation and termination, with no inherent switching losses. Further, the current rate of rise is much higher than in the resistance-limited case, where the voltage is a falling exponential. This is illustrated in Figure 10-35, where the two cases are compared for equal supply voltage.

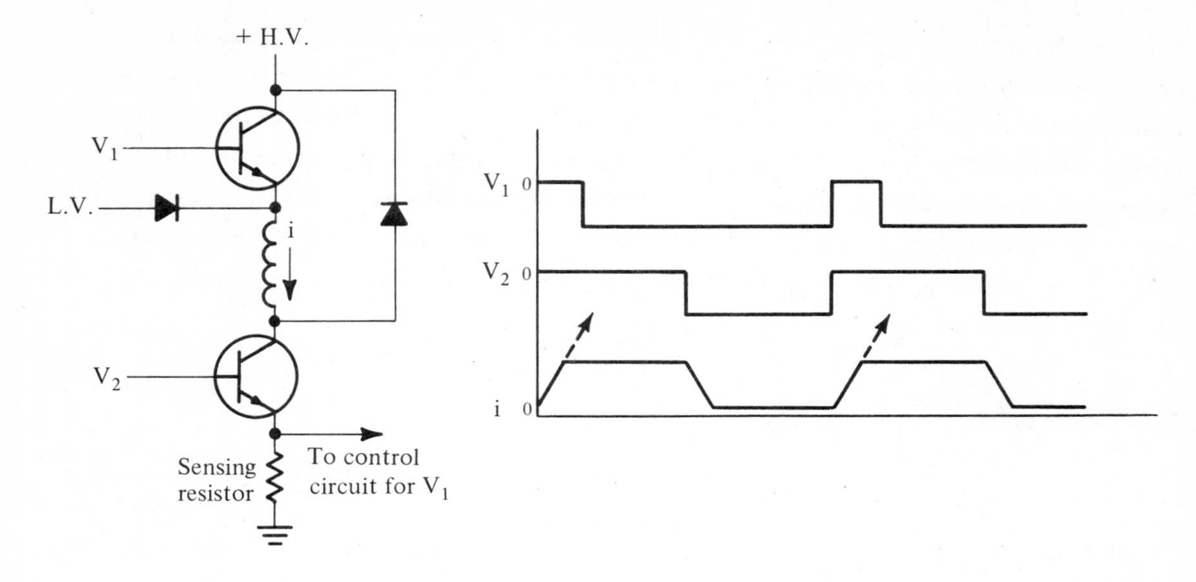

Figure 10-34. Bilevel drive.

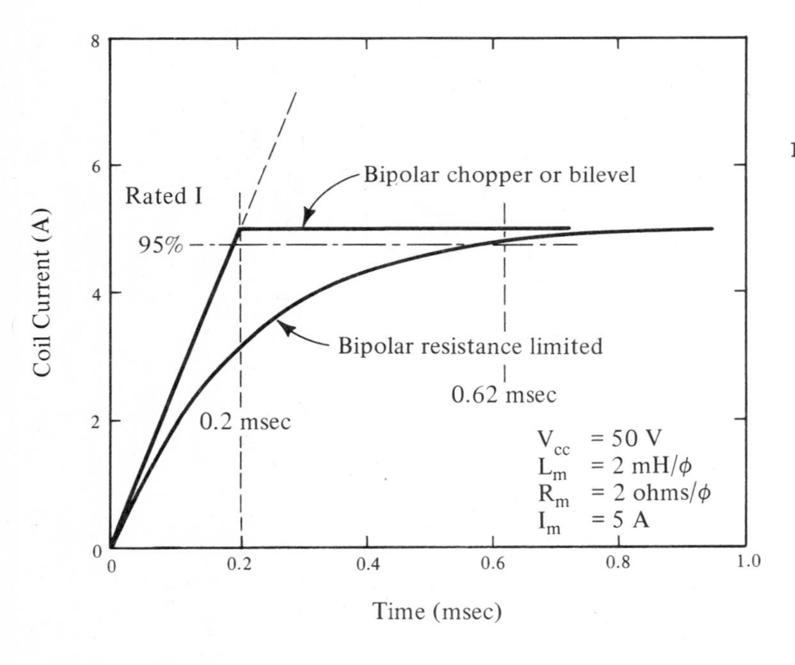

Figure 10-35. Comparison of current rise times.

Current Fed Drive Circuitry

Regardless of the basic drive system (unipolar or bipolar) high driver efficiency is attainable if the voltage source can be programmed to deliver the required current at all stepping rates. A convenient method of programming voltage efficiently is in the initial conversion from the power line, using such methods as ferro-resonant transformers, saturable reactors, and SCR phase control. All of these methods are inherently slow in response (≈ 50-200 milliseconds), since they are limited by the necessity of filtering line frequency (a three-phase system is considerably faster than the conventional single-phase approach because of the inherent reduction in carrier lag). This discussion will focus on the SCR phase control method, but it is generally applicable to the other schemes.

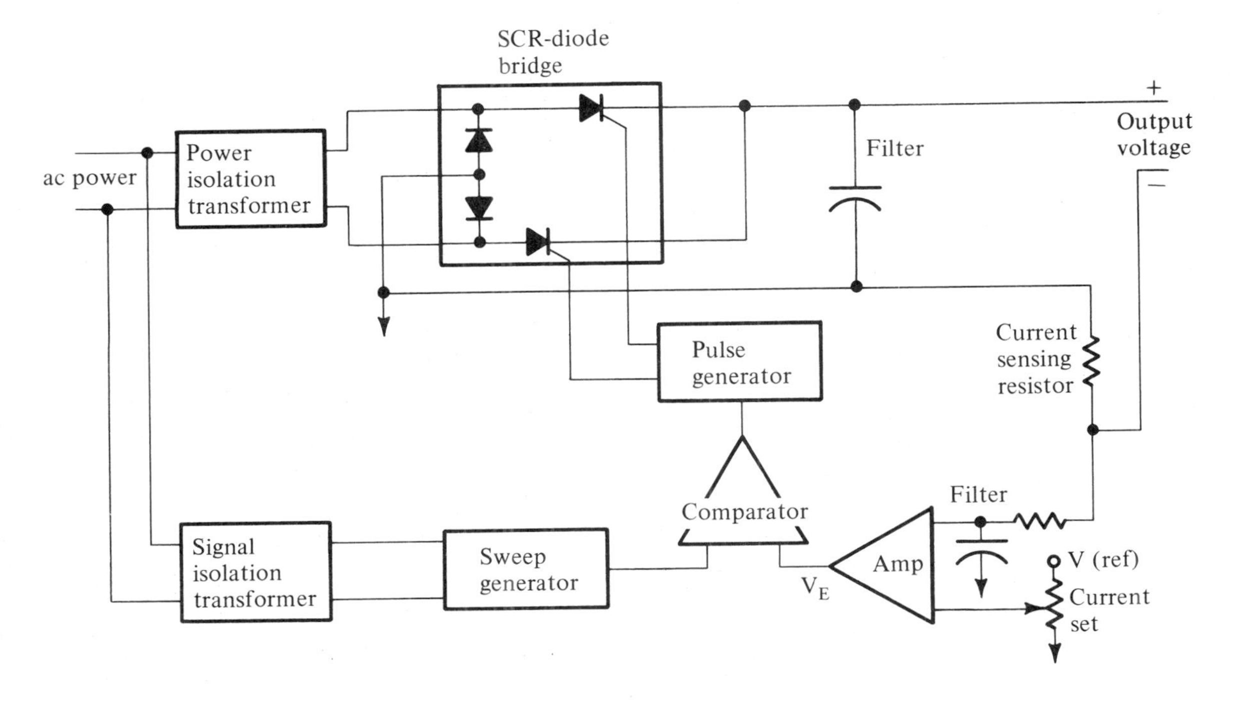

Figure 10-36. Block diagram of SCR-controlled constant-current source.

A basic block diagram of an SCR phase control system is shown in Figure 10-36; response times to a current step are also indicated. This type of motor drive system has the following advantages and limitations:

1. Relatively long response times limit the start-stop performance. When the system is at rest, a low percentage of the supply voltage is applied to the driver. For instance, in a system that will deliver 100 volts at high speeds, the voltage may be 3 volts when the motor is stopped. Since the voltage can rise only at a relatively slow rate, start-stop operation is severely limited. However, this limitation may not present much difficulty in a ramped system, where the motor is accelerated to a high speed after starting at a low speed. In this case, the time to accelerate to high speed is impaired little by the driving system, since the response of motor and load inherently limits acceleration.
2. Care must be exercised in system design to prevent excessive current flow in the switching system. This may occur if the system is suddenly switched

to a low speed after a period of high-speed operation. In this case, the system voltage is high to allow high-speed running, and the filter capacitor is fully charged. If the motor is suddenly switched to low-speed operation, the full capacitor charge will be dumped through the switches, with possible excessive current flow or switch dissipation. In normal ramped operation, of course, the motor speed is decelerated gradually and this problem does not arise.

3. This mode of operation is highly favorable with regard to resonance control. Since only enough voltage is applied to force rated current through the stepping motor windings, the problem of rotor overshoot and resonant modes is greatly reduced. Motor operation is smooth, and it is not necessary to avoid or rapidly traverse critical resonance regions. Another result of this mode of operation is that the low frequency torque enhancement described does not occur, since high overdrive does not exist at low speeds. In the block diagram of Figure 10-36, it is important to note that the average current to the drive system is the factor that is regulated. Because of this, as the frequency of operation is increased, the average voltage is likewise increased to continue forcing the set level of current through the motor winding, until the voltage limit of the supply is reached. At high speeds, an appreciable fraction of the motor current is returned to the supply by the recirculating diodes, but this factor is effectively ignored by the regulating system which continues to adjust the voltage to the system on the basis of average drain from the supply. The result of this type of operation is that the high speed torque is enhanced, but the RMS current through the windings, the winding RMS voltage, and consequent motor dissipation, are increased during high-speed operation. This type of operation results in the greatest power output from a given stepper, but thermal factors, including duty cycles, motor construction, heat sinking, and ambient temperatures, must be considered carefully in the design of a system of this type.

Chopper Drive Circuitry

Another high-efficiency system that offers certain advantages in stepping motor driving is the so-called "chopper" system for current limiting by means of voltage modulation. With this arrangement, the full high voltage supply (10 to 20 times motor voltage) is applied to the motor winding until the correct current level is reached (Figure 10-37). The voltage is then switched off, and the current is allowed to circulate in the motor winding. When the current decays to a predetermined level, voltage is again applied to drive the current back to the correct level. This cycle is continued throughout the driving pulse time. At the termination of the

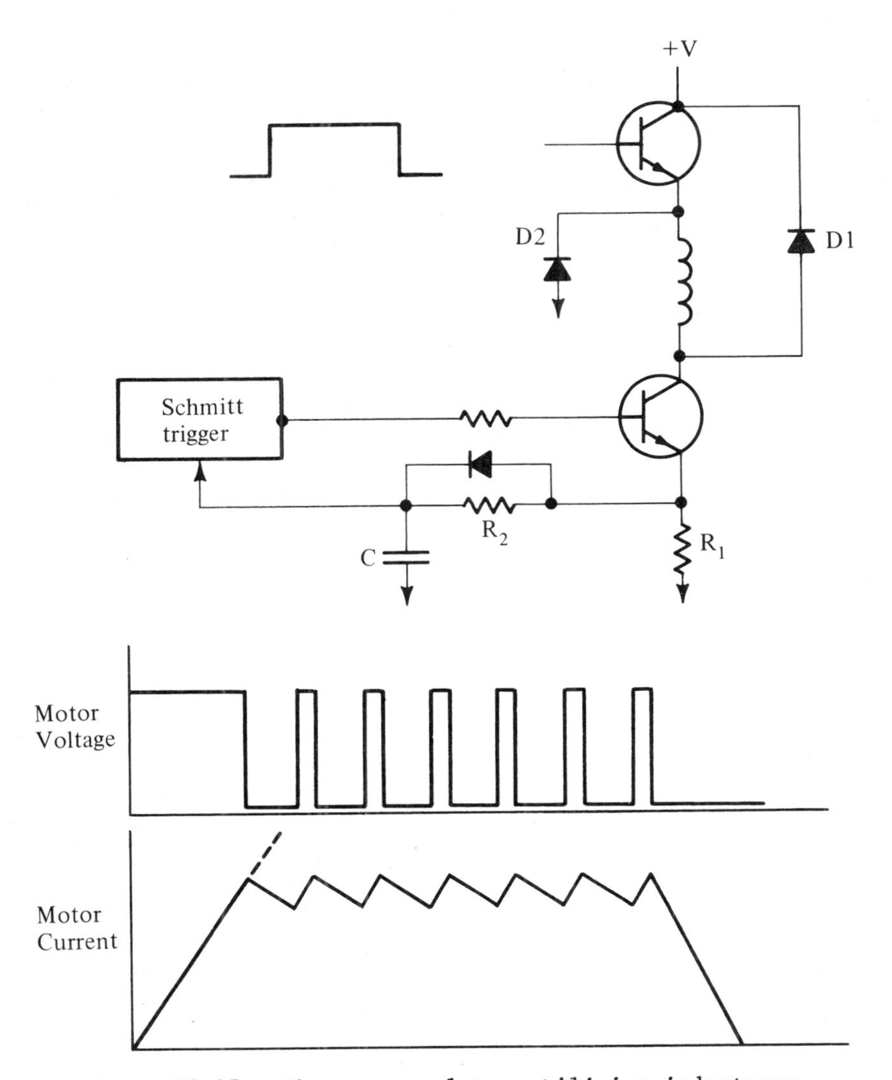

Figure 10-37. Chopper regulator utilizing inductance.

driving pulse, the motor winding current is recirculated rapidly to the high voltage supply. Essentially, the operation involves a high voltage to initially charge the motor winding, indicating a low average voltage (achieved by time modulation) to sustain the current, and high voltage return to discharge the motor winding inductance. In this case, the effective motor voltage is increased at high stepping frequencies, since the voltage is switched on to the motor for a high percentage of the time. The current sensing system used, as in the current-fed system, does not take account of the current circulating in the motor winding. Therefore, when high stepping frequencies are used, the RMS voltage, current, and motor power increase as discussed in the following sections.

Chopper operation is inherently well suited to start-stop applications - high voltage is available to accelerate the rotor as rapidly as possible, with a maximum of low frequency torque. For ramped applications, resonance problems must be considered, and resonant regions avoided. It is generally possible to accelerate through resonance regions, but it may not be possible to achieve stable steady state operation in such regions, particularly the major resonance at about 1 KHz in Series 20 motors.

Performance Results

Measured results using one particular motor (20-4270), and high voltage supply (65V) are shown as Figure 10-38. With the chopper drive the low frequency torque is high and start-stop characteristics are good (but note the resonant regions). A bilevel drive would give approximately the same results. The current-fed driver is smoother in operation, with poorer start-stop characteristics and low frequency torque. All three types of high efficiency drive would give the same results at high stepping rates (> 1.5KHz). This is true because they present identical waveforms to the motor in this region, where they all become voltage sources switched by a transistor bridge. The difference between the drives is only in the method used to limit motor current at low stepping rates. For comparison purposes, the curve of the same motor and power supply with a conventional unipolar resistance-limited driver is included (Figure 10-39) - the difference in power output and systems efficiency is rather striking.

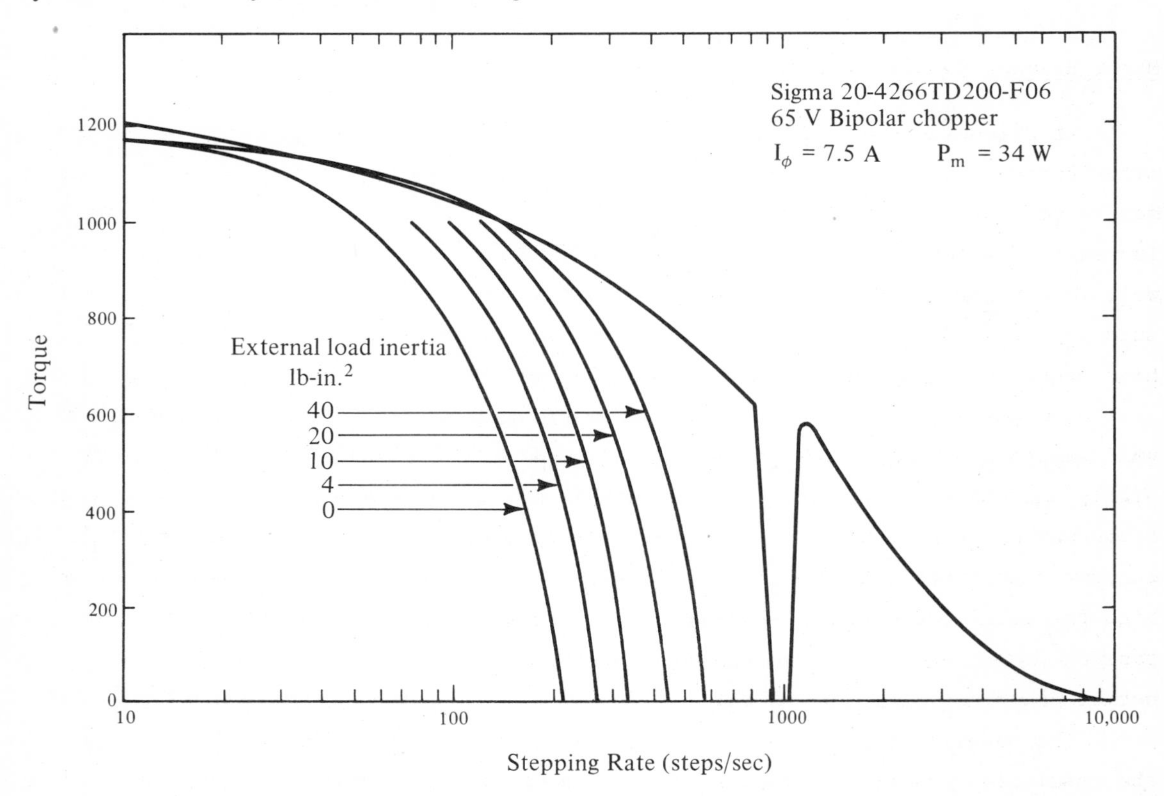

Figure 10-38. Speed-torque curves with bipolar chopper.

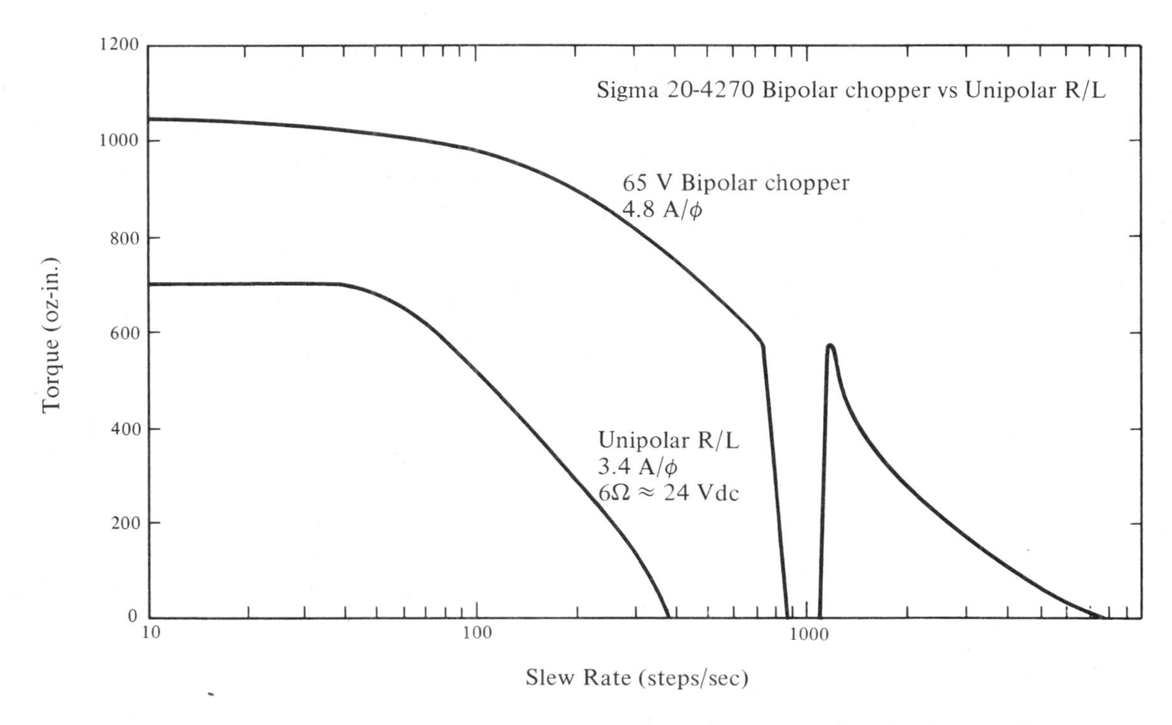

Figure 10-39. Speed-torque curves with bipolar chopper and unipolar resistance-limited driver.

Motor Heating Considerations

As discussed previously, high efficiency drives result in an increase in RMS motor current and voltage. Although an appreciable fraction of the power increase may be delivered as output power in a loaded motor, there is still a considerable increase in motor dissipation at high speeds under high overdrive conditions. It is well to note that an increase in motor dissipation also occurs in inefficient drives, such as resistance-limited systems - the dissipation is basically the result of the high overdrive required for high speeds, however it is applied.

The key to high-speed motor performance has been to use higher and higher voltages to approximate a current source in order to drive the current into the highly inductive motor winding. A number of schemes are used to limit the low frequency current, ranging from series resistors to ramped voltages or switching current regulators ("choppers"). Regardless of how the current is controlled, at high frequencies the voltage across the windings approaches the supply voltage, causing increased switching losses in the rotor and stator, which can exceed the normal power rating of the motor.

The power rating of a motor is limited by the maximum allowable temperature of the insulation system used in the motor. In accordance with NEMA specifications, a motor with Class B insulation has a maximum winding temperature rating of 130°C.

Therefore, the allowable temperature rise of a Class B motor cannot exceed 130°C minus an allowance for "hot spots," and minus the maximum expected ambient. Accordingly, the motor power rating is equal to that total input power which produces a winding temperature rise compatible with the insulation system used. In general, the ratings for most stepping motor manufacturers are based on an unmounted, stopped or slowly running motor for an expected 55°C or 60°C maximum ambient.

If the current or voltage ratings (derived from the power rating) are used at low frequencies and low ratios of power supply voltage to motor voltage, referred to as overdrive, the motor temperature rise will be safe. However, if very large overdrive ratios are used, the input power to the motor can rise above rated power.

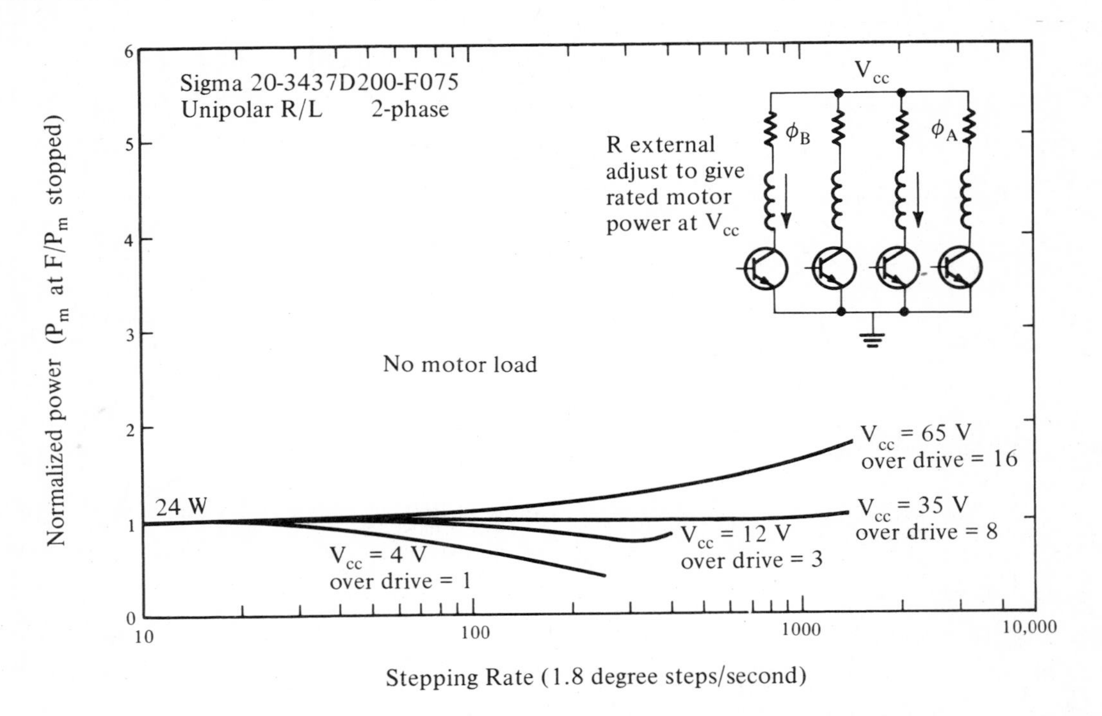

Figure 10-40. Normalized motor power as a function of frequency and power supply voltage.

For example, Figure 10-40 shows the normalized motor power for an unloaded motor as a function of frequency and power supply voltage for the common R/L method of driving stepping motors. The motor is excited by applying rated current to the motor in the stopped condition, with the circuit as shown in Figure 10-40. The motor used was a Sigma 20-3437D200-F075, rated at 24 watts.

From Figure 10-40, the power input to the motor is always equal to or less than the DC power input for overdrive ratios up to 8. For ratios over 8, the motor power increases with frequency: Specifically, for a ratio of 16, the power at 1500 sps is 1.8 times the low frequency value. Consequently, motor overheating could be a problem if the motor had to run continuously at this frequency. The reason for this power increase is as follows: Although the current is decreasing with frequency, and hence the I^2R heating is dropping, the stator and rotor switching losses are increasing with frequency. These represent power dissipated in the motor. Furthermore, the RMS voltage across the winding increases with frequency, adding a further increase in core losses. The switching losses increase with increasing overdrive because the voltage across the motor windings eventually reaches V_{cc} at high speeds.

Another method of driving motors for higher performance with resistor current limiting consists of driving all the windings with bidirectional currents, or two-phase bipolar driving. Figure 10-41 shows the normalized motor power with bipolar R/L driver. The motor is the same as used in Figure 10-40. In the examples cited, the motor power increases dramatically with frequency. The maxima are created when the I^2R losses are still large while the core losses are increasing. Eventually, the I^2R losses decrease faster than the switching losses. At very high frequencies, the losses are almost all core loss. At frequencies above 10K sps, the motor power continues to rise.

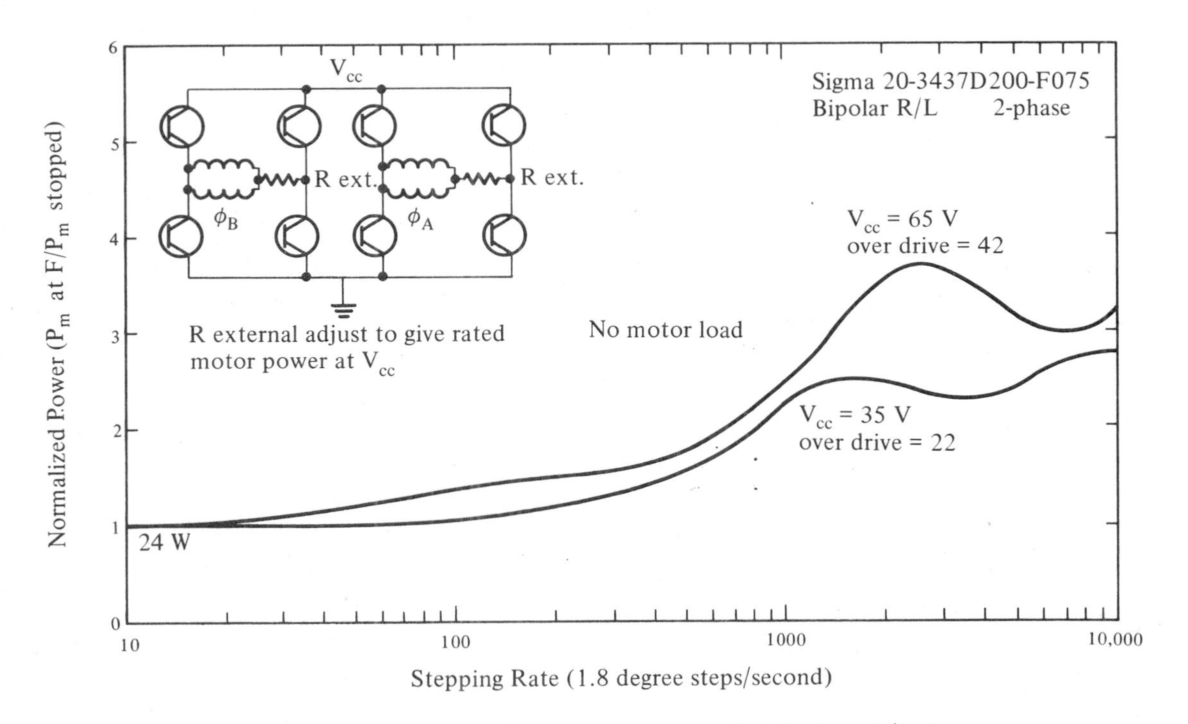

Figure 10-41. Normalized motor power with bipolar R/L driver.

It should be noted that bipolar switching doubles the effect of the power supply voltage because the peak-to-peak swing across a motor winding is twice the supply voltage. In addition, the rated motor voltage is less when two windings are connected in parallel because the resistance level is halved but the current in the pair only increases by $\sqrt{2}$. Thus the effective overdrive ratio is increased for the same supply voltage. Similar increases in motor power occur with "constant current" drives. Figure 10-42 is the power-frequency plot of a ferro-resonant constant

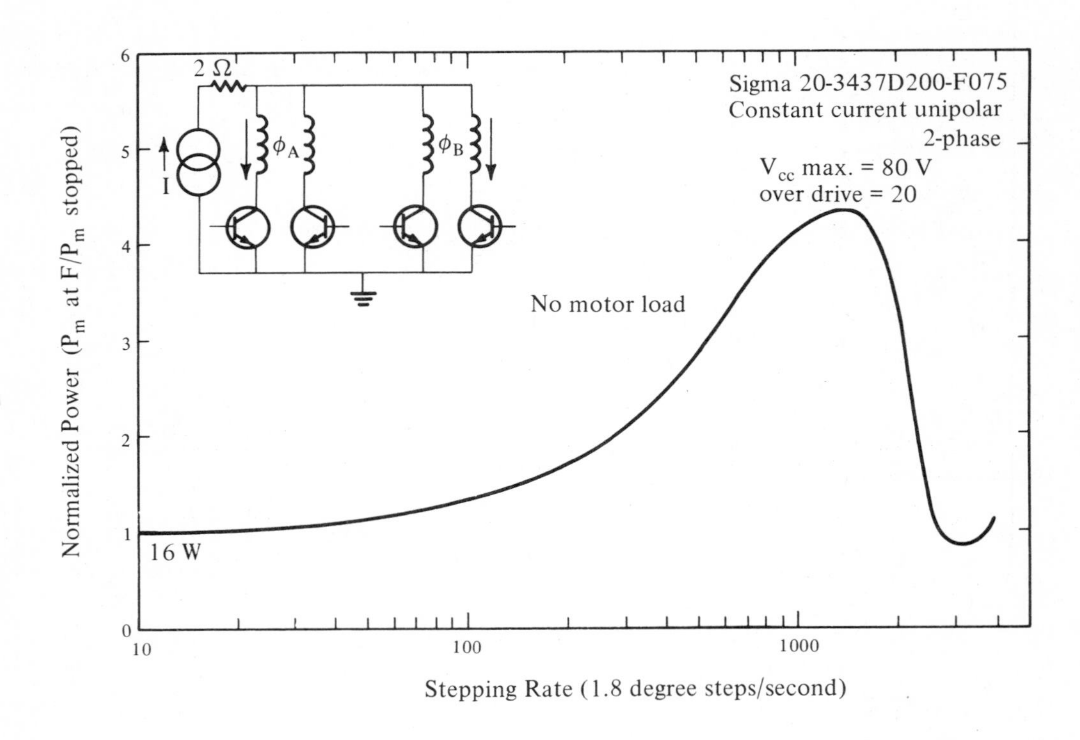

Figure 10-42. Power-frequency plot of a ferro-resonant constant-current source.

current source, which adjusts the supply voltage upward as the motor current tends to decrease with speed. With this driver, the available output voltage varies from 12 volts to 80 volts. The driver is basically a unipolar driver as in Figure 10-40. The motor is the same as in Figure 10-40. As indicated, the motor dissipation rises to more than 400% of the rated motor power.

Another constant current driver uses a switching regulator to control the motor current. Essentially, a winding is connected by a 4-transistor bridge to a voltage supply and the winding current is sensed in the common side of the bridge. The bridge is turned on until the current reaches rated value and then turned off. During the off interval, the winding current is recirculated to the power supply through diodes. The basic driver is a bipolar driver as in Figure 10-33. The power-frequency relationship is shown in Figure 10-43. The curve for the 65 volt circuit is shown for reference purposes only. It is not recommended for use with the small motor used in the previous examples. It does illustrate the effect of very high voltage overdrive conditions.

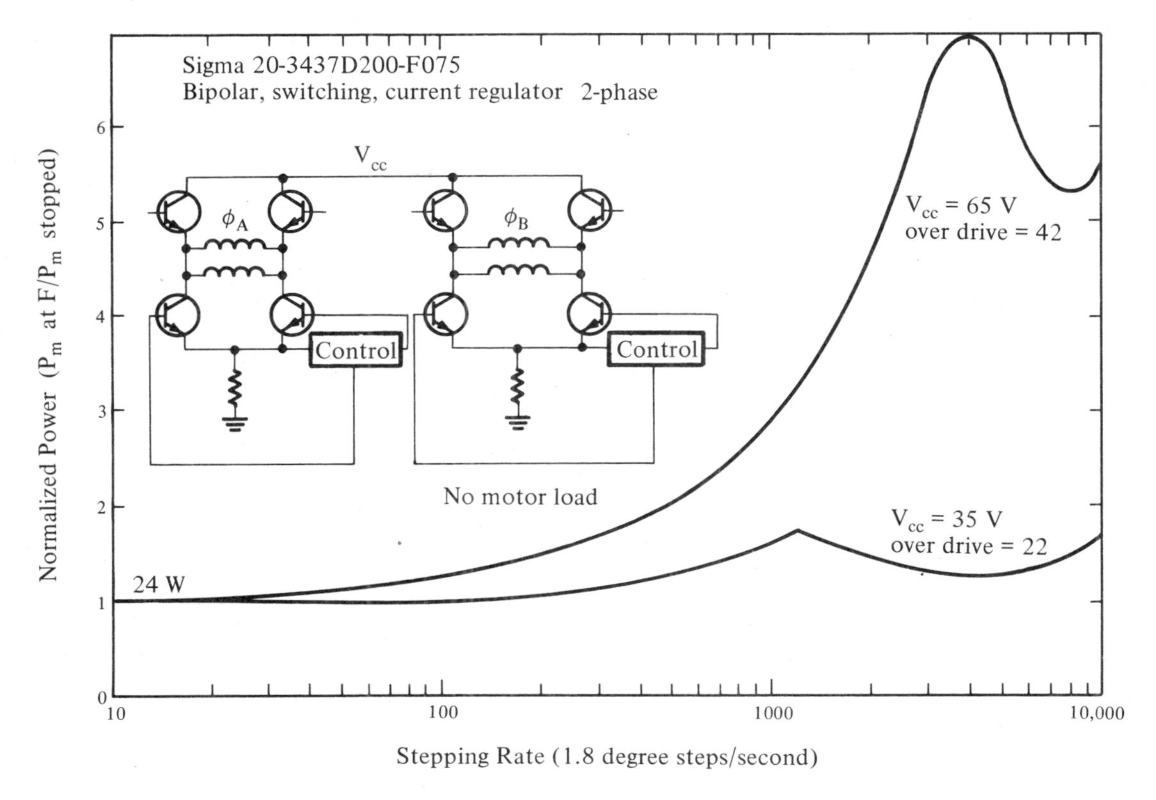

Figure 10-43. Power-frequency plot with bipolar driver.

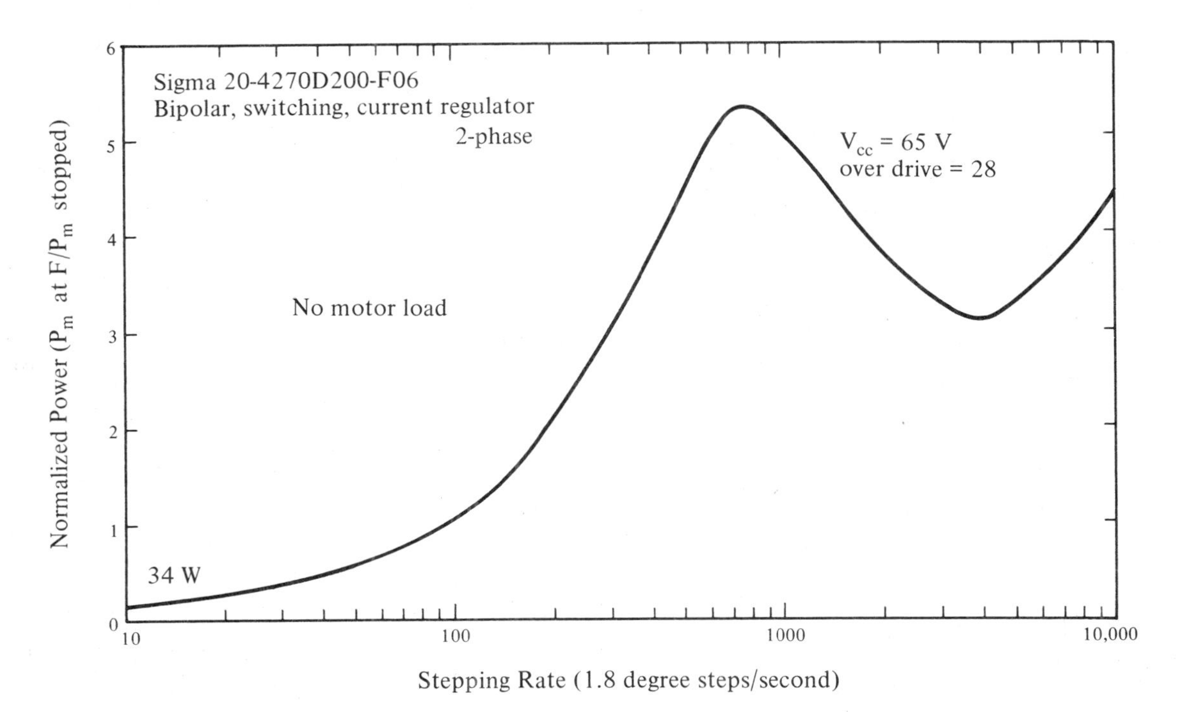

Figure 10-44. Power-frequency plot with bipolar driver.

For purposes of comparison, a much larger motor was run on the 65 volt bipolar switching regulator. This combination will deliver a peak output power of about 1/4 HP at 2000 sps. The motor is rated at 34 watts and is Sigma 20-4270D200-F0.6. Figure 10-44 is a power-frequency plot of this motor and driver, with the motor unloaded. The increase in power input is again dramatic.

The improvement in step motor performance has been achieved at a hidden cost - heat in the motor. The motor ratings must be used with an eye to the type of driver used. When high-performance drivers are used, the duty cycle of the application must be considered. A driver capable of running a motor at 10,000 steps per second with useful torque may have regions where the input power at lower frequencies is many times the power rating of the motor.

All these tests were made on an unloaded motor in order to show the trends more clearly. In general, a loaded motor will have slightly lower losses. In the case of resistance limiting, the through-put power tends to reduce the motor voltage at particular frequency, reducing switching losses. Furthermore, the rotor develops a lag angle which tends to reduce losses. However, the decrease in motor losses caused by loading rarely amounts to more than 20% of the unloaded losses.

Summary of Drive Circuits

The simplest and least expensive drive circuit is usually the resistance-limited unipolar circuit; however, it fails by a wide margin to yield the maximum output available from a given stepping motor, and is inefficient from the standpoint of power requirements and system heating.

The bipolar resistance limited system requiring only a modest increase in complexity, considerably improves the results of the unipolar circuit. The available improvement can be taken as improved torque (20%-40%) and speed (≈ 30%-50%) or reduced system power (50% reduction from unipolar case).

The high efficiency systems (bilevel,current-fed, chopper) allow dramatically improved motor performance and system efficiency, at the expense of complexity and cost.

Obviously, the designer must weigh his performance requirements carefully before deciding upon a particular motor-driver combination.

10-4 APPLICATION CONSIDERATIONS IN STEPPING MOTOR SYSTEMS

Occasionally, stepping motor systems applications are simple and straightforward. Perhaps the load is very light and well controlled, the speed requirements are low, and the environment constant and agreeable. In such a case, probably the only concern would be the lowest priced motor and driver available that would fit in the space allocated. Or, a heavy machine tool load is to be driven as fast as possible, because reduced cycle time is extremely valuable. Here, the largest motor and best driver available might be well justified.

Unfortunately, most applications tend to lie between these extremes, and the designer must make a choice of systems based on a careful comparative study. First and foremost, the load requirements must be studied thoroughly - not only to determine the load parameters as well as possible, but also, if possible, to consider modifications to the load to optimize the system for stepping motor drive. For instance, in general, stepping motor systems are more sensitive to inertial loads than to friction loads. Therefore, if fast response and quick settling are desired, the most favorable load configuration will probably be the one that minimizes inertia at the stepping motor, and effort spent in this direction is likely to be most rewarding. The friction load is the other basic load parameter. Friction and inertia, combined with the required load speed and acceleration, determine the basic requirements on the stepping motor system, since friction determines the power output, and inertia and speed define the amount of kinetic energy that must be put into the system upon starting and removed upon stopping.

In some cases, required speed and resolution determine the step angle in a stepping motor system, but in others a choice is available, since an effective gear ratio may be obtained through a timing belt, lead screw, linkage, or other means. In these cases, it may be useful to compare motors of differing step angles. Usually, the simplest approach is to consider shaft speed in RPM, or some other convenient measure. For instance, a 200 step/rev (1.8°/step) device at 200 steps/sec is moving at 1 RPS, or 60 RPM. A 24 step/rev (15°/step) motor achieves this output speed at 24 steps/sec.

In making comparisons involving stepping motor systems with differing step angles and speed-torque characteristics, it is useful to plot speed in RPM against torque and power output. Power output in watts can be calculated from the formula P_{watts} = Speed (steps/sec) × Torque(in-oz) × 0.0444 ÷ steps per rev. For various step angles, the formulas are

$$P_{watts} = 2.22 \times 10^{-4} \times \text{Torque (in-oz)} \times \text{Speed (steps/sec) for } 1.8^\circ \text{ step angles}$$

$$P_{watts} = 1.84 \times 10^{-3} \times \text{Torque (in-oz)} \times \text{Speed (steps/sec) for } 15^\circ \text{ step angles.}$$

Power output curves on stepping motors reveal a broad maximum (Figure 10-40) which arises from the nature of the power output relationship - at very high speeds, zero torque (and zero power) is delivered; at zero speed, torque is high but power delivered is necessarily zero. If other system factors allow, consideration should be given to operating at the peak of the power output curve in order to obtain maximum performance from the motor-driver combination.

Operating point selection and motor driver choice must take into account thermal conditions in the motor. As discussed earlier, high performance drivers frequently lead to increased motor dissipation at high speeds. Since motor thermal lag is generally fairly long (thermal time constant of 20 minutes or more), it is frequently possible to allow high motor dissipation for short periods of time if the net effect is not one of excessive motor temperature rise. This is basically a classic duty cycle problem - the criterion is final temperature rise considering the effects of ambient temperature, motor heat-sinking, and operating times. It should be emphasized that the limitation on power output of most step motors is simply temperature rise - more torque is available at higher than rated currents if the final temperature can be held to its rated value.

The best way to determine if a motor will overheat is to measure the temperature rise of the motor winding by measuring winding resistance after it has operated at the worst duty cycle for a period of 3 or 4 hours. This long period is required because the thermal time constant of most motors is in excess of 20 minutes. If the winding resistance is measured when the motor is at room temperature and again at the end of a period of operation at the worst duty cycle, the temperature rise of the winding is approximately 2.5 times the percentage resistance increase from the room temperature resistance. This temperature rise should be limited to a safe value recommended by the manufacturer based on the maximum ambient expected. Heat radiators or fan cooling are sometimes required.

In general, most control systems have only limited periods where high speeds are required. In these, heating is not a severe problem. Great numbers of successful applications are in the market place that combine high-performance drivers and stepping motors to achieve fast, precise positioning with excellent reliability.

In making a choice of driver-motor combinations, resistance-limited drivers frequently offer adequate performance when used in conjunction with a suitable motor. However, serious consideration must be given to the power dissipated in the limiting resistors, and the effects of the resultant temperature rise on the remaining system components. In modern systems, it is quite probable that limiting resistor dissipation will represent a significant percentage of total system power. Considerations of this kind not infrequently lead the designer to consider a more efficient driving system, even though the actual system drive requirements could be met adequately by the simpler but less efficient resistance-limited drive.

Another category of problems arises when the stepping motor system is used as a high-speed positioning device, and angular accuracy is critical. Here, both start-stop speed and damping characteristics must be considered. Unfortunately, drives that yield high start-stop rates tend to be poor in settling characteristics. Reverse pulse damping on the last pulse of a train and/or current-fed drive may be satisfactory answers for damping problems. Mechanical dampers are sometimes used, but their life span may be inadequate for high reliability systems.

Mechanical design for stepping motor systems must take into account fits, mounting dimensions, etc., but it is often sensible to plan the design for a possible future change in motor size. It often happens that it is desirable to improve the performance in speed or torque of a system after the preproduction or production stage is reached. This can occur because the actual load was not as estimated, or it may be that the product will be more marketable if its capability can be increased. In any event, the design of a stepping motor family typically will include a number of different length motors with given frame diameter. This is done not only for economic reasons (much of the tooling on a motor is related to diameter, not length) but also to maintain a favorable torque/inertia ratio. Torque increases directly with the stack length of a motor, as does rotor inertia. However, as a first order approximation, torque increases directly as diameter, but the inertia goes as the square of the diameter. Therefore, to obtain high performance motors, it is preferable to make a long motor (related to diameter) rather than a short, fat motor. Obviously, this approach reaches a limit when it becomes impractical to hold tight tolerances on air gap as the rotor and stator lengths are increased. However, the point is that it is frequently possible to upgrade system performance rather simply by increasing motor length - if provision has been made for such a change. Another important mechanical consideration is system backlash or "tightness". Stepping motors are inherently a transient device, and system oscillations, with consequent poor performance, are readily excited by the pulsing torque delivered by a stepping motor. The ideal mechanical system for stepping motor drive has good fits and low backlash characteristics, and, preferably, the lowest possible inertia.

Coupling of stepping motors to the load is frequently treated rather casually - but it is the source of considerable problems. The ideal coupling is somewhat compliant, but not sloppy or subject to deterioration with time and operation. Attachment of couplings to stepping motors with set screws is satisfactory only for the smaller sizes. Larger motors should be pinned or keyed - the discontinuous stepping torque will soon loosen less positive fasteners over even short periods of time.

In the broad sense, probably the most practical way to apply step motor systems is with a healthy dose of conservatism, particularly as regards load conditions. Many loads change appreciably during system life, inevitably, it seems, for the worse. Further, it is desirable many times to handle transient load conditions that occur only infrequently. Adequate performance margins are the realistic approach to such problems.

10-5 OTHER MODES OF APPLICATION FOR STEPPING MOTORS

Step motors are frequently used in modes other than straightforward digital load positioning. Although basic motor operation is the same in all modes, it may be useful to consider these modes of operation separately.

Servo Motor Mode

In many applications, open-loop stepping motor systems are competitive with closed-loop servos. However, stepping motors are sometimes used as the mover in servo systems. In this application, the step motor is generally in the digital, or stepping, mode. Therefore, the system is limited in resolution to the value represented by one step. In essence, this amounts to a bang-bang servo where an error signal results in a quantized response, and the deadband necessary for stability is represented by the distance between steps. This type of system has resolution limited by the step size (as seen at the load through any appropriate gearing) but it does have some interesting features. Holding torque is always applied to the load; almost full torque can be generated by the motor with a shaft position error of less than the deadband, i.e., no system error signal. Speed of response is also independent of the load characteristics: the load is driven to the system null at the same speed regardless of the magnitude of the error. Overshoot problems are easily minimized: in effect, the stepping motor system controls overshoot on every step. The stepping motor does not have the brush problems of DC servos, and is suited for operation over a wide range of frequencies compared to conventional two-phase AC servomotors.

Synchronous Servo

A method of using a stepping motor in an inexpensive servo system is shown as Figure 10-45. The synchronous mode of operation requires a minimum of extra hardware to provide a reversible drive. This operating mode utilizes the low synchronous speed of the stepping motor when operated directly from line frequency (72 RPM for a

Figure 10-45. Circuit diagram of synchronous servo.

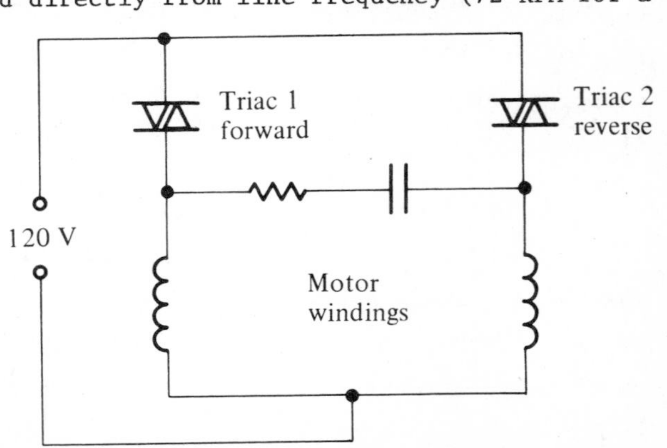

200 step/rev motor operated at 60 Hz). In this type of operation, low speed and relatively high torque are available without a gear box. The motor is well damped for positive start-stop characteristics, and can be stalled indefinitely without incurring damage.

Speed Control

The stepping motor with appropriate driver offers the possibility of virtually infinite speed control range, since the low speed limit can be as low as desired. Obviously, the output is stepped and not smooth, but in many applications this is not objectionable. Further, the output speed is unaffected by load up to the torque limit of the motor, and the speed can be programmed or controlled from an external source in either analog or digital fashion.

Encoder Operation

Shaft position encoders have been used in conjunction with stepping motors for two principal reasons:

1. Acceleration Improvement

   If the shaft position can be determined accurately as a function of time, it is possible to control drive pulse timing to obtain maximum acceleration. Since the dynamic shaft position depends upon load, a coupled shaft encoder takes into account load characteristics to allow maximum possible acceleration. The encoder chosen must be of such design as to allow determination of direction (usually accomplished with quadrature tracks), since overshoots of shaft position otherwise could be interpreted incorrectly as a forward pulse by the decoding circuitry a considerable amount of sophisticated logic is required to use an encoder, but it does offer improvement of acceleration that is difficult to obtain in any other way.

2. Proof of Step Indication

   An encoder, again with two quadrature tracks to decode direction information, can be used to ensure that all command steps have been taken by the motor. This system basically guards against transient load conditions, or momentary malfunctions in the motor driver or logic system.

11 CLOSED-LOOP CONTROL OF STEP MOTORS

B. C. Kuo
Department of Electrical Engineering
University of Illinois at Urbana-Champaign
Urbana, Illinois

11-1 INTRODUCTION

Open-loop control of step motors suffers from the disadvantage that the motor may not be able to follow the input pulse train so that the top speed which a motor can run is limited. Also, the speed of a step motor under open-loop control may have wide fluctuations. These difficulties can be overcome by using positional feedback to the step motor to determine the proper rotator positions at which phase switchings should occur. With the closed-loop control, one cannot only achieve much higher speed and more stability in speed, but also more versatility in many other aspects of the control of the step motor.

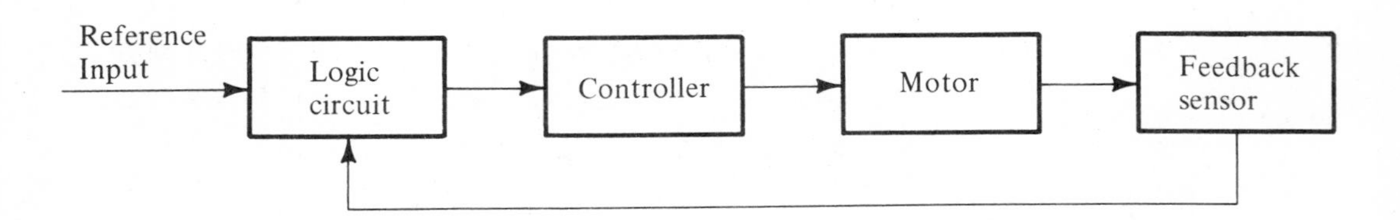

Figure 11-1. Block diagram of closed-loop control of step motor.

A block diagram illustrating such a closed-loop control scheme is shown in Figure 11-1. The feedback sensor in this case could either be a photoelectric device or a magnetic pickup device which would give a pulse for every step of motion. The motor is started initially with one pulse from the controller and subsequent pulses are generated from the feedback sensor assembly. Figure 11-2 shows a feedback sensor arrangement using a slotted disk which is mounted on the motor shaft and two photo-sensors. It will be shown that the positions of the photosensors with respect to the step position are very important in the feedback control of the step motor.

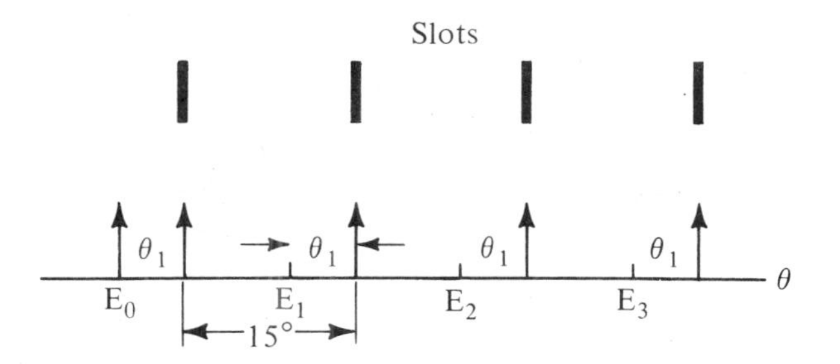

Figure 11-2. A feedback sensor using a slotted disk and two photoelectric sensors.

11-2 THE LEAD ANGLE AND THE SWITCHING ANGLE

The encoder disk has uniformaly spaced slots which are equal in number to the number of steps per revolution of the motor. Each photoelectric sensor unit consists of a light source and a photoelectric cell. The basic operation of each sensor is straightforward. Each time a slot passes through the beam from the light source, a pulse is generated by the photocell and sent to the controller. The motor is started initially with one pulse from the input command, and subsequent pulses are generated from the encoder assembly.

The position of each sensor unit may be adjusted to a fixed angle, often called the "lead angle" or the "switching angle", with respect to the equilibrium or detent positions of the motor. The relationship between the switching angle and the equilibrium positions is illustrated in Figure 11-3 for a 7.5-degree-per-step motor.

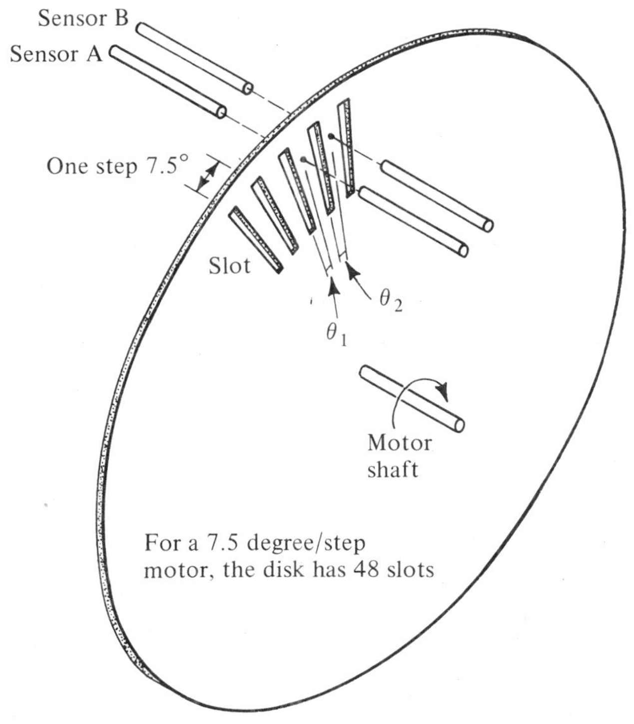

Figure 11-3. Definition of lead angle or switching angle of closed-loop motor control.

Let θ_1 denote the switching angle of one set of the photoelectric sensors. In this case, $0 \leq \theta_1 \leq 7.5°$. In Figure 11-3, the equilibrium or detent positions are designated by E_0, E_1, E_2, ... The first pulse is applied when the motor is at the equilibrium position E_0. Since the sensor is positioned at θ_1 degrees from the equilibrium position, subsequent pulses generated by the sensor are at an angle θ_1 degrees from each of the subsequent equilibrium positions.

The lead angle is an alternative way of expressing the switchings for closed-loop control of step motors. The lead angle refers to the number of degrees in advance of (in lead of) a particular equilibrium position that the corresponding phase is switched. For instance, the final equilibrium position due to the initial pulse and the first pulse generated by the sensor is E_2 (two steps of travel). Therefore, as shown in Figure 11-3, the lead angle, as the name implies, is defined as $15° - \theta_1$. In practice, both the switching angle and the lead angle are used. However, in this chapter for simplicity, we shall make reference mostly to the switching (switch) angle θ_1.

Variation of the switching angle has an important effect on the acceleration and deceleration, and especially the final steady-state speed of the step motor. Figure 11-4 illustrates a typical steady-state speed versus switching angle curve of a step motor with closed-loop control.

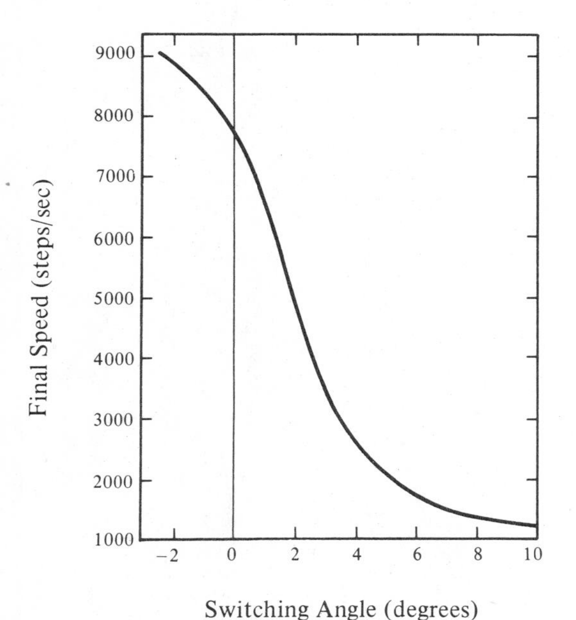

Figure 11-4. Steady-state speed versus switching angle for step motor.

Although the mounting of the sensor places physical limitations on the actual range of the switching angle for a given motor, by injecting extra pulses into the input pulse train to the controller, negative values of the switching angle θ_1 and values greater than one motor step in degrees may be attained.

The phenomenon of extra-pulse injection and negative switching angles will be discussed in subsequent sections of this chapter.

The second set of feedback sensors are for the purposes of direction sensing, or when different switching angles are required for acceleration and deceleration of the motor. For direction sensing purposes, the two feedback sensors are adjusted so that their respective output pulses overlap, as shown in Figure 11-5. This pulse overlap is sent to the direction-sensing circuit of Figure 11-6 which insures that only one pulse results from the two feedback pulses which are produced each time the step motor advances one step and that the motor rotates in the proper direction.

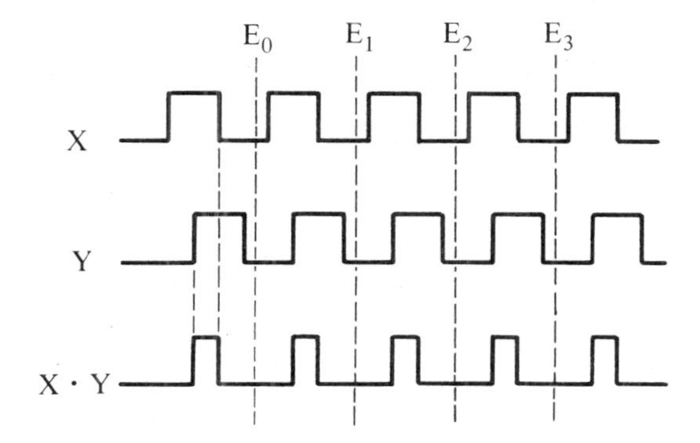

Figure 11-5. Feedback pulse overlap for direction sensing.

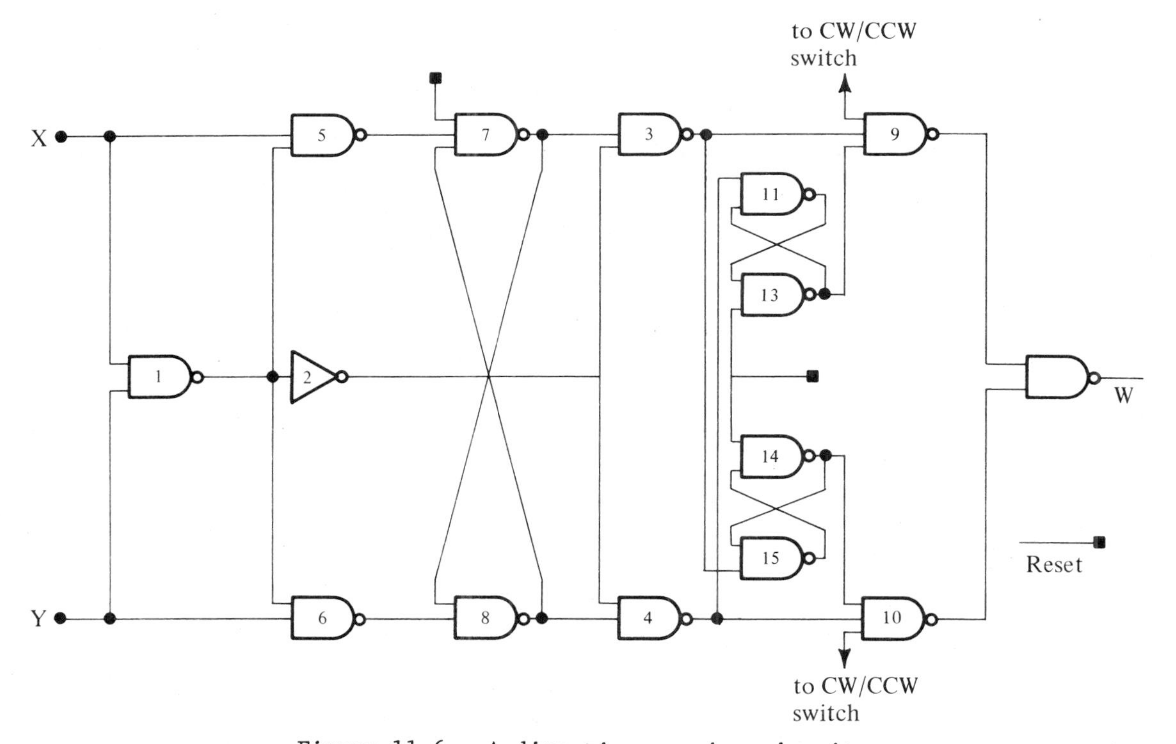

Figure 11-6. A direction sensing circuit.

Basically, the circuit divides the overlapping pulses into the three regions shown in Figure 11-7. As long as the sequence 1, 2, 3 or the sequence 3, 2, 1 occurs, the pulses are directed to the forward or reverse outputs, respectively. An incomplete or mixed sequence will either generate no output pulses or one in the forward and reverse outputs, depending on the oscillation of the motor.

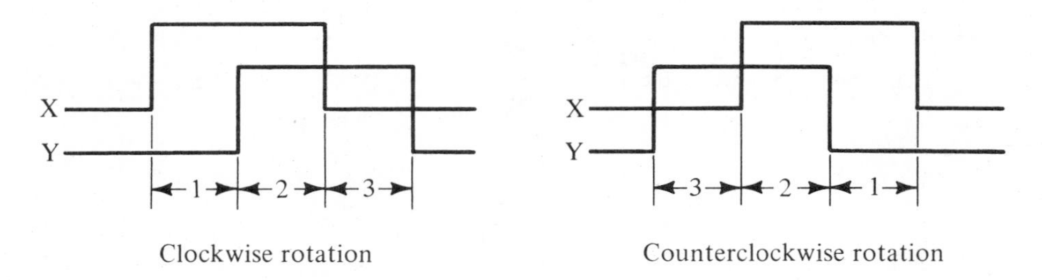

Figure 11-7. Feedback pulse overlap.

11-3 CLOSED-LOOP SWITCHING

The static torque curves of a step motor are useful in gaining understanding on how the closed-loop control scheme works. Shown in Figure 11-8 are the idealized (sinusoidal approximation) static torque curves of a three-phase 7.5 degree-per-step motor. The approximate value of the angle at which the motor should switch from one torque curve to the next (or one phase to the next) to maintain positive or negative torque, can be determined by using the curves of Figure 11-8. Suppose that a maximum

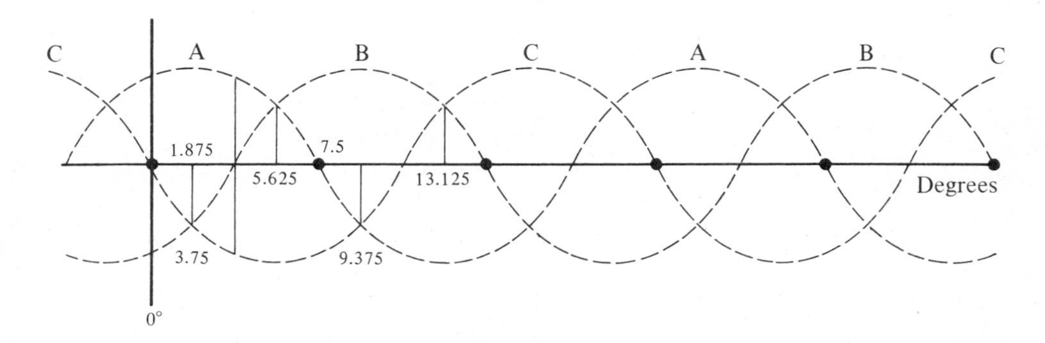

Figure 11-8. Static torque curves of a three-phase 7.5-degree/step step motor.

positive torque is to be maintained, and also that at the instant of switching, the current in the deenergized phase decays instantaneously, and that the current in the newly excited phase builds up instantaneously. A switching angle of 5.625° (or lead angle of 9.375°) would then give the maximum torque, since this is the point at which the torque curves overlap. For example, let the motor be stationary at 0° with a steady-state current in phase C. The first forward pulse would be applied at 0°, the next at 5.625°, and the next at 13.125°, which is 7.5° beyond 5.625°, and all subsequent pulses at 7.5° intervals. However, in reality, the currents do not build up or decay instantaneously, and thus the switching process is dependent upon the rise and decay times of the current as well as the position. The time constants for these two processes are dependent upon the resistances of the control circuit and the motor, the self and mutual inductances of the windings, as well as the back emf of the motor. Also, the inductances of the windings are functions of the rotor position and the motor current since the magnetic material of the motor is subject to saturation. Consequently, it becomes difficult analytically to solve for the optimal switching angle exactly. However, it has been found from experimental results that in a typical step motor the switching process from one torque curve to the next may take up to several milliseconds of buildup time. In order to switch in a manner that maintains a maximum positive torque, the pulses must be applied at an instant corresponding to the buildup time before the 5.625° position. At low speeds this buildup time might require that the switching take place at a point less than 1° before the 5.625° angle. At high speeds the time build up might require that the switching take place more than 7.5° before the 5.625° angle. Therefore, the controller must use both positional and velocity feedback in order to optimally control the motor.

If only the positional feedback is employed, as is done in most practical applications, the angle at which the motor is switched from one torque curve to another must be chosen so that the motor maintains a high average torque, and is capable of starting from a stationary position. It is evident that this angle must be less than 5.625° for the motor portrayed, since any angle greater than this would give a lower average torque. Also, it has been found that if the motor initially starts to move from the equilibrium point, a switching angle of less than one degree may cause the motor to reverse its motion and oscillate about the -7.5° equilibrium position. From the static torque curves of Figure 11-8, it can be seen that the first feedback pulse after the equilibrium point will switch the supply voltage from phase A to phase B which will place a negative torque on the motor until the shaft

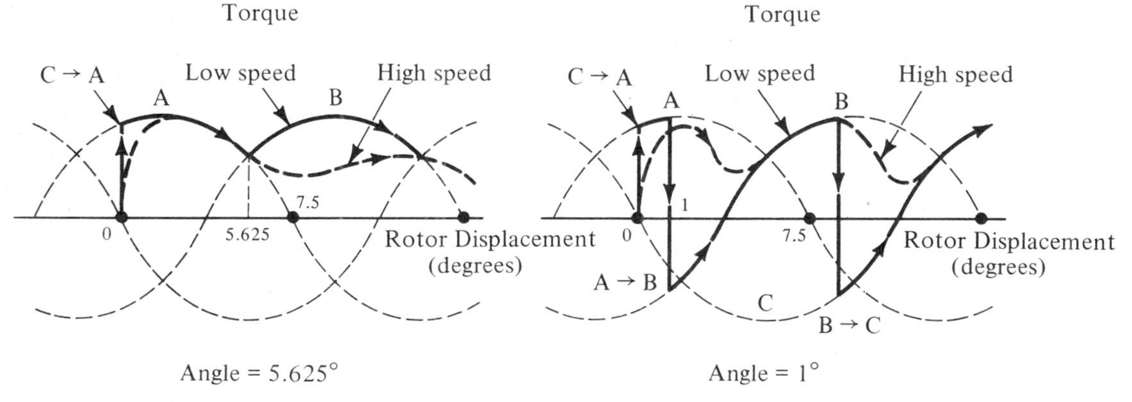

Figure 11-9. Torque curves showing the effect of the switching angles at low and high motor speeds during acceleration.

has passed the 3.75° point. If this negative torque is applied long enough to reverse the direction of motion, the motor will move in the opposite direction to the phase B equilibrium position at -7.5°. Thus, for a 3-phase motor with positional feedback alone, the switching angle must fall between 1° and 5.625°. Figure 11-9 shows these two extreme cases. At low speeds the angle corresponding to current buildup is small, and angles close to 5.625° give the greatest average torque. At higher speeds this angle is greater (due to constant buildup time), and the torque is reduced, as shown by the dotted lines. On the other hand, angles close to 1° give low torque at low speeds, as shown by the solid lines, but give much greater torque at high speeds (dotted line), due to the effect of the buildup time.

Similar constraints are also present for the switching angle used to apply a maximum negative torque. Figure 11-10 shows the two limiting cases for this angle.

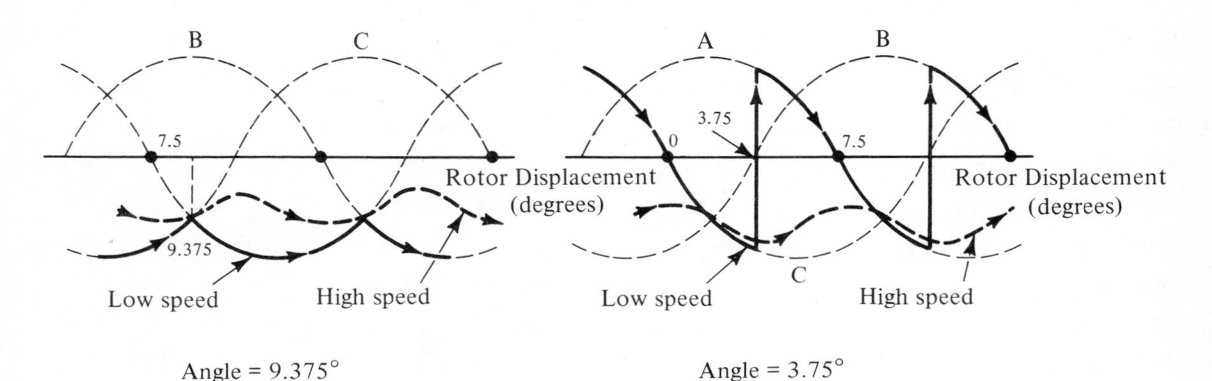

Figure 11-10. Torque curves showing the effect of the switching angles at low and high motor speeds during acceleration.

It is assumed here that two pulses have been injected to initiate the deceleration. At very low speeds where the time constants for the current are negligible, the maximum negative torque is given by a switching angle of 9.375°. At progressively higher speeds the effective switching angle increases, thus resulting in a decrease

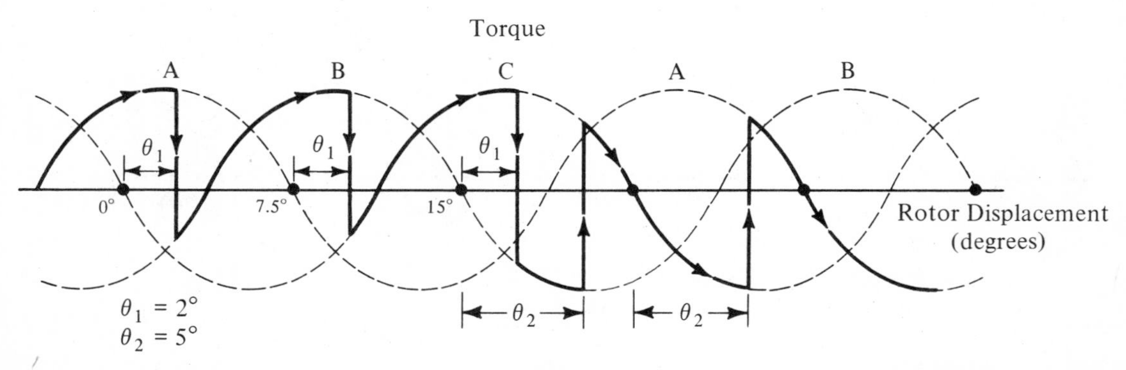

Figure 11-11. Torque curves showing the switching of the motor from acceleration to deceleration.

of torque. as shown by the dotted lines. Any switching angle greater than 9.375° will evidently give less torque for all speeds. The switching angle of 3.75° shown in the second set of curves in Figure 11-11 will provide a higher average torque at higher speeds than does 9.375°, due to the effect of the buildup time. However, as the speed approaches zero this angle will give less torque. Note that if the switching angle were any smaller than 3.75°, and the motor were rotating at a low enough speed, it will see a positive average torque. Thus, the switching angle for deceleration purposes must lie between 3.75° and 9.375°, where the smaller angles provide the greatest deceleration at high speeds, and the larger angles provide the greatest deceleration at low speeds.

At some point in the control scheme the motor must be switched from an acceleration mode to a deceleration mode. If the switching angle for the acceleration mode, θ_1, is smaller than the switching angle for the deceleration mode, θ_2, which is usually the case, either two forward-sequence pulses must be given at θ_1 or one pulse in the reverse sequence must be given at θ_1 to cause the motor to go into a deceleration mode. This is shown in Figure 11-11 for $\theta_1 = 2°$ and $\theta_2 = 5°$. A limiting case of the above situation is when the same angle is used for both acceleration and deceleration ($\theta_1 = \theta_2$). When the motor changes modes, either two forward pulses or one reverse pulse occurs at θ_1 and one forward pulse occurs at θ_2 (the new switching angle). Thus, as θ_1 approaches θ_2, either three forward pulses can occur at θ_1 or one reverse pulse and one forward pulse can occur at θ_1. Both cases are equivalent to not switching at all at θ_1. This special case is shown in Figure 11-12.

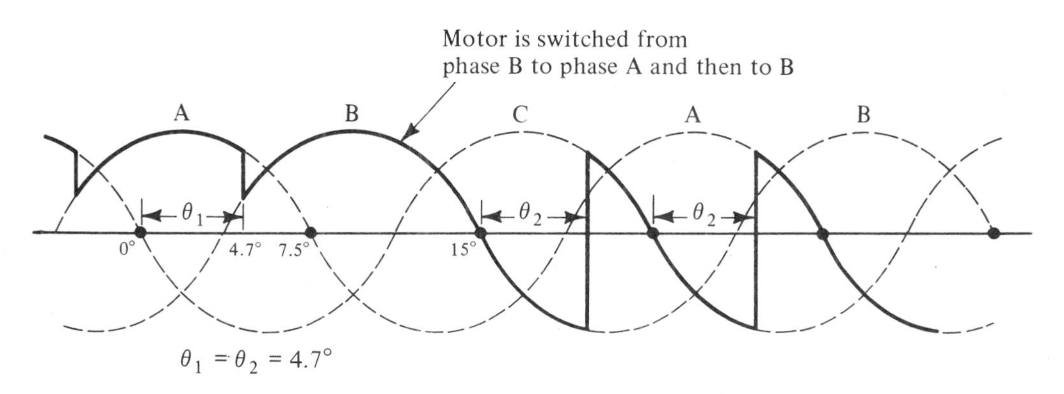

Figure 11-12. Torque curves showing the switching of phase from acceleration to deceleration when $\theta_1 = \theta_2$.

It is shown in Figure 11-4 that small switching angles will result in a high steady-state motor speed. Unfortunately, the switching angle θ cannot be too small because this particular range of values does not generate enough positive torque to start the motor. However, operation at these values may be obtained by injecting an extra pulse into the train of feedback pulses going into the controller. Extra pulse injection in effect adds a negative angle equivalent to one step to the initial switching angle θ (or increases the lead angle by one step).

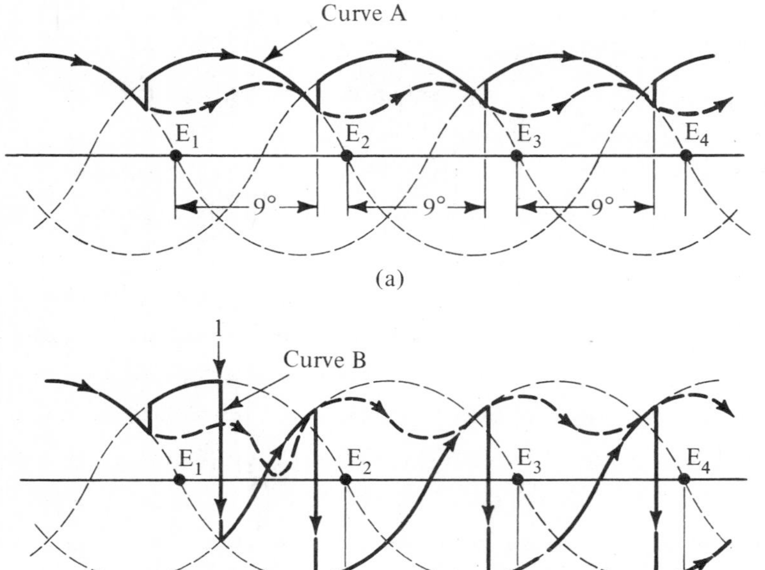

Figure 11-13. Torque curves showing the effect of extra pulse injection for θ = 9°.

Figure 11-13 shows the static torque curves of a three-phase 10-degree/step motor. Curve A (Figure 11-13a) shows the typical low-speed torque curve that the motor would follow for a switching angle of 9° (or a lead angle of 11°). Curve B (Figure 11-13b) shows the low-speed torque curve after an extra pulse has been injected at point 1. Observe that for Curve A switching takes place 9° after each equilibrium point, and for Curve B switching takes place 1° before each equilibrium point. Thus, if one considers each equilibrium point as the zero-degrees point, then, Curve B implies a negative switching angle of -1°(or a lead angle of 21°). In each of the figures, the high-speed torque curve is shown in dotted lines. This extra-pulse for high speed phenomenon is found to occur only for relatively large values of θ (greater than 7.0°). For values of θ in the range $1° \leq \theta \leq 7.0°$, extra pulse injection results in a rapid deceleration to zero steady-state speed. However, if the speed of the motor is sufficiently high, the motor may accelerate to a higher speed rather than decelerate to zero speed. Figure 11-14 shows both the high speed (dotted line) and low speed (solid line) torque curves for a switching angle of 5°, with an injection of an extra pulse at point (1). In the case of the low speed torque curve, the motor will decelerate to zero speed because of the negative torque being applied. For the case of the high-speed torque curve, the motor will accelerate because of the positive average torque being applied to the motor.

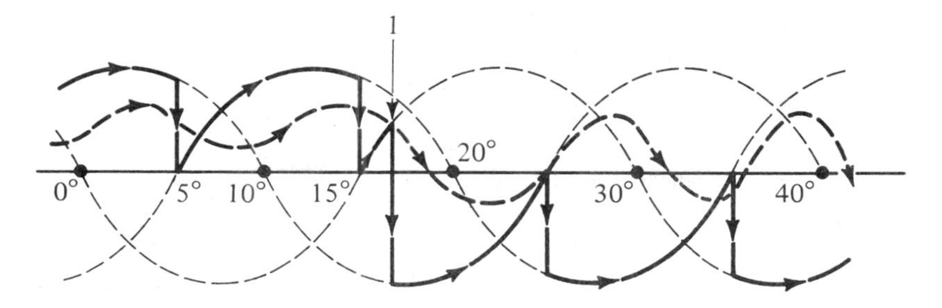

Figure 11-14. Torque curves showing the effect of extra pulse injection for $\theta = 5°$.

In order to achieve switching angle characteristics for $\theta > 10°$, which is equivalent to one full step in this case, two extra pulses are injected into the feedback path. Since the step motor considered is three phase, injection of three pulses results in no change in the torque characteristic. Injection of one extra pulse results in a change in θ of $-10°$. Injection of two extra pulses results in a change in θ of $+10°$. Figure 11-15 shows the resultant low-speed torque after two pulses have been injected for an initial switching angle of 2.5 degrees.

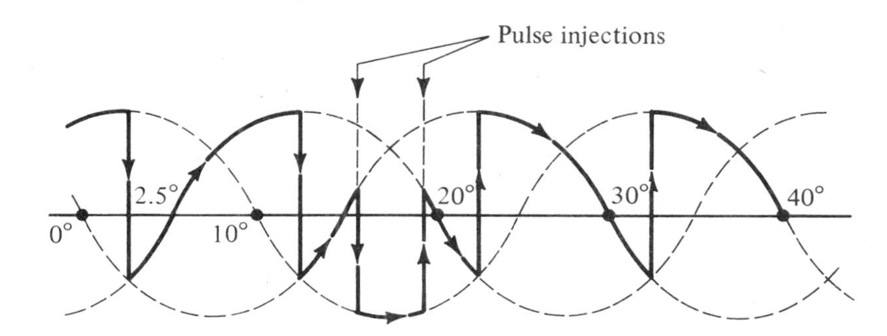

Figure 11-15. Torque curves showing the effect of the injection of two extra pulses for $\theta = 2.5°$.

11-4 CLOSED-LOOP CONTROL OF STEP MOTORS WITH TIME-DELAYED FEEDBACK

Thus, the discussion on the switching angle of the feedback sensor has been based on the assumption that the sensor angle is adjusted physically. However, in practice, the switching angle is regulated by the position of the photoelectric sensor and it cannot be adjusted easily, especially while the motor is operating. We shall now discuss a method of introducing a time delay in the feedback path electronically which is equivalent to varying the switching angle.

Consider the block diagram shown in Figure 11-16. A time-delay unit is introduced between the feedback sensor and the controller. Therefore, effectively, a fixed-time interval must occur between the time the switching signal is given and the time the actual phase switching occurs. The introduction of this time delay can also

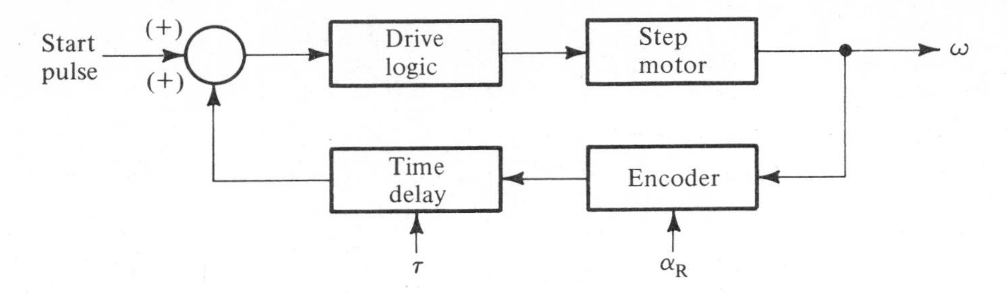

Figure 11-16. Closed-loop operation with time delay.

be interpreted as an additional angle delay, α_R, so that the effective switching angle is

$$\alpha = \alpha_R + \alpha' \quad . \tag{11-1}$$

If α' represents the angle traversed by the motor during a time interval T, at a speed w, it is clear that this angle is a function of both the time delay and the motor speed; i.e.,

$$\alpha' = wTr \tag{11-2}$$

where r is the number of degrees corresponding to one step of the motor.

For example, a time delay of 0.1 msec at 1000 steps per second represents a delay angle of 0.75 deg for a 7.5 degree-per-step motor. At 500 steps per second, the same time delay results in a delay angle of only 0.375 deg.

It should be noted, however, that the use of time delay in the closed-loop control can only increase the switching angle. Consequently, it is necessary to set the reference angle α_R so that the maximum desired speed is less than the speed that can be obtained with α_R acting alone ($\alpha' = 0$), since the speed decreases with the increase of the switching angle when the angle is within its "normal operating region."

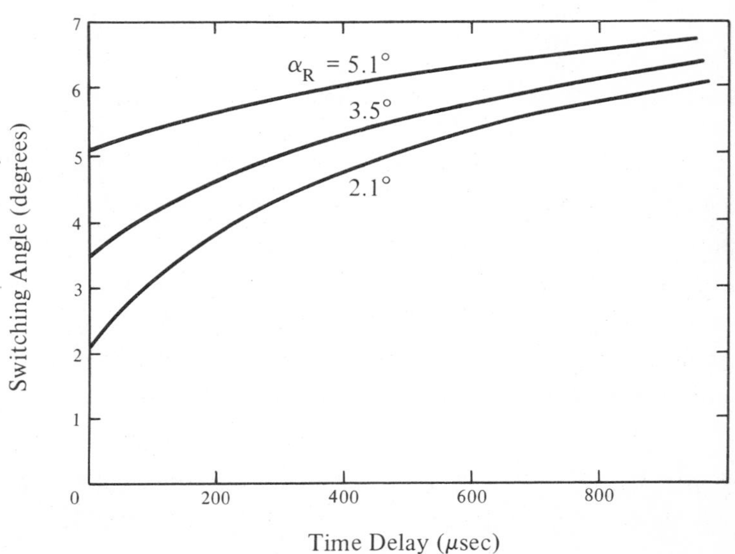

Figure 11-17. Effective switching angle versus time delay.

Given the desired final speed, we can now determine the amount of time delay T required for a desired effective switching angle α, for a given reference angle α_R. The relationship is plotted as shown in Figure 11-17.

As a sample calculation, consider that α_R = 2 deg, and the desired speed is 1000 steps per second. Then, from Figure 11-4, α = 5.1 deg, and using Equation (11-1),

$$\alpha' = \alpha - \alpha_R = 5.1 - 2 = 3.1° \quad . \tag{11-3}$$

Therefore, the amount of time delay required for a 7.5 degree-per-step motor is obtained from Equation (11-2),

$$T = \frac{3.1}{(1000)(7.5)} = 0.413 \text{ ms} \tag{11-4}$$

For the same motor described above, using Equation (11-2), the final speed of the motor can be determined as a function of the time delay. Figure 11-18 illustrates the relationship for several values of α_R.

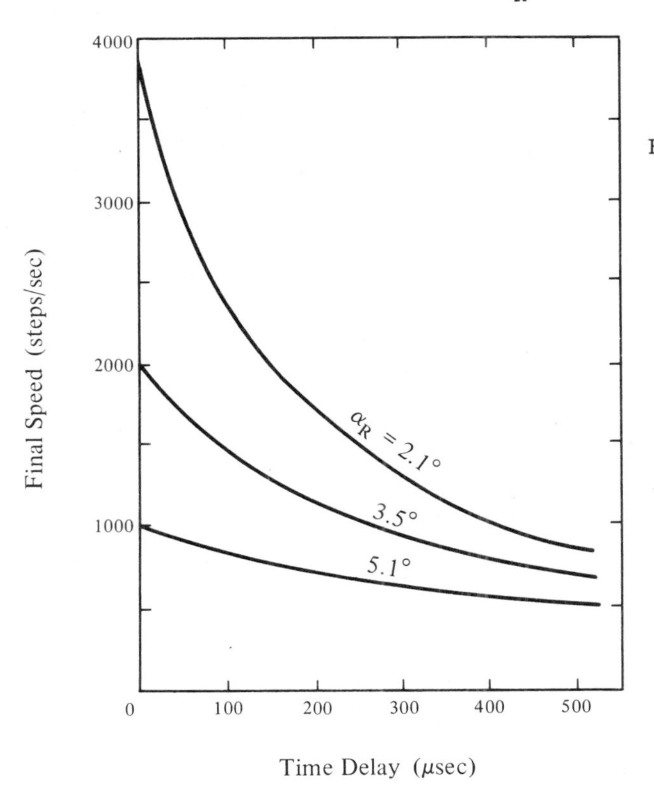

Figure 11-18. Steady-state speed versus time delay for step motors.

Now let us illustrate the effects of the time-delayed feedback on the speed response of step motors by some simulation results. The computer simulation is conducted on a Warner Electric SM048-AB motor.

We have established earlier from Figure 11-18 that when the reference switching angle α_R is 2 deg, a time delay of 0.41 ms is necessary for the motor to reach a final steady-state speed of 1000 steps per second. The simulation result for this condition is shown in Figure 11-19 along with the speed response with only a switching angle of 5.1 deg and no time delay. We can clearly see that the rise time of the speed response with the time delay is shorter than that of the response with a pure switching angle. The reason for this difference is due to the speed dependence of the angle delay α' which is attributed to the time delay.

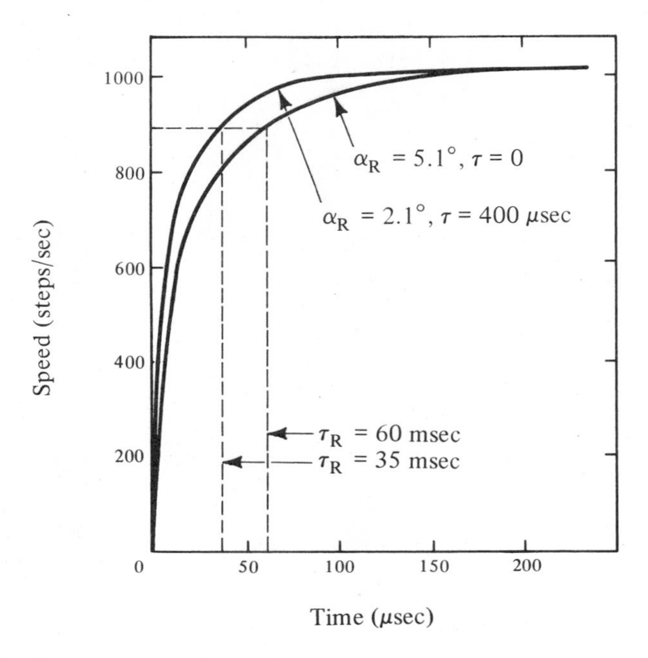

Figure 11-19. Time-delay effect on transient speed response.

During the initial acceleration of the motor when the speed of the motor increases from zero to around 500 steps per second, the effective switching angle is between 2 deg and 3.5 deg. From Figure 11-20 it can be seen that the angle 5.1 deg gives large values of torque only at low speeds. For speeds greater than 325 steps per second in Figure 11-20, the smaller angles of 2 to 3.5 deg produce a higher average torque that accelerates the motor more rapidly. This explains the fact that the motor with a speed-dependent switching angle has a better acceleration characteristic at the start. As the motor speed approaches 1000 steps per second, the effective switching angle increases to 5.1 deg and the system with time delay and the system without time delay settle at the same final speed.

Figure 11-21 shows the change in speed due to a sudden increase in the delay time once a steady-state speed has been reached. With α_R = 2 deg and an initial delay of 0.18 ms, the effective switching angle is 4 deg and the final steady-state speed is 1450 steps per second. As shown in Figure 11-21, a change of time delay to 0.72 ms after the steady-state speed is reached results in a new speed of 720 steps per second within 30 ms with an effective switching angle of 6 deg. This is compared

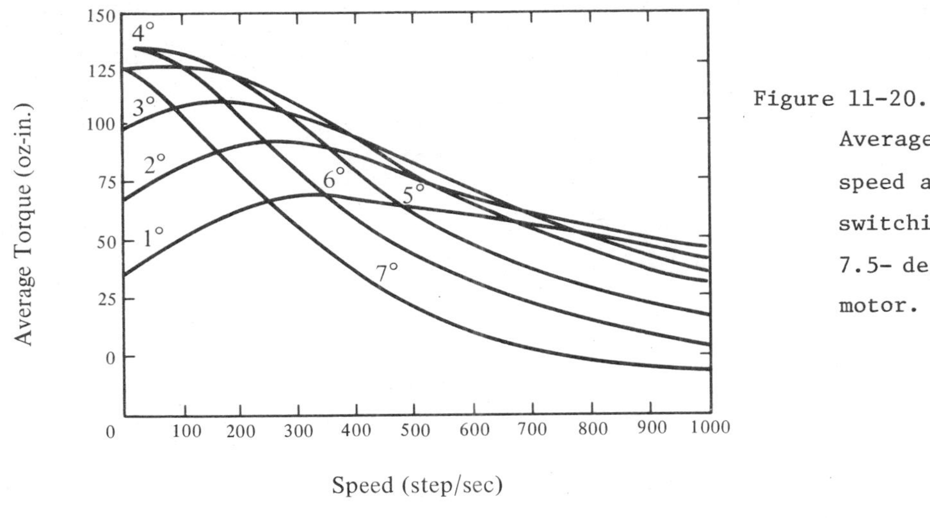

Figure 11-20. Average torque versus speed at constant switching angle for a 7.5- degree/step step motor.

with a response which was obtained by using an exact switching angle of 4 deg and then switching it to 6 deg. Notice that the system with the time delay results in a much shorter transient time.

Figure 11-21. Effect of sudden change in time delay on speed response. Also, difference between speed responses with and without time delay.

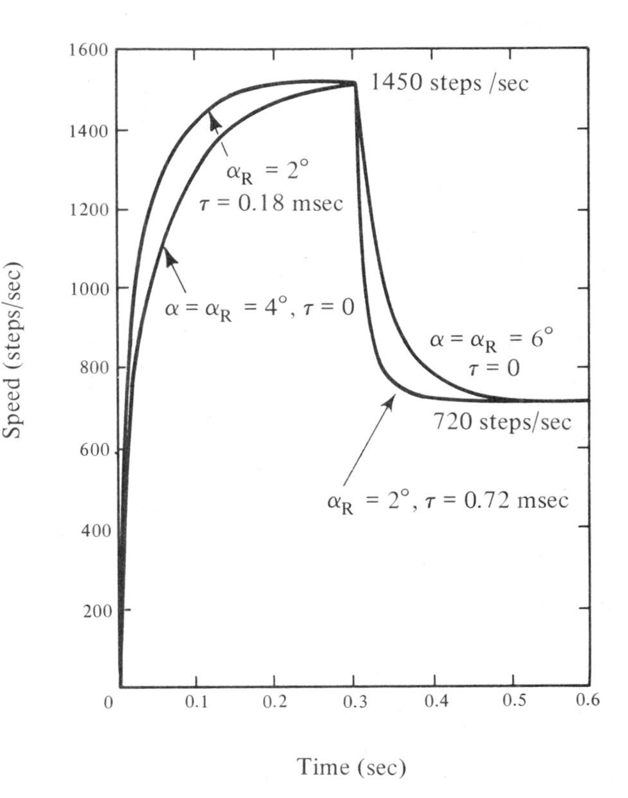

The reason for the improved transient response can again be seen with reference to Figure 11-22. When the time delay is shifted from 0.18 ms to 0.72 ms at 1450 steps per second, the effective switching angle also decreases until at 720 steps per second the effective switching angle is 6 deg and a small positive torque maintains this constant speed. When no time delay is used and the switching angle is changed from 4 deg to 6 deg, the motor develops a small negative torque which decelerates it more slowly than the previous 9 deg angle.

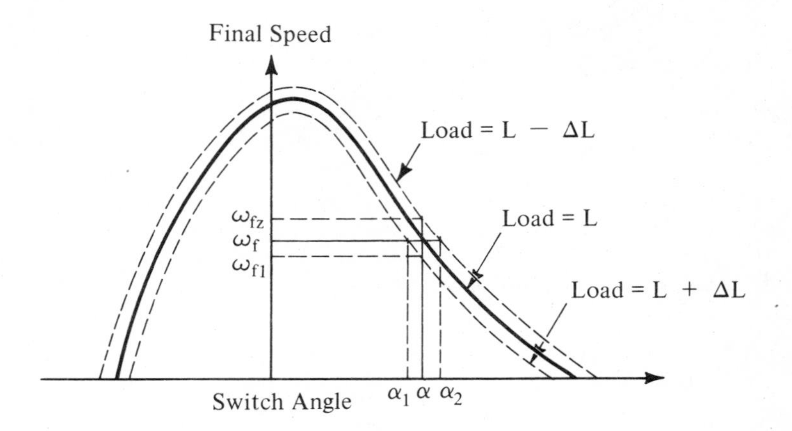

Figure 11-22. Speed variation as a function of the switch angle for small load variations.

We shall now show that the time delay has the effect of decreasing the sensitivity of the final speed of the motor due to load variations. For small load variations, the final steady-state speed of the motor as a function of the switching angle α is as shown in Figure 11-22. It is shown that for a given load L, the motor speed is designated by W_f when the switching angle is α. An increase in load to L + ΔL will result in a decrease in speed to W_{f1}, and a decrease in load to L - ΔL will result in an increase in speed to W_{f2}, if the switching angle is maintained at α. Figure 11-22 shows that when the load is increased, to maintain the same steady-state speed, the switching angle must be decreased, and the reverse is true for a decrease in load.

Consider a desired final speed of 1000 steps per second, and a load of 5 oz-in. It was shown earlier that this requires a switching angle of 5.1 deg. Now if the

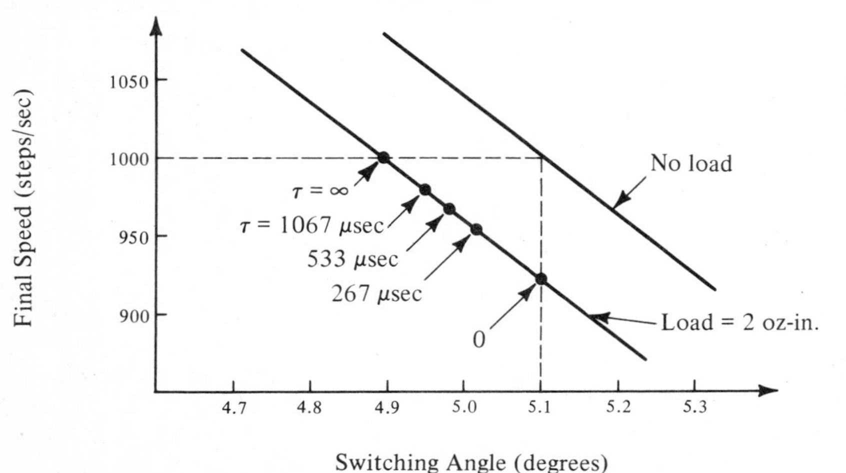

Figure 11-23. Linearized speed curves showing effect of time delay on load sensitivity.

load is suddenly increased to 7 oz-in, the speed is dropped to 920 steps per second. Figure 11-23 shows the final speed versus switching angle curves linearized at the speed of 1000 steps/sec. For the loads of 5 and 7 oz-in, notice that when the load is increased from 5 to 7 oz-in, in order to maintain the speed at 1000 steps per second, the switching angle must be decreased from 5.1 to 4.89 deg. Now consider that a time delay is introduced at a reference switching angle of 1.1 deg. Then, $\alpha' = 4$ deg, and the required time delay for a speed of 1000 steps per second is

$$T = \frac{4}{(7.5)(1000)} = 0.533 \text{ ms} \tag{11-5}$$

As the load is increased, the speed would normally decrease to a new speed W_{f1}. However, the effective switching angle at the speed W_{f1} which is given by

$$\alpha' = W_{f1}Tr = (0.533)(7.5)W_{f1} \tag{11-6}$$

is now less than 4 deg. Therefore, the final speed will be greater than W_{f1}. In the present case, the final speed with time delay will be greater than 920 steps per second. Similarly, if the load decreases, the switching angle α' increases, thus causing the final speed to be less than the value when there is no time delay. The speed response of the motor is shown in Figure 11-24 for a load increase from 5 oz-in to 7 oz-in with time delays of T = 0, T = 0.267 ms, and T = 0.533 ms.

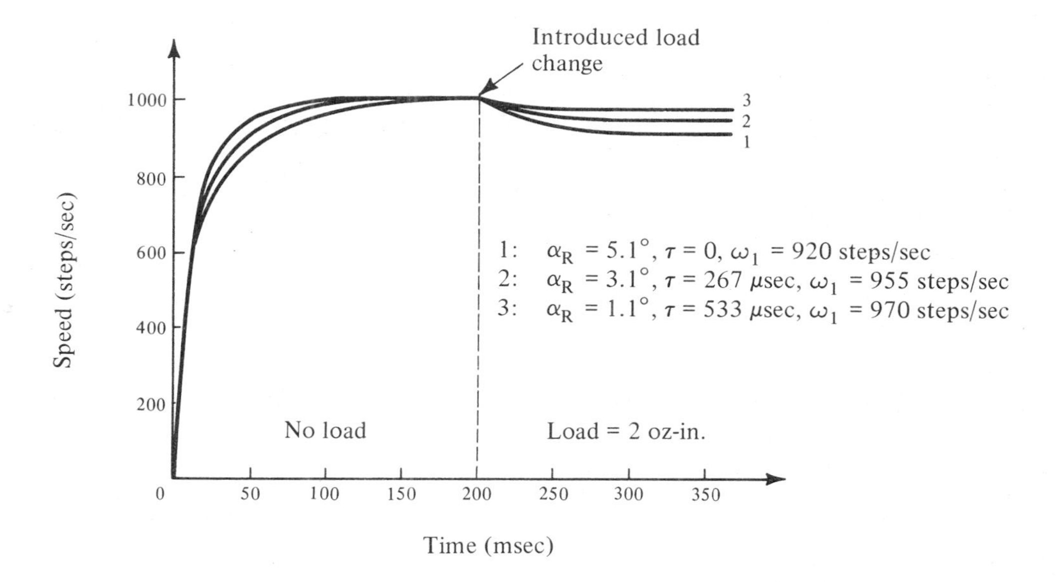

Figure 11-24. Effect of time delay on speed response with load change.

Using Figure 11-23 it is possible to calculate the final steady-state speed when the load changes. As an illustrative example, consider that the reference switching angle α_R is 3.1 deg, and T = 0.267 ms. This gives a switching angle of 5.1° and a speed of 1000 steps per second at the nominal load of 5 oz-in. When the load is increased to 7 oz-in, we know that the final speed will be less than 1000 steps per second.

This final speed can be calculated with the use of Equations (11-1) and (11-2), and Figure 11-24. About the nominal speed of 1000 steps per second, the final speed W_f may be expressed as a function of the switching angle, α, and the load, L, in oz-in,

$$W_f = B_1 + B_2\alpha \tag{11-7}$$

where B_1 is given by

$$B_1 = C_1L + C_2 \quad . \tag{11-8}$$

B_2 and C are constants which may be obtained from Figure 11-24:

$$B_2 = -382$$

$$C_1 = -40.0$$

$$C_2 = 3148.$$

Then,

$$W_f = C_1L + C_2 + B_2\alpha \quad . \tag{11-9}$$

Substituting Equation (11-1) into Equation (11-9), we have

$$W_f = C_1L + C_2 + B_2(\alpha_R + \alpha') \quad . \tag{11-10}$$

Using Equation (11-2), the last equation becomes

$$W_f = C_1L + C_2 + B_2\alpha_R + B_2W_fTr \tag{11-11}$$

or

$$W_f = \frac{C_1L + C_2 + B_2\alpha_R}{1 - B_2Tr} \tag{11-12}$$

At the new load of 7 oz-in, the final speed is found to be 955 steps per second.

Figure 11-25 illustrates the final speed of the motor as a function of the reference angle and time delay when the motor is initially at a speed of 1000 steps per second, and the load is increased from 5 oz-in to 7 oz-in. Since the motor is at an initial speed of 1000 steps per second, the reference angle and the time delay are not independent variables but are algebraically related by Equation (11-2). The lowest point on the graph corresponds to the case where no time delay is used and the motor speed drops to a value of 920 steps per second. The second point on the graph corresponds to the above case where T = 0.267 ms and the speed only drops to 955 steps

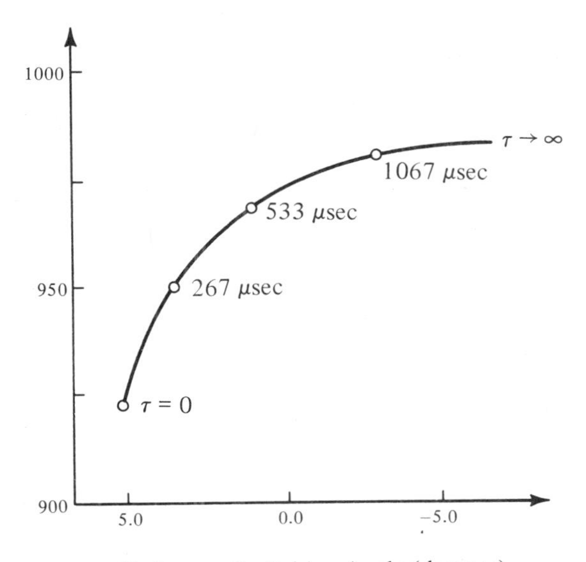

Figure 11-25. Steady-state speed of step motor with load for a system with time delay.

per second. If the reference angle is shifted further back and a larger value of time delay is used to obtain the initial speed of 1000 steps per second, the change in motor speed will be less and the final speed will be higher. In other words, when the time delay is increased, the sensitivity of the final speed to load variation decreases.

In the limit, as T approaches infinity, the sensitivity reaches zero. However, in practice, as the delay time T becomes large the closed-loop feedback system itself becomes ineffective as changes in speed are not compensated for immediately, but only after a time T. As a result, the speed would oscillate even without load variations. In fact, as T becomes infinite, the closed-loop system becomes an open-loop one, and it is well known that the open-loop speed response of a stepping motor is usually quite oscillatory. In practice, a compromise should be reached in determining the proper value of T. In the present case, Figure 11-25 shows that very little is gained by using a time delay greater than 1 ms.

11-6 EXPERIMENTAL RESULTS ON CLOSED-LOOP STEP MOTOR CONTROL

In this section experimental results are given to show the effects of variation in sensor angle setting and shaft loading on the operation of the step motor in the closed-loop system. Various inertial loads with moments of inertia equal to some multiples of the moment of inertia of the rotor of the Warner Electric SM-036 step motor ($5 \times J$, $10 \times J$, $20 \times J$, $J = 0.000354$ oz-in-sec^2) are used for the tests. For the unloaded case, ($J = 0$), the data were taken with only a tachometer mounted on the shaft.

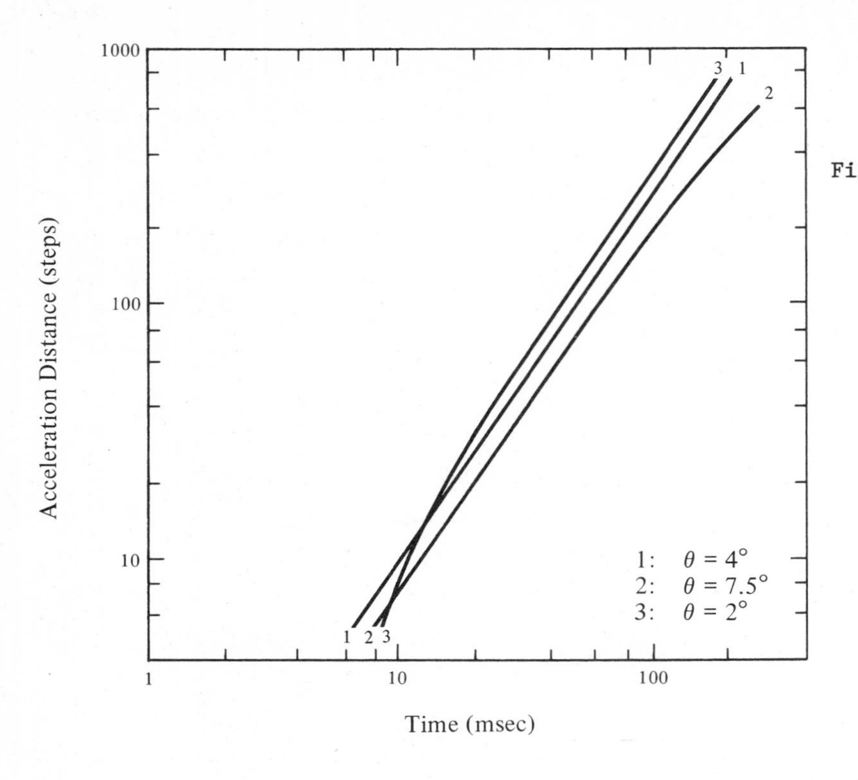

Figure 11-26. Relationship of acceleration distance and time for unloaded case.

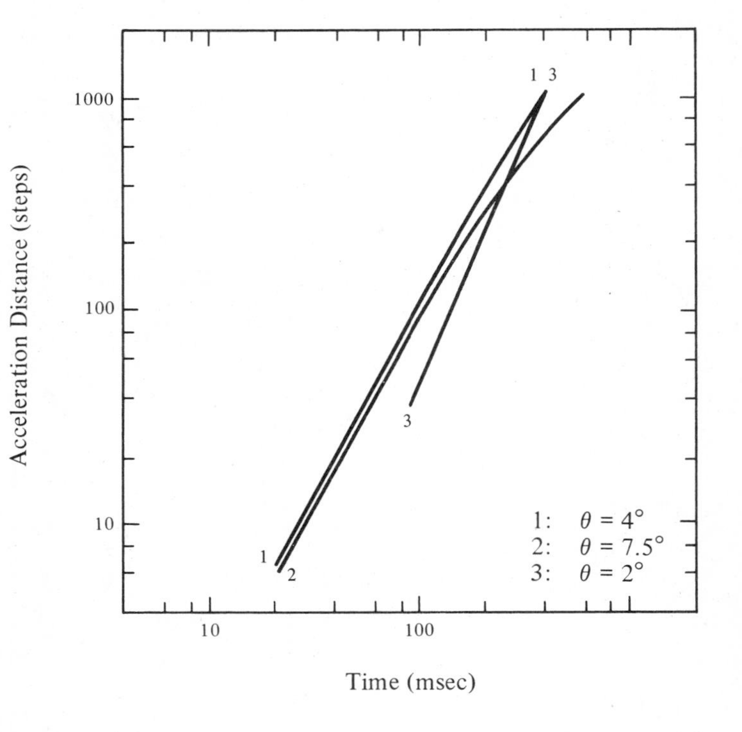

Figure 11-27. Relationship of acceleration distance and time for load = 5 × J ($J = 0.000354$ oz-in-sec^2).

Figures 11-26 and 11-27 show the effect of sensor angle variation on the acceleration of the step motor. These data were taken by allowing the motor to accelerate for a given number of steps and observing the time required to travel that distance. For the unloaded case, a setting of $\theta = 2°$ produced the best acceleration for distances greater than 16 steps. This phenomenon occurs because at distances greater than 16 steps, the step motor is operating at relatively high speeds (greater than 2000 steps/sec), and thus, the settings of $\theta = 4°$ and $\theta = 7.5°$ do not allow the current in the stator windings sufficient time to build up to maximum value. Consequently, the average torque is not maintained at its highest value and a slower acceleration results. For distances less than 16 steps, a value of $\theta = 4°$ produces the best acceleration because the motor is traveling at slower speeds (500 to 2000 steps/sec) in this distance range, and the values of $\theta = 2°$ and $\theta = 7.5°$ provide phase switching too early and too late, respectively, to produce maximum average torque. With a shaft load of five times the rotor inertia, the motor required approximately 1000 steps to achieve a relatively high speed (2000 steps/sec, see Figure 11-29). Thus, a sensor setting of $\theta = 4°$ provided the best acceleration for distances less than 1000 steps for the same reasons stated in the unloaded case. Figure 11-28 illustrates the effect of load variation on step motor acceleration. It may be seen quite readily that the larger loads require more torque to accelerate the motor, and, therefore, require longer time intervals to travel the same given distance.

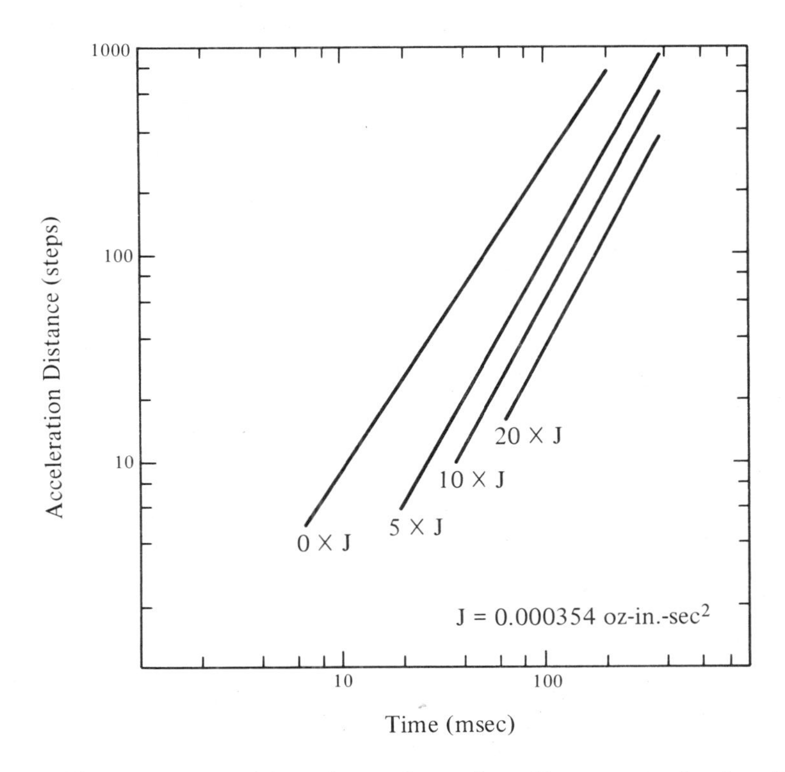

Figure 11-28. Relationship of acceleration distance and time for $\theta = 4°$.

Figures 11-29 and 11-30 show the relationship of maximum rotor speed to acceleration distance. For distances greater than 4 steps, the values of $\theta = 2°$ and $\theta = 4°$ produced the higher maximum speeds. This characteristic occurs because the value of $\theta = 7.5°$ does not allow the current sufficient time to build up in the windings, resulting in low average torque values, and, hence, low maximum speeds. Similarly, the maximum rotor speed varies inversely with the shaft loading because the larger loads require more torque to accelerate the motor to a given speed, and, therefore require longer distances to reach that speed.

Figure 11-29. Relationship of maximum rotor speed and acceleration distance for unloaded case.

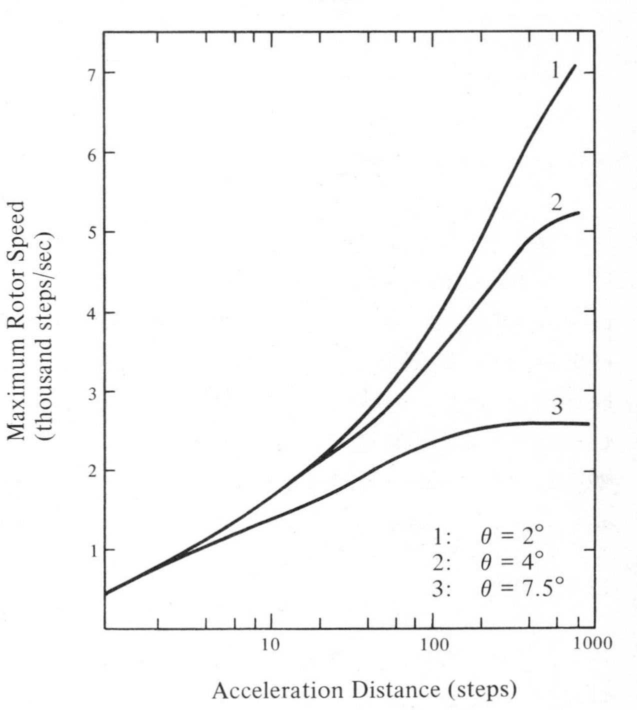

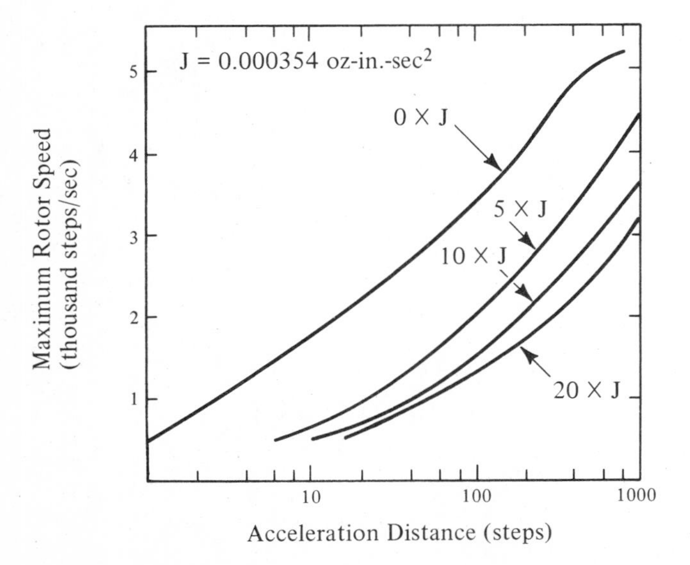

Figure 11-30. Relationship of maximum rotor speed and acceleration distance for $\theta = 4°$.

12 APPROACHES TO STEP MOTOR CONTROLS

J. P. Pawletko
IBM System Products Division
Endicott, New York

12-1 INTRODUCTION

Step motors have been enjoying increasing application and this is due, in part, to the wide selection of motor control systems that are currently available. The choice of a particular control scheme depends largely on whatever performance criteria and economic factors that have to be met in a specific application. Once an initial selection has been made between a variable reluctance (VR) and permanent magnet (PM) motor the next step is to determine the best combination of drive system and motor parameters to obtain the desired dynamic response.

12-2 THE BASIC CONTROL SYSTEM

Step motor control systems can be generally classified into two groups: the closed-loop and open-loop systems. In these basic categories there are several control variations, each of which has its own characteristics and applications.

In general, however, any step motor control system can be represented by a simplified system as shown in Figure 12-1. Here, the command source, depicted by

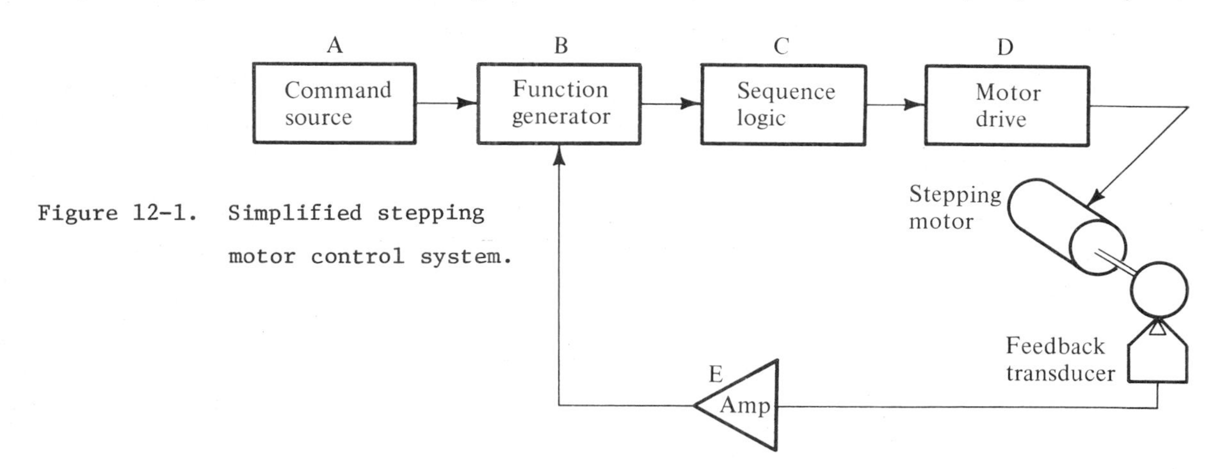

Figure 12-1. Simplified stepping motor control system.

block A, may be a manual or local control or part of a controller in a larger system. The start-stop and direction commands are generated from this area. The function generator, which we will discuss in detail, is the source of the stepping motor advance pulses which are manipulated in such a manner so as to achieve the desired motor dynamics. The sequencing logic, block C, provides for proper driver switching sequences. This logic is usually implemented in latches or triggers. The motor drive circuits consist of solid state devices capable of sufficient current carrying capacity and voltage breakdown protection to handle worst case operating conditions. If closed loop operation is desired then an optical, magnetic or capacitive feedback device is used in conjunction with an amplifier to supply feedback signals to the function generator. Each of these areas with the exception of the command source will be examined.

12-3 STEPPER MOTOR FUNCTION LOGIC - OPEN LOOP SYSTEMS

The function generator logic is the source of motor advance pulses to the sequencing circuits. Manipulation of this logic allows one to tailor the motor characteristics to fit the desired response. Either open loop or closed loop feedback control can be used. The simple open loop control methods will be discussed in this section. Two phase energization provides the most favorable detenting accuracy and best damping, therefore we will confine our discussions to this basic motor scheme except for the "bang-bang" drive scheme.

The Simple "Bang-Bang" Stepping Motor Drive

A frequent requirement of an incrementer is to advance one step at a time. Single step response, as we all know, is very oscillatory and various mechanical means have been employed to reduce that kind of behavior. By properly controlling the field switching sequences of the stepping motor a damped response can be achieved.

In order to understand this kind of system more clearly a simplified stepping motor is shown in Figure 12-2. The motor diagram shows the relative pole and winding relationships. The pulse sequence applies to this model for a counterclockwise (CCW) rotation. When winding A is turned on and winding B turned off the rotor moves toward A at a rate determined by the system's $\ddot{\theta}$. When the windings are returned to their original condition (B on), a negative torque is generated thus causing the system to decelerate. Since no rotation reversal occurs, the rotor will approach the desired pole A with minimum velocity. Just before the rotor stops, winding A is energized again and this holds the rotor at that position. Two phase drives should be used whenever possible because of the higher damping associated with this approach.

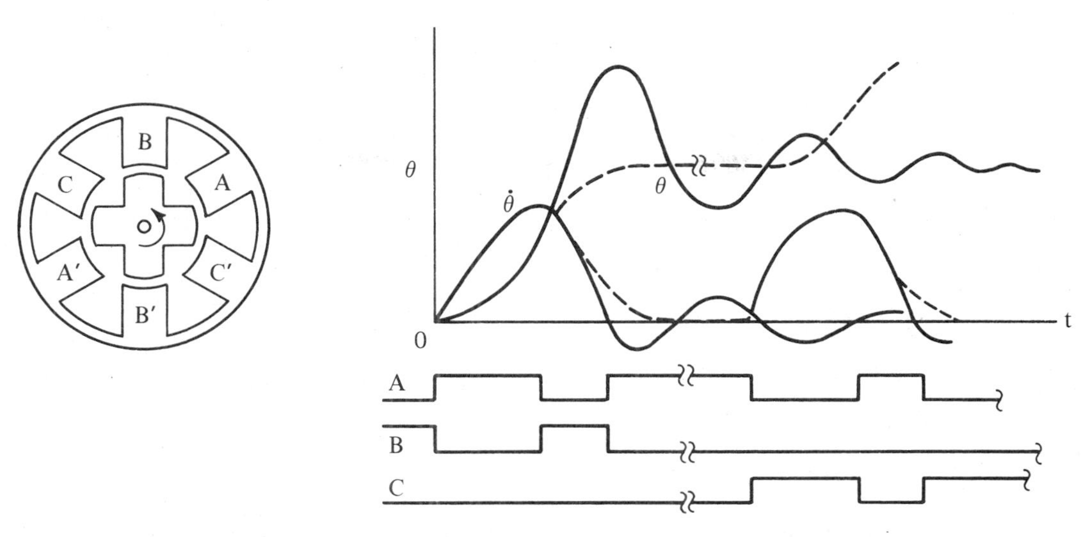

Figure 12-2. Simplified step motor, switching sequence and dynamic response.

Single-Shot Drive Sequence

For continuously incrementing motion, a group of cascaded single-shots can be used to generate a nonoscillatory displacement profile. A time displacement profile shown in Figure 12-3 is a composite of a three-step response. The switching at T_1 results in the typical single-step response: $\theta_1(t)$ and $\dot{\theta}_1(t)$. The T_2 pulse, which

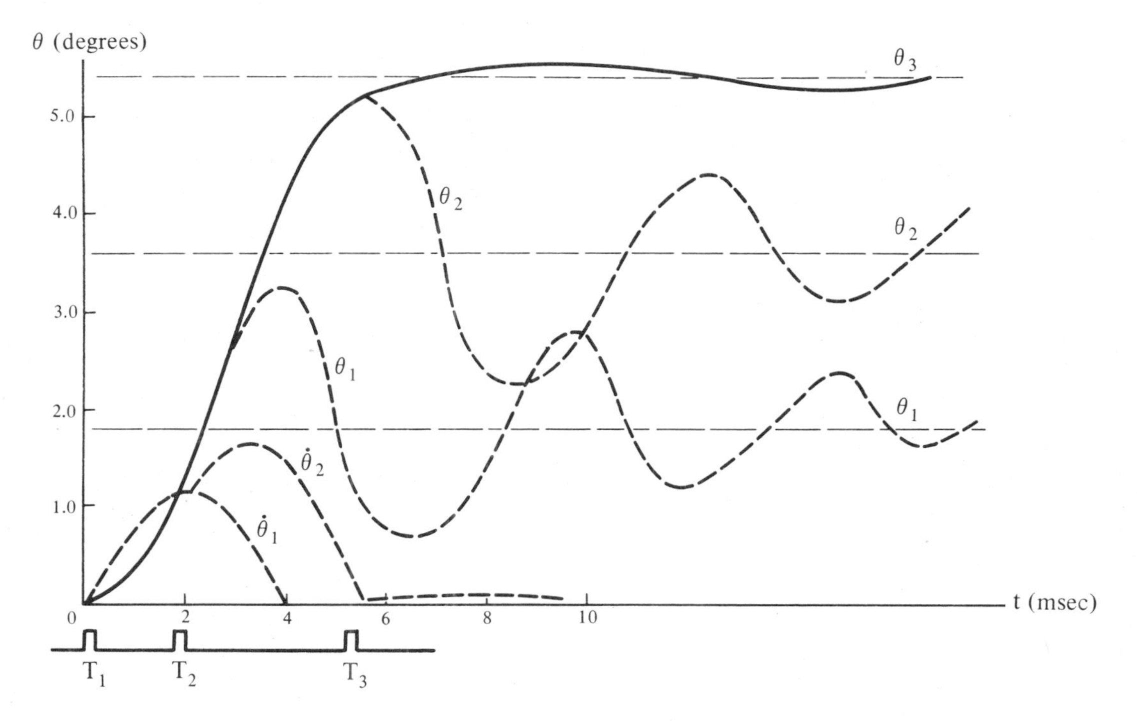

Figure 12-3. Typical three-step response.

occurs just past the peak of $\dot{\theta}_1$ accelerates the rotor toward the θ_2 position. Because of the higher angular velocity the rotor overshoots to within a few percent of θ_3. The T_3 pulse should occur just prior to $\dot{\theta}_2 \simeq 0$. When the rotor is at θ_3 negligible perturbation of the rotor will occur while it detents at θ_3. A properly designed and adjusted system can have a nominal settling amplitude of only a few minutes of a degree.

Gated Oscillator

A simple gated oscillator as shown in Figure 12-4 is sufficient to control the motion of a stepping motor in most noncritical applications. This system will work quite satisfactorily provided the stepping rate of the motor is not exceeded. Start, stop and reversal can be accomplished by external means such as optical sensors or just plain microswitches. The last step is controlled by a single-shot whose timing is initiated by a turn-off signal that is adjusted to coincide with zero velocity condition of the system.

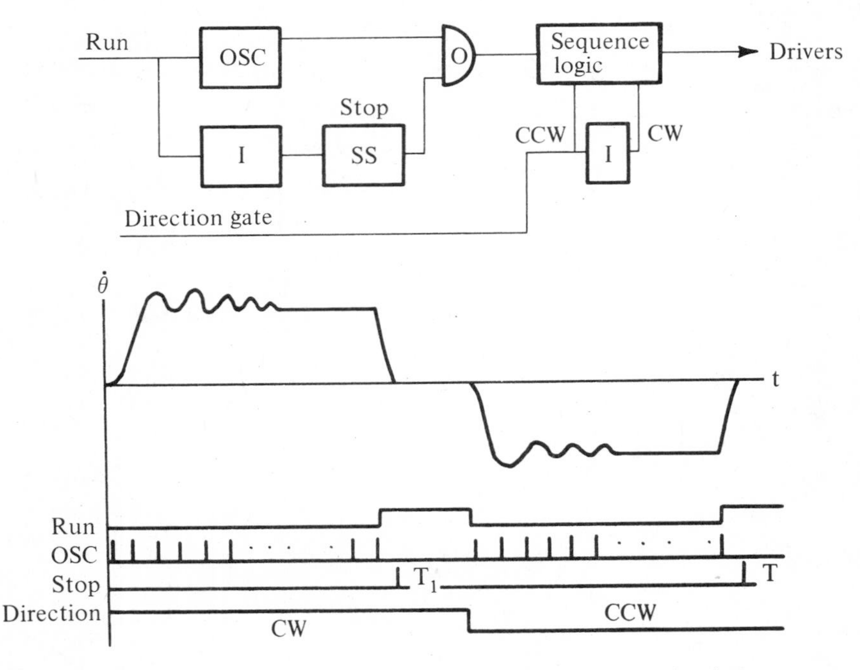

Figure 12-4. Gated oscillator logic and timing.

Variable Frequency Function Logic

In a more sophisticated extension of the gated oscillator, a variable frequency oscillator can be used in open loop drive schemes. In the case where the desired operating frequency is in the slew region of the stepping motor, gradual acceleration to that speed must be achieved. The rate at which the frequency converges to the desired frequency depends on the acceleration capability of the motor in the system:

$$\ddot{\bar{\theta}} = \frac{T_{avg.} - T_{friction}}{J_{total}}$$

Typically, a voltage controlled oscillator (VCO) (as shown in Figure 12-5) is used in these applications because the control ramp can be easily generated to suit the particular requirements.

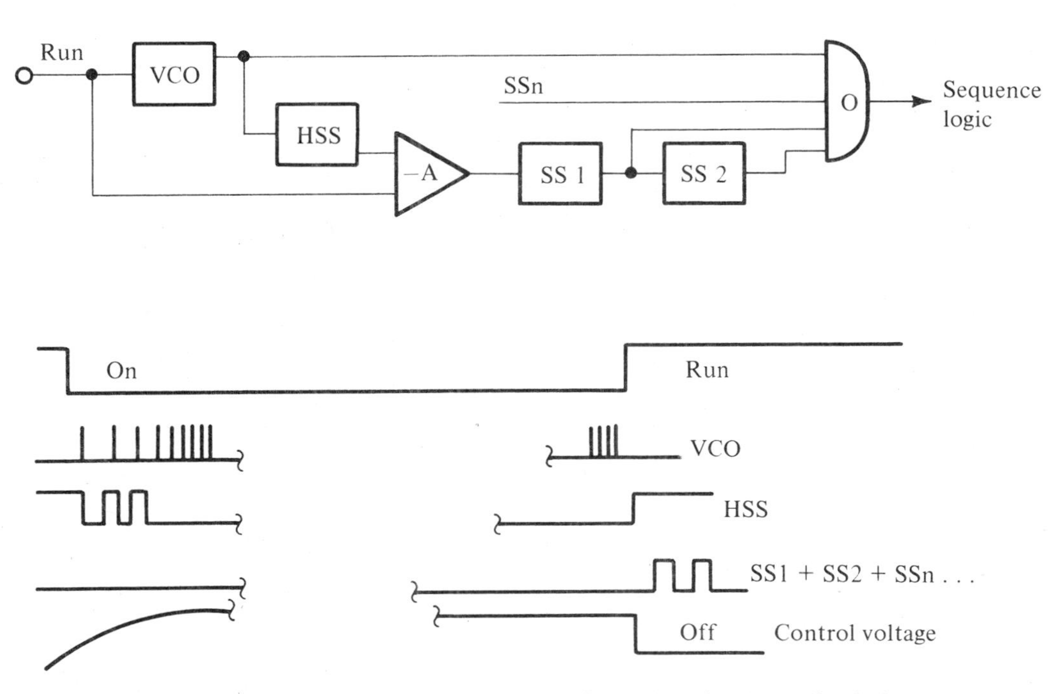

Figure 12-5. Variable-frequency function logic and timing.

A major problem in open loop drive systems after constant velocity motion has been achieved is the generation of a proper stopping response. Stopping sequence pulses must be directly related to the oscillator pulses. No problem exists if counters are used to turn the oscillator off, but if external means are used, such as

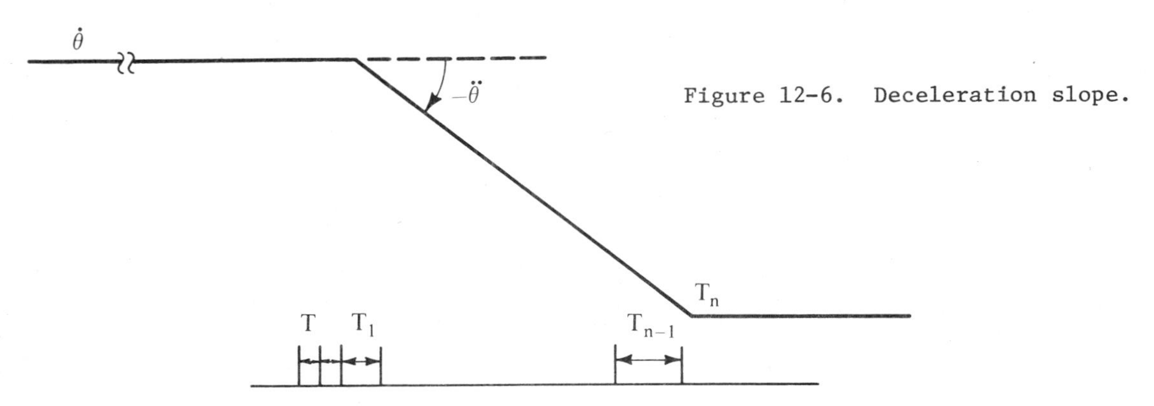

Figure 12-6. Deceleration slope.

photocells, microswitches, etc., it is very difficult to achieve the proper timing. To obtain a nonoscillatory stop from any velocity, two basic conditions must be met: The deceleration slope (see Figure 12-6) must be such that both zero velocity and maximum overshoot are achieved with an integer number of steps. Therefore, the first decelerating pulse must be sufficiently delayed to allow the rotor to lead the field by the amount necessary to meet the $-\ddot{\theta}$ condition. Subsequent pulses are not as critical with the exception of the last pulse. The timing interval between T_{n-1} and T_n is determined by the system ω_n. From previous discussion of the three-step increment, a nonoscillatory stop should occur at the maximum overshoot condition.

Since

$$\omega_n \simeq \left(\frac{K_r}{J}\right)^{1/2}$$

where K_r is the slope of the holding torque curve in oz/radian.

J_T - Total system inertia in oz sec^2,

then

$$4T_n = \frac{2\pi}{\omega_n}$$

Therefore,

$$T_n \simeq \frac{\pi}{2}\left(\frac{K_r}{J}\right)^{-1/2}$$

assuming a low system friction.

Three different approaches to a proper stopping response are summarized in Figure 12-7 which shows timing conditions of a typical system. In Case I, if a

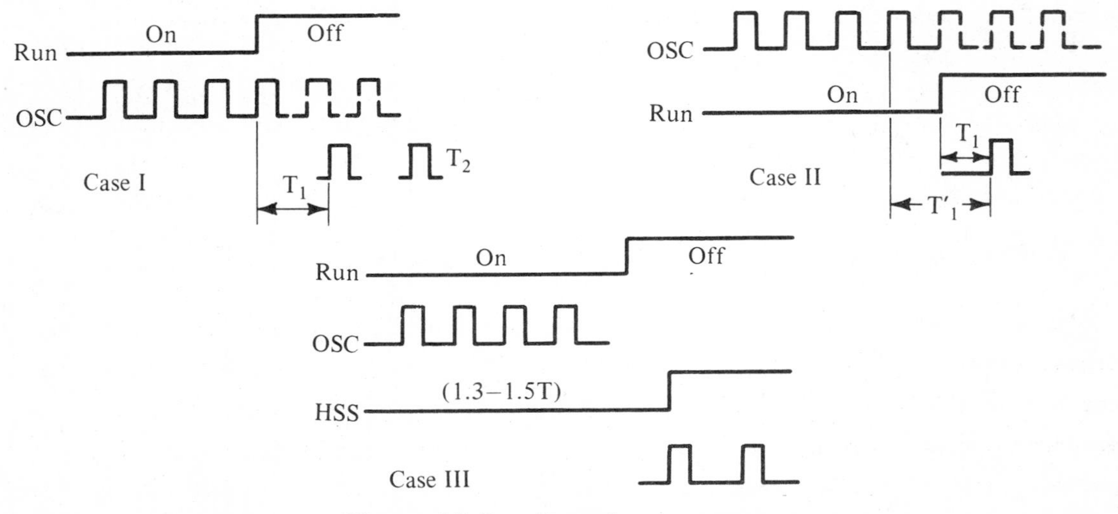

Figure 12-7. Stopping responses.

counter is used to turn the oscillator off, a direct relationship exists between the last oscillator pulse and the first deceleration interval and no additional conditions are needed. If external sensing is used, then the turn-off can happen anytime within the oscillator period as shown in Case II.

To establish the first deceleration interval in Case II, a holdover single-shot is used. The turn-off of the run gate initiates the deceleration single-shot pulse at T_1.

In Case I the run gate turned off shortly after the oscillator pulse and $T_1 \geq$ 1.3T of an oscillator pulse. This will generate the proper slope.

Case II shows the run gate turn-off somewhere within the oscillator interval with the resultant T'_1 approaching 2T. This condition will result in loss of control of the motor.

Case III shows the holdover single-shot extended to 1.3 - 1.5T and it makes the turn-off gate adjustment much less critical.

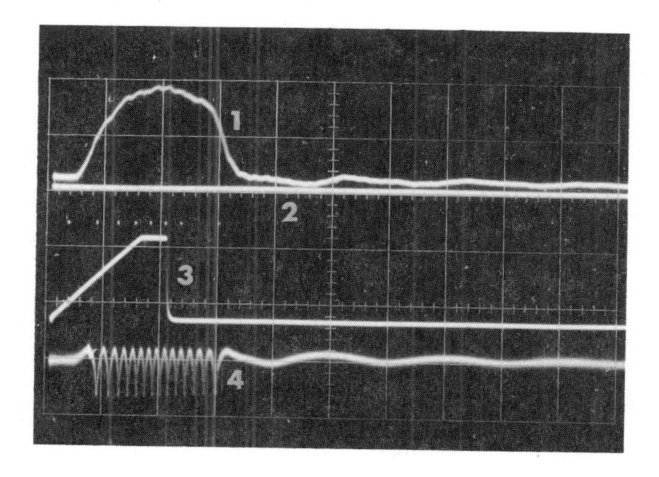

Figure 12-8. Dynamic behavior of a voltage-controlled oscillator driven system with the stopping sequence discussed in the text.

12-4 STEPPER MOTOR FUNCTION LOGIC - CLOSED LOOP FEEDBACK SYSTEMS

Closed loop operation is the most versatile approach to stepping motor drive systems. There are several feedback methods, all of which use a displacement transducer having a one-to-one correspondence between the number of steps and the transducer output frequency. There are two basic transducer types: optical and magnetic. Both of these transducers or encoders generate a square wave output. The DC signal is used as a logic switching signal in DC coupled logic or just the positive and negative transitions can be used in AC coupled logic. The basic advantages of feedback drive are:

Freedom from hunting

High acceleration

Constant velocity

High torque

Simple deceleration methods

Maximum motor performance and various velocity profiles can be obtained by properly controlling the feedback angle. This section discusses several methods that can be implemented to obtain the desired results.

Typical Feedback System

A simple example of a feedback system is one that uses a slotted disc and optical sensor arrangement with a fixed feedback lead angle (Figure 12-9). The disc, which is driven by the stepping motor, provides moving windows for light to the optical sensor. Each feedback signal from the sensor advances the field in the motor by one pole. The typical application finds the resolving slit on the sensor initially in the dark zone. An external advance pulse must be applied to the sequence logic to advance the field one pole and start the motor turning. During the motion either a positive or negative signal transition will occur due to the edge of the next feedback window. The signal transition will advance the field to the next pole. In a typical motor the field will be leading the rotor by 1.25 steps. The rotor will advance each step, continuously accelerating until external parameters impose a limiting condition on the final velocity of the motor.

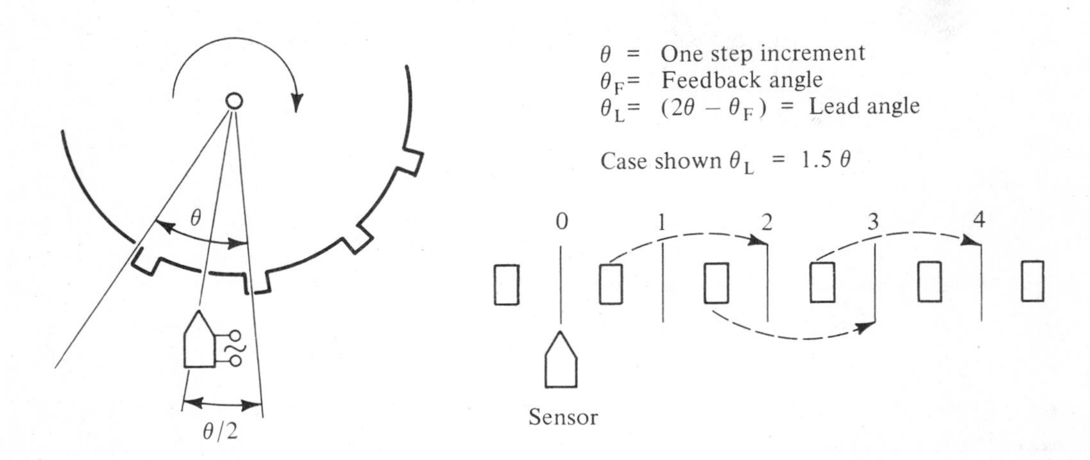

Figure 12-9. Typical feedback system (optical sensor).

Feedback Lead Angle Modulation

Different performance curves can be obtained by modulating the feedback lead angle. The following paragraphs explain three methods to obtain a lead angle change.

In order to obtain a lead angle advance, a signal inversion and a specific level transition can be used to switch the drivers. Using AC coupled logic, positive transitions of the feedback signal causes the motor winding drive circuits to advance. By switching over to the out-of-phase signal the positive transition occurs 1/2 step earlier and a new lead angle is established. This sequence can be repeated one step later with a corresponding increase in slewing speed.

Continuous lead angle modulation can be achieved by using a function generated ramp or by external modulation of the reference voltage. Figure 12-10 shows the logic necessary to do this.

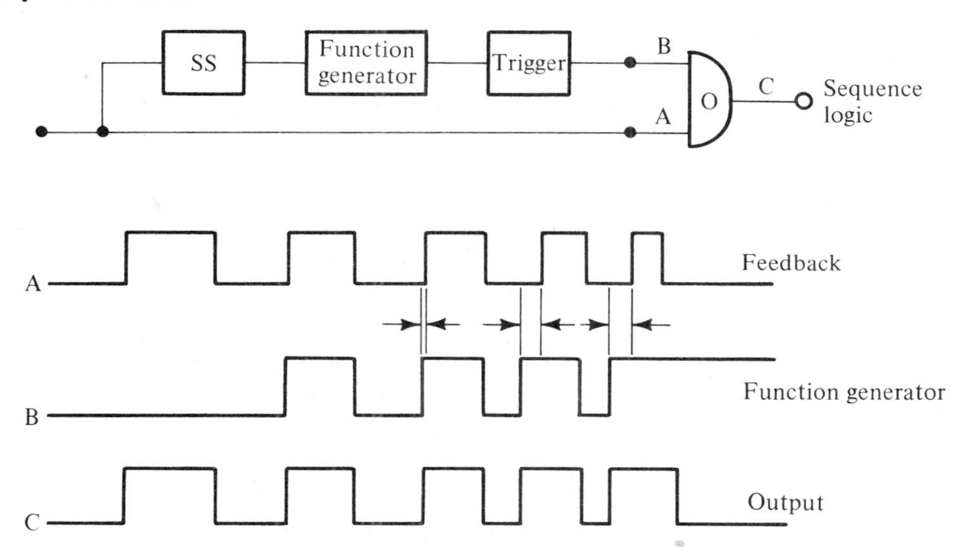

Figure 12-10. Continuous lead angle modulation logic and timing.

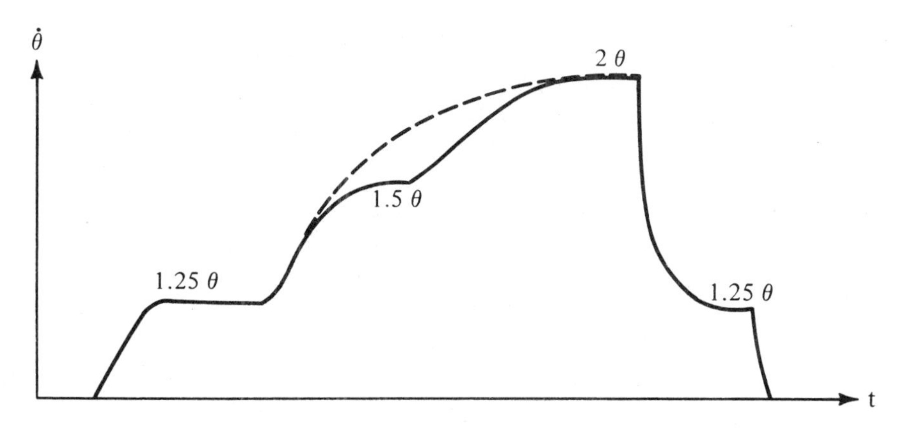

Figure 12-11. Velocity as a function of lead angle.

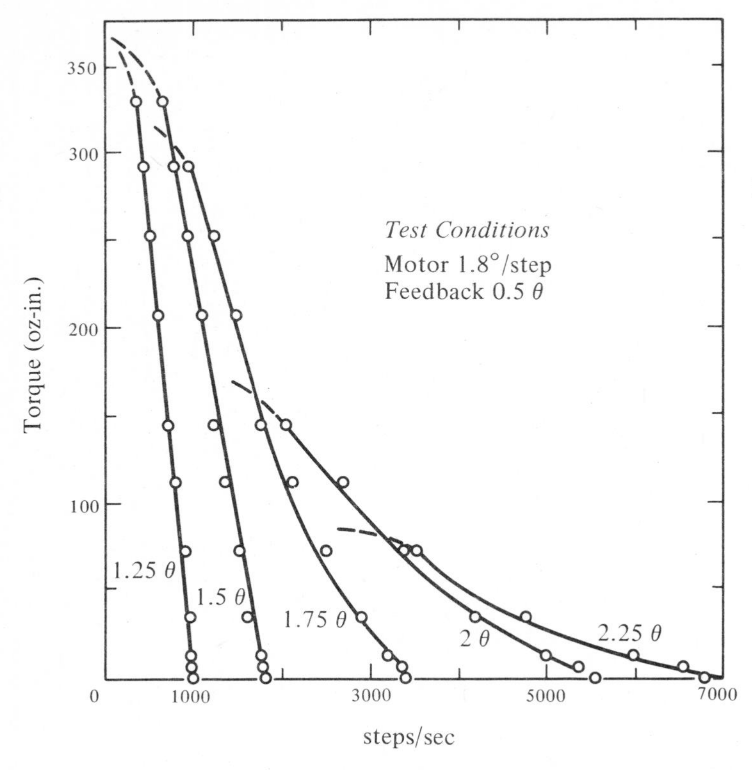

Figure 12-12. Torque-speed for various lead angles.

If higher speeds are desired, an external pulse can be injected between two feedback intervals. This causes a full θ feedback advance which results in a very high slew rate.

A typical velocity profile as a function of lead angle is shown in Figure 12-11 and the effect of lead angle change on the torque-speed slope is shown in Figure 12-12. Observe the very steep or "stiff" characteristic with the minimum lead angle. The upper slew rate is approximately 1000 pps. An additional lead of 0.25θ increses the slew rate to approximately 1.8 of the original value; however, the load line is shallower. Additional lead angle increases the upper rate by another 1.8 of the previous one. The 2θ lead angle follows the same ratio (0.75θ lead angle increase will cause approximately a $(1.8)^3$ slew speed increase). A 2θ lead cannot be obtained initially because this is an ambiguous condition; however, once a specific slew rate is reached one can then advance the lead angle further.

Stepping Motor Deceleration With Feedback Control

Methods to obtain motor deceleration resemble the acceleration techniques except that in this case a lagging field angle is desired instead of a leading angle. Three typical deceleration methods are explained in the following paragraphs.

To initiate deceleration, an externally controlled single-shot can be used to blank any desired number of feedback pulses. The blanking gate will cause a lead angle change in discrete increments of θ. If a typical lead angle is 1.5θ the blanking of two feedback intervals will result in a 0.5θ lag and a corresponding $-\ddot{\theta}$ deceleration rate as shown in Figure 12-13a.

Deceleration by field reversal can be accomplished by generating a CW-CCW signal which causes the driver sequence to be reversed. By injecting the desired number of pulses within a feedback interval, the angle can be changed instantaneously which will cause deceleration to occur (see Figure 12-13b).

In unidirectional drive systems, the second approach would not be feasible. A simpler and more flexible approach is pulse injection. When deceleration is desired, the proper number of pulses are injected in rapid succession. This advances the driver sequence logic without causing the drivers to respond. As shown in Figure 12-13c, the interval can occur at a predetermined point in time and does not require bidirectional sequence logic as in the case of a unidirectional system.

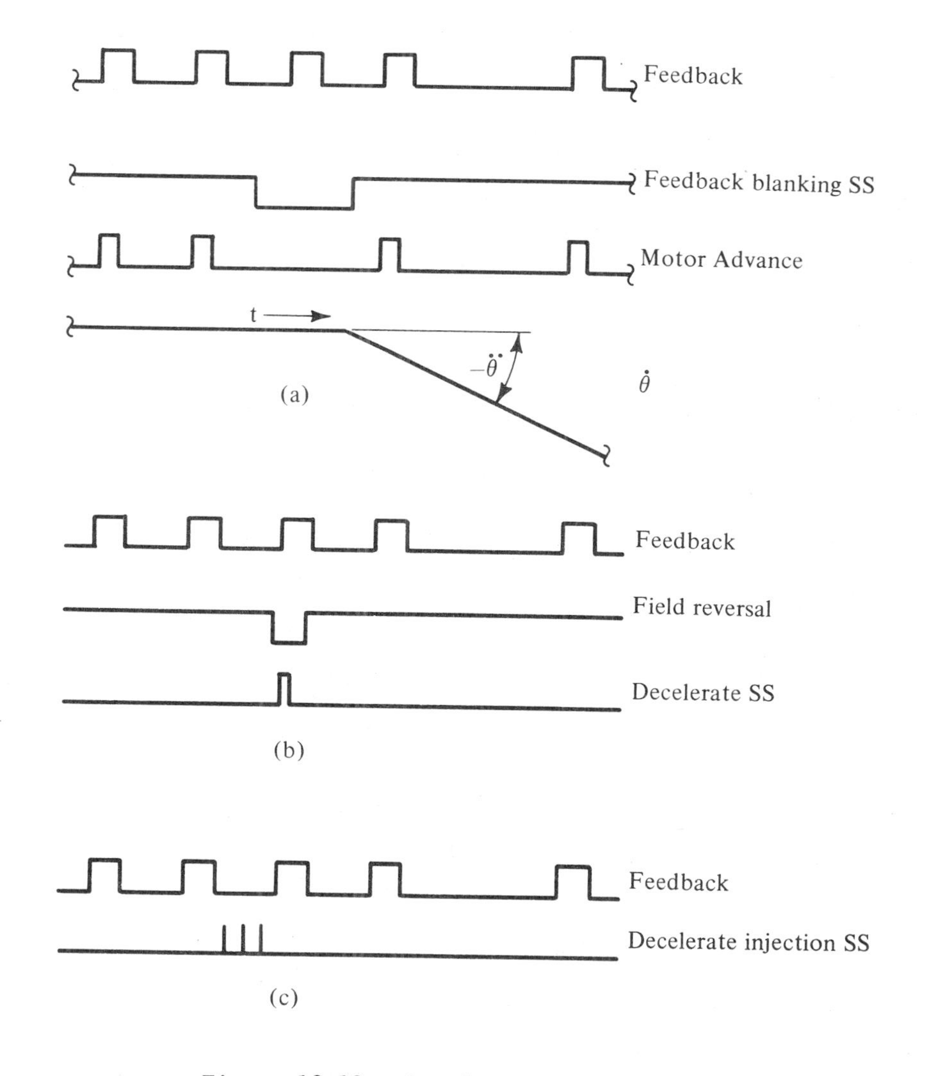

Figure 12-13. Deceleration techniques.

Constant Velocity Operation

If a constant and absolute motor velocity is required, a phase-locked voltage controlled oscillator can be used in the feedback loop. Basic operation does not differ very much from standard phase-locked oscillator applications; however, there are a few basic requirements that must be satisfied:

1. The feedback signal must be symmetrical
2. The drive circuit parameter values must be chosen such that the upper slew rate is above the oscillator frequency
3. The maximum load must not exceed the pull-out torque available at the operating point.

A basic phase-locked circuit configuration is shown in Figure 12-14. The motor is accelerated with feedback in the usual manner and at every feedback transition the oscillator is reset. The up-to-speed comparison logic compares the oscillator to the feedback interval. The oscillator itself runs at the nominal frequency assured by the standard reference voltage. The up-to-speed latch blocks any error voltage signal from appearing at the output of the phase comparator and also blocks the oscillator pulses from appearing at the winding advance OR block. Once the feedback interval becomes shorter than that of the oscillator the up-to-speed latch turns off allowing the oscillator to take over. Since the oscillator is continuously reset by the feedback it is in phase with the feedback and a smooth transition occurs at that point (see Figure 12-15). The feedback deceleration method is also shown in this oscilloscope trace; however, the last stopping pulse was missing and as a result some small oscillations exist. The upper slew rate was 5000 pps.

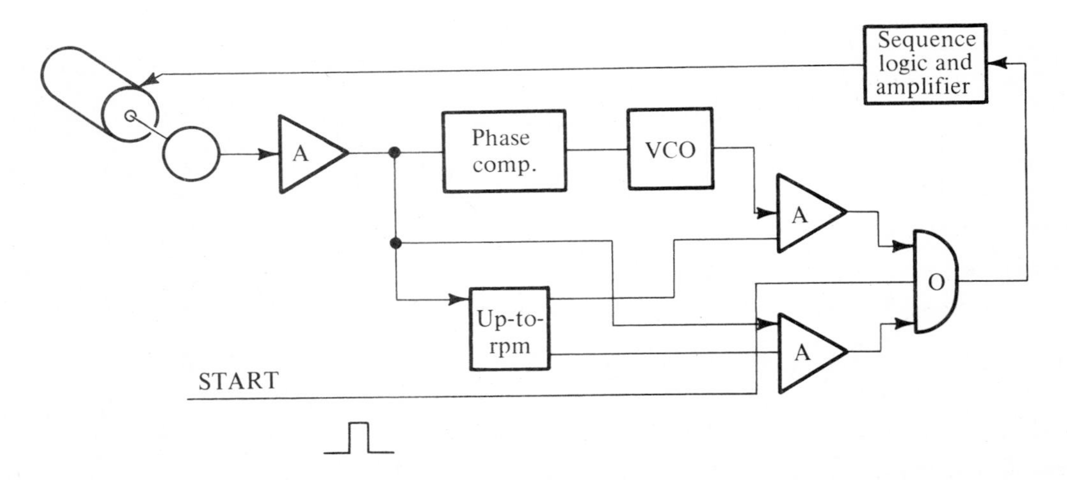

Figure 12-14. Basic phase-locked circuit.

Figure 12-15. Motor velocity profile with phase-locked oscillator.

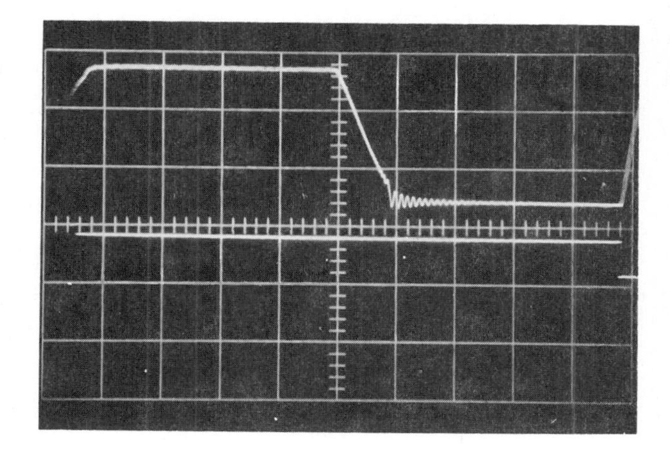

12-5 SEQUENCING LOGIC

The sequencing logic used in most stepping motor applications is either DC logic or triggers. The main purpose of the sequencing logic is to provide the proper phasing sequencing when a series of input pulses or transitions occur.

12-6 MOTOR DRIVE CIRCUITS

The most common motor drive circuits use either PNP or NPN transistors with voltage and current ratings to fit the motor. For most any given application, the desired torque speed curve can be achieved by selecting the proper motor and by manipulating the external drive circuit parameters. These parameters are:

1. L/R ratio of the motor windings
2. Operating voltage and current
3. Overdrive conditions for faster acceleration
4. Commutating capacitor values
5. Feedback lead angle modulation.

The results of manipulating the pullout torque of a stepping motor are shown in Figure 12-16. Curve A is typical for a common connection of a high inductance winding to the rated supply voltage. This circuit is shown in Figure 12-17a. An overdrive condition is shown in Figure 12-16b as curve B and the respective circuit is shown in Figure 12-17b. Curve C shows the torque characteristics for a split phase connection and a low L/R ratio at rated current. The corresponding circuit is shown in Figure 12-17c. An overdriven case used for acceleration only is represented by Curve D and Figure 12-17d.

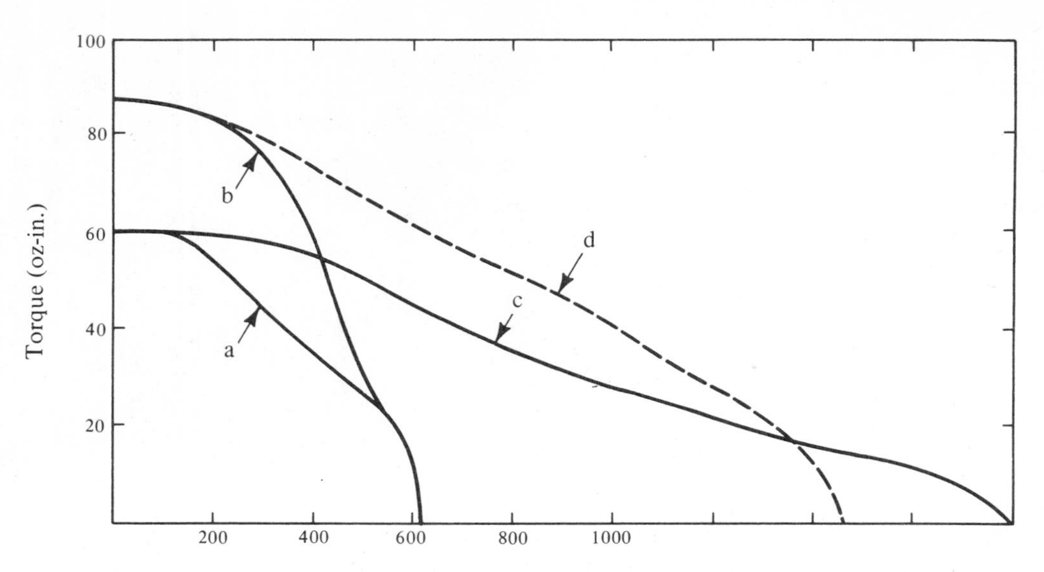

Figure 12-16. Pull-out torque (typical) as a function of drive circuit parameter.

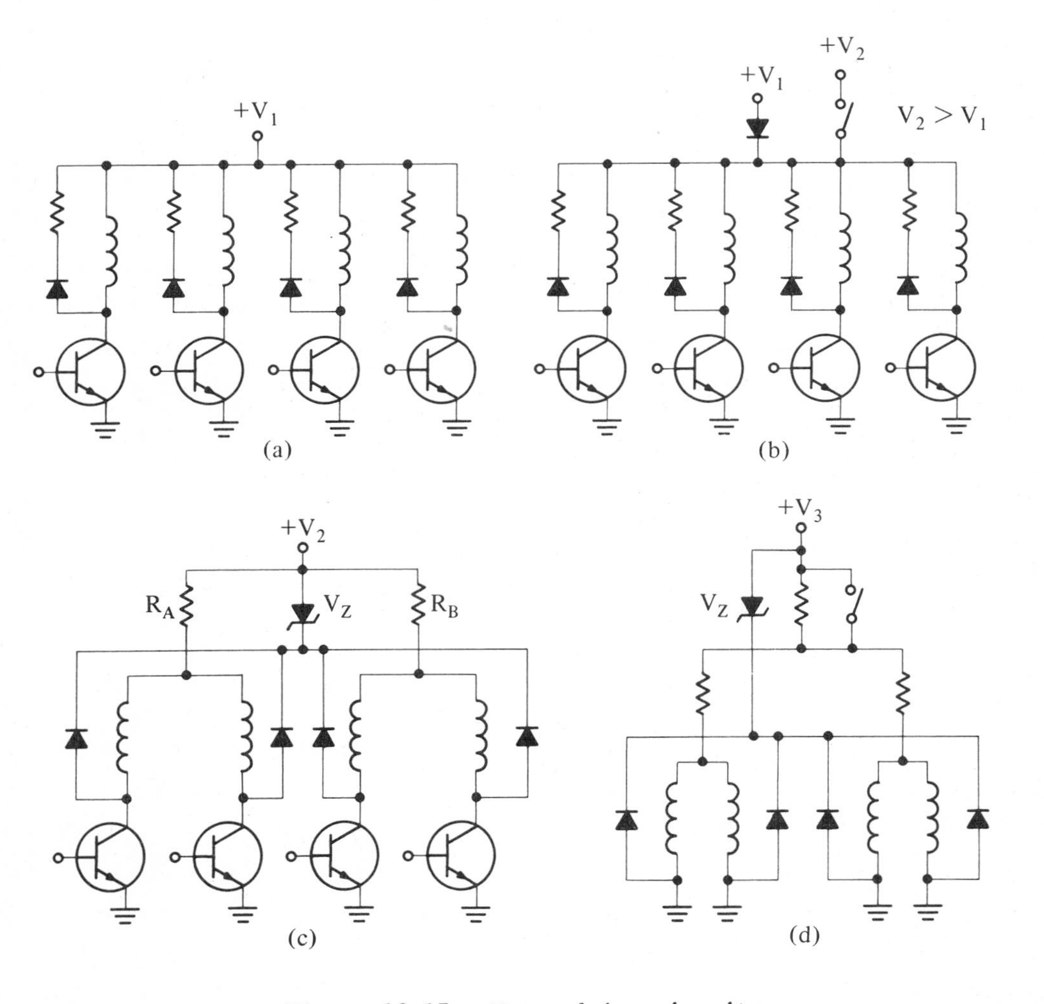

Figure 12-17. Motor drive circuits.

One of the most effective means of shaping the load characteristics of a stepping motor is by the use of a commutating capacitor in the drive winding. This capacitor is effective in both open and closed loop operation. A typical drive circuit employing a commutating capacitor is shown in Figure 12-18. The following advantages are realized by using this capacitor:

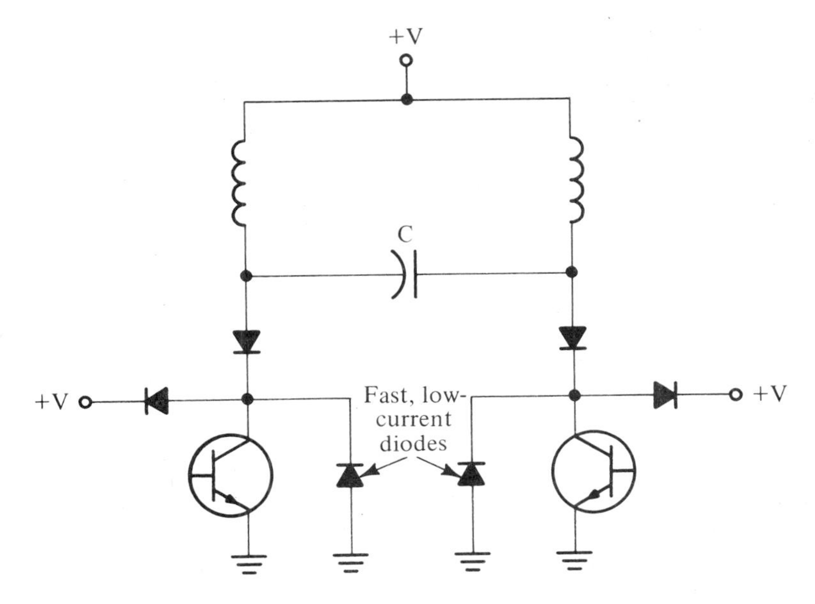

Figure 12-18. Drive circuit with commutating capacitor.

1. Higher power efficiency due to the intermediate storage of energy.

2. Higher current rise time in a resonant circuit as opposed to an L/R ratio of the same motor.

3. Collector protection; V_C limited by C value.

4. No high current spikes, therefore less RFI.

5. Lower harmonic components, i.e. lower eddy currents at any slewing rate.

The torque and average phase current versus stepping rate for a motor with a commutating capacitor in the drive circuit is shown in Figure 12-19. Torque speed characteristics for different values of a commutating capacitor in the drive circuit are shown in Figure 12-20.

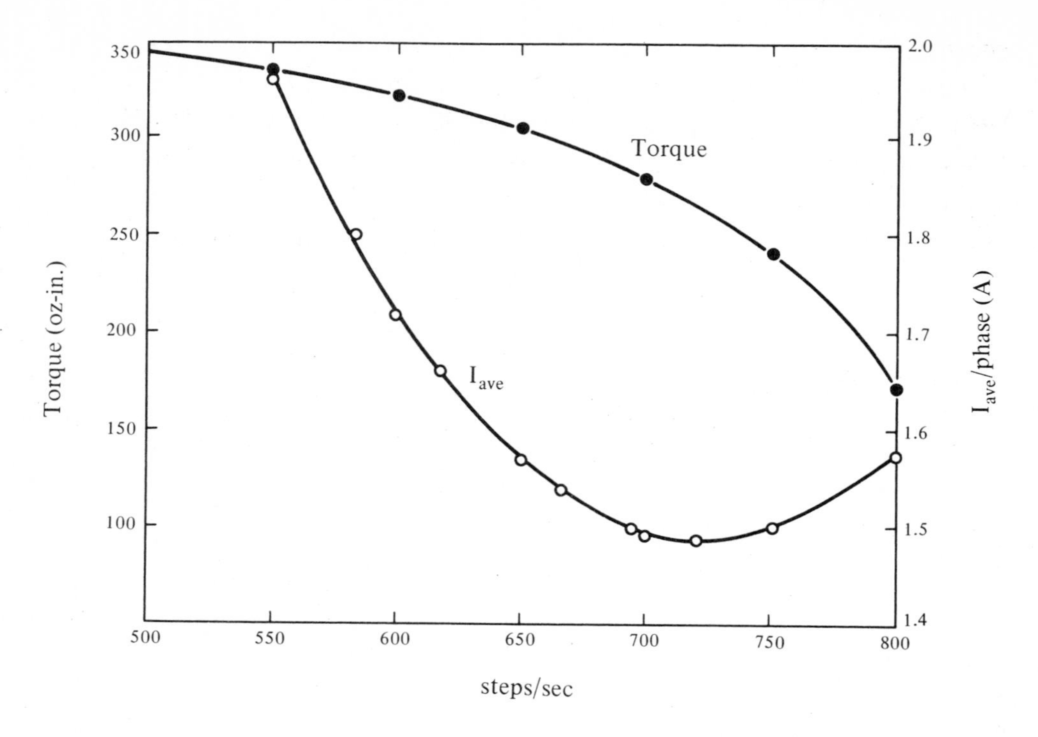

Figure 12-19. Torque and average phase current versus stepping rate.

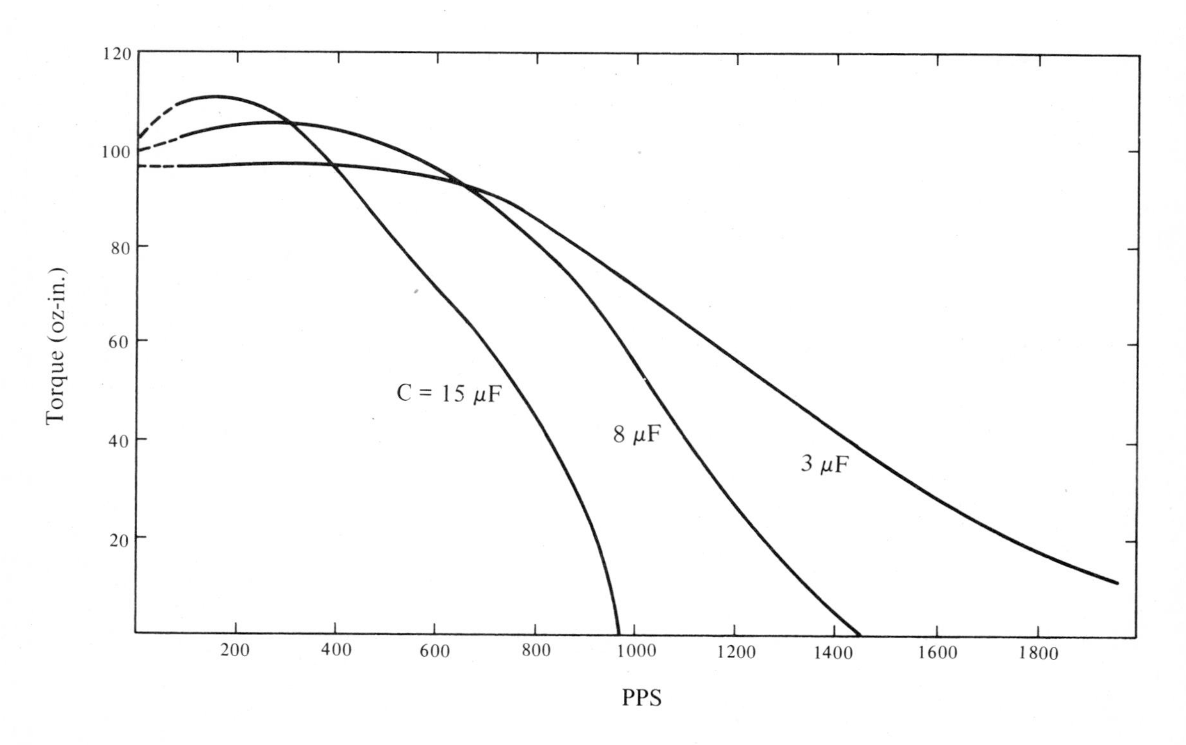

Figure 12-20. Torque-speed characteristics as a function of commutating capacitance.

12-7 CONCLUSION

From the brief discussion of stepping motor control methods, it should be apparent that a whole array of methods is available to the applications engineer. The particular approach depends not only on systems requirements, economic factors, type of stepping motor used but also on the experience of the designer.

13 DESIGN OF DIGITAL CONTROL SYSTEMS WITH STEP MOTORS

T. R. Fredriksen
T. Roland Fredriksen & Associates
Palo Alto, California

13-1 INTRODUCTION

The rapid evolution of digital circuit elements has made it economical to literally chop the classical analog circuits to pieces. Almost daily, a new digital device appears which performs a subfunction that previously required a number of separate components. The result is that today it is less expensive to buy a mini-computer than to construct process sequencers, even if only part of the capabilities of the computer are utilized.

Stepping motor control has likewise benefitted from the digital circuit growth and has indeed kept pace. Beginning in earnest about a decade ago [1 - 9], the step motor made the claim that even mechanical motion should be digitized. Adding the closed-loop principle [10, 11, 13 - 15] increased motor performance and reliability by a factor of ten. Since then, new open-loop techniques have appeared and, in general, there is room and requirement for both open and closed-loop stepping motor control systems.

In this paper, open and closed-loop control of stepping motors is discussed from a practical design point of view. The digital computer is considered to be part of the overall system and control schemes are worked out accordingly. Emphasis is placed on correctly identifying the major control elements rather than to guarantee complete interconnect accuracy and logic details are only given when needed to clarify some critical point.

13-2 COMPOSITION OF A DIGITAL POSITIONAL CONTROL SYSTEM

A digital positioning system is the natural subordinate of a digital computer. In a most realistic way, the stepping motor serves as the physical working arm of the computer and this combination is very analogous to those sections of the human control system associated with basic motion. The movement of an arm, the clenching of a fist, operate from a stored program in the brain which was learned and debugged by experience.

The composition of a digital positioning system must have the same structure as the human arm, where all motions are executed on a sense and control basis. In order of priority levels, the gross elements are:

1) Stored program which lists the sequence of desired motions.

2) A set of control points which by selective combination give rise to all physical motions necessary to fulfill the program.

3) A set of sense points which indicate the physical states necessary for unique motion control.

Computer Interface Selection

Since all elements, apart from coil drivers and sensors, are standard binary logic, there is theoretically a choice where the total system is divided into PROCESSOR and MOTION CONTROLLER. Until this division is made, the input versus output, with respect to sense and control, cannot be defined. As far as the stepping motor is concerned, the INTERFACE between controller and processor can be drawn at many levels. In general, the more logic included in the controller, the less frequent it will need attention from the processor. Referring to Figure 13-1, the decision is largely one of economics, optimizing cost of memory for the stored

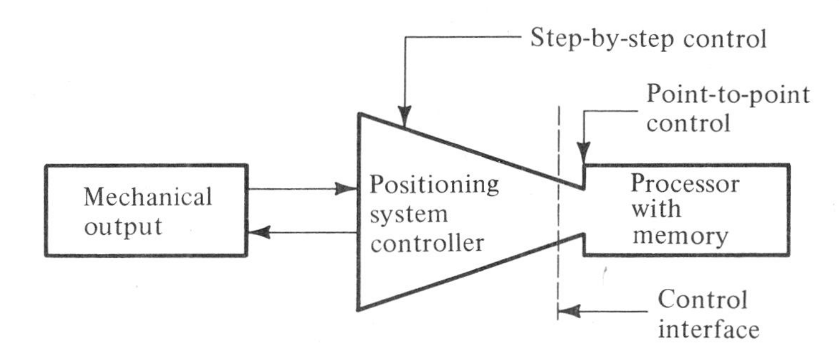

Figure 13-1. Controller-processor interface selection.

program versus cost of logic in the controller. Regardless of how many other external devices are to be time shared, each controller must receive adequate attention from the processor so that the motions can be executed reliably [12].

There is also normally three parties involved in this type of equipment selection; the USER, who's process is to be mechanized by the total control system, the COMPUTER (PROCESSOR) MANUFACTURER and the POSITIONAL CONTROL SYSTEM SUPPLIER. The name of the game for the user is invariably to control a process adequately at a minimum cost, while the equipment suppliers must sell his unit making an adequate profit. A practical solution to this situation is to standardize the interface and assign the process options and flexibility as programming tasks within the computer. This allows both supplier and user to largely eliminate special engineering, which is costly and restrictive from all points of view.

For the purpose of this paper, it is assumed that mini-computers have been standardized with 16-bit input-output channels. That is, they "send" 16-bit signal combinations out to an external device which initiates a motion and receives a 16-bit signal combination back as input which reports the dynamic state of the motion. The processor interface must also include an address channel which permits one particular external device to be selected for operation and finally, a set of clock signals is required which initiates the transfer of information between the processor and motion controller.

Mechanical Output Criteria

The productive output function of a positional control system is mechanical motion of some form. This motion is restricted by physical bounds within the process equipment, and it is not only necessary to control the shaft motion of the stepping motor, but also relate it to geographical limits within the actual equipment. To achieve optimum reliability, uniqueness of position must be ensured. This essentially means the exclusion of analog elements.

Using the stepping motor as basic reference, all motions are defined in terms of steps. In this paper, it is assumed that a stationary position point is synonomous with a locked step of the motor. These step positions must now be related to unique positions within the bounds of the processing equipment, and moreover, the processor is required to be able to establish this reference without human intervention. Thus, the overall criteria for digital positional control system is total automation, giving the computer the ability to power up the system, as well as setting it up for a fixed reference position.

Control Elements

A number of well defined control elements execute commands from the interface into mechanical motion. Many of the elements are similar for both open and closed-loop control, but generally, more electronics are required for the closed-loop system.

Referring to Figures 13-2 and 13-3, a linear load is presented with drive motor and electronic limits (boundary detectors). The step motor is driven from a BASIC MOTOR CONTROL which consists of motor power drivers and some appropriate logic. For

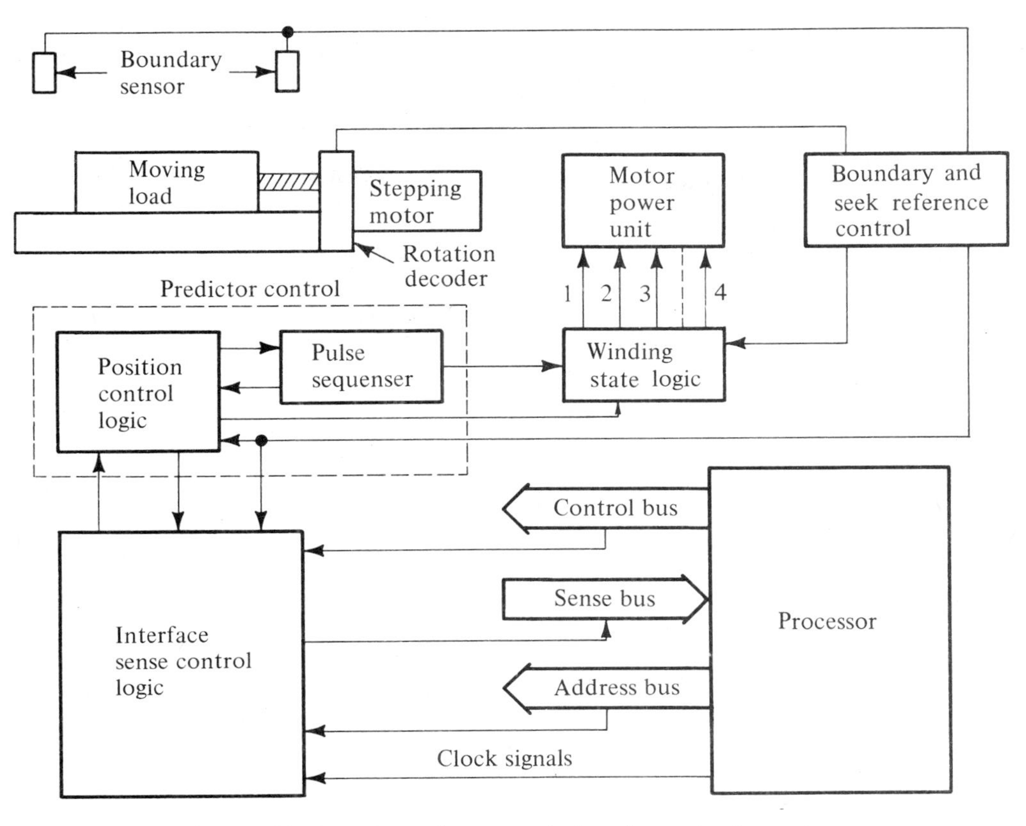

Figure 13-2. Open-loop positional control system.

the open-loop system (Figure 13-2), the WINDING STATE LOGIC accepts inputs in the form of pulses and direction signals, while in the closed-loop system (Figure 13-3), the MINOR LOOP LOGIC operates on ON-OFF command sending actual step pulses back to the position control.

The open-loop PREDICTOR CONTROL converts the new position command into a direction signal and a train of pulses from which motor action is produced by the basic motor control. The closed-loop system works as a servo and the control logic starts the motor via the minor loop. Then, as motion takes place, pulses representing actual measured steps are fed back to the position control, reducing the error towards zero.

The seek reference control (SRP) is very similar for both open-loop and closed-loop systems. When the computer commands SRP, the positional control forces the motor to seek a predetermined reference step regardless of present position. The reference step is approximately one-half rotation of the motor shaft from one of the electronically detected limits. In the SRP mode, both systems are open loop when searching for the reference step, and both are essentially closed loop when locking in on the reference step.

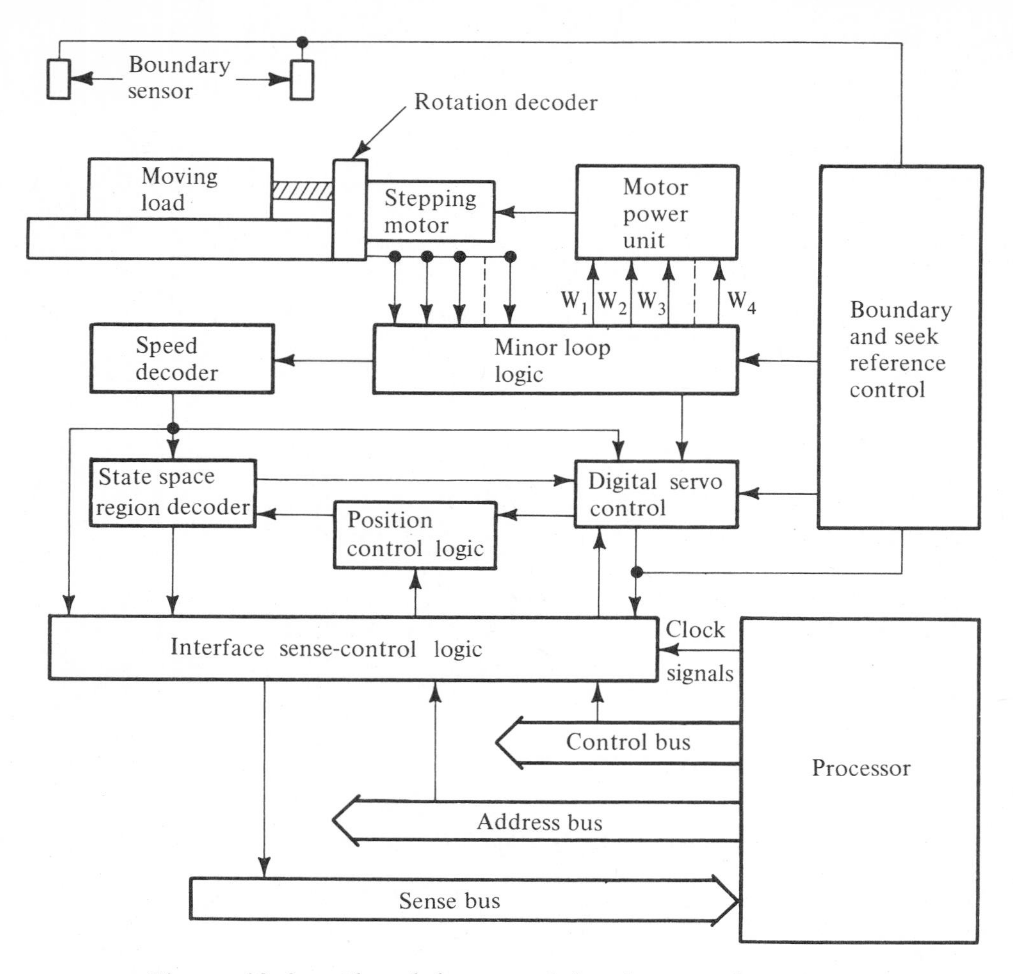

Figure 13-3. Closed-loop positional control system.

13-3 INTERFACE DESIGN

The interface is designed to match the computer, but even more important, it must give the computer access to all necessary control inputs and state variables of the positioning system. In other words, the computer must be able to monitor (sense) the state of the system in sufficient detail to always make adequate control decisions.

The interface logic operates on the regular computer input-output channels and establishes communication when the computer selects the positional control system via the address code. Care must be taken here to stay within channel loading specification, otherwise, very subtle transmission reflections can create errors that are very difficult to locate.

Control Point Definitions

The control functions are the same for both open and closed-loop systems.

Defining U and V as two sequential 16-bit data words, we have:

$$U = \begin{bmatrix} u_1 \\ u_2 \\ \cdot \\ \cdot \\ \cdot \\ u_{16} \end{bmatrix} = \{u_1, u_2 \ldots u_{16}\} \tag{13-1}$$

and

$$V = \begin{bmatrix} v_1 \\ v_2 \\ \cdot \\ \cdot \\ \cdot \\ v_{16} \end{bmatrix} = \{v_1, v_2 \ldots v_{16}\} \tag{13-2}$$

where $U \equiv |\text{motion in motor steps}|$

$v_1 \equiv \text{sgn}|\text{motion}|$

$v_4 \equiv$ halt motion

v_5

v_6 $\equiv$ maximum speed selection

v_7

$v_8 \equiv$ seek reference point

and v_2, v_3, $v_9 \ldots v_{16}$ are unused.*

Within U and V we have included the basic motion requirement of a positioning system, namely; how far to move the load from present position (U), direction of move (v_1), and speed of motion (v_5, v_6, v_7). In addition, two diagnostic control points are included which enable the processor to stop all motion (v_4) regardless of previous commands and the seeking of reference point (v_8) which enables the computer to restart the system from a known geographical point within the travel boundary.

For the purpose of explanation, it is easier to think of V as five separate control inputs, even though the computer transfers them all at the same time. Thus, we have

---

* In a two axis system, these bits can be applied to the second axis of control.

while the other components of V retain their individual notation since they are single control points.

$$V' = \begin{bmatrix} v_5 \\ v_6 \\ v_7 \end{bmatrix} = \{v_5, v_6, v_7\} \qquad (13\text{-}3)$$

Open-Loop Sense Point Definitions

The necessary set of sense points for a predictor control is derived directly from the state plane trajectories shown in Figure 13-4. Here, the points

$$P = (e, o), \qquad (13\text{-}4)$$

where

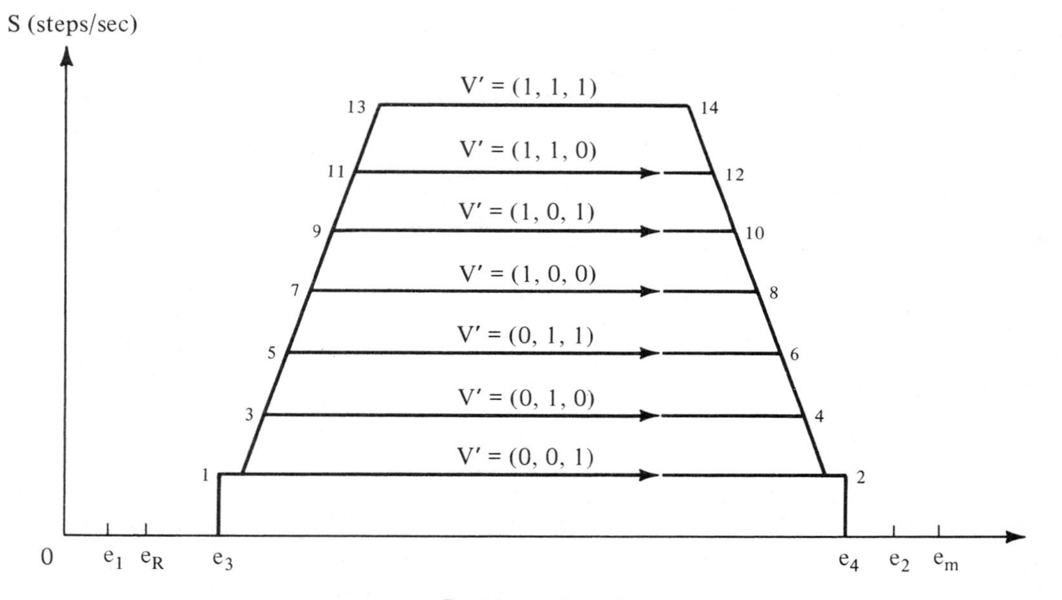

Figure 13-4. Open-loop trajectories in state plane.

$$e = 0, 1, 2 \ldots e_m$$

represent the absolute maximum set of stable terminal positions. The practical boundaries (e_1 and e_2) are placed inside the physical limits to prevent damage to the equipment. Thus, it is necessary to detect the two limit zones and define two sense points X_1 and X_2 where

$$X_1 = 1, \; 0 < e \leq e_1$$

$$X_1 = 0, \; e_1 < e \qquad (13\text{-}5)$$

and

$$X_2 = 1,\ e_2 \leq e < e_m \qquad\qquad (13\text{-}6)$$
$$X_2 = 0,\ e < e_2$$

The point $P = (e_R, 0)$ is defined as reference position and detected by a third sense point, X_3 where

$$X_3 = 1,\ e = e_R \qquad\qquad (13\text{-}7)$$
$$X_3 = 0,\ e \neq e_R$$

The dynamic trajectories from e_3 to e_4 are responses to the control input U and V where the speed component of V, V', takes on the binary format as shown. These trajectories are predetermined by the predictor control and the stepping motor is slaved to following a particular pattern. The whole open-loop control fails if the motor does not follow, thus, the only required sense point is X_4 which indicates whether a pulse train is being generated. If $t = 0$ at e_3 and $e_4 - e_3$ pulses have been produced by $t = t_1$ then

$$X_4 = 1,\ t \geq t_1 \qquad\qquad (13\text{-}8)$$
$$X_4 = 0,\ 0 \leq t < t_1$$

Closed-Loop Sense Point Definition

A typical closed-loop set of trajectories are shown in Figure 13-5. The previous definitions of X_1, X_2 and X_3 hold for this situation as well, but the dynamic side is now controllable. The system will not necessarily fail if e_4 is

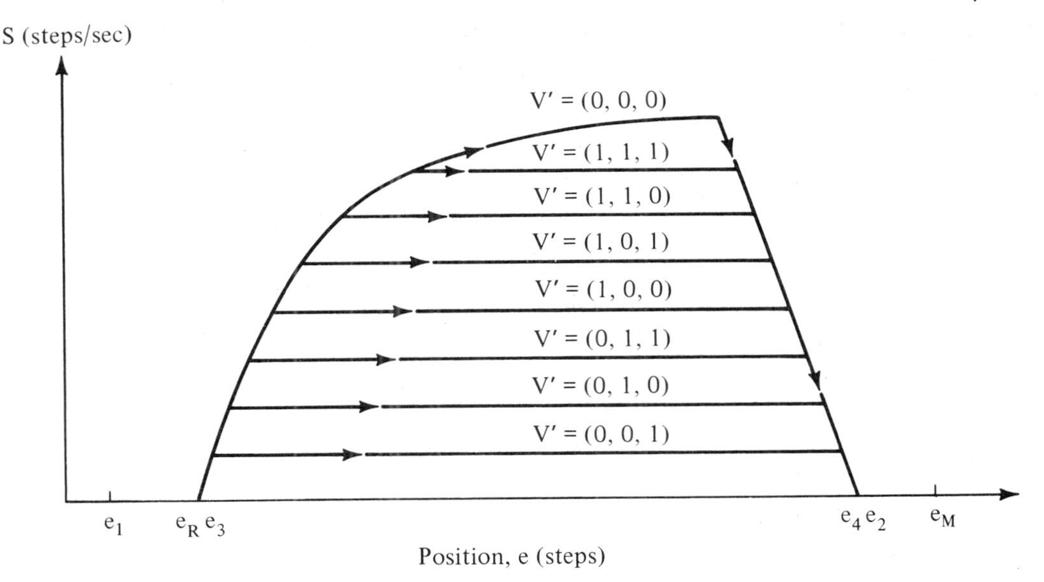

Figure 13-5. Closed-loop trajectories in state space.

reached at other than s = 0, in fact, simple load changes could make this happen. It is therefore necessary to add a sense point for zero velocity detection, but since there is in the real world no true zero sensor, X_5 is defined as:

$$X_5 = 1, S < S_1$$
$$X_5 = 0, S \geq S_1 \qquad (13\text{-}9)$$

The "motion complete" sense point X_4, has a different meaning in closed-loop control. Rather than being pretimed, X_4 now depends only on whether the target has been reached, thus:

$$X_4 = 1, e = e_4$$
$$X_4 = 0, e \neq e_4 \qquad (13\text{-}10)$$

The closed-loop stepping motor is not, of course, a true second order system. Other state variables could be added for more optimum control; however, our objective is "adequate function at minimum cost" and that rules out all but two dimensional control.

Addressing Definition

The interface is not complete without giving the processor the ability to select a particular device for sense-control operation. Referring to Figures 13-2 and 13-3, the device needs to respond to two combinations of the address function A' and A''. Either A' or A'' is sufficient to connect the sense lines to the device, while A' and A'' channel the control data from the processor into the U and V receivers, respectively. If W is the control data from the processor, R is sense data returned to the processor and B is the address word from the processor, then:

$$U = Wk_1$$
$$V = Wk_2 \qquad (13\text{-}11)$$

and

$$R = Xk_3$$

The constants k_1, k_2 and k_3 are generated by the device address detector, such that:

$$k_1 = 1 \text{ when } B = A'$$
$$k_1 = 0 \qquad B \neq A'$$
$$k_2 = 1 \text{ when } B = A''$$
$$k_2 = 0 \qquad B \neq A''$$
$$k_3 = 1 \text{ when } B = A' \text{ or } B = A''$$
$$k_3 = 0 \text{ when } B \neq A' \text{ or } B \neq A''$$

Interface Hardware

The block diagram of the interface sense-control logic is shown in Figure 13-6. Here, the components of the column matricies W, U, V, R, X and A now correspond to parallel sets of signal lines. The address detector contains three comparitors where

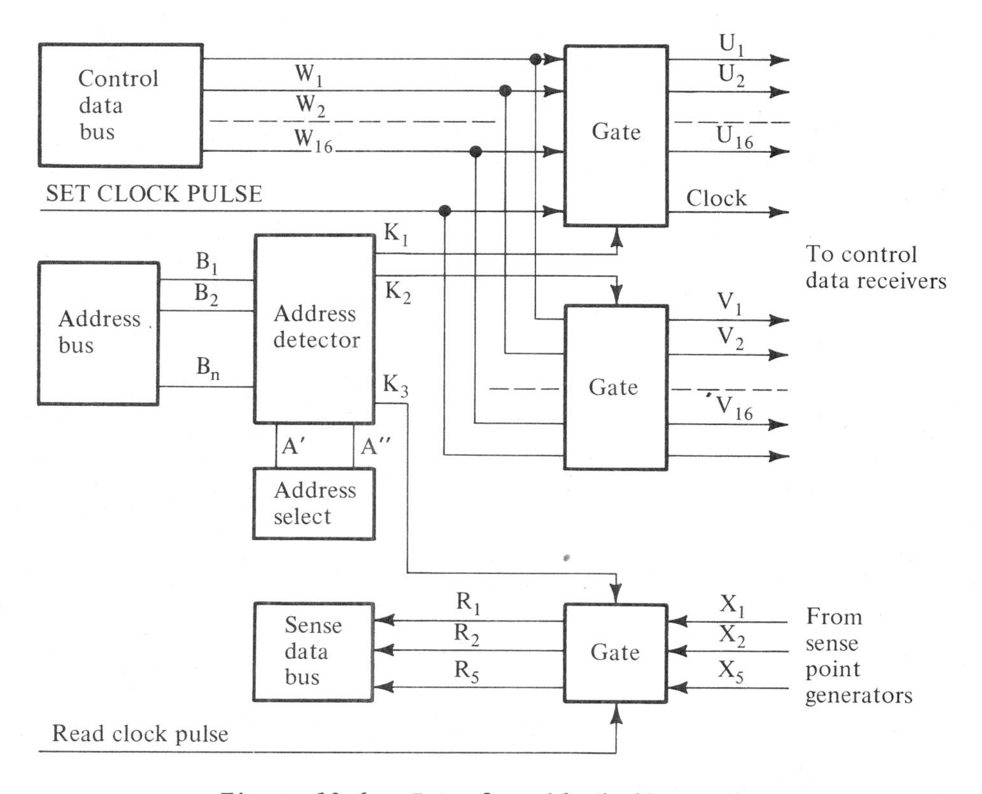

Figure 13-6. Interface block diagram.

A' and A'' are matched to the B input components. The k_1, k_2 and k_3 outputs are 1 or 0, depending on the composition of B. The set clock pulse from the computer is used to strobe the data in the appropriate counters or registers within the control elements. Similarly, the read clock pulse gates the sense word onto the sense bus at the appropriate time in the computer input cycle. The read pulse need not necessarily be sent to the device, since this gating function is often performed within the processor.

System Operation

The actual operations conducted by the computer are sequential chains of:

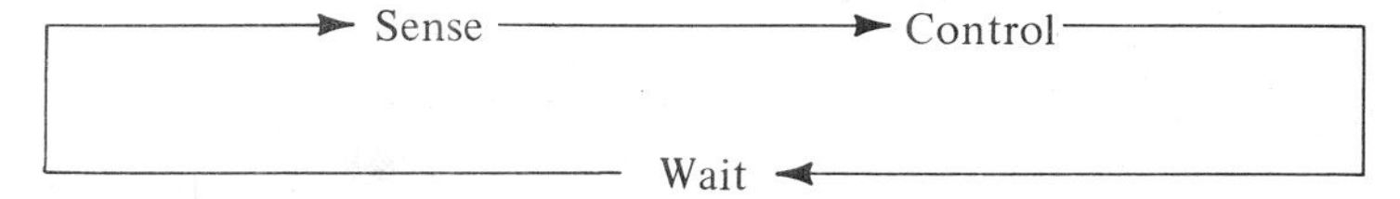

During "wait", the computer concerns itself with other devices or tasks. Then, when

the previous motion should be completed, the sense points are checked on a more frequent basis until, a) expected state comes true or, b) system is overdue and probably malfunctioned. The first condition results in a new control being executed, while a set of diagnostic routines are consulted for the second condition. The diagnostic solution may simply say, "try again", but it can also be demanding complete restart (SRP), or if nothing succeeds, sounding alarms and calling for the operator to intervene.

For the systems described here, the sense function consists of reading the X word until it matches the expected combination (i.e., 0, 0, 0, 1 for open loop and 0, 0, 0, 1, 1 for closed loop). Then the control data is sequenced as given in Figure 13-7.

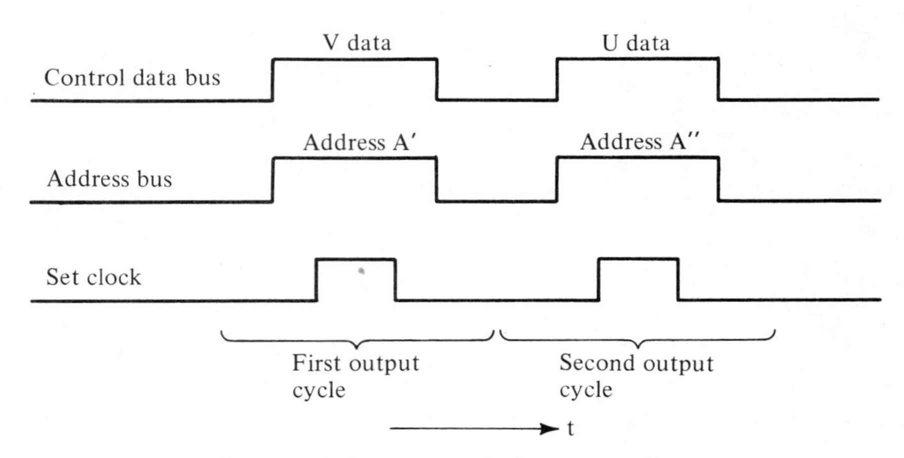

Figure 13-7. Control data transfer.

13-4 MECHANICAL OUTPUT CONFIGURATION

The stepping motor is occasionally directly coupled to a rotary load, but by enlarge, some form of mechanical lineage is required to give the desired load displacement. When these are carefully designed to give good linearity with negligible backlash, it will not be necessary to measure relative position at the load, but rather use the motor steps as reference.

Precision ground lead screws are frequently used to translate motor rotation into linear motion. Combining these with 200 step/revolution motors yields very acceptable resolutions and speeds. For example, a 5-pitch lead screw will give a minimum step increment of 0.001", and 8 - 10"/sec. maximum linear speed for a commonly used closed-loop motor. A linear accuracy of ±0.0001 inch/inch is reasonable to expect and each step position is easily repeatable within ±0.00005 inch.

With such performance, it is very difficult to justify elaborate position sensing at the actual load. Moreover, the unavoidable spring coupling between motor and load can often set up severe stability problems. The ability of the stepping motor to lock into a step with very high torque stiffness gives the load excellent stability in a very absolute sense. In fact, the inaccuracies in instrumentating the load position alone would be more difficult to cope with than this final settling motion of the load. Also, in most practical designs, it is more economical to construct the system somewhat faster than needed and allow some fixed time for settling at the termination point, than to worry about damping it out by control measures.

All positional objectives are very simply defined in terms of motor steps. With present day digital electronic reliability, it is better to construct positional error by step counters than by applying transducers to the actual load.

Limit control must be implemented by sensors mounted directly at the load. A very useful and reliable device for this purpose is the LED (light emmitting diode)-phototransistor combination. This small, single package is easy to mount and simple shutter arrangements attached to the load give very flexible and reliable limit control signals.

The reference point detection (e_R in Figures 13-4 and 13-5) is generated as a combination of motor step and limit control signals. To guarantee the location of such reference, a step detector is added to the motor shaft which can sense a particular step within a rotation. The limits are then arranged so that they are reached when the motor shaft is 180° away from sensing the reference step. Finding the reference then becomes a control sequence of moving the load to the limit and backing up until the FIRST reference step is detected.

13-5 BASIC MOTOR CONTROL

Stepping motors are constructed to facilitate solid state drive control. That is, each winding requires current to be passed in one direction or no current at all. The most common bifilar construction [6, 8, 14] provides a choice of four windings arranged in two phase pairs. Having a permanent magnet rotor, the four windings energize the stator so that the rotor advances one step at a time if the correct sequence or combination is used.*

---

* See references [10, 11, 14] for more details on sequence and motor construction.

For the purpose of this paper, we shall treat the step motor as a black box and be concerned only with input-output characteristics. Figure 13-8 shows a complete set of drive circuits for a bifilar motor using some of the new power switching devices available. At rated voltage, the motor is severely restricted as

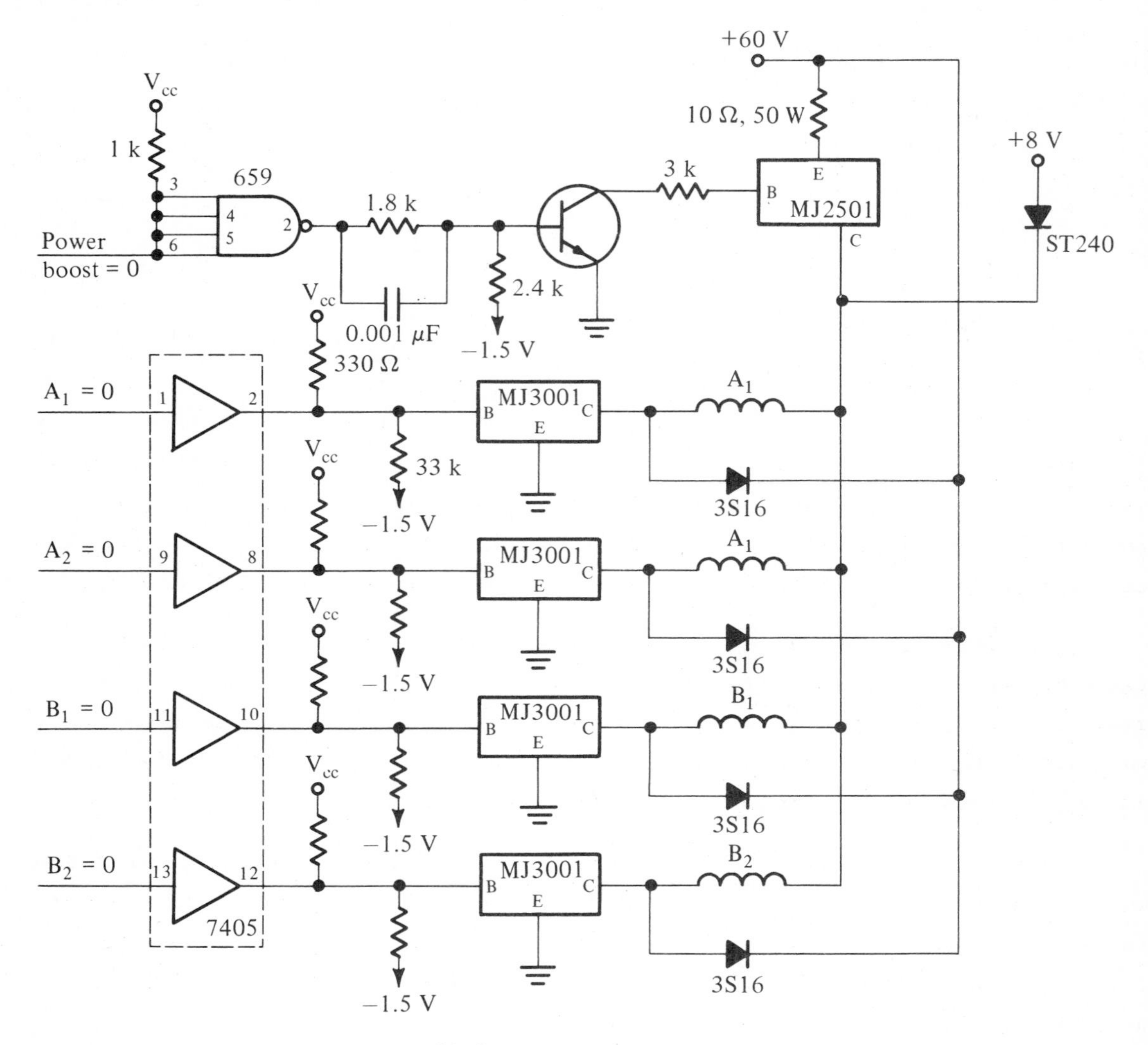

Figure 13-8. Power driver circuits.

to stepping rates. To improve this, a higher voltage source is switched in during operation, allowing the current to rise faster by lowering the L/R constant. The high voltage source is disconnected by the power boost (PB) switching element when the motor is at standstill to prevent excess heating of the series resistor, R.

The five power switches are driven by conventional predriver circuits, which in turn require regular logic level inputs. Combining the motor and power drivers as one unit, the input-output for a 1.8° step motor correlates as listed in Table 13-1.

TABLE 13-1

Winding Energized For Various Steps

STEP NOS.	A_1	A_2	B_1	B_2	RELATIVE POSITION (degrees)	DECODER 1 2 3 4
1	0*	1	0	1	0 + 7.2n	1 0 0 0
2	0	1	1	0	1.8 + 7.2n	0 1 0 0
3	1	0	1	0	3.6 + 7.2n	0 0 1 0
4	1	0	0	1	5.4 + 7.2n	0 0 0 1
1	0	1	0	1	0 + 7.2(n + 1)	1 0 0 0

* NOTE: Logical "0" means winding is energized.

Since there are only four distinct winding combinations energized, it is convenient to label these as step #1, 2, 3, and 4, and define 1, 2, 3, 4, 1 as a clockwise (CW) sequence, and 4, 3, 2, 1, 4 as a counterclockwise (CCW) sequence.

For open-loop control, the motor drivers are simply sequenced in a CW or CCW pattern. Provided the rate of change is within certain limits, the motor will respond by incrementing to corresponding positions. If the rate is fast, the motor will not really step, that is, come to a halt at each steady state position, but rather, follow at a continuous speed as any regular motor.

When closed-loop control is implemented, a set of sense points are generated which correspond to the four cyclic steps 1, 2, 3, 4. The construction of such a shaft encoder (step discriminator) has been covered in many previous paper [10, 11, 12, 14] and will not be repeated here in detail, however, Table 13-1 includes the decoder sense words for the various steps.

Open-Loop Winding State Logic

The winding state logic shown in Figure 13-9 consists of two J-K flip flops (IC1, IC2), which change states according to the gate transfer of IC3. The AND-OR logic in this latter element operates on the present state of the windings and the direction inputs (CW = 1 and CCW = 1) so that the correct sequence defined in Table

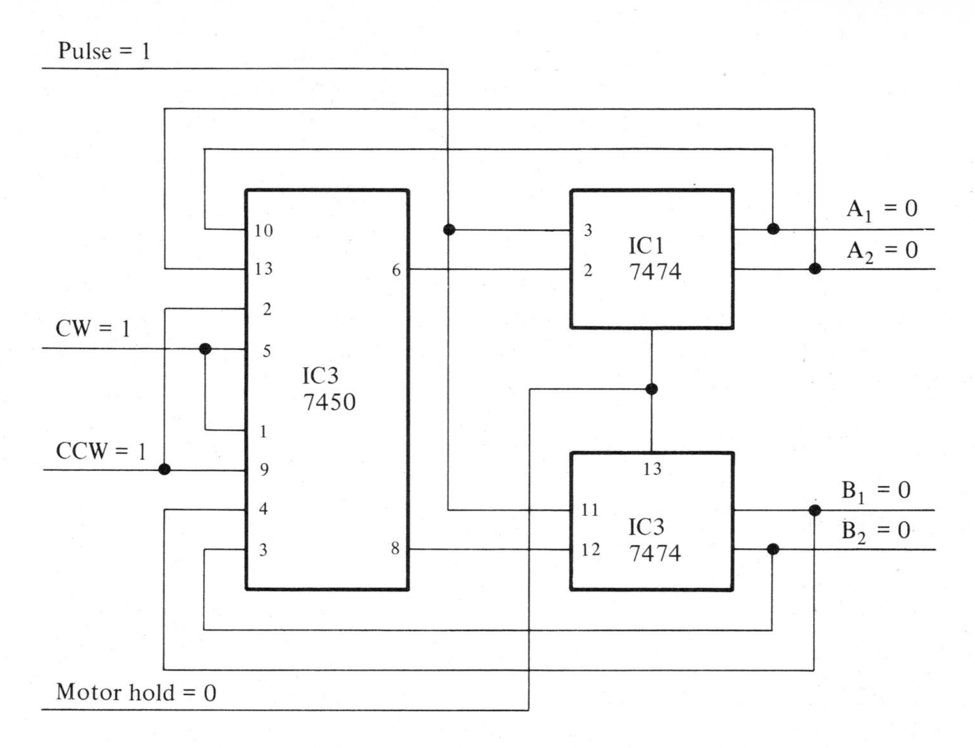

Figure 13-9. Open-loop winding state logic.

13-1 is incremented each time a pulse is received from the predictor control (see Figure 13-4).

The MOTOR HOLD input is used to force step #1 winding combination, regardless of pulse and direction states. This control input is used in conjunction with the seek reference control which forces the motor to hang on to the reference step until absolute stability has been reached.

Closed-Loop Control Logic

The minor loop logic operates on the basic sense-control points defined in Table 13-1. Previous papers [10, 11, 12, 14, 15] give a variety of logic details, therefore, Figure 13-10 only shows the input-output signals required. The minor loop performs two functions; a) provides winding sequencing signals to the motor power control, and b) generates actual step detection signals to position control and speed detectors. The four control inputs; Stop, Speed 1, CW and CCW relate to the winding sequence as listed in Table 13-2. The translation is strictly a logic function which uniquely drives the motor coils like a commutator would on an ordinary D.C. motor. The main exception is, of course, that the closed motor can be stalled, as well as run at several speeds which a regular D.C. motor cannot normally do.

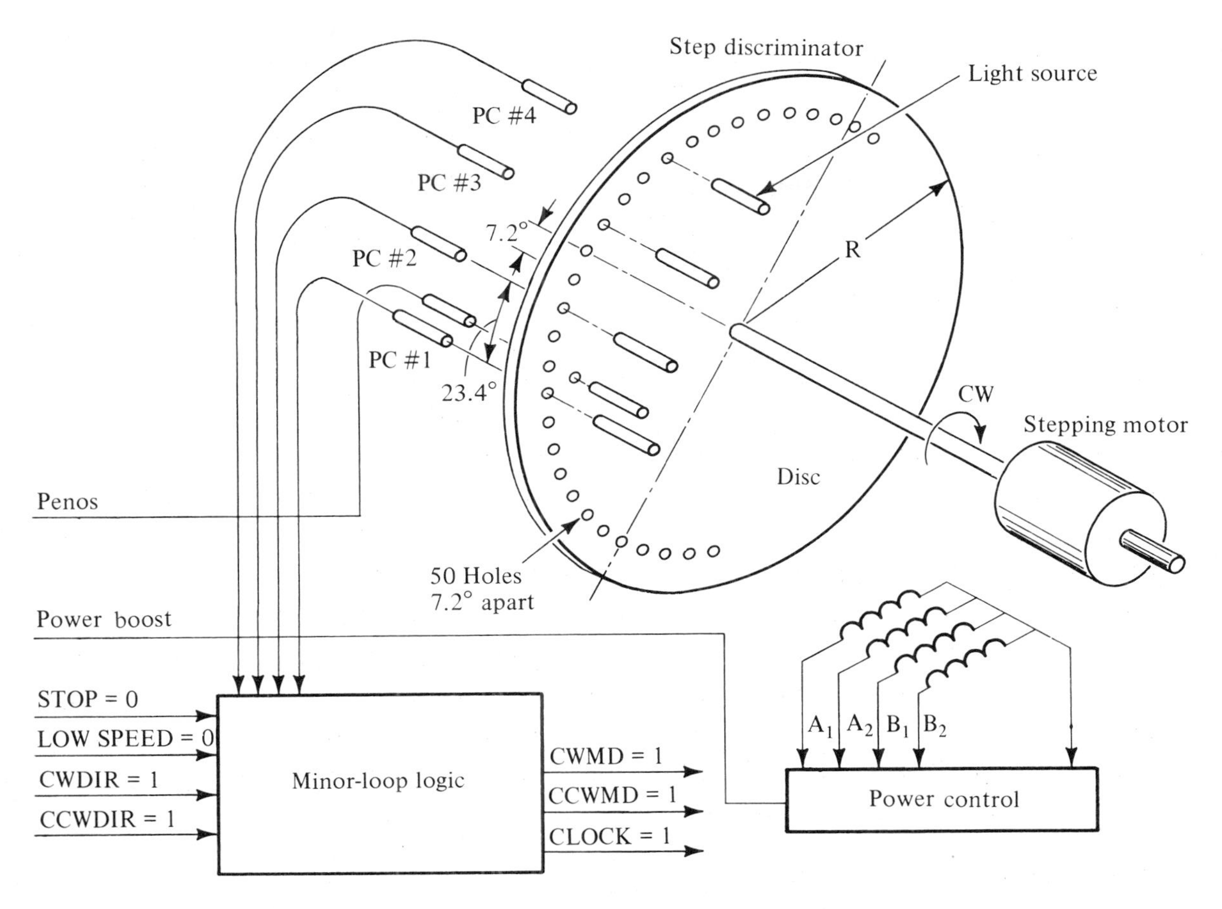

Figure 13-10. Minor-loop control.

TABLE 13-2

Minor Loop Motor Drive Translation

CONTROL INPUT	PC. # 1 ON	PC. # 2 ON	PC. # 3 ON	PC. # 4 ON
STOP	0, 1, 0, 1	0, 1, 1, 0	1, 0, 1, 0	1, 0, 0, 1
CW + LOW SPEED	0, 1, 1, 0	1, 0, 1, 0	1, 0, 0, 1	0, 1, 0, 1
CCW + LOW SPEED	1, 0, 0, 1	0, 1, 0, 1	0, 1, 1, 0	1, 0, 1, 0
$\overline{\text{LOW SPEED}}$*	1, 0, 1, 0	1, 0, 0, 1	0, 1, 0, 1	0, 1, 1, 0

* $\overline{\text{LOW SPEED}}$ ≡ not LOW SPEED

The winding inputs listed in Table 13-2 correlates with Table 13-1 and at any step position, the command input gives the definition:

STOP ≡ The motor is energized to hold the step it is on.

CW + LOW SPEED } ≡ The motor is energized to move one step in the CW direction.

CCW + LOW SPEED } ≡ The motor is energized to move one step in the CCW direction.

$\overline{\text{LOW SPEED}}$ ≡ The motor is energized to move two steps in either direction.

The STOP command is applied whenever deceleration or standstill is required. Starting from standstill condition, either LOW SPEED inputs will drive the motor at a relatively low constant rate* in the specified direction. Once the motor is rotating, the LOW SPEED input can be removed (i.e., $\overline{\text{LOW SPEED}}$). The motor will now sharply accelerate to a maximum speed, 10 to 20 times the low speed value.

A new technique has been developed which enables the maximum speed to be selected anywhere between the low speed value and maximum. The low speed input is modulated by a pulse train so that the low speed input is enabled during the pulse and not between pulses. By varying the pulse width, and to some extent the frequency, the speed can be varied very effectively. To spread the range even further, POWER BOOST is not applied for the lower speeds, which set the basic low speed even further down. Figure 13-5 shows typical response curves for the seven controlled and the one (0, 0, 0) free running maximum top speed input.

The step detector output signals are tightly controlled and screened to give a one microsecond CLOCK pulse each time the motor has actually moved one step. If the step was detected as a clockwise rotation, the CWMD signal switches to a position level prior to the clock pulse being issued. If the following steps are in the same direction, the CWMD will continue to stay positive, but if the next step is detected in the counterclockwise direction, the CWMD output will go to zero while the CCWMD output will switch to a positive (1) level. Thus, CWMD and CCWMD signals indicate the actual motor direction during the last step taken, while the accumulated number of pulses give the total distance travelled. The rate of the pulses give a measure of motor speed. This is not really instantaneous speed, but rather the average speed during the last step motion.

13-6 PREDICTOR CONTROL DESIGN

The open-loop stepping motor can follow a variable pulse train, providing the change of rate is within the torque available for holding synchronism, that is, for

* Smoothness of speed depends on physical tolerances of discriminator parts.

each step the motor must move one step and be able to accept the next winding change of state.

The parameters involved are:

Start Rate ≡ Constant pulse frequency which the motor can follow from standstill.

Ramp Slope ≡ Rate of change in pulse frequency which the motor can follow either accelerating or decelerating.

Max. Frequency ≡ Max. rate at which motor will drive load without losing synchronism.

A typical predetermined open-loop trajectory is shown in Figure 13-11.

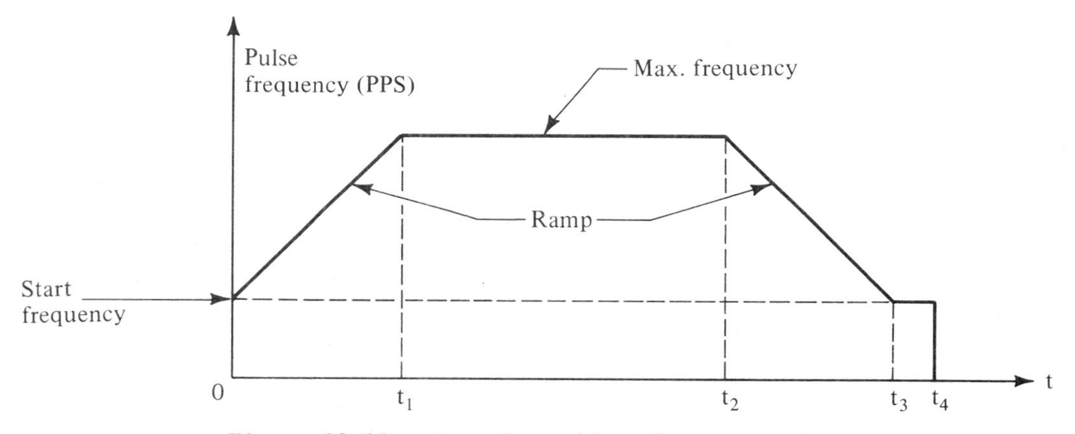

Figure 13-11. Open-loop pulse frequency plot.

At t = 0, the oscillator is started and begins pulsing at the start frequency rate. The rate increases linearily until t_1, when the maximum frequency is reached. The positional control initiates deceleration at t_2 by commanding the oscillator to change back to the low frequency. At t_3, the start frequency has been reached and will be maintained until the full distance has been travelled.

The variable oscillator must be logically controlled by the inputs shown in Figure 13-12.

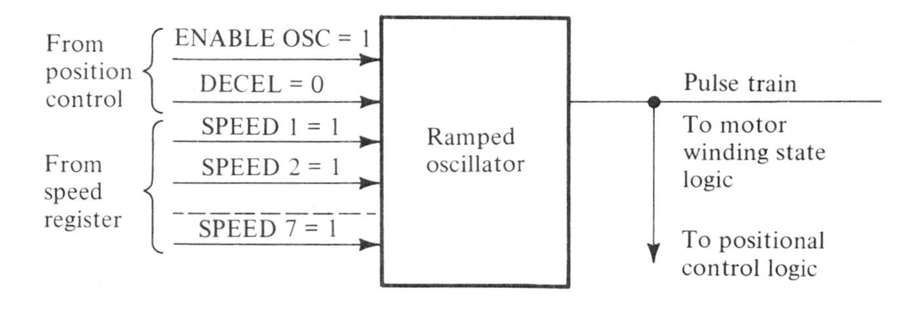

Figure 13-12. Oscillator input-output definitions.

Referring back to Figure 13-11, the control signal sequence is:

	Enable OSC = 1	
at $t = 0^+$	Decel = 1	(no decel commanded)
	Speed 7 = 1	
	Enable OSC = 1	
at $t = t_2$	Decel = 0	
	Speed 7 = 1	
	Enable OSC = 0	
at $t = t_4$	Decel = 0	
	Speed 7 = 1	

Thus, the top speed is preselected and remains constant, both acceleration and deceleration termination (t_1 and t_3) depends on the internal RC networks, while start-stop and beginning of change of ramp is externally controlled. A typical workable ramped oscillator circuit is detailed in Figure 13-20.

Incremental or Absolute Position Control

Before discussing the various position controls, it is necessary to dispose of the controversial absolute control versus incremental control. Since day one of servo control, it has been general practice to command the new position in an absolute manner, that is, all inputs were referred to a specific zero position. Using analog circuitry and potentiometer type position transducers, this was indeed a logical solution. The same can be accomplished in digital servos, but instead of using differential amplifier and analog voltage, the true error must now be calculated by digital subtracting logic. Such logic is already part of the computer, therefore, the digital positional servo mechanism will receive the initial error (distance to go) and direction rather than a new absolute position. Hence, the positional control for both open and closed-loop systems has a counter which is preset by the initial error and the target is reached when this counter has been decremented to zero.

Open-Loop Position Control

The open-loop position control consists of presettable counter which will be decremented by the pulse input from the ramped oscillator. As shown in Figure 13-13, a zero error detection logic resets the ENABLE OSCILLATOR LATCH while the DECEL-POINT DETECTOR sets the DECEL-LATCH. Both of these flip-flops are initiated by the clock pulse from the interface control.

The deceleration detector must allow sufficient steps before target to allow the oscillator to reach the start frequency before terminating. Since the actual steps needed to accomplish this depends on the maximum speed this logic is conditioned by the speed selected.

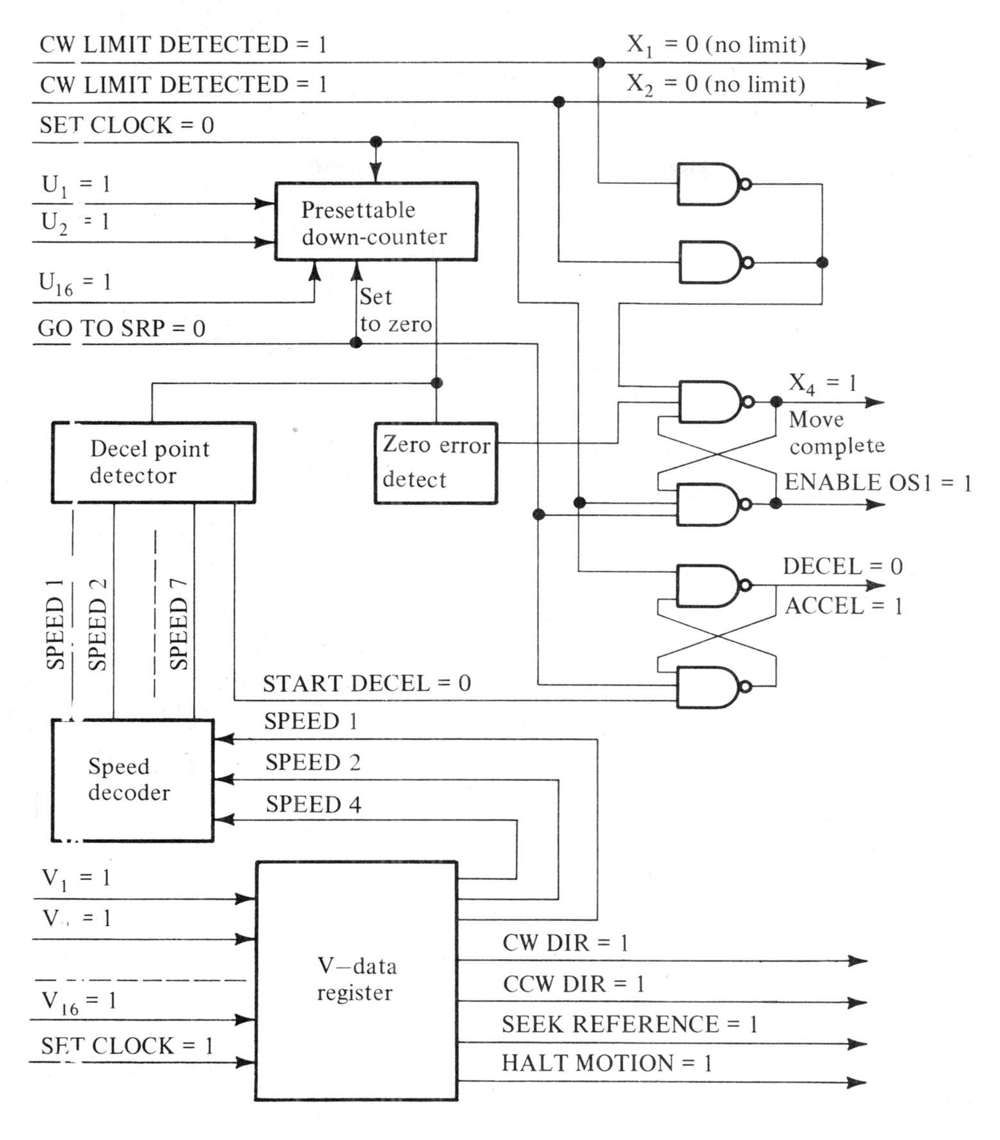

Figure 13-13. Open-loop position control.

The open-loop position control also contains a register which stores the direction, speed, halt and seek reference commands given by the processor via control word V. This word is loaded in first so that when the counter is preset with the contents of control word U, both the oscillator and motor control will operate on the first pulse in the correct manner. The rest of the operation follows readily. The oscillator runs through its preset pattern and at t_4 (Figure 13-11), the counter is zero. The oscillator is stopped and this halts the motor. Assuming the motor was able to follow the pulse train, the positioning job is complete.

The sense point X_4 gives a true signal whenever the enable oscillator latch is reset (see Figure 13-13). However, as shown in Figure 13-13, the latch is also reset by entering the limit zone, thus, the computer must receive the sense word X = 0, 0, 0, 1 or X = 0, 0, 1, 1 to register a satisfactory "move complete" sense.

13-7 OPEN-LOOP SEEK REFERENCE CONTROL

The logic in Figure 13-13 shows how the seek reference control point set by the computer forces the counter to zero but also clamps the oscillator to continuous start frequency. Figure 13-14 gives the logic details of the actual seek reference

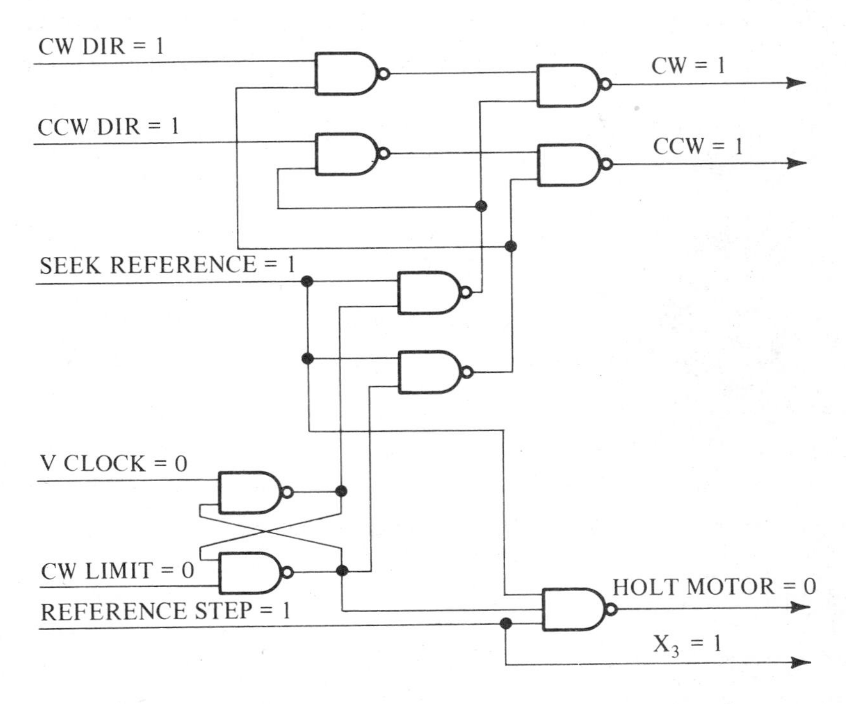

Figure 13-14. Open-loop seek reference control.

control. Here, the initial direction is always towards a specific limit, but as soon as the limit is reached, the direction is reversed and the motor proceeds to back out from the limit. When the reference step is reached, the HALT MOTOR signal is set to zero, which clamps the correct winding input (see Figure 13-9) and the motor settles down at the reference step, while the oscillator is shut off. No further action can now take place until the seek reference control point is reset. This is done by the computer after sensing that the reference step has indeed been found.

13-8 DIGITAL SERVO DESIGN

The closed-loop position control logic is almost identical to the open loop up to the point of having counted down to zero error. However, the decel signals are generated by true state variable conditions and the clock pulses from the minor loop are used to decrement the counter rather than a prefixed oscillator. Also, at the first zero error, the mode is changed to a proper feedback control where if the system overshoots the target and the counter goes to 9999, 9998, 9997, etc., a reversing control takes over which forces the motor to decelerate then reverse and move back to the target.

Positional Bookkeeping Control

The initial command given by the computer to the digital servo sets the counter to the initial error and the initial direction is stored in the V word register. Figure 13-15 shows how the minor loop signals, CWMD and CCWMD are combined with the

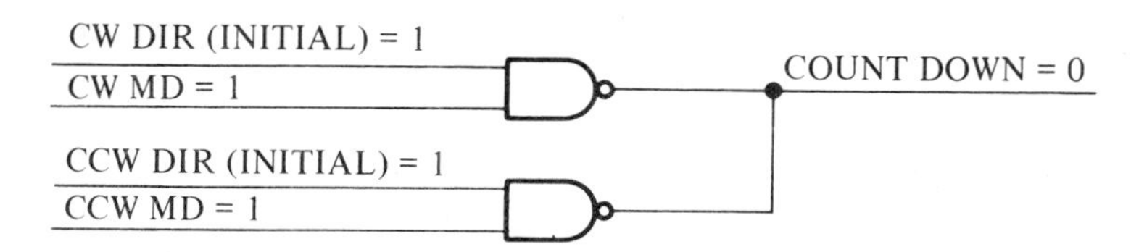

Figure 13-15. Normalizing counter decrementing.

initial direction to set the input to the up-down counter for "down" count as long as the actual steps taken by the motor coincide with the initial direction. If the motor is stepping in the oposite direction, the counter will count "up". Thus, provided the motor is actually commanded to reverse upon overshooting the counter will reach zero each time the actual target step is encountered.

Direction Control

When the counter first reaches zero, a latch is set, called "overflow". A direction control is immediately enabled which now considers the upper 60% of the counter (i.e., 9999 to 4000) as an overshoot condition, while the lower 40% of the counter is considered undershoot. As detailed in Figure 13-16, the initial direction

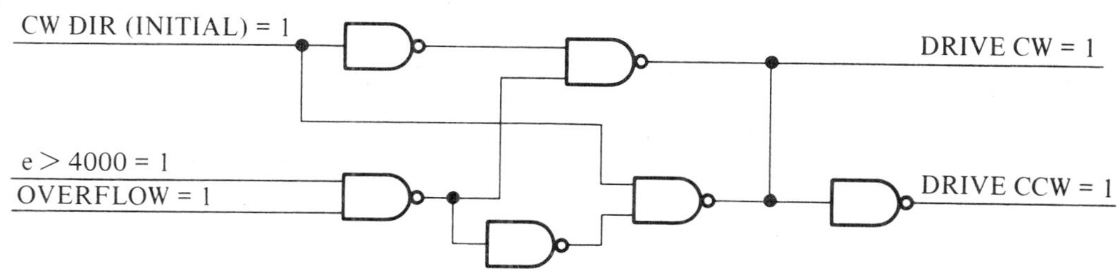

Figure 13-16. Motor reversing logic.

signal is intercepted by the servo direction logic which reverses the initial direction fed to the minor loop logic if OVERFLOW and OVERSHOOT is detected. In addition, the overflow condition also forces low speed, thus, from a stability point of view, a single overshoot is all that will occur since the low speed lock-in at the target assures absolute stability (see Figure 13-17).

Time Optimal Trajectory Control

The logic shown in Figures 13-15 and 13-16 is sufficient to assure proper servo control, and the resulting trajectory will be as drawn in Figure 13-17. Obviously, the time consuming slow speed return to the target can be much improved by providing a control which will detect a model of the time optimal trajectory and guide the motor into the target along the ALTERNATE curve shown in Figure 13-17.

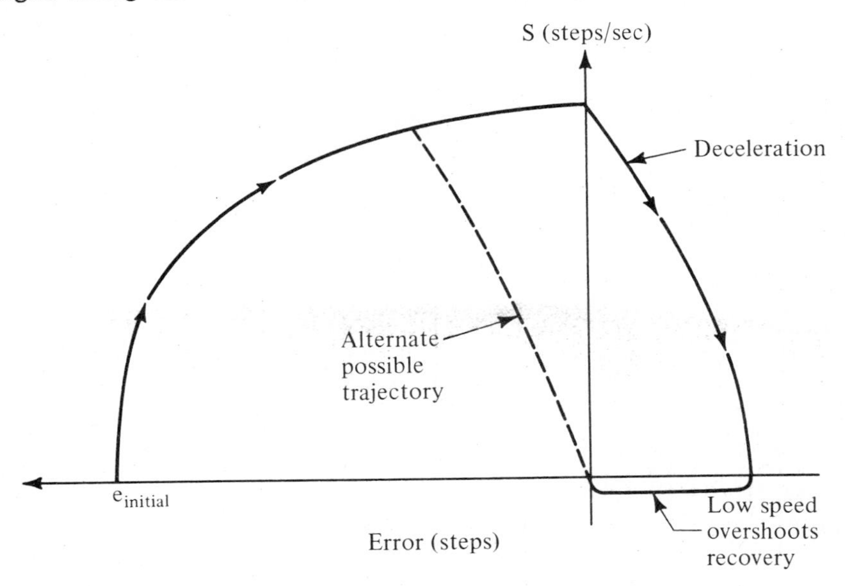

Figure 13-17. Closed-loop state space trajectory.

The implementation of such control is made easier by designing the minor loop input so that STOP takes priority over all other inputs. Thus, to cause deceleration and terminal halt, the control only has to force the STOP condition.

Ideally, the true optimal trajectory could be tracked by a stepping rate-step error function with one step resolution. In a practical system, this is too costly and really not necessary. By making a three step staircase approach, as shown in Figure 13-18, a near time optimal response can be obtained. [If the system is not performing fast enough on this approach, it is always good to remember that a more powerful motor costs infinitely less than a computer.]

To implement the staircase control in Figure 13-18, the speed need only be detected as $S \geq S_1$, $S \geq S_2$ and $S \geq S_3$. Similarily, the error need only be detected as $e = 0$, $e \leq e_1$, $e \leq e_2$ and $e \leq e_3$. If the corners of the staircase are selected close to the optimal trajectory, all trajectories will eventually join the minimum time trajectory. Since most of the correction is done at higher speed, the percentage loss is not too great.*

* See reference [15] for a more detailed discussion of the problem.

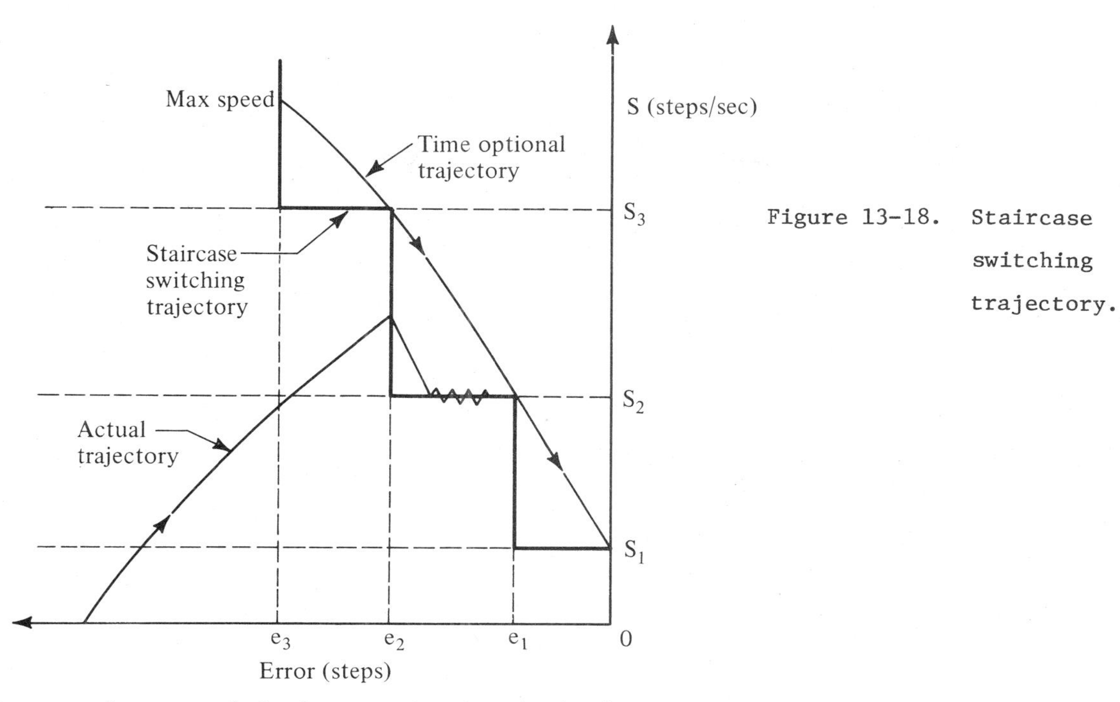

Figure 13-18. Staircase switching trajectory.

The control implementation is simply forcing STOP in the regions to the right of the staircase and accelerate (not stop) under the curve. The terminal stability is sensed by $s < s_1$ and $e = 0$, which gives true dynamic null. However, since the speed is not instantaneously measured nor really zero, the recommended practice is to read the sense word at least twice to assure permanent stability.

13-8 CLOSED-LOOP SEEK TARGET CONTROL

The closed-loop stepping motor can seek the reference target at much higher speeds than the open loop, but otherwise, the principle remains the same. When the seek target is initiated, zero error is maintained by keeping the counter reset to zero but the minor loop input is forced at maximum speed towards the limit. When the direction is reversed, the low speed is implemented and the motor backs out from the limit. When the reference step is detected, the forced minor loop input is dropped and since the reference step now remains as zero error, the system will servo to this step if overshoot occurs.

13-9 HYBRID STEPPING MOTOR CONTROL

The open-loop control does give a smoother constant speed than the closed loop when the frequency is kept constant. However, the closed loop gives more secure and normally faster positional control. The best feature of both can be incorporated in a hybrid control scheme where tight constant speed is required between two positions. Figure 13-19 shows a trajectory for such a system and indicates where the motor is switched into open loop and back to closed loop.

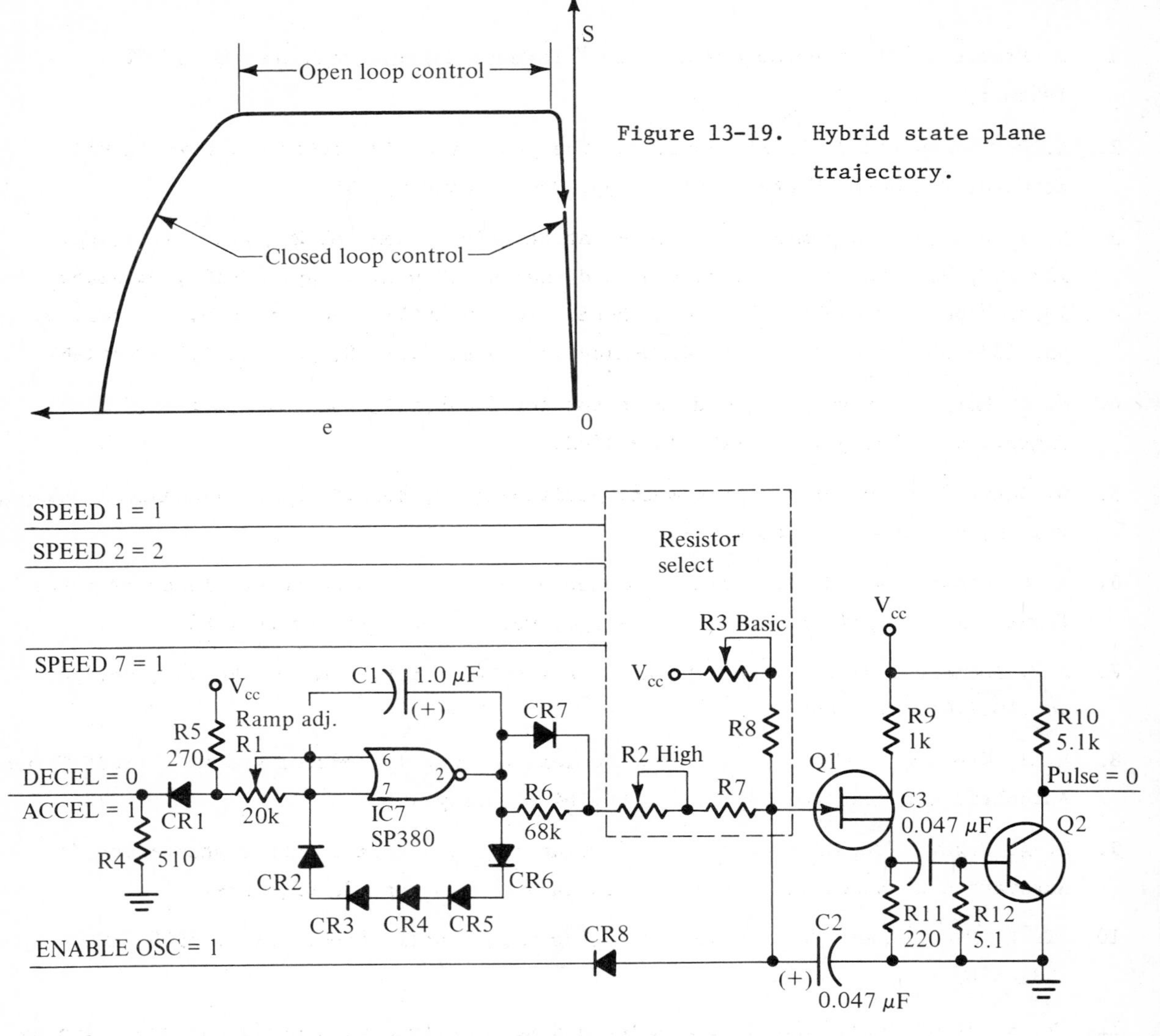

Figure 13-19. Hybrid state plane trajectory.

Figure 13-20. Ramped oscillator circuit.

13-10 CONCLUSION

In this paper, both open and closed-loop control systems have been discussed. Which type to choose for a particular application remains a cost versus function decision. However, there is one performance characteristic which should be remembered when making a decision: If the motor is removed from the open-loop system, the controls will still execute and the computer will never know the difference. If the motor is stalled in the closed-loop system during an operation and held for a period of time, then released, the motor will continue and complete the objective without error. Hence, the cost of failure in the process will, in most cases, dictate whether to use open or closed-loop systems.

REFERENCES

1. J. Proctor, "Stepping motors move in," Product Engrg., vol. 34, pp. 74-78, February 1963.

2. A. G. Thomas and J. F. Fleischauer, "The power stepping motor - a new digital actuator," Control Engrg., vol. 4, pp. 74-81, January 1957.

3. S. J. Bailey, "Incremental servos - Introduction," Control Engrg., vol. 7, pp. 123-127, November 1960; "Operation and analysis," vol. 7, pp. 97-102, December 1960; "Applications," vol.8, pp. 85-88, January 1961; "Industry survey," vol. 8, pp. 133-135, March 1961; "Interlocking steppers," vol. 8, pp. 116-119, May 1961.

4. N. L. Morgan, "Versatile inductor motor for industrial control problems," Plant Engrg., vol. 16, pp. 143-146, June 1962.

5. G. Baty, "Control of stepping motor positioning systems," Electromechanical Design, vol. 9, pp. 28-39, December 1965.

6. A. E. Snowden and E. M. Madsen, "Characteristics of a synchronous inductor motor," Trans. AIEE (Applications and Industry), vol. 8, pp. 1-5, March 1962.

7. J.P. O'Donahue, "Transfer function for a stepper motor," Control Engrg., vol. 8, pp. 103-104, November 1961.

8. R. B. Kieburtz, "The step motor - the next advance in control systems," IEEE Trans. Automatic Control, vol. AC-9, pp. 98-104, January 1964.

9. B. A. Segov, "Dynamics of digital routine control systems with a step motor," Automation of Electric Drives (in Russian), no. 4, pp. 38-61, 1959.

10. T. R. Fredriksen, "Closed loop stepping motor application," Proc. 1965 JACC, pp. 531-538.

11. T. R. Fredriksen, "New developments and applications of the closed-loop stepping motor," Proc. 1966 JACC, pp. 767-775.

12. T. R. Fredriksen, "Direct digital processor control of stepping motors," IBM J. Research and Develop., vol. 11, March 1967.

13. J. L. Douce and A. Refsum, "The control of a synchronous motor," presented at the 3rd IFAC Cong., London, England, June 20, 1966, paper 4A.

14. T. R. Fredriksen, "Stepping motors come of age," Electro-Technology, vol. 80, pp. 36-41, November 1967.

15. T. R. Fredriksen, "Applications of the closed-loop stepping motor," IEEE Trans. Automatic Control, vol. AC-13, number 5, October, 1968, pp. 464-474.

14 LINEAR STEP MOTORS

J. P. Pawletko and H. D. Chai
IBM System Products Division
Endicott, New York

14-1 INTRODUCTION

There are many applications in which a controlled linear motion is required. The step motor with its inherent stepping and detenting properties and its ability to slew at a high velocity provides the necessary controlled linear motion. It is extremely trouble free. It requires much simpler electronic drive and control than the d-c servomotors. It requires low current (implying low input power and low heat dissipation) thus avoiding the cooling problems often encountered with printed-circuit motors.

However, in most of today's applications, the linear motion is provided by a rotary step motor that is mechanically connected to a lead screw, which provides the linear motion. Although this approach provides the desired results, it is not an optimum solution. It requires a mechanical interface to transform the rotary motion to linear, and it consumes valuable hardware space - both, adding to product cost. Consequently, it seems logical to use a linear motor for linear motion. This eliminates the mechanical interface to transform the rotary motion into linear. We eliminate the lead screw and replace it with a tooth shaft as a stator. The armature which moves along the shaft carries the working device.

One version of a linear step motor is presented in Reference 1; another version - a three-phase, variable reluctance, linear step motor - is described in this paper. The basic physical configuration of the motor, its operation, necessary energization schemes, and measured data from experimental hardware are presented.

14-2 DESCRIPTION

Figures 14-1 and 14-2 show the simple construction of the motor and its schematic diagram, respectively. The stator is a rectangular laminated shaft. A pair of laminated E cores make up the armature, which moves along the shaft. Prewound coils are easily inserted around the E core legs. The opposing cores are rigidly held by a mechanical support which contains a set of roller bearings and a means for adjusting the pole gap.

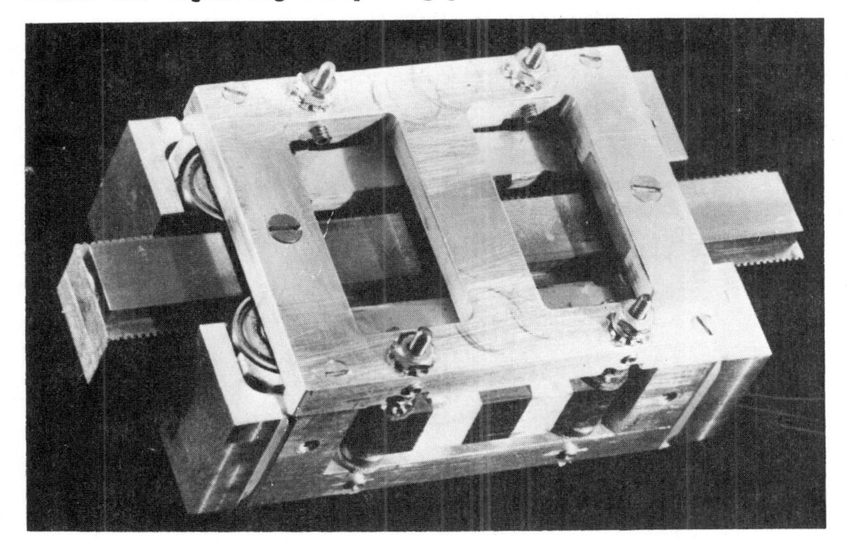

Figure 14-1. Experimental three-phase linear step motor.

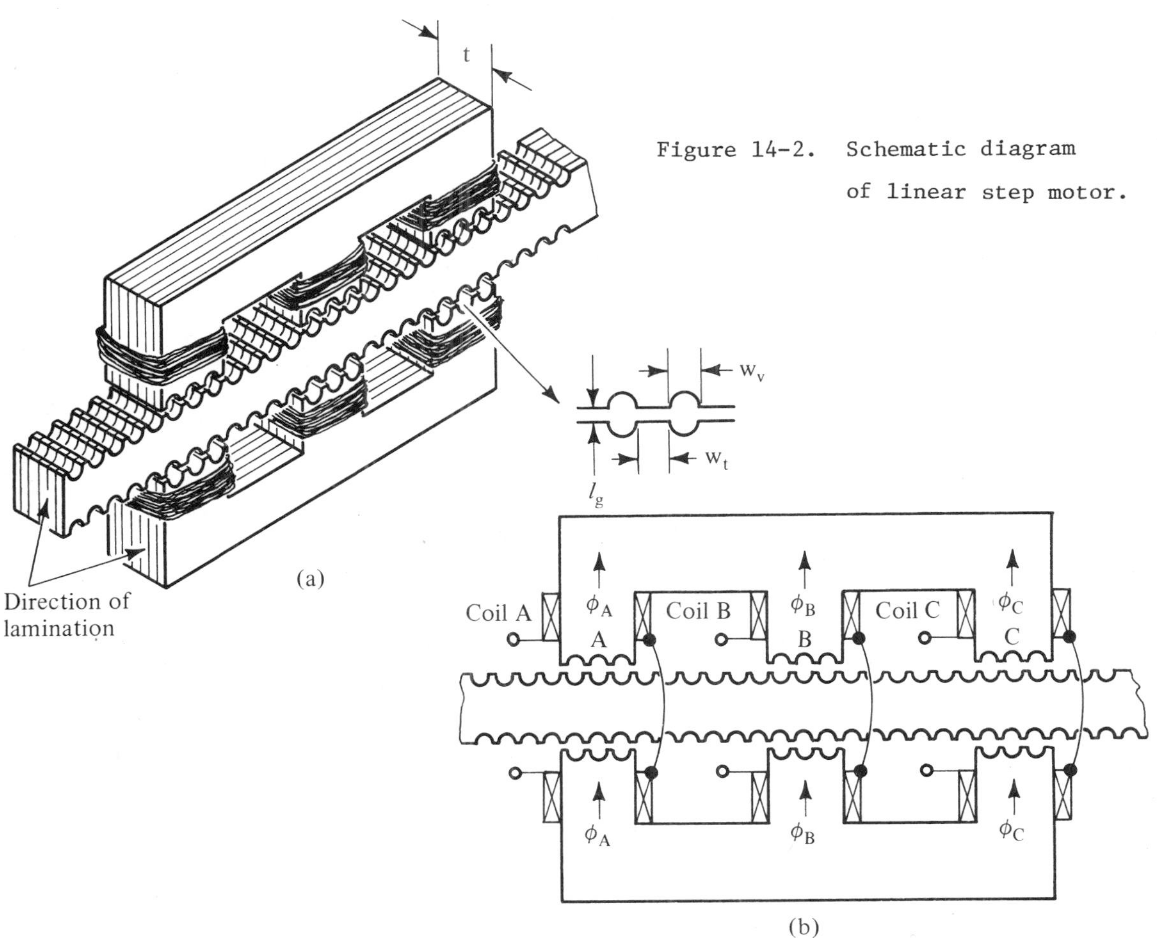

Figure 14-2. Schematic diagram of linear step motor.

The rectangular construction enables the stator and armature to be laminated, thus improving the high-speed response by minimizing eddy currents in the magnetic circuit. The stator and armature are easily assembled from laminations that are punched out of commercially available electrical iron sheets. After assembly, the valleys are filled with a non-magnetic plastic material to prevent accumulation of foreign particles.

14-3 FORCE EQUATION

Figure 14-3a shows the magnetic circuit describing the motor; Fig. 14-3b reduces this circuit to a simpler form. Here we are only considering the gap

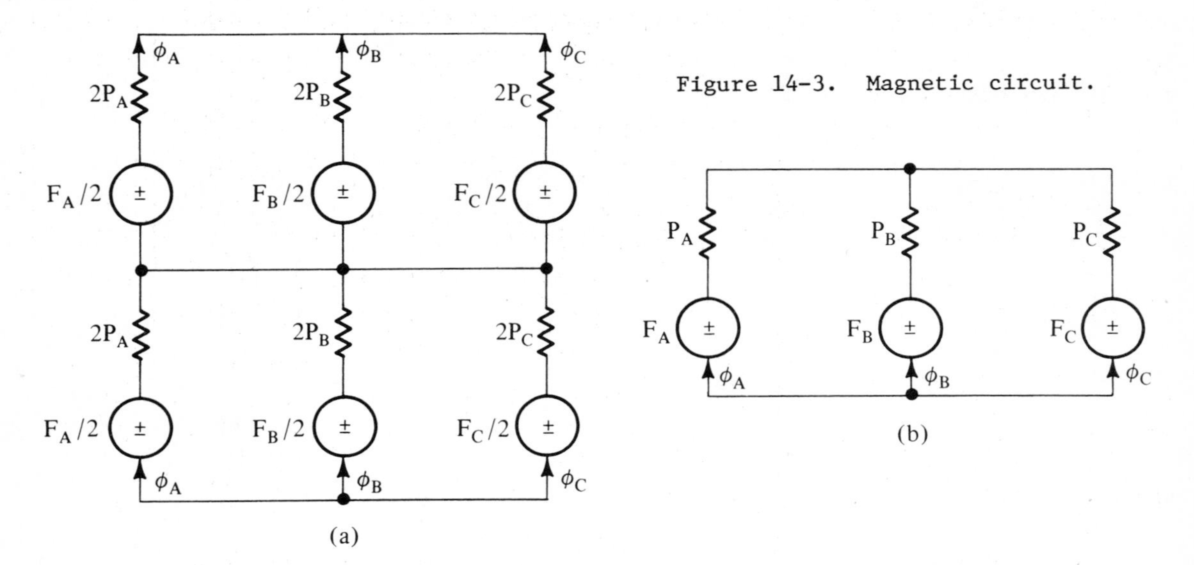

Figure 14-3. Magnetic circuit.

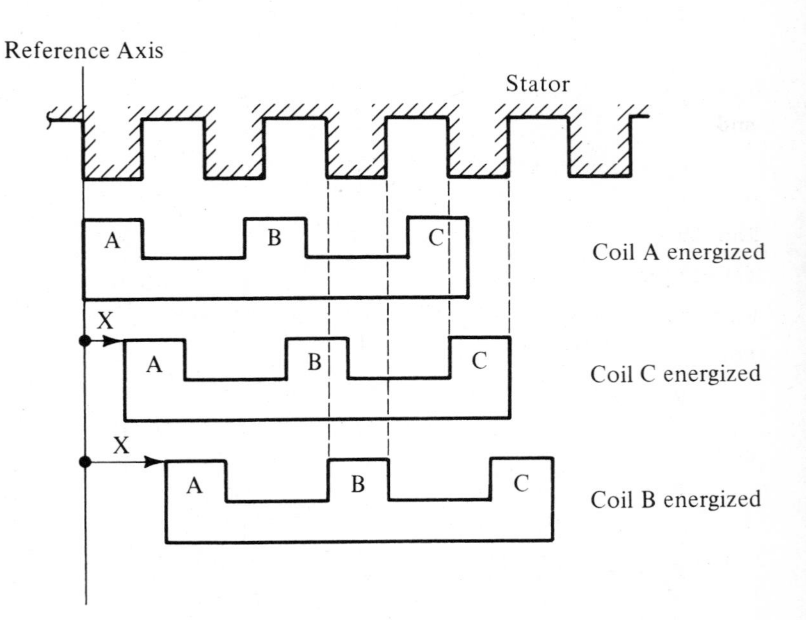

Figure 14-4. Detent positions with single-phase energization.

permeances; permeances in the iron are neglected. Using the reference axis shown in Fig. 14-4, the gap permeances are approximately given by

$$P_A = P_0 + P_1 \cos\theta_e$$

$$P_B = P_0 + P_1 \cos(\theta_e - 120^o)$$

$$P_C = P_0 + P_1 \cos(\theta_e - 240^o), \qquad (14\text{-}1)$$

where

P_0 = d-c component of the permeance

P_1 = magnitude of first harmonic component of permeance

θ_e = electrical angle, $2\pi\left(\frac{x}{w_t + w_v}\right)$

x, w_t and w_v are respectively the armature displacement, tooth width and the valley width.

The reluctance force on the armature (See Chapter 5) is:

$$F = \frac{1}{2}\sum_{i=1}^{3}(F_i - F_o)^2 \frac{\partial P_i}{\partial x}$$

$$= \frac{\pi}{w_t + w_v}\sum_{i=1}^{3}(F_i - F_o)^2 \frac{\partial P_i}{\partial \theta_e}, \qquad (14\text{-}2)$$

where

$$F_o = \frac{\sum_{i=1}^{3} P_i F_i}{\sum_{i=1}^{3} P_i}, \qquad (14\text{-}3)$$

and

F_i = excitation mmf F_A, F_B, F_C.

The index i in Eqs. (14-2) and (14-3) signifies subscripts A, B, and C shown in Eq. (14-1) and Fig. 14-3. These subscripts are used to avoid confusion between the first harmonic component P_1 and the gap permeances.

For

$$F_A = NI$$

and

$$F_B = F_C = 0$$

$$F_A = -K_t(NI)^2 \sin\theta_e + \Delta F_A, \qquad (14\text{-}4)$$

where

$$K_t = \frac{\pi P_1}{w_t + w_v}\left[\frac{8}{9} - \frac{1}{18}\left(\frac{P_1}{P_0}\right)^2\right]$$

and

ΔF_A = higher harmonic components.

Similarly, for

$$F_B = NI$$

and

$$F_A = F_C = 0$$

$$F_B = -K_t(NI)^2 \sin(\theta_e - 120^\circ). \qquad (14\text{-}5)$$

And for

$$F_C = NI$$

and

$$F_A = F_B = 0$$

$$F_C = -K_t(NI)^2 \sin(\theta_e - 240^\circ). \qquad (14\text{-}6)$$

Note that the force is proportional to $(NI)^2$.

This is only true for a low value of NI. When NI becomes large, the mmf drop in the iron becomes significant and the force reaches saturation condition.

For single-phase energization, the detent positions are at θ_e = 0, 120 and 240° (Fig. 14-4). It takes three steps to travel one pitch ($x = w_t + w_v$). For the motion to the right, the sequence of excitation is A-C-B-A. For the motion to the left, the sequence of excitation is A-B-C-A.

14-4 ENERGIZATION SCHEMES

Single-phase energization is accomplished by a simple driving scheme (Fig. 14-5a) in which each coil is sequentially energized to accomplish an incremental motion. The timing diagram in Fig. 14-5a shows the necessary pulsing sequence for the motion of the armature to the right (Fig. 14-4). Each pulse causes the armature to move $1/3(w_t + w_v)$.

In rotary step motors, to obtain a high torque, two-phase energization with bifilar windings is used. A similar scheme can be applied to the linear step motor. Coils in each pole are bifilar wound. At each instant two coils are on; this gives

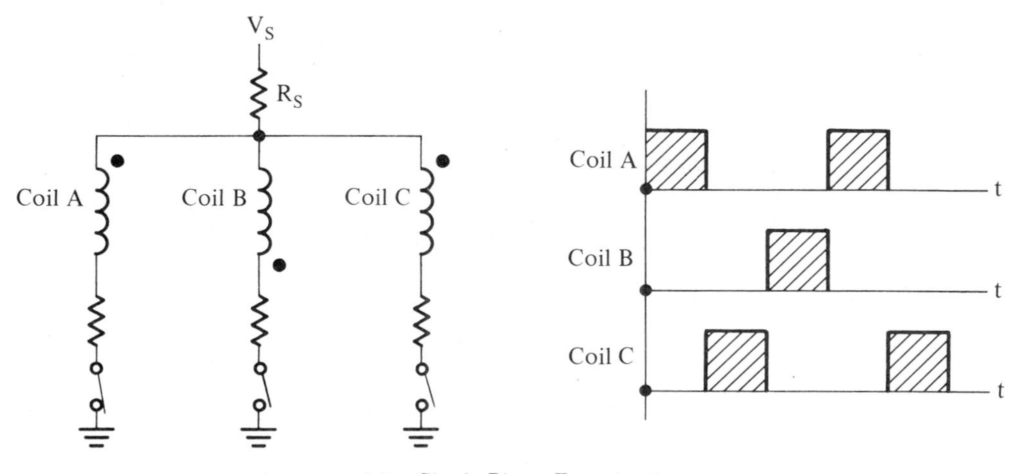

(a) Single Phase Energization

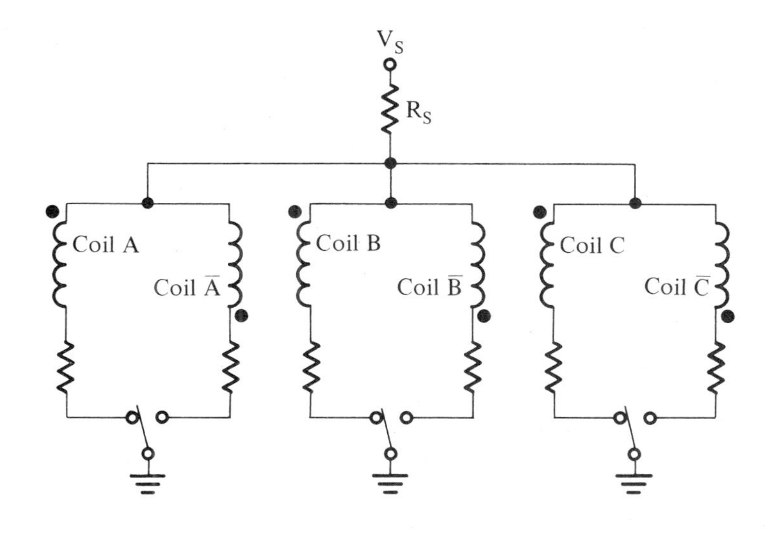

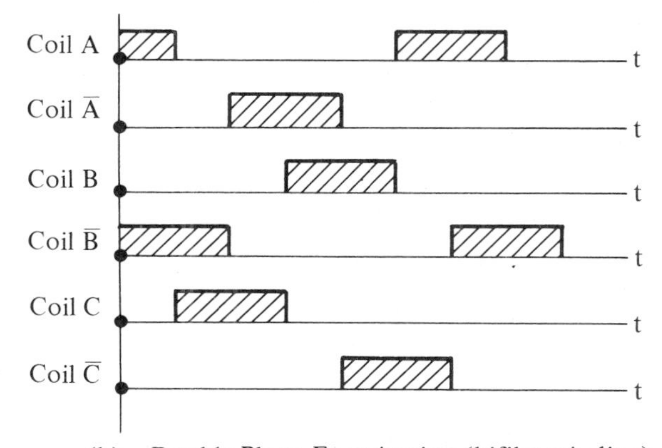

(b) Double Phase Energization (bifilar winding)

Figure 14-5. Energization schemes.

a higher force. Also since each coil is turned on for twice the length of the single-phase energization for a given step rate, the detrimental effect of current rise is reduced, and the high-speed performance is improved. One drawback to this scheme is the increased mass of the armature. Consequently, the choice of a particular energization depends on the application in which the motor is used.

Figure 14-5b shows the bifilar arrangement and the necessary timing diagram for the motion to the right. It should be noted that a complete pulsing cycle requires six steps instead of three. In this cycle, the armature travels $2(w_t + w_v)$.

14-5 DATA FOR THE EXPERIMENTAL VR MOTOR

The following are the data for the three-phase variable reluctance linear motor that was built and tested.

Physical Dimensions

Armature mass including the housing = 0.7 lb

Tooth width = 25 mil

Valley width = 25 mil

Pole width = 0.375 in

Number of teeth per pole = 8

Gap length = 3 mil nominal

Lamination thickness = 14.5 mil electrical iron

Armature and stator thickness = 0.542 in

Test Data

Step increment = 16.67 mil nominal
8.33 mil min (1/2 step)

Rated current = 3 amp/phase

Holding force = 6 lb

Single step response = 4.5 msec

Maximum slew rate = 3000 pps

Additional data are given in Figs. 14-6 through 14-9.

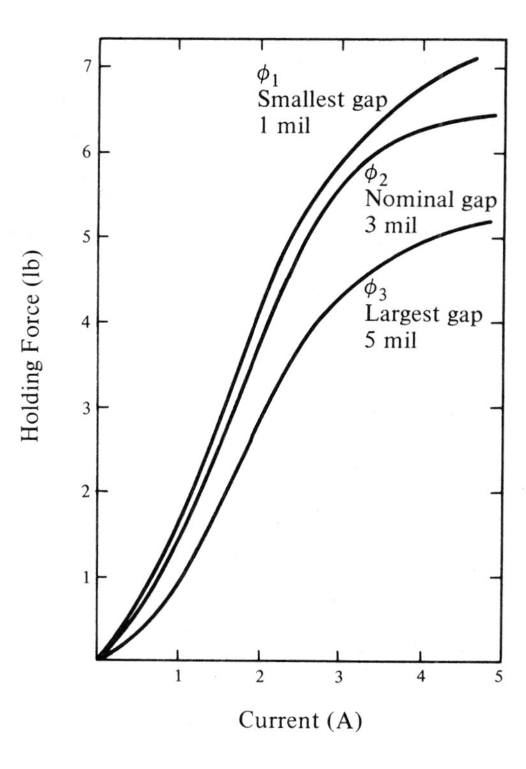

Figure 14-6. Holding torque versus current for single-phase energization.

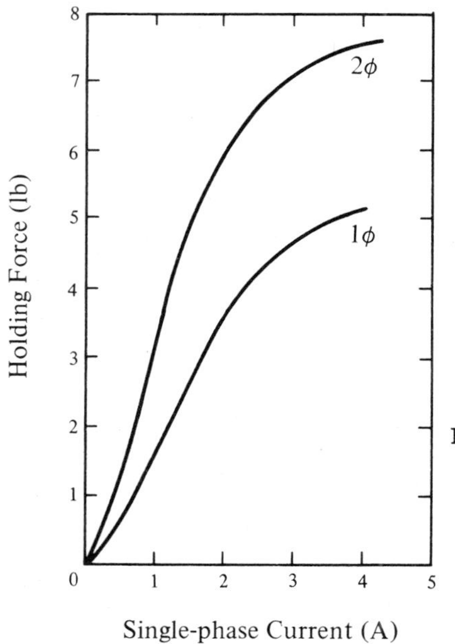

Figure 14-7. Holding torque versus current (5 mil gap).

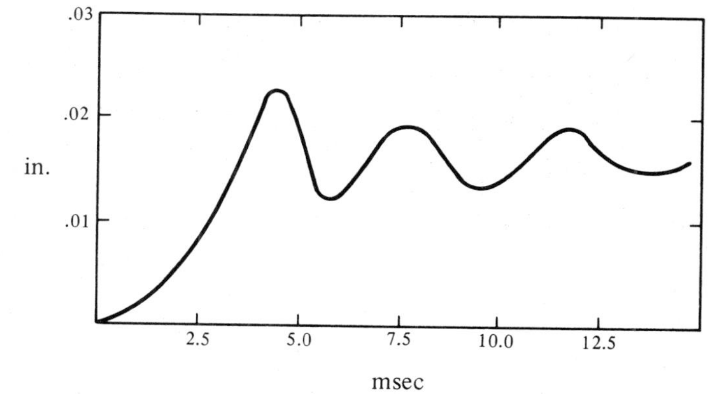

Figure 14-8. Single-step response.

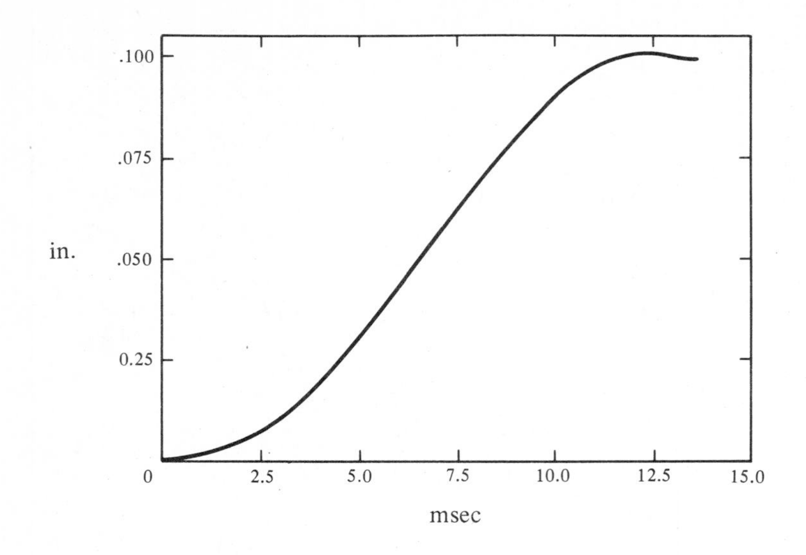

Figure 14-9. Six-step response.

14-6 DISCUSSION AND CONCLUSION

From the description, it is apparent that a versatile linear motor of this type of simple construction can be built. Various step increments are possible depending only on tooth size and pitch. Many force ranges can be achieved by lateral expansion of the stack. The stator bar, when properly reinforced, serves also as the structural member.

Laminated construction has a dual purpose:

1. Reduction of eddy currents
2. Easily assembled simple strips rather than solid machined bars.

The armature assembly is totally self-contained and, because of its structural integrity, the payload can be directly mounted on the motor.

A permanent magnet version of this unit is also achievable. By mounting the magnets parallel to the stator bar, d-c flux biasing is achieved. Less power is needed to steer the flux through the different legs than in the VR version, but, since the magnets are also part of the movable mass, a reduced response is the result.

1-2 mode operation is possible resulting in 1/2-step increments similar to the rotary version. Operation in a high performance application requires six drivers. The prototype discussed has a minimum 1/2-step size of 8.33 mil. There is a drawback, however, in using small-tooth motors since a small change in air gap length has a significant effect on force developed. To reduce this dependency, a larger tooth size is more desirable. By properly connecting the coils, a minimum number of flux reversals are generated for a given increment sequence. Aluminum wire can be used where the highest force to mass ratio is necessary. Winding directly on the armatures results in a lower winding mass, lower d-c resistance, lower cost, and lower leakage flux.

Flexibility and adaptability can be further increased by closed loop operation. By mounting a multitoothed magnetic pickup above the stator bar, the stator teeth become the position sense signal which, when properly processed, can drive the sequence logic.

Optical sensing can be accomplished in a similar manner. By using the reflective properties of the stator teeth or the absorptive property of the tooth valleys, a proper signal can be obtained to perform the desired function. An absolute encoder film pattern can be bonded to the stator box, and the proper signals can be obtained in that manner.

In open loop operation all the approaches given in References 2 and 3 can be fully applied.

14-7 SUMMARY OF ADVANTAGES AND DISADVANTAGES

Advantages

Direct linear motion

Favorable force-to-mass ratio

Simple drive schemes

Stator is also a supporting structure

PM as well as VR possible

High slew rates achievable

Open and closed loop operation possible

Stator bar also transducer grid

Disadvantages

Finite structural mass value
No gearing or spatial motion transformers possible
Finite step size due to physical limitations and manufacturability
Umbilical cord problems
Six drivers needed for high-performance operation

It is apparent that the linear motor is a versatile device. Its simplicity should result in a reasonably priced device with a large number of possible applications.

REFERENCES

1. H. D. Chai and J. P. Pawletko, "Magnetic Actuators Using Reluctance Torque Characteristics," 1971 Design Engineering Conference, April 20, 1971.

2. J. P. Pawletko, "Approaches to Stepping Motor Controls - Part I," Electromechanical Design, November 1972, p. 18.

3. J. P. Pawletko, "Approaches to Stepping Motor Controls - Part II," Electromechanical Design, December 1972, p. 26.

15 THE SAWYER LINEAR MOTOR

W. E. Hinds and B. Nocito
XYNETICS, Inc.
Conoga Park, California

15-1 INTRODUCTION

The need for high efficiency motors and positioners that have a linear instead of a rotary output grows steadily with the increasing use of automatic machine tools, remote actuators, and the entire field of automatic control systems. The Sawyer Linear Motor, invented in 1969, overcame most of the limitations of such linear motors and actuators as voice coils, pneumatic or hydraulic actuators, lead screws and linear induction motors.

The Sawyer Motor provided improved long motion capability, freedom from mechanical wear, and the unique ability to position itself precisely in space without the need for a closed-loop control system.

This paper discusses the mechanics and the principle of operation of the Sawyer Motor.

15-2 THE SAWYER LINEAR MOTOR

The Motor Mechanization

The simplest mechanization of a Sawyer Motor is shown in Figure 15-1.

The mechanism is called a "forcer." It consists of a permanent magnet, PM, two electromagents, EMA and EMB, and a platen made of a ferromagnetic material.

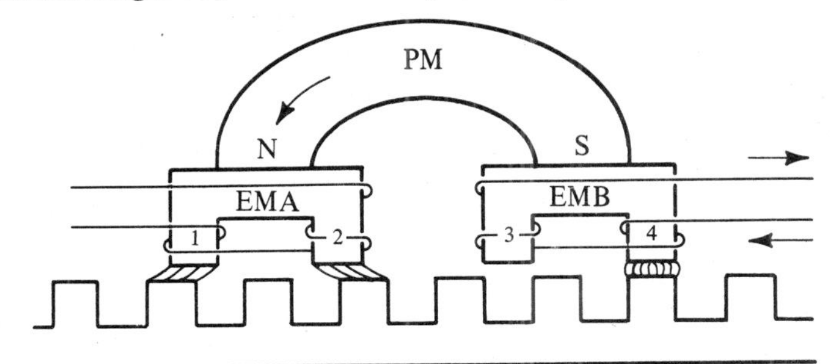

Figure 15-1. Mechanization of the Sawyer motor.

The permanent magnet flux closes its path through the electromagnet stacks, the air gap between stack and platen, and the platen.

In the absence of currents in the electromagnet coil, the magnet flux flows through both stack legs as shown in electromagnet EMA. Full coil current excitations switch, or commute, the flux entirely into one leg of the electromagnet, as shown in EMB. That brings the air gap density in that leg to a maximum, while the flux density in the other leg is brought down to a negligible value.

The appropriate coil flux commutation, in conjunction with the relative position of pole and platen teeth, produces forces, perpendicular to the teeth and parallel to the platen plane, that drive the motor assembly along one axis of motion.

These forces are called tangential to distinguish them from the normal force or pull between motor and platen surfaces caused by the magnetic field.

The motor assembly is floated on the platen by a gas bearing.

To understand more about the functioning of this novel motor, Figures 15-2a to 15-2d illustrate how a full current cycle produces forces between motor and platen teeth that drive the motor by discrete increments of one quarter of the platen pitch for each current switching. When the motor operates in this fashion, it is the analog of a step motor. But if a cosinusoidal current is applied to Phase A, and a

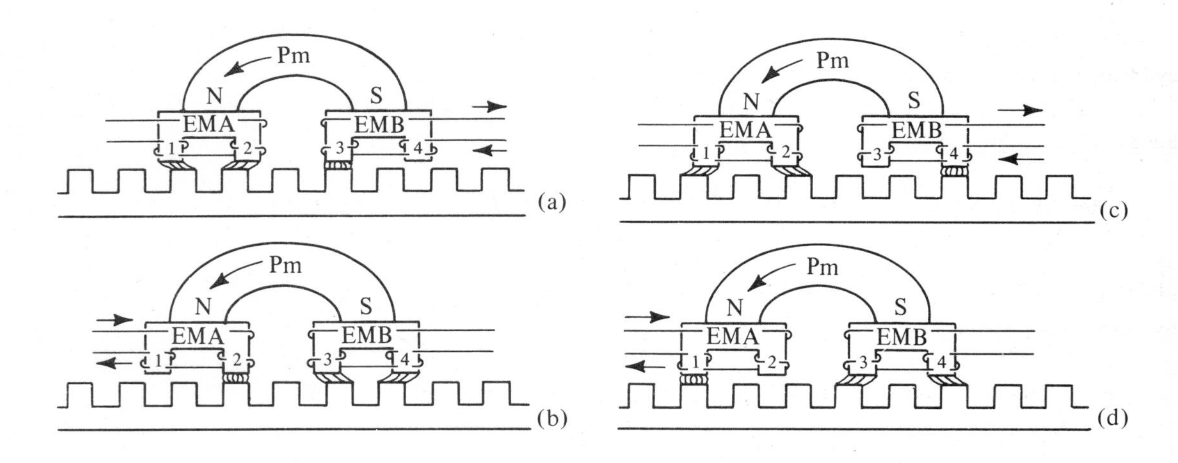

Figure 15-2. Diagrams illustrate how the Sawyer motor works.

sinusoidal current to Phase B, the motor moves by any increment corresponding to an incremental change of the exciting currents. Thus, the step resolution can be made as small as desired.

Two axis motion can be obtained when two forcers are assembled on a single motor frame, with their axes of action perpendicular to each other. In this case, a waffle platen with sets of perpendicular slots is used (Figure 15-3).

Three-axis motion, combined with rotation, can be achieved by merely adding a third motor platen assembly to the two-axis motor as shown in Figure 15-4.

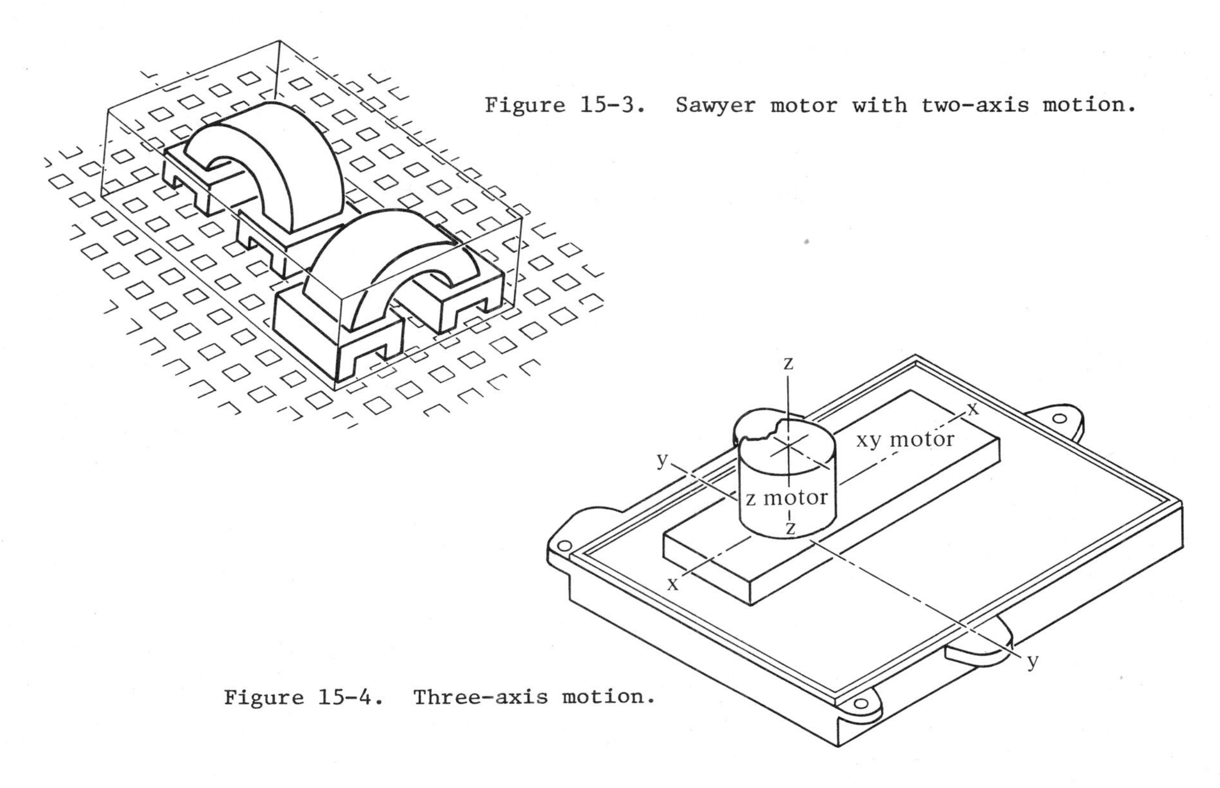

Figure 15-3. Sawyer motor with two-axis motion.

Figure 15-4. Three-axis motion.

It is interesting to note that rotary motion can be achieved by using a cylindrical platen that rotates in a cylindrical forcer.

Basic Motor Equations

The basic mechanization of a 2-phase single axis motor is shown in Figure 15-5. It is important to note that poles 1 and 3 are in phase; poles 2 and 4 are in phase, but they lag poles 1 and 3 by $\tau/2$; poles 5 and 7 are in phase but lead 1 and 3 by $\tau/4$; and, finally, 6 and 8 are in phase but lag 1 and 3 by $\tau/4$.

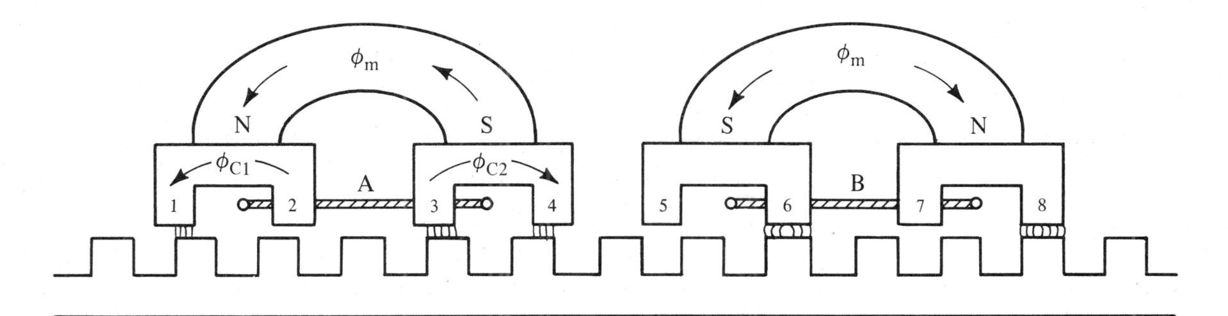

Figure 15-5. Basic mechanization of a two-phase single-axis motor.

The coils A and B have the function of commuting the permanent magnet flux. With full current, as in the case of coil A, all the magnetic flux circulates through poles 1 and 3. Meanwhile, poles 2 and 4 carry negligible flux. To derive the motor force equation, we will start by deriving the force acting on pole 1 for a given flux circulating across the air gap. To simplify the problems, the permeability of the iron is assumed to be large enough to make the reluctance of the iron path negligible. Leakage flux will also be neglected.

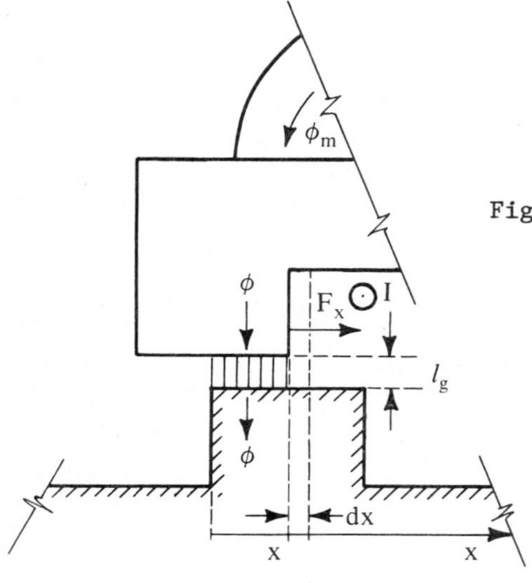

Figure 15-6. Pole 1 of the Sawyer motor.

With these two assumptions, pole 1 can be represented as in Figure 15-6.

The magnetic energy stored in the air gap is equal to:

$$E = \frac{B^2}{8\pi} V = \frac{B^2}{8\pi} h l_g \tag{15-1}$$

E = air gap magnetic energy (ergs)

B = air gap flux density (gauss)

h = electromagnet stack length (cm)

x = pole engagement (cm)

l_g = air gap length (cm)

If an incremental displacement dx of the pole pieces is allowed, the tangential force F_x will perform work equal to the air gap energy change:

$$F_x dx = dE = \frac{B^2}{8\pi} h l_g dx$$

then, (15-2)

$$F_x = \frac{dE}{dx} = \frac{B^2}{8\pi} h l_g$$

This can be expressed as a function of the total flux ϕ crossing the air gap as follows:

$$F_x = \frac{B^2}{8\pi} h l_g \left(\frac{hx^2}{hx^2}\right) = \frac{B^2 h^2 x^2}{8\pi} \frac{l_g}{hx^2} \tag{15-3}$$

but

$$\phi = Bhx$$

then

$$F_x = \frac{\phi^2}{8\pi} \frac{l_g}{hx^2}$$

The air gap reluctance is:

$$R = \frac{l_g}{hx}$$

then

$$\frac{\partial R}{\partial x} = -\frac{l_g}{hx^2} \qquad (15\text{-}4)$$

or

$$\frac{l_g}{hx^2} = -\frac{\partial R}{\partial x} \qquad (15\text{-}5)$$

Substituting Equation (15-5) in Equation (15-4), we have

$$F_x = -\frac{\phi^2}{8\pi} \frac{\partial R}{\partial x} \qquad (15\text{-}6)$$

The minus sign in Equation (15-6) comes from the fact that F_x is in the direction in which the reluctance R decreases. The flux across pole 1 gap is

$$\phi = \frac{\phi_c}{2} + \frac{\phi_m}{2} = \phi_m \qquad (15\text{-}7)$$

where

ϕ_c = coil flux (maxwells)

ϕ_m = permanent magnet flux.

The air gap reluctance function for a platen pitch τ, neglecting higher harmonics, is equal to

$$R = R_o(1 - K\cos\frac{2\pi x}{\tau}) \qquad (15\text{-}8)$$

where

R_o = average reluctance over a pitch

K = reluctance per unit variation

R is minimum when the pole pieces are aligned and x = 0 for this condition.

Substituting (15-7) and (15-8) into (15-6), for each pole, with due consideration to the relative pole phase shift, and adding the following expression, the total tangential force acting on the motor is obtained:

$$F_x = \frac{kR_o\phi_m\phi_c}{2\tau}\left(\phi_{ca}\sin\frac{2\pi x}{\tau} + \phi_{cb}\cos\frac{2\pi x}{\tau}\right) \qquad (15\text{-}9)$$

For excitations of peak value, ϕ_c is equal to:

$$\phi_{ca} = \phi_c \sin(wt + \psi)$$

$$\phi_{cb} = \phi_c \cos(wt + \psi) \qquad (15\text{-}10)$$

and taking into account that the motor advances a pitch, τ, in a current cycle, we have the following relationship:

$$v = \tau f$$

$$x = vt = \tau ft \qquad (15\text{-}11)$$

$$w = 2\pi f$$

Substituting (15-10) and (15-11) in (15-9), we have

$$F_x = \frac{kR_o\phi_m}{2\tau}\left[\phi_c\cos(wt + \psi)\cos\frac{2\pi}{\tau}\tau ft - \phi_c\sin(wt + \psi)\sin\frac{2\pi}{\tau}\tau ft\right] \qquad (15\text{-}12)$$

which reduces to:

$$F_x = \frac{kR_o\phi_m\phi_c}{2\tau}\sin\psi \qquad (15\text{-}13)$$

where ψ is the force angle or phase shift between the flux and the reluctance function.

Equation (15-13) indicates that the output force, neglecting saturation and iron losses, is constant and proportional to the product of the magnet flux times the coil flux, and inversely proportional to the motor and platen pitch.

Effects of Iron Saturation

The need for high force versus motor weight ratio in order to achieve high accelerations, dictates high flux densities which reduce the iron permeability to the point where the reluctance of the iron path is no longer negligible with respect to the air gap reluctance at coil peak current intensities. This effect introduces a distortion in the flux waveshapes. Coil current and total resulting air gap flux, for no iron saturation and for iron saturation, are shown in Figures 15-7 and 15-8, respectively, for the first quadrant.

With no iron saturation, the force contribution of both phases is due to sinusoidal flux variations. If the currents are increased by discrete angular

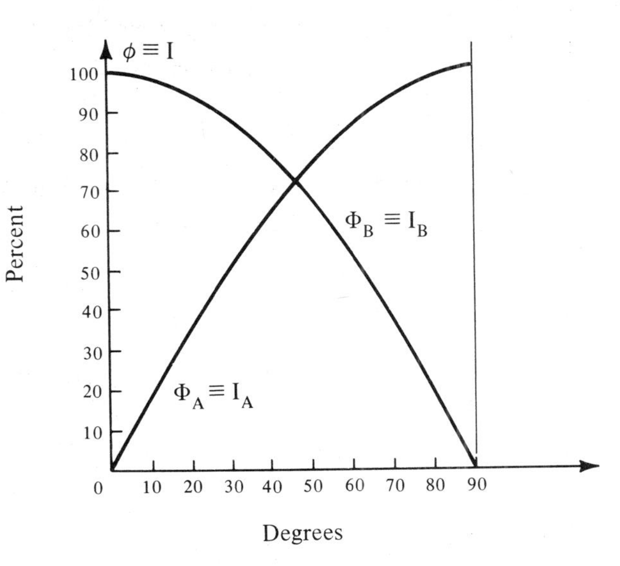

Figure 15-7. Coil current and total resulting air gap flux, no iron saturation.

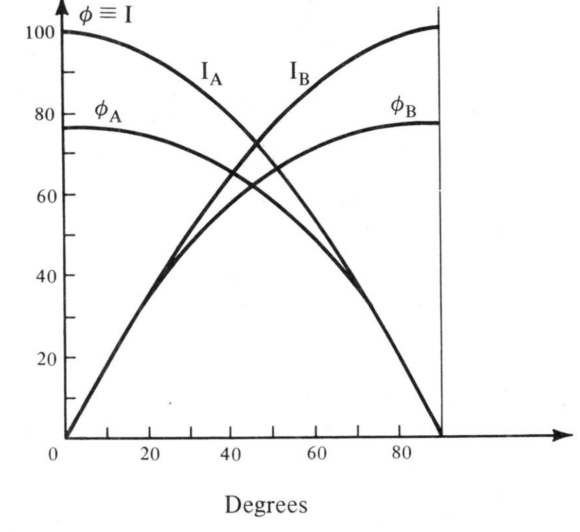

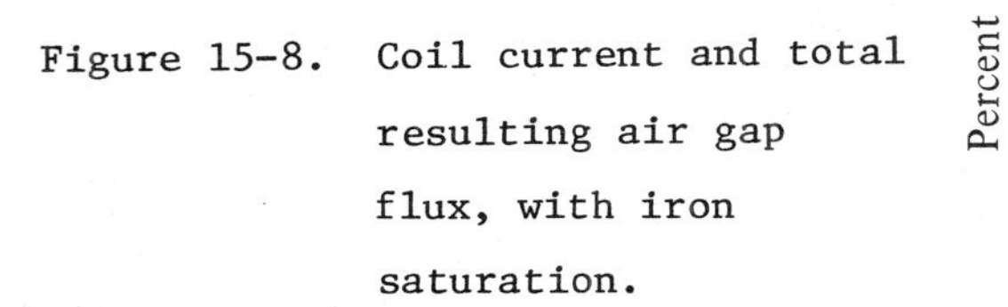
Figure 15-8. Coil current and total resulting air gap flux, with iron saturation.

increments, the motor advances by increments proportional to the angular increments of the currents:

$$x = \frac{\tau}{2\pi}\,\theta$$

since

$$\theta = t_g^{-1}\left(\frac{\phi_b}{\phi_a}\right) = t_g^{-1}\left(\frac{I_b}{I_a}\right) \tag{15-14}$$

However, if saturation exists, equation (15-14) no longer holds ture and, in general,

$$t_g^{-1}\left(\frac{\phi_b}{\phi_a}\right) \neq t_g^{-1}\left(\frac{I_b}{I_a}\right) \tag{15-15}$$

which implies that an angular increment of current does not correspond to an equal angular increment of flux. With saturation, $\theta = t_g^{-1}(\phi_b/\phi_a)$ is equal to $\theta_c = t_g^{-1}(I_b/I_a)$ which is the commanded angular displacement, only at the origin, at 45°, and its multiples; that is,

$$\theta_c = t_g^{-1}\left(\frac{I_b}{I_a}\right) = 45N°$$

where N = 0, 1, 2...

This introduces a fourth harmonic error position as shown in Figure 15-9.

For a two-phase motor with a pitch $\tau = 0.040$ inch, the peak position error is of the order of 0.0015 to 0.002 inch. The four-phase motor was developed to correct the fourth harmonic position error. It can be thought of as a single mechanical frame assembled with two two-phase motors, shifted mechanically by $\tau/4$ and with their respective two-phase excitations shifted by 45 electrical degrees. The linearity of a four-phase motor is of the order of 0.0002 inch.

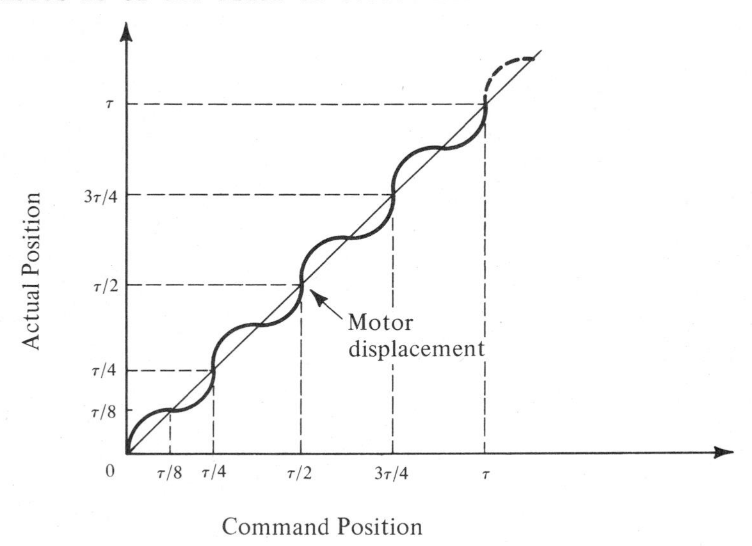

Figure 15-9. Fourth harmonic error in motor position.

Effect of Platen Iron Losses

Though the force equation

$$F = \frac{kR_o\phi_m\phi_c}{2\tau}\sin\psi \tag{15-16}$$

shows the motor force as being independent from motor velocity, this is only because the platen losses were neglected.

When the Sawyer Motor is accelerated to high velocities, 30 in/sec or more, platen losses can no longer be neglected.

The motor magnetic field sweeps the platen as it moves. Each platen element, being made of ferromagnetic conductive material, is the seat of eddy current and hysteresis losses. These platen losses translate into a drag force opposing the motor motion, proportional to the velocity.

The drag forces, per unit volume of platen, can be derived as follows:

$$v = \tau f \tag{15-17}$$

where

v = motor velocity

τ = motor pitch

f = excitation frequency

The losses per unit volume are

$$P_v = \eta B_m^n f + \xi B_m^2 t^2 f^2 \tag{15-18}$$

where

η = hysteresis coefficient

B = peak flux density

f = frequency

ξ = eddy current coefficient

t = platen thickness

The drag force per unit volume is then expressed by

$$F_v = \frac{P_v}{v} \tag{15-19}$$

$$F_v = \eta B_m^n + \xi B_m^2 t^2 f \tag{15-20}$$

and the total drag force is equal to the volume integral

$$F_d = \int_v P_v dV \tag{15-21}$$

The integration has to be carried over the platen volume directly under the motor.

The total motor tangential force available for acceleration at a given velocity is given by

$$F = \frac{kR_o \phi_m \phi_c}{2\tau} \sin\psi - F_d \tag{15-22}$$

where F_d is the drag force at that particular velocity. The maximum accelerating force available at a given velocity is

$$F = \frac{kR_o \phi_m \phi_c}{2\tau} - F_d \tag{15-23}$$

Figure 15-10 represents the maximum constant acceleration ramps as a function of the maximum velocity that can be achieved with a standard single axis four-phase motor with 2.50-inch stacks for a given payload and a laminated steel platen.

Figure 15-11 represents the performance of the same motor but with a 1010 cold-rolled steel platen, while Figure 15-12 corresponds to the combination of a two-axis, two-phase motor with 2-inch stack over a two-axis 1010 cold-rolled steel platen.

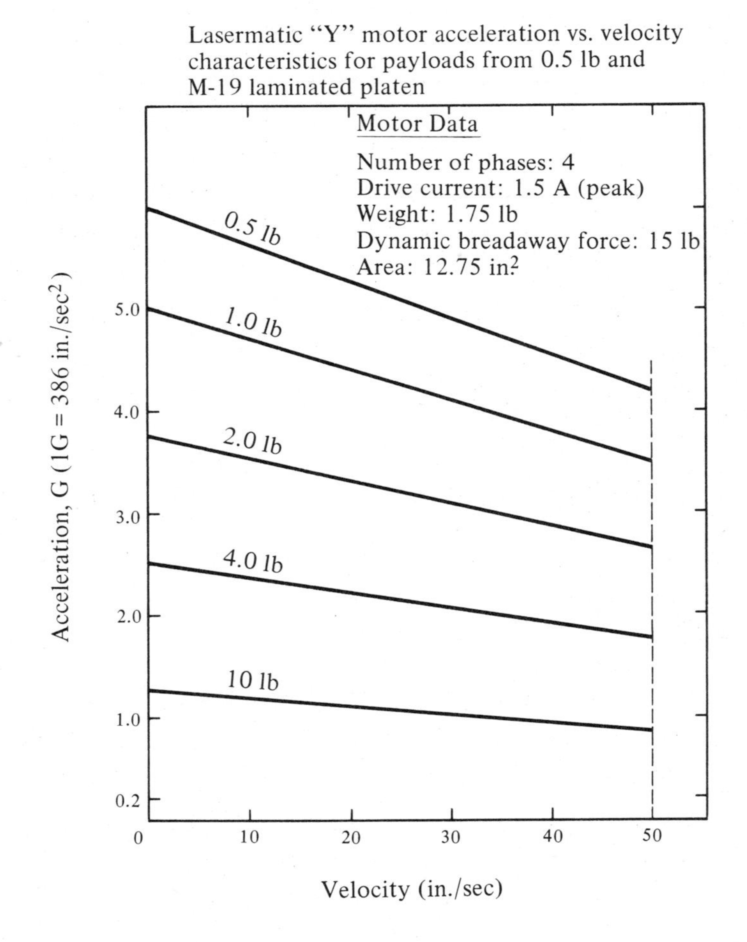

Figure 15-10. Lasermatic "Y" motor acceleration versus velocity characteristics for payloads from 0.5 lb to 10 lbs and M-19 laminated platen.

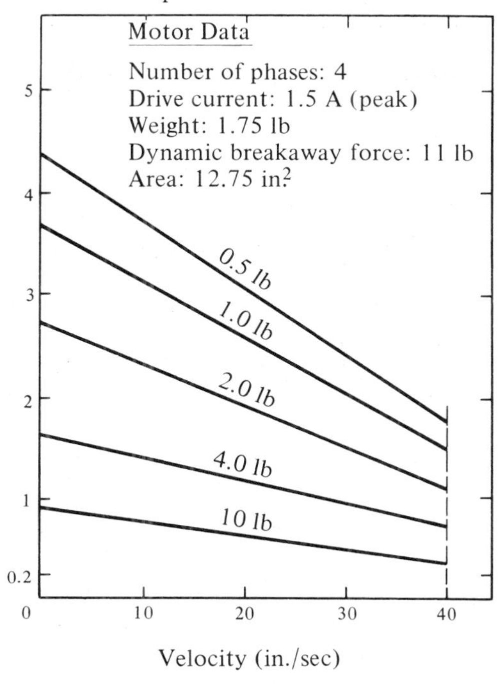

Figure 15-11. Lasermatic "Y" motor acceleration versus velocity characteristics for payloads from 0.5 lb and 10 lbs and 1010 steel platen.

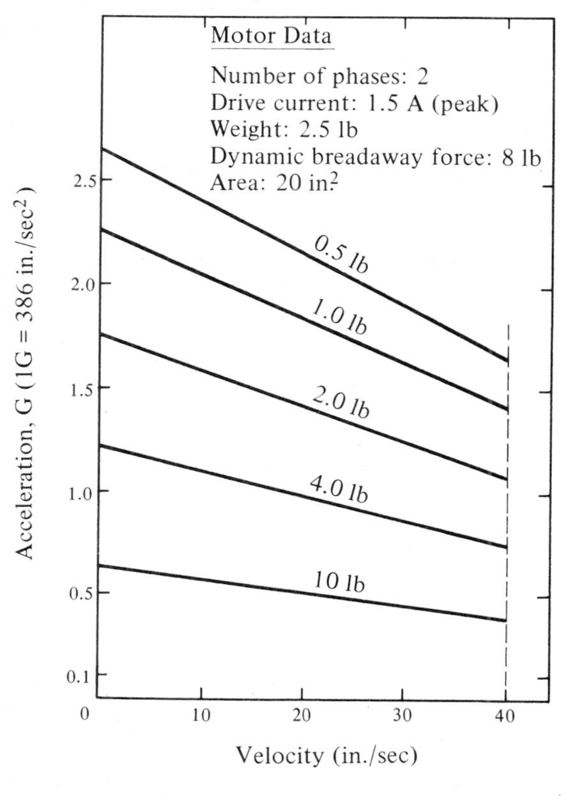

Figure 15-12. High-force XY motor acceleration versus velocity characteristic for payloads from 0.5 lb to 10 lbs and 1010 steel platen.

15-3 CONCLUSION

Since its invention in 1969, the Sawyer Linear Motor has proven to be an invaluable linear positioner for open-loop computer or NC-controlled systems. Linear induction motors, voice coils, hydraulic or pneumatic actuators require closed-loop control systems for precise positioning, and suffer wear. Lead screw systems driven by step motors suffer from wear and backlash. The Sawyer Linear Motor with its unique ability to position itself precisely without the need for a closed-loop control system, its wear-free operation and its high force-to-mass ratio is an ideal motor for automated systems where precise positioning and long life are required.

To conclude this chapter, we will present three typical applications developed by XYNETICS, INC. The company, since its founding, has devoted its efforts to developing Sawyer Linear Motors for specific industrial applications.

Drafting Head Positioner

One of the first applications of the Sawyer Principle was in the XYNETICS Automated Drafting System (Figure 15-13). The drafting head and motor are an

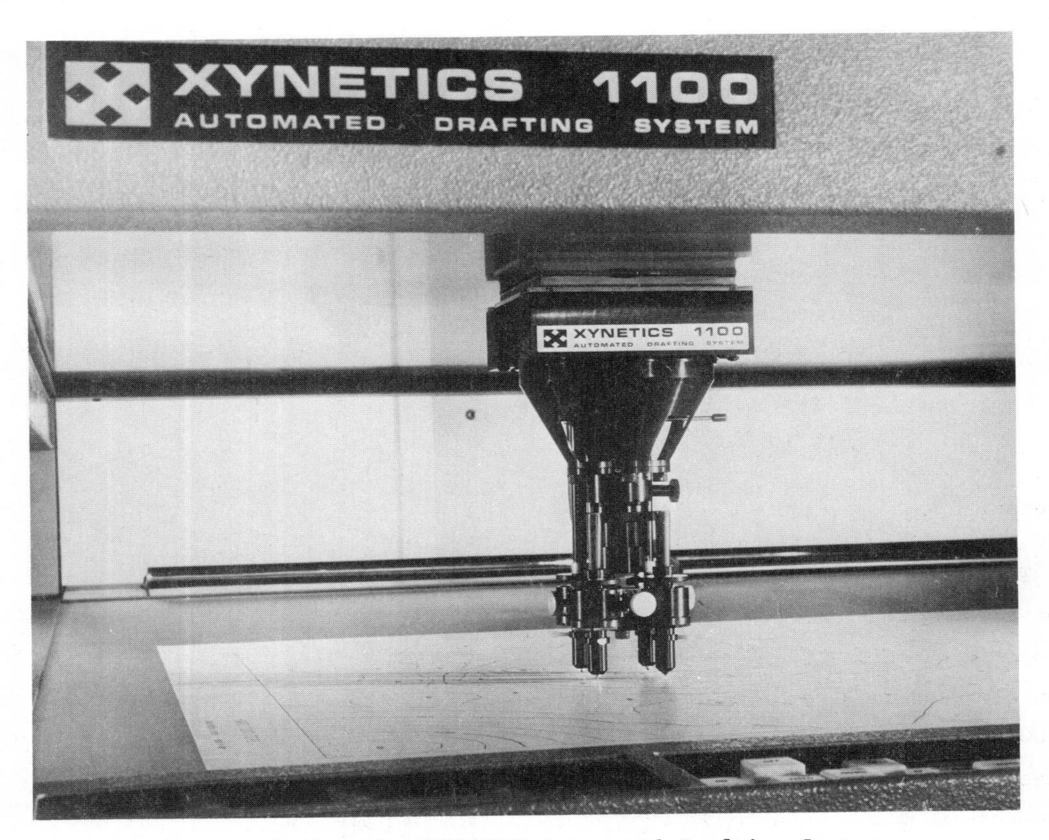

Figure 15-13. The XYNETICS Automated Drafting System.

integral unit riding on an air bearing under an X-Y platen. The positioning head, which carries various drafting tools, can move along any selected vector with a resolution of ± 0.001 inch, at accelerations up to 1g and speeds up to 50 ips. The head is controlled over the 5-foot by 8-foot drafting area by a small digital computer.

Laser Beam Positioners

The high speed and acceleration of the Sawyer Linear Motor made it possible to design an efficient laser beam fabric cutting system for use in apparel manufacturing, as shown in Figure 15-14. The positioning head, with laser focusing optics, rides on a Y-beam which is driven by X motors over X beams. The laser optics can be positioned anywhere within a cutting area of 5 feet by 10 feet, at speeds over 40 ips and accelerations over 1g, with a resolution of ± 0.001 inch. The system is controlled by a Hewlett-Packard 2100 computer.

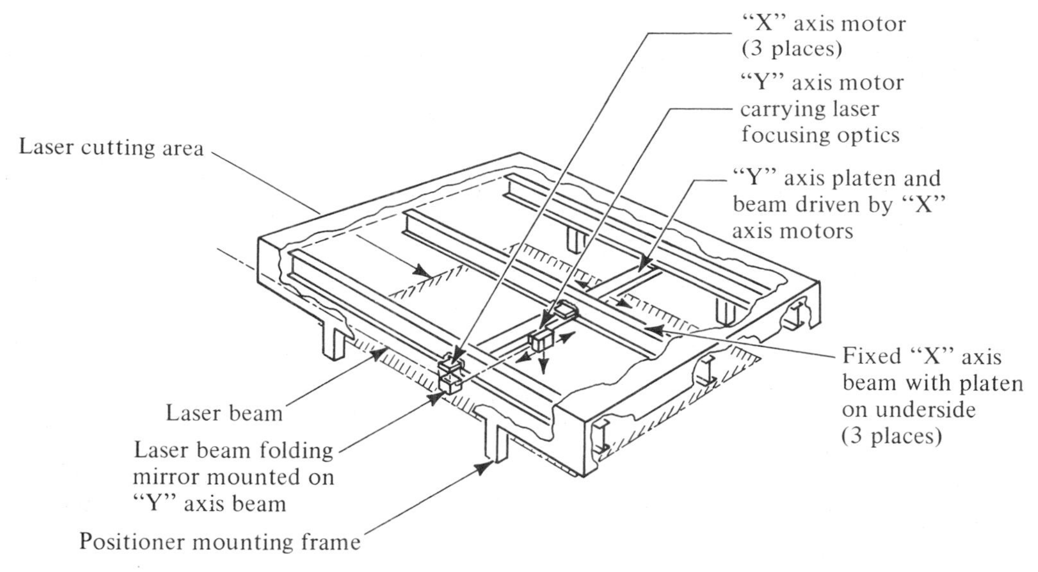

Figure 15-14. A laser beam fabric cutting system.

Micropositioner

To provide quick and highly accurate three-axis positioning of a semiconductor wafer under fixed probes, XYNETICS has incorporated the Sawyer Principle into a micropositioning system (Figure 15-15). The X-Y motor, carrying the Z stage, rides on an air bearing over a combined single-axis platen. Acceleration up to 1g with ± 0.0005 inch accuracies are achieved in all three axes over an operating area of 5 inches by 10 inches. The Z axis motor, which carries the semiconductor wafer on a vacuum chuck, has a vertical travel of 1/2 inch. The micropositioner is numerically controlled.

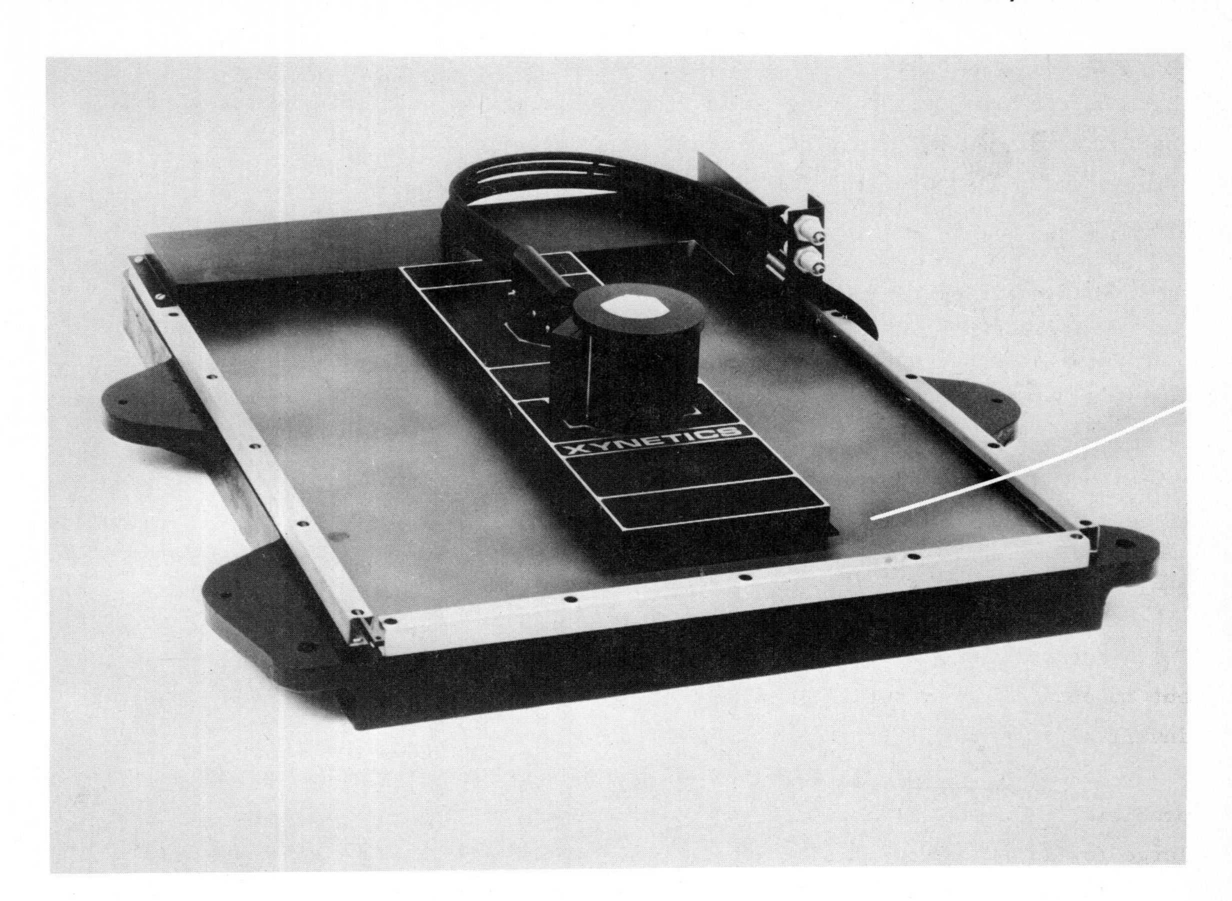
XYNETICS

16 ON THE CONTROL OF LINEAR MOTION STEP MOTORS

T. R. Fredriksen
T. Roland Fredriksen & Associates
Palo Alto, California

16-1 INTRODUCTION

Rotary step motors have become widely used in discrete positioning controls, but to obtain linear motion some form of mechanical translator is required. This invariably introduces friction and wear problems.

Recently, a linear step motor was developed eliminating the mechanical wear problems. However, from practical consideration the normal digital step size is too large for many high resolution applications.

This paper describes how this can be overcome by analog step interpolation. By partial current control to the phases a 0.005-inch step motor is controlled to give a 0.0005-inch step resolution.

Due to the high velocity capability of the linear motor special care must be taken to create a ramped pulse train. Concepts are presented for a totally digital pulse generator with burst control to give near perfect electronic damping.

The velocity profiles given illustrate typical performance characteristics of the linear step motor.

16-2 ANALOG STEP INTERPOLATION OF A LINEAR STEP MOTOR

The Sawyer motor discussed in Chapter 15 can be looked at as a regular step motor with the two phases, coils A and B, operated in a single-phase mode. Thus, we have four possible natural steps as shown in Table 16-1.

Table 16-1 - Cardinal Steps

Step #1	A = +	B = 0
Step #2	A = 0	B = +
Step #3	A = -	B = 0
Step #4	A = 0	B = 1

For a typical small step motor the length of these steps are 0.005-inches and give a tooth spacing of 0.020-inches. Figure 15-1 can be used to illustrate the action.

To get finer increments it is necessary to define states in between the natural or cardinal positions. Logically this would be to apply partial current in each phase coil, and the relation turns out to be sinusoidal. If we assume a 0.005-inch step to be 90°, then the current distribution in the coils are as shown in Figure 16-1.

If the required step resolution is 0.0005-inch, the current values are required to be generated only at every 9° interval. The reset point is defined in Figure 16-1, and from then on a "positive" step command will change the current levels to the motor progressively along the curves from left to right. A "negative" step command will change the current levels 9° to the left along the curves.

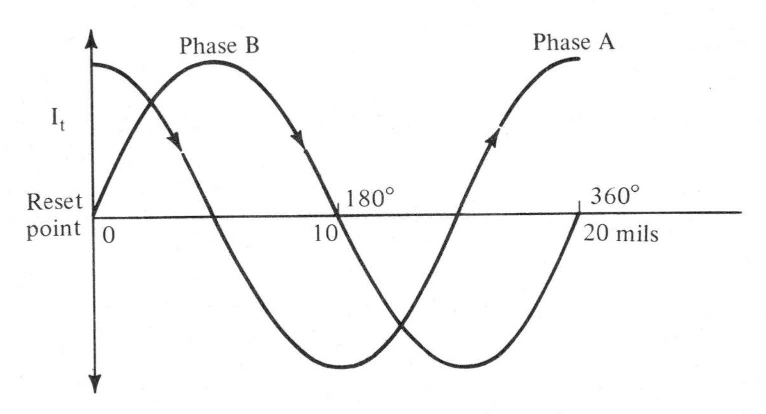

Figure 16-1. Steady-State Phase Currents.

Four-Phase Motor Configuration

By adding a second two-phase motor and displacing it 45° with respect to the first unit, some nonlinear force effects are cancelled. This motor is current-driven along curves also displaced 45° with respect to those in Figure 16-1. This means that the theoretical step positions are the same but deviations from predicted points are 180° out of phase. It follows that the error is largely a fourth-harmonic phenomenon. It follows also that unless the two motors are matched in characteristics and alignment, new positional errors may be introduced.

The linear step motor manufactured by Xynetics, Inc. consists of a permanent magnet/electromagnet assembly which skims over a toothed base. The moving member (Motor) has two sets of two-phase forcers per axis which are aligned with specific patterns on the base (or platen). Each forcer has two phase coils which are terminated to ground through a small feedback resistor as shown in Figure 16-2.

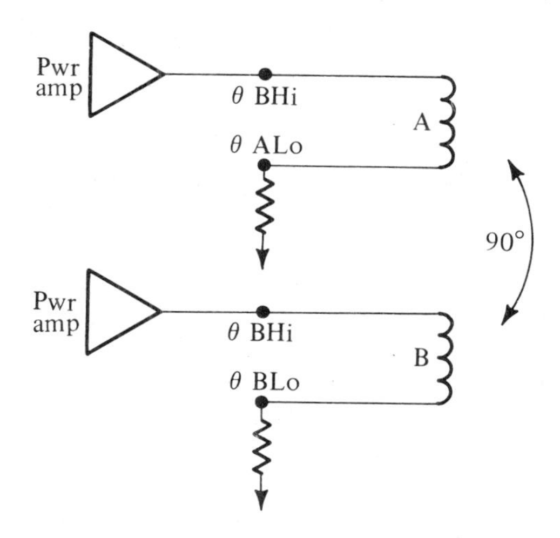

Figure 16-2. Power Driver Configuration.

The "Hi" inputs swing from +28 to -28V maximum in an effort to maintain constant current levels. Maximum current level is 1.5 amp and the actual steady-state phase currents are:

$$I_A = 1.5 \cos \theta$$
$$I_B = 1.5 \sin \theta \qquad (16\text{-}1)$$

The angle θ relates to linear position with 360° equal to 0.020" or as center-to-center tooth distance.

The second forcer has phase C and D which are displaced 45° from the A and B. Basic phase relations are:

$$I_C = \frac{1}{\sqrt{2}}(I_A + I_B)$$
$$I_D = \frac{1}{\sqrt{2}}(I_B - I_A) \qquad (16\text{-}2)$$

Figure 16-3 shows a vector diagram for Equation (16-2).

Reset condition sets θ to zero, which gives

$$I_A = 1.5$$
$$I_B = 0$$
$$I_C = \frac{1.5}{\sqrt{2}} \quad @ \quad \theta = 0° \qquad (16\text{-}3)$$

(Reset condition)

$$I_D = \frac{1.5}{\sqrt{2}}$$

Figure 16-3. Vector Diagram.

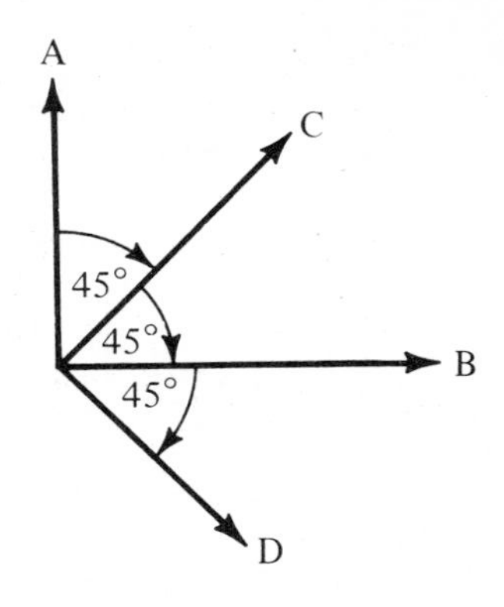

Power Amplifier Hardware Implementation

The hardware implements Equations (16-1) and (16-2) in three translations. First, the resolution of 0.0005-inches is created by recognizing 40 values for θ as

$$\theta = 9n \quad (n = 0, 1, ----, 39).$$

The number "n" is generated by a two-section up/down counter which is incremental each time an input pulse is received. Next, the lower digit (0, 1, ----, 9, 10) generates analog voltages

$$|V_A| = 10|\cos(9m)| \quad m = 0, 1, ----, 9, 10$$

$$|V_B| = 10|\sin(9m)|$$

while the upper digit sets

Sign V_A as +1 for 0 and 1

-1 for 2 and 3

Sign V_B as -1 for 0 and 1

+1 for 2 and 3

The lower digit counter is controlled by the upper digit so that the natural sequence is: --1, 0, 1, 2, 3, 4, 5, 6, 7, 8, 9, 10, 9, 8, 7, 6, 5, 4, 3, 2, 1, 0, 1--

Similarly, two analog voltages V_C and V_D are generated by operating on V_A and V_B, i.e.,

$$V_C = (V_A + V_B)\frac{1}{\sqrt{2}}$$

$$V_D = (V_B - V_A)\frac{1}{\sqrt{2}}$$

Finally, the analog voltages are translated into driving currents and

$$I_A = K_p V_A$$

$$I_B = K_p V_B$$

$$I_C = K_p V_C$$

$$I_D = K_p V_D$$

Where

$$K_p = 0.15 \text{ amps/volt}$$

16-3 DAMPING WITH PULSE BURST CONTROL

The Sawyer motor is highly underdamped since practically no mechanical friction exists. Therefore, when the final step is reached excessive ringing is likely to be sustained longer than desirable.

Another consideration is that because of the interpolated step size only partial torque is set up by one step. Assuming sinusoidal torque distribution and 9° steps, maximum torque should be at a 10-step phase advance.

In general, during acceleration the motor position is trailing the pulse train position; during constant velocity the motor position is more or less in phase with the pulse train; while during deceleration the motor position is leading the pulse train.

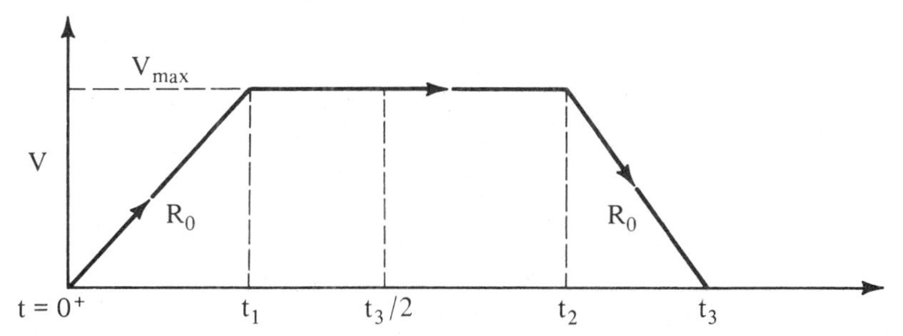

Figure 16-4. Velocity Profile and Definitions.

If the motor control is fed an increasing rate of pulses as described in Figure 16-4, the motor will adjust and fall into a phase-lag or lead condition. However, a much closer approximation to a constant acceleration and sharp acceleration changes can be generated by instantly changing the motor winding phase to correspond with the desired action. Using a mechanical analogy the situation is similar to pulling an inertia load with a spring. From a standstill the spring will have to be elongated until the required force is applied to the load to accelerate at the desired rate.

Similarly, to create constant speed the spring is relaxed to near normal position while deceleration is executed under compression.

The pulse-burst concept is implemented to instantly create the correct force condition in relation to position.

Starting from standstill a burst of pulses shifts the motor phase to give maximum force, (i.e., up to 10 pulses) in the desired direction of motion. At the end of the ramp the same number of pulses are subtracted giving essentially zero net torque. When deceleration is called for, a second reverse set of pulses shifts the motor phases to create maximum reverse force. Finally, at the desired terminal position the phase is stepped ahead to create practically zero force.

In a practical control, it is better to set the burst size to create zero force at the zero position. This type of action is similar to the well-known predictor control illustrated in Figure 16-5. Here, the oscillatory curve is the ringing or settling expected by the underdamped system when the stepping stops at position 0. Since the system overshoots to +n steps, it will have reached zero velocity and maximum reverse force proportional to n steps. If the desired position is now stepped to +n the force is reduced to zero and sitting at zero velocity the result is theoretically a deadbeat response as described by the horizontal dotted curve. While the approach to the zero velocity point is different, the condition at the zero velocity position is the same for the linear step motor following the predicted ramp. Thus, we get a substantially reduced ringing. (Perfectly deadbeat response is only possible if the phase lag is an integral number of steps.)

If we assume the ideal burst size to be n, then at point $t = 0^+$ in Figure 16-4, the burst size is +n; at t_1, −n; at t_2, −n; and at t_3, +n.

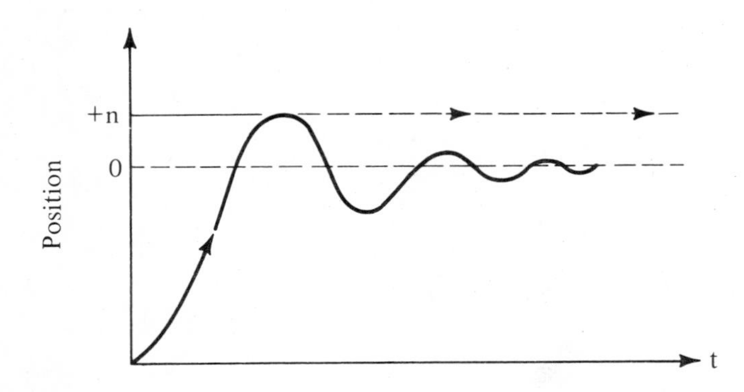

Figure 16-5. Predictor Control.

16-4 DIGITAL RAMP CONTROL

The ramp control is divided into two major functional blocks:

a) Zone generator

b) Pulse generator

Referring to Figure 16-4, the nomenclature defines the parameters for both functions.

V = velocity steps/sec

$a = \frac{dv}{dt}$ = acceleration steps/sec^2

R_o = ramp length = $\frac{V_{max} t_1}{2} = \frac{V_{max}}{2} (t_3 - t_2)$

D_o = index distance = $V_{max} t_2$

Phase 1 = $0^+ \leq t < t_1 \rightarrow$ acceleration phase

Phase 2 = $t_1 \leq t < t_2 \rightarrow$ const. velocity phase

Phase 3 = $t_2 \leq t < t_3 \rightarrow$ deceleration phase

Phase 0 = $t_3 \leq t \rightarrow$ countout, (ready)

Zone Generator Concepts

The zone generator has the following functions:

a) Receives and stores indexing commands, i.e., distance, direction, maximum velocity and acceleration magnitude.

b) Receives step pulses from the pulse generator and sends back information when to change the velocity rates.

c) Provides manual extensions of indexing functions.

The principal components in the zone generator are shown in Figure 16-6. During the load cycle the indexing distance, D_o, is preset into the D_1 - counter, and D_2 - counter; the ramp length R_o is preset into the R_o - register and R_1 - counter; and the ramp slope A_o is stored in the A_o register.

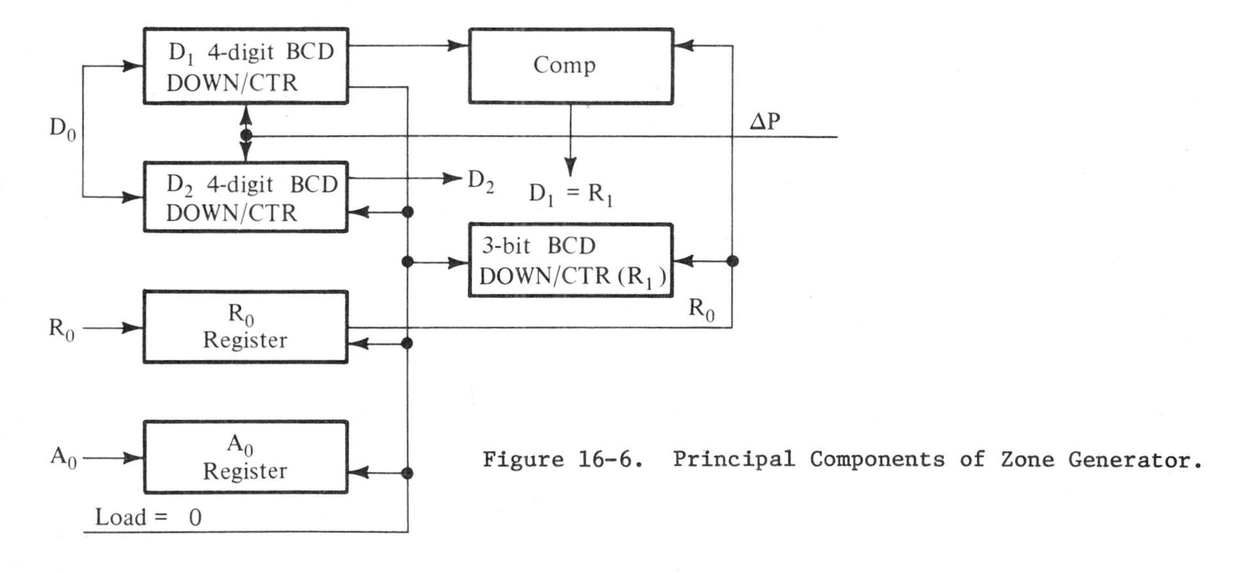

Figure 16-6. Principal Components of Zone Generator.

As the input pulses ΔP count the three counters down the following sequential events are detected:

1) Load pulse sets acceleration (Phase 1)

2) End of acceleration

a) $R_1 = 0$, $D_2 > 0$ in Phase 1 — Set constant velocity (Phase 2)

b) $D_2 = 0$, $R_1 \geq 0$ in Phase 1 — Set deceleration (Phase 3)

3) End of constant velocity (Phase 2)

$D_1 = R_1$ — Set deceleration (Phase 3)

4) $D_1 = 0$ — Set ready state (Phase 0)

In addition to the functional blocks shown in Figure 16-6 we need:

1) Detect when $R_1 = 0$

2) Detect when $D_2 = 0$

3) Detect when $D_1 = 0$

4) A two-Bit phase register

A common clock will assure that the changes of phase will not create a race condition, since the phase register changes state when clock drops.

Pulse Generator Concepts

The pulse generator must respond to the signals: a) Phase 0, b) Phase 1, c) Phase 2, d) Phase 3; and produce: a) ΔP, pulse train (one pulse/step), b) General clock, c) Burst pulses to motor.

Step Pulse Generator

Assume first that the velocity or stepping rate for the motor is represented by a number, V. Then from a fixed clock frequency, f, we want to generate a pulse train ΔP where

$$\Delta P = F(V, f) \tag{16-4}$$

A simple mechanization scheme of Equation (16-4) is shown in Figure 16-7.

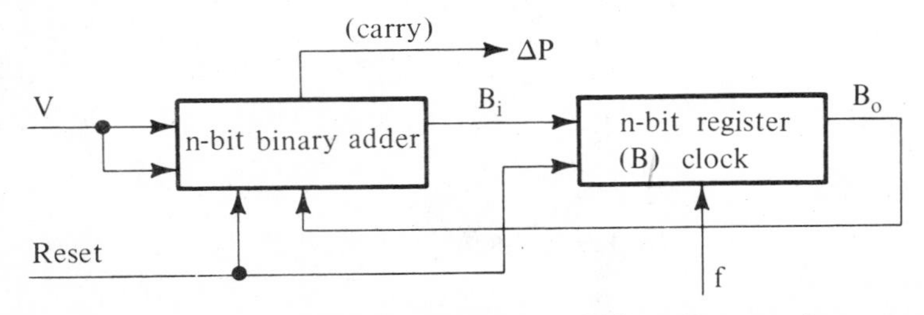

Figure 16-7. Velocity-to-Step Pulse Conversion.

Assume that both the adder and register are reset initially (t = 0) which sets $B_o = 0$ and $B_i = V$.

When the first f - pulse comes, the V - number is stored in the register and at the falling edge,

$$B_o = V$$

and

$$B_i = 2V$$

In general, if $B_i \leq 2^n - 1$

$$B_o = kV$$

and

$$B_i = (k + 1)V$$

after the kth pulse has updated the B register. Suppose after the kth pulse

$$(2^n - 1) - V < B_i \leq (2^n - 1)$$

then the (k + 1)th pulse will cause the adder to overflow. A carry pulse, ΔP, is generated. While B_o and B_i changes to

$$B_i = (k + 1)V - 2^n$$

$$B_o = (k + 1)V$$

If $V << 2^n$, the ΔP pulse rate is

$$P = \frac{V}{2^n}f \tag{16-5}$$

<u>Scaling the Step Pulse Generator</u>

The only real time parameter in Equation (16-5) is ΔP; the other values are selectable within practical ranges.

Assume first that $f = 1.0 \times 10^6$ pps and n = 16 (i.e., a 16 - Bit register); then for a max. $P = 2.0 \times 10^4$ pps

$$V_{max} = \frac{P2^n}{f} = \frac{(2.0 \times 10^4)(63,436)}{1.0 \times 10^6}$$

$$V_{max} = 1,268.72$$

A more convenient binary number would be $V_{max} = 511$ (i.e., 9-Bit register) or from a packaging point of view, even $V_{max} = 255$ giving an 8-Bit register. Changing to $V_{max} = 255$ would require $f = 2.57 \times 10^6$ pps or n to be lowered.

Since the adders are normally packaged in 4-Bits, n = 12 would be the next lower sized adder. In this case

$$2^n = 2^{12} = 4096$$

$$\frac{f}{\Delta P_{max}} = \frac{2^n}{V_{max}} = \frac{4096}{255} = 16$$

and

$$f = \frac{(2.0 \times 10^4)(4096)}{255} \approx 3.2 \times 10^5 \text{pps}$$

A ratio of 16 will give a maximum of 6.25% dither in the ΔP generation or 3.12μs. As a rule of thumb it should be less than 1.0μs. We now have a choice of increasing f and scaling ΔP down or increasing f and n.

To get better ramp generation we choose to select n as 16 and f as 1.06×10^6pps*. Thus, the final calculation gives

$$V_{max} = \frac{\Delta P 2^n}{f} = \frac{(2.0 \times 10^4)2^{16}}{1.06 \times 10^6} = 1200$$

Velocity Generation

The velocity changes as shown in the profile of Figure 16-4. A hardware implementation of acceleration and velocity parameters is detailed in Figure 16-8.

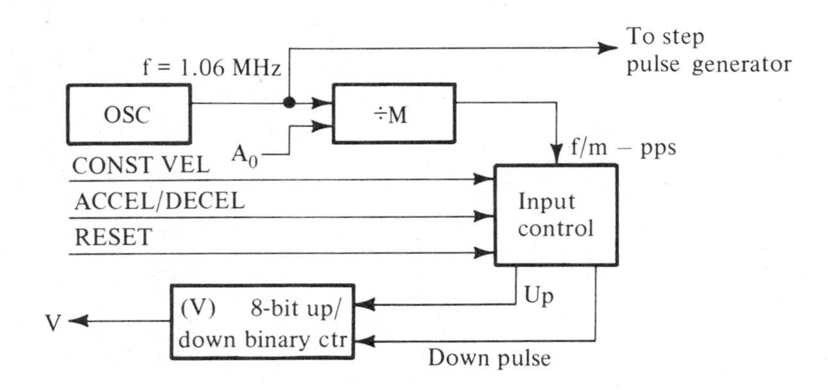

Figure 16-8. Velocity Generation.

*This frequency is a convenient harmonic of a 4.19304MHZ crystal controlled oscillator.

The oscillator (OSC) produces the required f - pulse train (1.06×10^6 pps), which are divided by M and sent to an up/down counter. Relating back to the phase control, we get the following input-control applied:

1) Phase 0: Counter is kept reset to zero.
2) Phase 1: Counter is allowed to count up (i.e., acceleration).
3) Phase 2: No pulses reach the counter (i.e., V = const.).
4) Phase 3: Counter receives down-pulses (i.e., deceleration).

Acceleration/Velocity Scaling

The V scale is $2.0 \times 10^4/1200$ and the desired ramp slope range is from 0.4 to 0.8×10^6 steps/sec^2. Then, maximum t_1 is

$$t_{1max} = \frac{2.0 \times 10^4}{0.4 \times 10^6} = 5.0 \times 10^{-3} \text{sec}$$

Thus, with scaling,

$$\frac{f_{min}}{M} = \text{Ramp slope pulse rate} = \frac{1200}{50 \times 10^{-3}} = 24 \times 10^3 \text{pps}$$

$$t_{1min} = \frac{2.0 \times 10^4}{18 \times 10^6} = 25 \times 10^{-3} \text{sec}$$

and

$$\frac{f_{max}}{M} = \frac{1200}{25 \times 10^{-3}} = 48 \times 10^3 \text{pps}$$

Thus

$$M_{min} = \frac{1.06 \times 10^6}{48 \times 10^3} = 22$$

and

$$M_{max} = 44$$

Mechanization of $(\div M) = F(A_o)$

The ramp slope is controlled by the input number A_o. If we use a similar arrangement to Figure 16-7,

$$\frac{f}{M} = \frac{A_o}{2^n} f$$

or

$$A_o = \frac{2^n}{M}$$

If n = 8

$$A_{omax} = \frac{256}{22} = 11.66$$

$$A_{omin} = \frac{256}{44} = 5.83$$

This allows the slope to be selected in 7 steps as A_o changes from 6 to 12. Assuming the frequency ratio of 22 to be adequate for this division application, the resulting hardware is shown in Figure 16-9. The A_o register is fixed at 4 Bits.

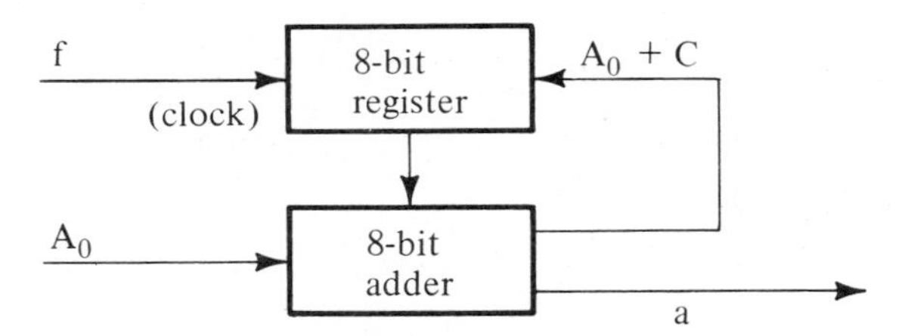

Figure 16-9. Variable Ramp Slope Mechanization.

Burst Pulse Control Concepts

The purpose of the burst is to introduce extra step pulses to the motor driver circuits either advancing or retarding the phase control with respect to the normal direction input. The basic elements of burst control are shown in Figure 16-10. At time $t = 0^-$ the up/down counter is preset by a selectable number N_o which ranges from 0 to 7. The "Load" pulse sets this condition.

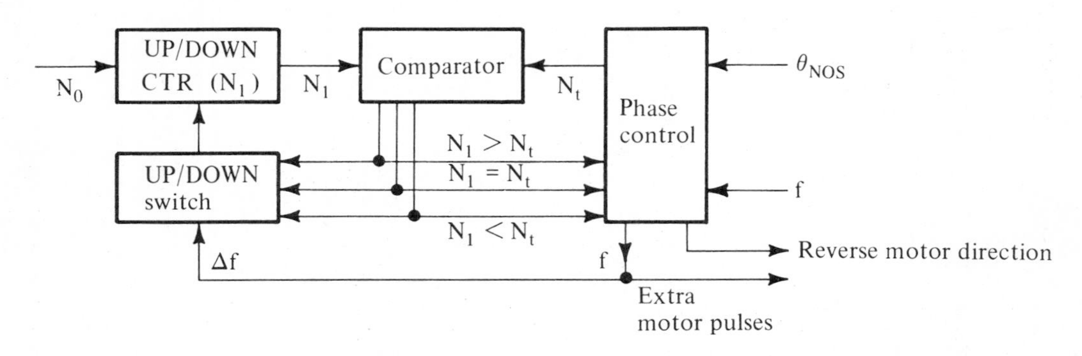

Figure 16-10. Burst Control Elements.

When Phase 1 is switched in at $t = 0^+$, the control number N_t is switched to $N_t = 0$, and the Δf - pulses allowed to exit from the phase control without calling for "reverse motor" direction. When $N_t = N_1 = 0$, the control shuts off the Δf pulses.

When Phase 2 is switched in $N_t = N_o$ and the comparison and phase control allows the Δf pulses to step the counter-up, reverse motor direction is now issued. At $N_1 = N_o$, the control shuts off Δf and inhibits reverse motor direction.

When Phase 3 is switched in, $N_t = 2N_o$, and depending on whether the previous phase was 1 or 2, either $2N_o$ or N_o reverse pulses are sent to the motor.

When Phase 0 is set, $N_t = N_o$, and the comparitor and phase control allows N_o forward pulses to exit to the motor. This returns the control to its original $t = 0^-$ state.

16-4 SYNCHRONIZING THE PULSE CONTROL

The crystal controlled clock frequency $f = 1.06 \times 10^6$ pps is generated by a 4.19304MHZ oscillator followed by a divide by four counter. Thus, the f-pulsewidth is approximately 250 ns.

Since all events are closely tied in with the clock f, any external command such as, Load = ⌉0⌈, must be converted to a set of trigger pulses which are synchronized with the free running clock.

A simple scheme for synchronizing the initialization procedure is shown in Figure 16-11. Here, the external load pulse sets the index command flip-flop which in turn allows the initial condition counter to count up to 9 of the following f-pulses.

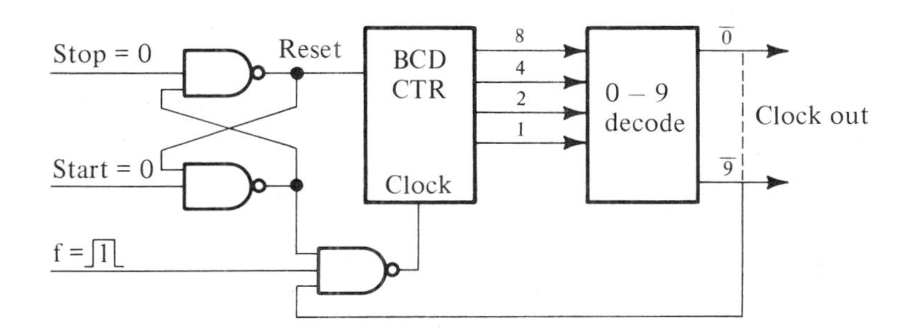

Figure 16-11. Synchronizing Clock.

The counter output is decoded and the sequential events now perfectly synchronized with the clock executes the resets, sets and start signals as listed in Table 16-1.

In general, mechanization care must be taken so that all changes of states such as the phases take place at the falling edge of the clock and that all sub-function pulses such as ΔP are clocked to the same basic f-pulsewidth. This makes a clean system free of race condition and allows the burst pulses to be slipped in very naturally with a minimum of 750 ns spacing to any active motor pulse.

16-5 CONCLUSION

The linear step motor represents another major advance in discrete positioning control. It has already found many applications in high speed automatic drafting machines, laser cutting equipment and precision semiconductor production equipment.

The control techniques described in this paper were developed for the linear step motor, but it seems very probable that the same methods can be applied to rotary step motors. This will combine the high speed capabilities of the large angle step motor with the fine angle resolution resulting from analog step interpolation.

REFERENCES

1. J. Proctor, "Stepping motors move in," Product Engrg., vol. 34, pp. 74-78, February 1963.
2. A. G. Thomas and J. F. Fleischauer, "The power stepping motor - a new digital actuator," Control Engrg., vol. 4, pp. 74-81, January 1957.
3. S. J. Bailey, "Incremental servos - Introduction," Control Engrg., vol. 7, pp. 123-127, November 1960; "Operating and analysis," vol. 7, pp. 97-102, December 1960; "Applications," vol. 8, pp. 85-88, January 1961; "Industry survey," vol. 8, pp. 133-135, March 1961; "Interlocking steppers," vol. 8, pp. 116-119, May 1961.
4. N. L. Morgan, "Versatile inductor motor for industrial control problems," Plant Engrg., vol. 16, pp. 143-146, June 1962.
5. G. Baty, "Control of stepping motor positioning systems, "Control of stepping motor positioning systems," Electromechanical Design, vol. 9, pp. 28-39, December 1965.
6. A. E. Snowden and E. M. Madsen, "Characteristics of a synchronous inductor motor," Trans. AIEE (Applications and Industry), vol. 8, pp. 1-5, March 1962.
7. J. P. O'Donahue, "Transfer function for a stepper motor," Control Engrg., vol. 8, pp. 103-104, November 1961.
8. R. B. Kieburtz, "The step motor - the next advance in control systems," IEEE Trans. Automatic Control, vol. AC-9, pp. 98-104, January 1964.
9. B. A. Segov, "Dynamics of digital routine control systems with a step motor," Automation of Electric Drives (in Russian), no. 4, pp. 38-61, 1959.
10. T. R. Fredriksen, "Closed loop stepping motor application," Proc. 1965 JACC, pp. 531-538.
11. T. R. Fredriksen, "New Developments and applications of the closed loop stepping motor," Proc. 1966 JACC, pp. 767-775.

12. T. R. Fredriksen, "Direct digital processor control of stepping motors," IBM J. Research and Develop., vol. 11, March 1967.

13. J. L. Douce and A. Refsum, "The control of a synchronous motor," presented at the 3rd IFAC Cong., London, England, June 20, 1966, paper 4A.

14. T. R. Fredriksen, "Stepping motors come of age," Electro-Technology, vol. 80, pp. 36-41, November 1967.

15. T. R. Fredriksen, "Applications of the closed loop stepping motor," IEEE Trans. Automatic Control, vol. AC-13, number 5, October, 1968, pp. 464-474.

16. T. R. Fredriksen, "Design of digital control systems with step motors," Symposium Proceedings, Incremental motion control systems and devices, University of Illinois, pp. 479-522, March 1972.

17 OPEN-LOOP CONTROL BY USING ELECTRO-HYDRAULIC MOTORS

M. I. S. Bajwa
Washington Scientific Industries, Inc.
Long Lake, Minnesota

17-1 INTRODUCTION

The fast expanding field of computer auxiliary equipment has been utilizing the unique characteristics of electrical step motors for the last few years. The only major disadvantage of electrical step motors is the lack of power that can be developed by such motors. Electrical step motors, being digital devices, are basically inefficient and any attempt in obtaining higher from these motors would be feasible only at the expense of high heat losses which, of course, cause many secondary problems. If a small electrical step motor could be used to control the power available from a hydraulic motor, the advantages of direct digital control could be spread over a very wide horsepower range. An electro-hydraulic step motor is a device which successfully exploits this concept.

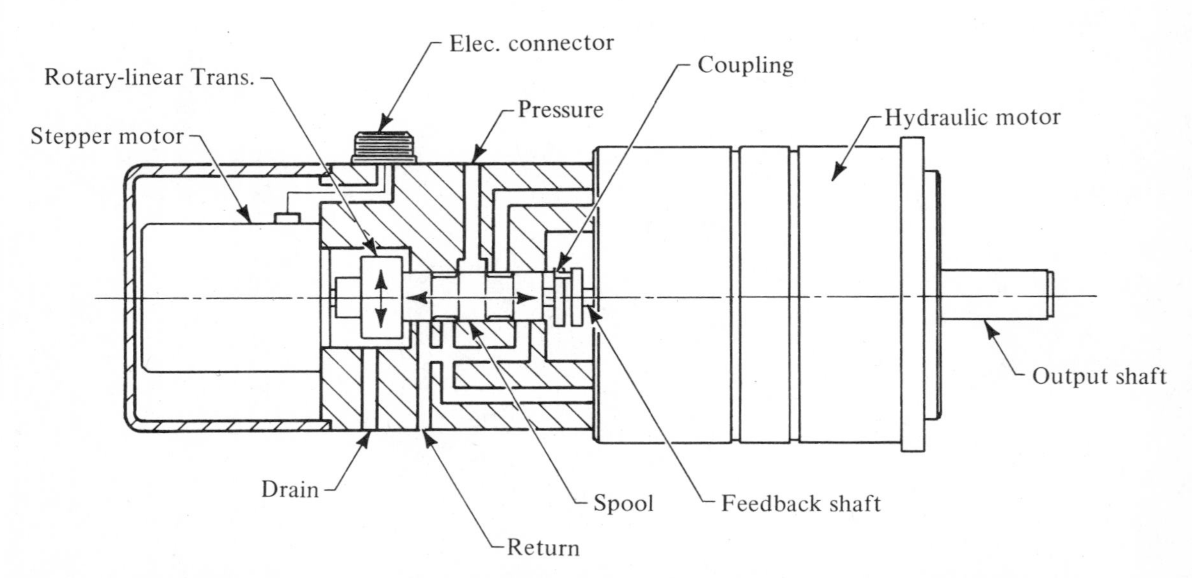

Figure 17-1. Schematic drawing of electro-hydraulic step motor (EHSM).

17-2 WORKING OF THE EHSM

The electro-hydraulic step motor has the following three components: Electrical Step Motor, Servo-Valve, and Hydraulic Motor. These three independent components, when integrated in a particular fashion, provide for the hydraulic motor to accurately follow the electrical step motor, but with a torque output which is several hundred times greater than the capabilities of the electrical step motor. The electric step motor is the mahout which controls the enormous power of the elephant--the hydraulic motor.

The electric step motor used in this package may be any commercially available off the shelf motor. This motor is directly coupled to the rotory linear translator of the servo-valve. The output torque of the electric motor must be capable of overcoming the flow forces in the servo-valve. The flow forces in the servo-valve are directly proportional to the rate of flow through the valve. Figure 17-2 gives the axial force the valve spool for various flows. The torque required to operate the rotory linear translator against this axial force is dependent on the flow gain

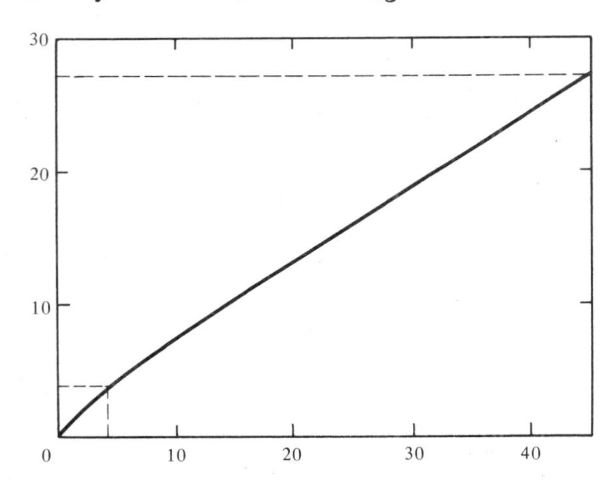

Figure 17-2. Axial force (thrust) on step motor shaft for various flows through servovalve.

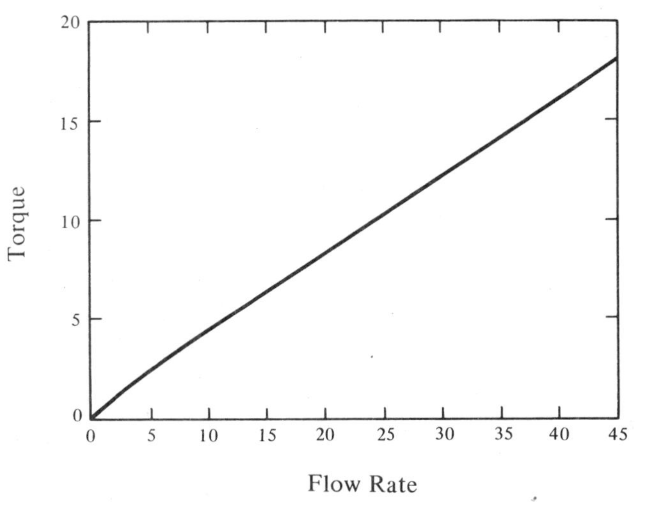

Figure 17-3. Torque required to drive servovalve spool for various flow rates.

in the servo-valve. Figure 17-3 gives the torque required for the operation of the servo-valve at various flow rates. This is the torque demanded of the electric step motor and as can be seen from the graph, a very small electric step motor can produce enormous output horsepower to be available at the hydraulic motor shaft. The electric motor is run submerged in oil. This is achieved by diverting the case drain flow of the hydraulic motor and servo-valve, through the can surrounding the electric step motor. This technique has the advantage of protecting the electric step motor from overheating due to the ability of the oil to act as a very efficient heat sink.

The mechanical input servo-valve performs the following two functions:

(1) The hydraulic motor comes to rest only when there is no positional error between the electric step motor and the hydraulic motor, thus the servo-valve makes the hydraulic motor reproduce exactly the position of the electric step motor. A closed-loop circuit exists between the electric step motor and the hydraulic motor, through the servo-valve. The hydraulic motor exactly reproduces the position of the electrical step motor, ignoring any minor mechanical imperfections.

(2) The amount of oil flow through the servo-valve is such that it tends to cancel out the lag between the electric motor and the hydraulic motor. Thus the speed and direction of rotation of the hydraulic motor would always try to reproduce the motion of the electric motor. The flow through the servo-valve is directly proportional to the lag between the

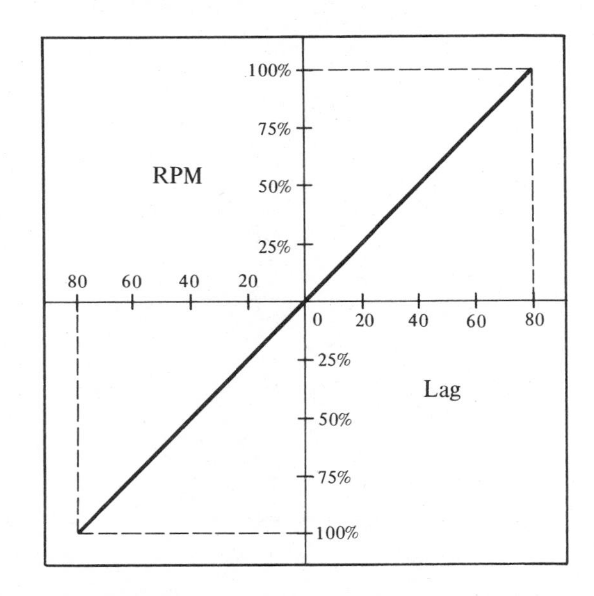

Figure 17-4. Servovalve flow versus lag between electric motor and hydraulic motor.

electric and hydraulic motors. Figure 17-4 gives the flow through the servo-valve for various pulses of lag between the electric step motor and the hydraulic motor. Thus, as the lag between the electric and the hydraulic motor increases, this causes additional flow to pass through the servo-valve which in turn speeds up the hydraulic motor. The hydraulic motor is faithfully trying to reproduce the position of the electric step motor and the servo-valve provides the feedback through mechanical linkage. The control of flow through the servo-valve is achieved by throttling (as in closed-loop servo-valve systems) and this produces heat because of wasted energy.

The hydraulic motor is the most important component of the EHSM package. The performance characteristics of the hydraulic motor would determine the performance that can be achieved from the whole package. The Washington Scientific Industries Inc. (WSI) hydraulic motor has certain very desirable characteristics for operation under these conditions. The motor is a fixed clearance, axial rolling vane motor and has a balanced rotor design as shown in Figure 17-5.

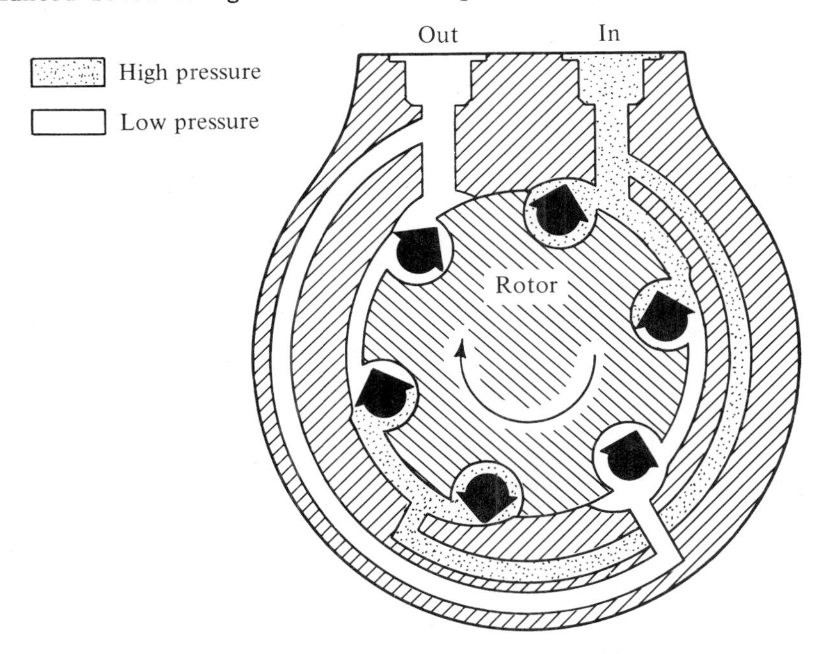

Figure 17-5. Schematic drawing of WSI hydraulic motor.

The fixed-clearance motor has the rotor always balanced on a film of oil and this produces a very low breakaway pressure (less than 10 PSI for all WSI motors) as well as very high response to an input signal. The rotor inertia is very low, e.g. the rotor inertia for the 7.0 cu. in. per rev. motor (capable of producing up to

53HP) is only 0.046 $lb\text{-}in\text{-}sec^2$. Due to the absence of any rubbing or sliding parts within the motor, the noise level of any WSI EHSM package is less than 83 dBA when operating anywhere within the speed and pressure range of the hydraulic motor. This is the noise level rating based on measurements at a distance of three feet from the EHSM package. The biggest advantage of the WSI hydraulic motor is the high torque available at stall. Due to an absence of "stick-slip" in the motor, the torque available at stall is better than 98 percent of the theoretical torque available from the motor. The torque speed curve of a 1.0 cu. in/rev. EHSM is shown in Figure 17-6 and for a 7.0 cu. in/rev. EHSM is shown in Figure 17-7.

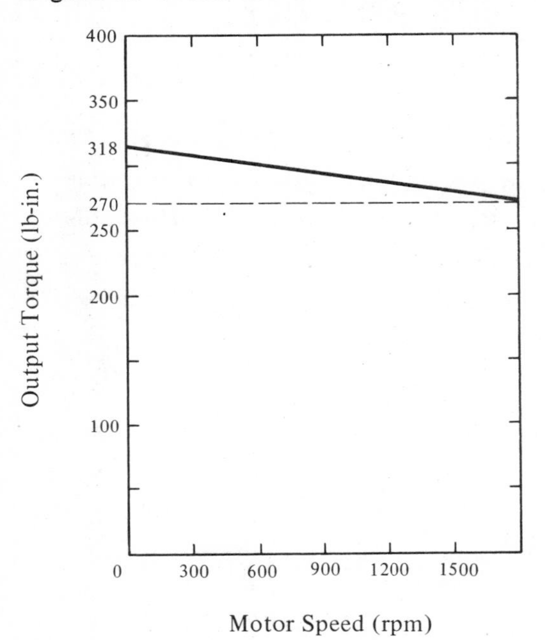

Figure 17-6. Torque-speed curve for the 1 cu-in/rev EHSM.

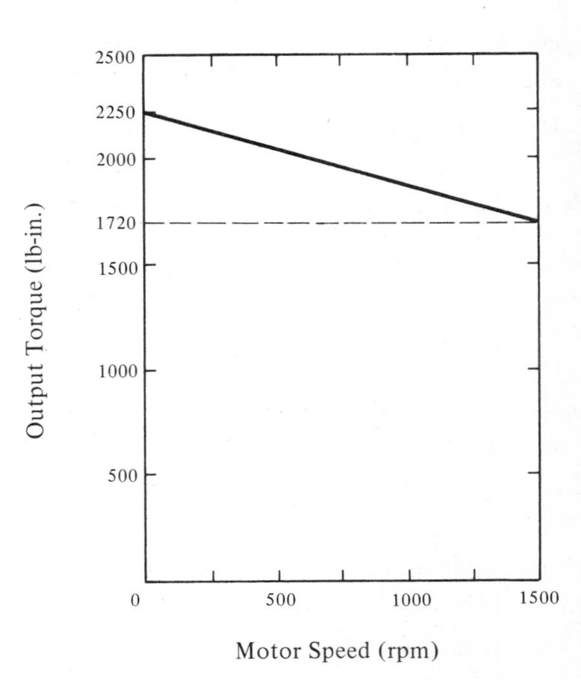

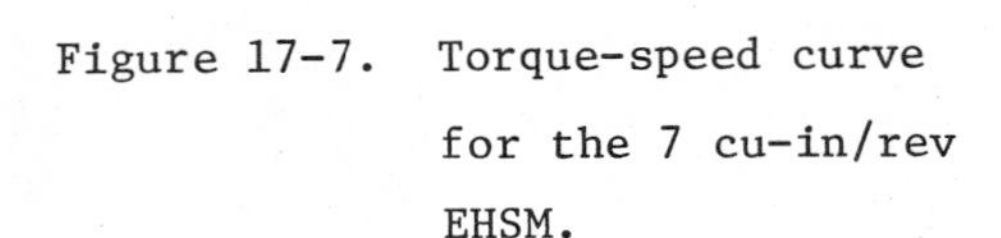

Figure 17-7. Torque-speed curve for the 7 cu-in/rev EHSM.

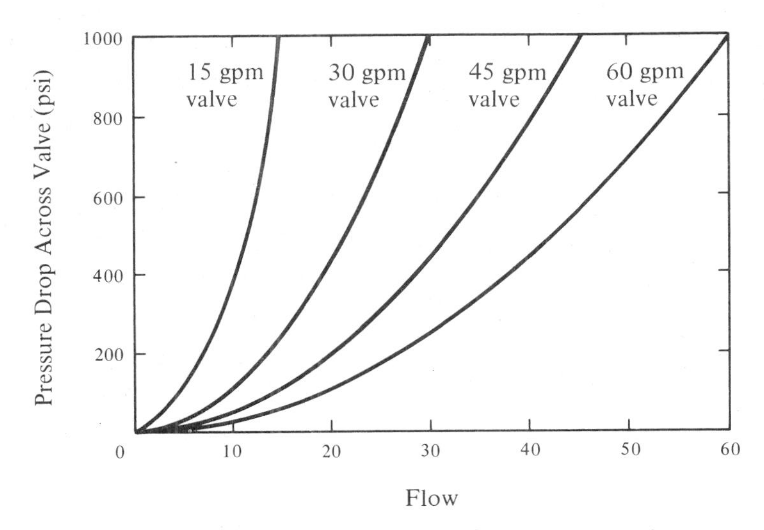

Figure 17-8. Flow versus pressure drop for various size servovalves.

The decrease in output torque at high speeds is due to the increase in pressure drop across the servo-valve at high speeds. This would result in lesser pressure being available to the hydraulic motor to do the useful work. The pressure drop across various servo-valves is shown in Figure 17-8.

17-3 RESPONSE CHARACTERISTICS

The main advantage cited in favor of hydraulic motor applications is that hydraulic motors can accelerate high inertia masses in the shortest time. This is because of the ability of hydraulic oil under pressure to have stiffness comparable with steel. At the same time hydraulic oil flows in the flow passages of the motor and provides for the maximum power that can be achieved from a minimum amount of enclosed space. In the EHSM, the high response of the package would depend upon the following parameters:

(1) The ability of the electric stepper to start at high pulse rates.

(2) The flow gain in the servo-valve.

(3) The ability of the hydraulic motor to accelerate at a high rate so that the lag between the electric step motor and the hydraulic motor stays within permissible limits.

In the WSI EHSM packages, the design of the servo-valve has been very carefully optimized so as to produce a high response and at the same time, not to introduce any stability problems. The high response of the EHSM is desired for the following reasons:

(1) To achieve short motions in the smallest time period.

(2) To minimize the lag between the electrical step motor and the hydraulic motor.

The mechanical response is such that the hydraulic motor is very close to being critically damped. In an underdamped or overdamped condition the motor takes a long time to come to rest. Higher inertia loads would increase the settling time for the EHSM but the good damping qualities of oil dampens out oscillations for normal applications in less than 100 msecs. A test was carried out for one of our customers to determine the response characteristics of the Model 10 EHSM. A load of 200 pounds was connected to the hydraulic motor through a one-pitch lead screw. This corresponds to an inertia of 0.02 $lb\text{-}in\text{-}sec^2$ when referred to the hydraulic motor shaft. It took 33 msec to complete 0.115 inches of motion of the table. Figure 17-9 shows the trace of the motion as was recorded by an oscillograph. The mechanical movement of the lead screw was recorded by obtaining a signal from a

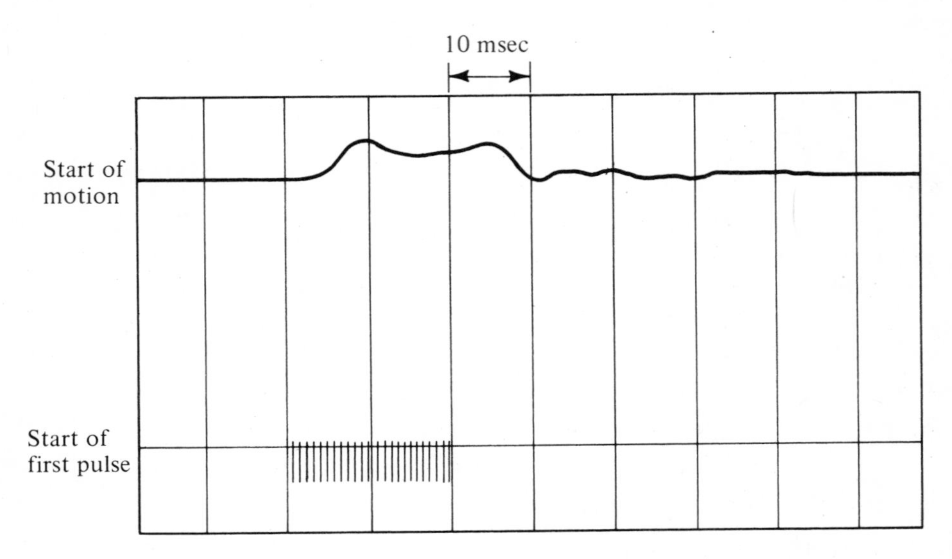

Figure 17-9. Oscillograph trace to determine the response characteristics of the motor.

tachometer directly coupled to the lead screw. For the same load, a distance of 26 inches was traversed in 1.5 seconds.

17-4 PROTECTION FROM HIGH INERTIA LOADS

The direct coupling of hydraulic motors to high inertia loads requires careful consideration. It is true that hydraulic motors have the capabilities of accelerating large inertia loads at high acceleration rates. The problem arises when these high inertia loads are decelerated at a high rate of deceleration; or worse still, when reversing high inertia loads. When a high inertia mass is moving at a certain speed, (linear speed or rotational speed as the case may be) it has enormous amounts of kinetic energy stored into it (1/2 mv^2). This energy has to be released, somehow, before the mass can come to rest. In the EHSM, when the electric stepper decelerates and stops, the servo-valve comes to the null position because of the mechanical feedback from the hydraulic motor shaft. At this null position, the flow passages between the servo-valve and the hydraulic motor are blocked by the servo-valve spool. The oil within the hydraulic motor is trapped and has no outlet. If a large inertia load were coupled to the hydraulic motor output shaft, this would have a tendency to overdrive the hydraulic motor beyond the commanded stop position as is dictated by the null position of the servo-valve. This tendency to be overdriven by the inertia load is resisted by the compression of the trapped oil within the hydraulic motor as the servo-valves passes through its null position. If the inertia load is too large this trapped oil may be compressed to generate dangerously high pressure spikes within the hydraulic motor. A single pressure spike generated in this manner is enough to cause permanent failure of the hydraulic motor. Some type of protection device must be incorporated between the servo-valve and the hydraulic motor to check the pressure from reaching dangerously high values. Figure 17-10 shows the relief

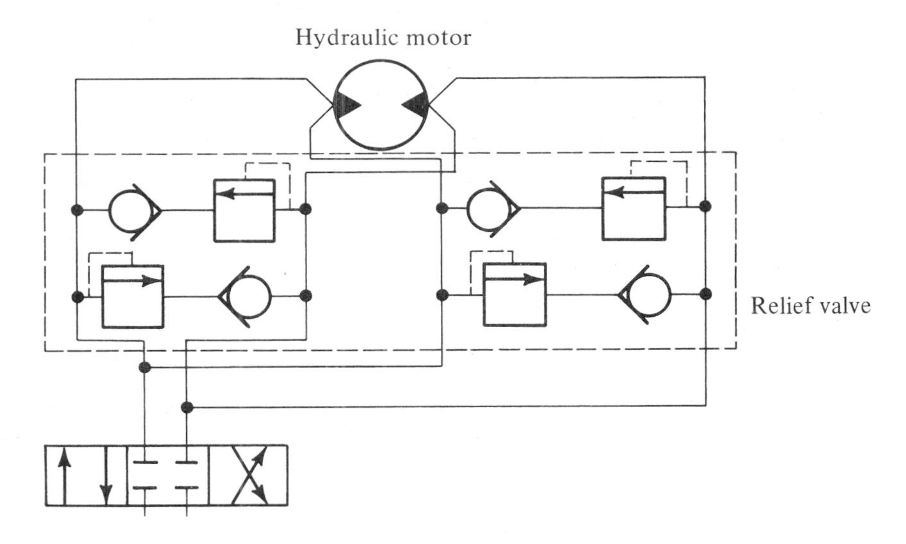

Figure 17-10. Dual relief protection for the hydraulic motor.

valve protection provided between all four parts of the EHSM. The relief valves are pre-set to operate at a pressure differential of 2000 PSI. This bleeds the oil from the high pressure side to the low pressure side of the hydraulic motor, thereby protecting the hydraulic motor from exposure to high pressure shocks. The advantage in EHSM applications is that any inertia or friction load applied to the hydraulic motor shaft is not reflected to the electric step motor. The load on the electric step motor shaft is a function of speed only, due to the increase in Bernoulli forces with increased flow through the servo-valve.

17-5 GAIN ADJUSTMENT

The inability to adjust the gain in EHSM packages is a disadvantage often cited when comparing closed-loop hydraulic servo-systems with an open-loop EHSM system. Since the motor is a fixed displacement hydraulic motor, the gain can only be adjusted by adjusting the flow gain of the servo-valve. The flow gain of the servo-valve may be changed by changing the gain in the rotary to linear translator. This refers to the axial travel transmitted to the valve spool per unit of rotary motion of the step motor shaft. Unfortunately, this gain cannot be readily adjusted in the field and the rotary linear translator would have to be replaced by one with a different gain. The flow gain however, is directly proportional to the system pressure and a limited variation of gain can be achieved by changing the system pressure. However, system pressure is generally dictated by load requirements and changing system pressure for obtaining optimum gain characteristics would affect system performance. Figure 11 shows an arrangement that can be employed for reducing the gain as experienced by the hydraulic motor. A portion of the flow from the servo-valve bypasses

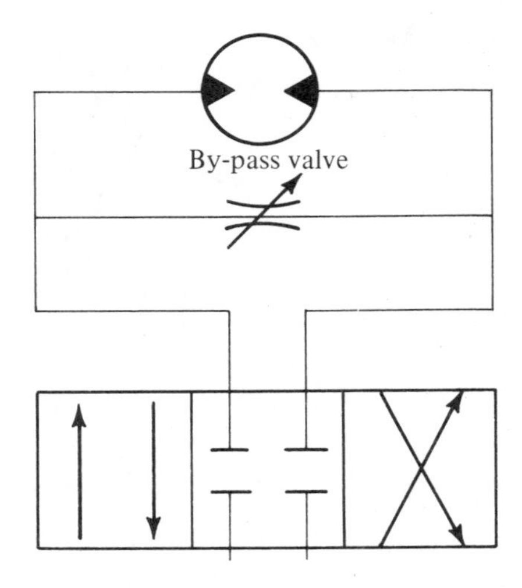

Figure 17-11. Gain adjustment by-pass valve.

the hydraulic motor and is fed into the return flow. The amount of bypass flow can be adjusted through a metering valve. This gives us the limited capability of being able to reduce the gain through a narrow range only. The disadvantage would be the loss of accuracy and also drift around the null position of the servo-valve.

Only the same EHSM line has a field adjustable mechanical arrangement of selecting between three or four fixed gain settings. This is achieved by changing the linkage length which rotates the rotary servo-valve, thereby changing the amount of rotation seen by the servo-valve for the same angular rotation of the step motor.

Accuracy

The step accuracy of the EHSM package is determined by four factors:

(1) The accuracy of the electric stepper.

(2) The accuracy within the servo-valve.

(3) The backlash in the rotary linear translator.

(4) The stick-slip or cogging tendency in the hydraulic motor.

The inaccuracy in electric step motors is due to the manufacturing tolerances in the rotor poles. Most good electrical step motors do stop within three percent of one step from the ideal position. This inaccuracy would be directly reflected at the hydraulic motor shaft also. The positioning accuracy obtained from the servo-valve would depend upon the very accurate inline configuration of the spool land and the sleeve port at the pressure port. Any overlap existing at the pressure port would deteriorate the step accuracy of the hydraulic motor. Any nderlap at the pressure port tends to cause hunting and instability. The inline configuration of the servo-valve ports is essential for high step accuracy and can only be achieved by repeated finish grinding of the valve spool, till it satisfactorily matches with a certain sleever. This trial matching of the valve spool and sleeve, though tedious and time consuming, pays off in extremely good accuracy achieved from the EHSM. The rotary linear translator mechanism must be torsionally rigid and have no axial backlash. In the WSI motor this is achieved through a patented preloaded arrangement which takes the backlash out of the mechanism. It is spring loaded and self compensating for wear. Finally, the balanced rotor of the hydraulic motor floating on a film of oil creates no stiction when the hydraulic motor seeks the null position. A combination of all these design features provides for very high step-to-step accuracy as well as repeatability. The repeatability is better than 10 mins. of shaft rotation for a 1.8 degree step motor.

17-6 APPLICATIONS

At present, EHSM applications are far more common in Japan and Europe than they are in the United States. The concept was originally conceived and successfully applied by Fujitsu Ltd. of Kawasaki, Japan. It has been so widely applied in the Machine Tool Industry in Japan that more than 90 percent of all machine tools built in Japan are equipped with electric or electrohydraulic step motors. France and England have also been applying EHSM's for a few years now. In the United States, only in the last five years, have we witnessed genuine interest in the application of EHSM's. We shall describe several applications of the EHSM in the following.

One of the applications of the EHSM is on a transfer line for the automotive industry, involving the adjustment of backlash.

In the assembly of rear axles, the adjustment of backlash in the differential gears has been done traditionally through the insertion of shims. The reason for the use of these adjustments is the slight non-uniformity that is present in the castings. The backlash between the differential gears is measured and the proper shim is selected to compensate this backlash and also to provide the required preload between the walls of the differential carrier. This method has the drawback of the inability to readjust in the field. Figure 17-12 shows the use of adjusting nuts to adjust the backlash.

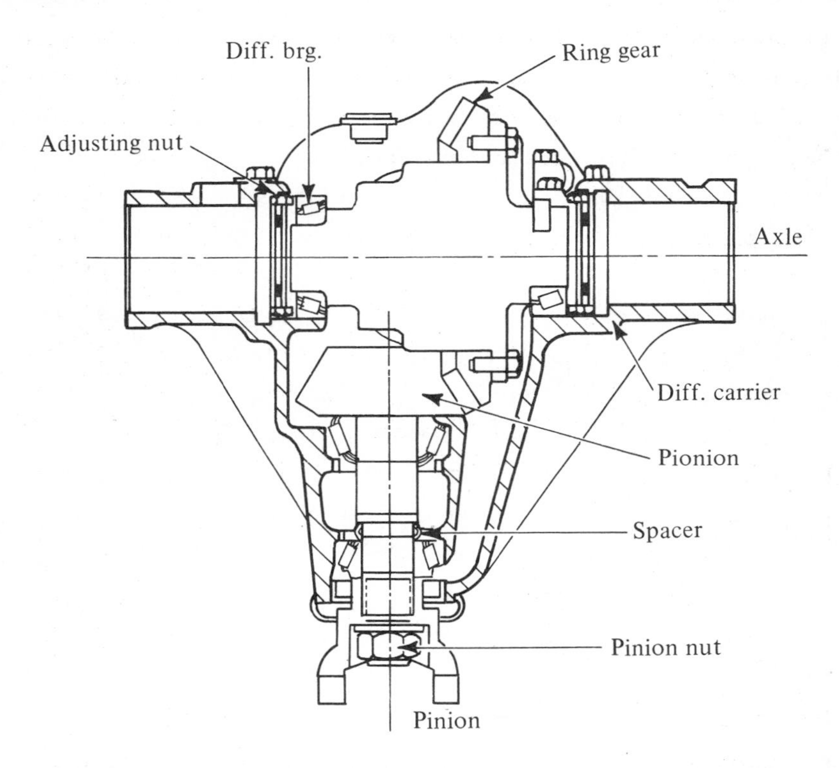

Figure 17-12. Differential backlash adjustment with adjusting nuts.

Adjusting nuts, though expensive when compared to shims, have the flexibility of field readjustment as well as reduce the assembly time for differentials. The adjusting nuts have holes into which the collet projections on the shafts of the drive motors come and engage. The drive motors are WSI electrohydraulic step motors. the drive motor shaft is hollow and through this goes the backlash measurement shaft. This shaft is connected to an electric step motor thru a large gear reduction. The backlash present in the gears is measured through the angular movement of this shaft. This measurement is then used to compute and trigger the amount of rotation to be given to the adjusting nuts to compensate the backlash and also to provide the predetermined amount of preload. The drive motors are then retracted and the part is sent to the next station in the transfer line. A new part comes in and the process is repeated.

Another interesting application involves running two motors in synchronism with an ability to manually retard one motor with respect to the other if, for external mechanical reasons the loads do not remain in synchronism. This has applications in the textile industry where cloth is fed through two sets of rollers during various printing operations. Due to slippage at the rollers the cloth can get skewed, and to keep the cross threads perpendicular to the longitudinal threads, one of the motors has to be retarded for a small time interval and then again allowed to proceed at the initial RPM to maintain the synchronism. The EHSM's used in such an application must possess high stiffness and the lag between the electric step motor and the hydraulic motor should stay fairly constant for small changes in load.

Figure 17-13 shows a system used for obtaining a constant surface speed while unwinding foil for a slitting application.

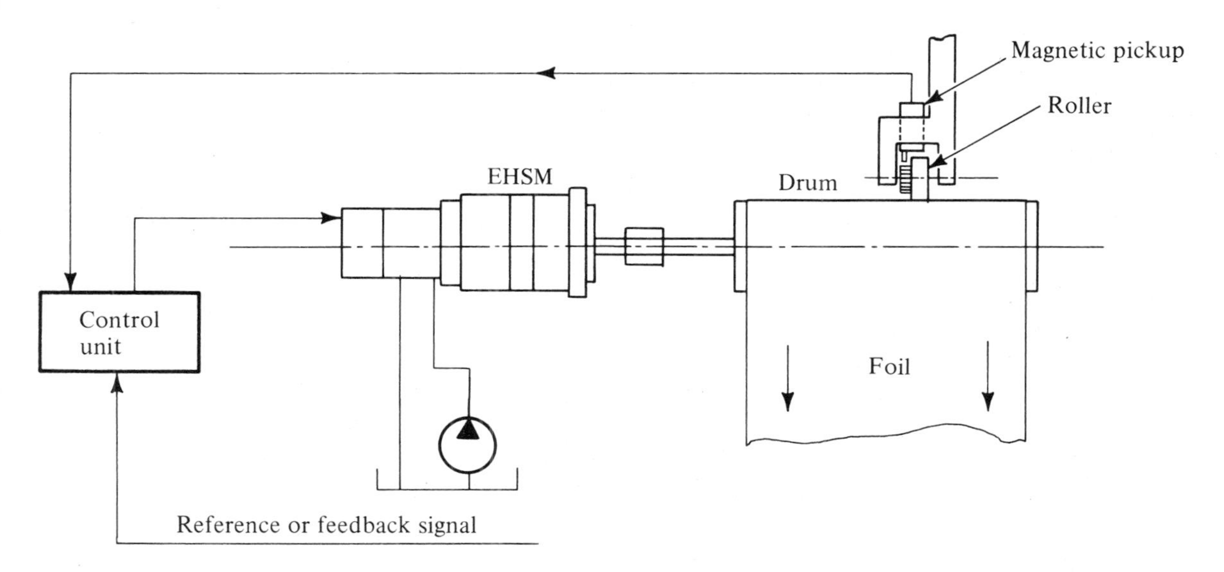

Figure 17-13. Arrangement for obtaining constant linear speed from a drum roll of changing diameter.

A gear and a magnetic pickup provide indication of foil velocity. The gear mounts on a small roller which rides on the foil surface. Pulses from the pickup are proportional to foil velocity. Comparing this pulse rate with the reference pulse rate for the EHSM the necessary correction is applied to the motor to obtain a constant surface speed during the unwinding of the roll. The concept may be applied for winding or unwinding for any material such as rubber, wire, fiber, paper or film.

An interesting application of the use of electro-hydraulics for precise incremental motion of a hydraulic cylinder is shown in Figure 17-14. The only difference between this device and an EHSM is that the hydraulic motor has been replaced by a linear actuator. The cylinder piston is connected to a lead screw, which is

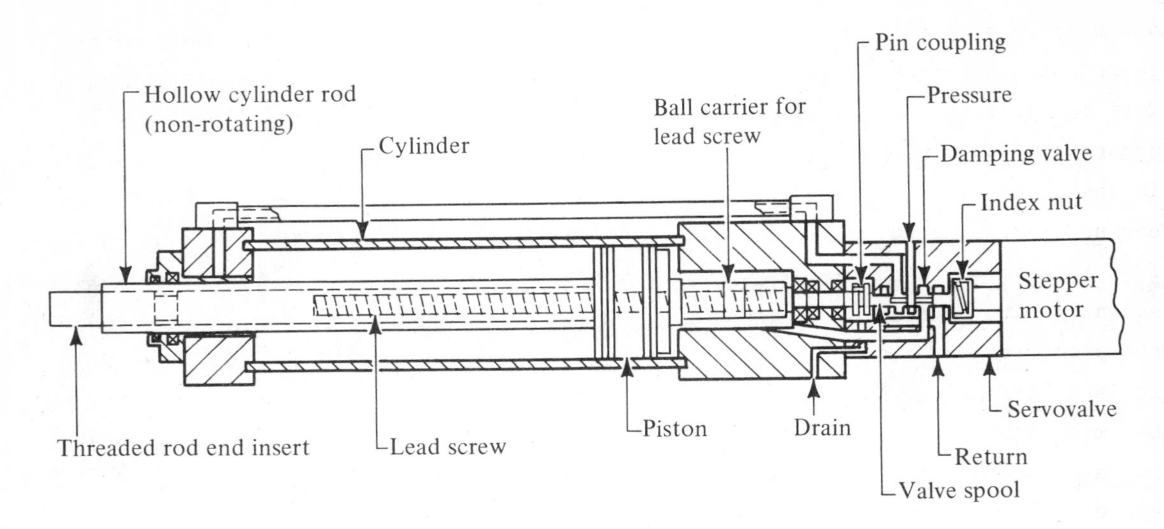

Figure 17-14. Electro-hydraulic linear actuator.

mechanically connected to the valve spool to provide the feedback between the electric stepper and the hydraulic cylinder.

The use of a 200 pulses/rev. motor and a five pitch lead would result in increments of 0.001 inches per pulse. The output force and speed of the cylinder piston would depend upon system pressure, piston dia and servo-valve capacity. A pulse rate of 5000 PPS would result in a cylinder speed of 300 inches per minute.

17-7 DESIGN OF HYDRAULIC POWER UNIT

In all the discussion on EHSM units till now, it was taken for granted that hydraulic power is available to the actuator. This power generated by a hydraulic power unit needs a very careful selection of components, and usually ends up being the most neglected aspect in the installation of the overall system. The hydraulic reservoir, pump, piping, filteration and heat exchanger; though simple devices, require many years of experience to tie them together in an efficient and neat manner. A few rules to be observed are given in the following paragraphs.

The reservoir capacity must be three to four times the capacity of the pump in gallons per minute. All lines to and from the reservoir must enter at the top so as to avoid draining the reservoir when disconnecting pipes. The return and suction lines should terminate well below the lowest level of the fluid and the space between these lines should be adequately baffled to make the return path of the oil to the pump inlet as long as possible. The reservoir should have ample surface area to allow for the bubbles to seperate from the oil. It should have heavy cast iron mounting surfaces.

The selections, mounting, and piping design for the pump is of great importance for the successful application of electro-hydraulic step motors. The use of a fixed displacement pump is advantageous from the point of view of continuous hydraulic power being available on demand. The great disadvantage being that during idle periods all this generated power is lost into heat when going over the relief valve. This necessitates a large heat exchanger to be used and the relief valve must be of proper size to minimize the noise. The relief valve being the noisiest component in the whole circuit, such a system would become annoying if long idle periods are encountered. The use of a variable volume pressure compensated pump is desirable from the standpoint of using power economically and sensibly but, the greatest disadvantage is in the loss of response, because of the excessive time it takes to make the pump respond to the load requirements. Due to this time required by the pump for stroking to provide the desired flow, the time required to move short distances becomes rather large. One way of overcoming this is by using accumulator circuits but accumulator application is tricky and can sometimes create more trouble than it solves.

Filteration is the most important aspect of a hydraulic system. Lack of proper filteration is overwhelmingly the largest single cause of premature failure in hydraulic valves and actuators. A ten micron filter ahead of the electrohydraulic motor is a good insurance against costly downtime for maintenance or premature failure of the actuator. A 33 micron filter in the return line is also recommended. The servo-valve in the EHSM is much less sensitive to contamination than the ordinary servo-valve used in closed loop applications. This is because of the fact that the servo-valve spool in the EHSM is rotating continuously at the hydraulic motor rpm (or electrical step motor rpm). This does not allow any dirt buildup or silting in the servo-valve, thereby retaining the high performance of the valve. This should not be taken as an implication that the lack of proper filteration would not hurt the EHSM and thereby neglect filteration precautions. Nothing is more important to the life of a hydraulic actuator than the use of clean hydraulic oil.

The hydraulic lines used to transmit hydraulic power could be pipe, tubing, or flexible hose. The choice would depend upon the system pressure and flow. The mean flow velocity through the pipe or tubing should not exceed 15 ft./sec. in any part of the hydraulic circuit. When using seamless tubing the tube should not be dented at the bends or collapsed at the point of attachment of fittings. A safety factor of at least 4:1 should be utilized in the use of tubing and hose. The use of manifolding in recent years has eliminated the need for a large amount of piping and fittings. This results in smooth hydraulic passages, neat appearances and enormous saving in space and time required to hook up the hydraulic circuit. Wherever hydraulic lines are used, they must be as short as possible and have the least number of bends. This would necessitate locating the hydraulic power supply as close to the hydraulic amplifiers as possible. Long lengths of pipe are expensive and create pressure drop due to friction to flow in the pipes. High pressure pipes can also be set into vibrations easily because of the pressure pulsations from the pump outlet. All high pressure lines should be securely fastened through supporting brackets. After all piping has been completed, the system must be flushed for several hours before the hydraulic actuator is connected to the circuit. Figure 17-15 shows a basic hydraulic power supply for a EHSM.

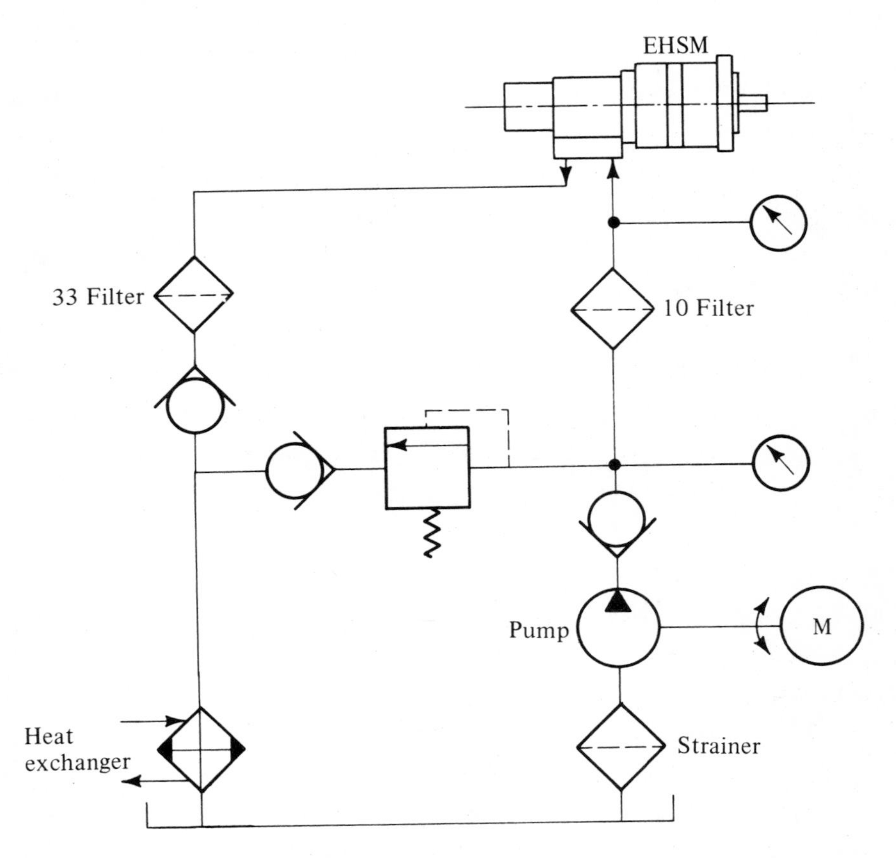

Figure 17-15. Simple hydraulic circuit for EHSM application.

17-8 OPEN LOOP OR CLOSED LOOP

Assuming that the decision to use hydraulics for a certain application has been made, the design engineers generally have two options:

(1) A closed-loop system.

(2) An open-loop system.

Figure 17-16 describes the two systems as applied to a machine tool control.

In a closed-loop system the velocity or position feedback from the machine is constantly compared with the input command signal, and the error signal is applied to the servo-valve that controls the hydraulic motor. In contrast, in an open-loop system, the EHSM is given a pulse train command and this moves the load in discrete increments to the desired position and at the desired speed. The feedback transducers are replaced by faith in the ability of the EHSM to respond accurately to the pulse train without the loss of any pulses. Both the systems have proven their performance regarding reliability and high accuracy. The closed-loop system is more

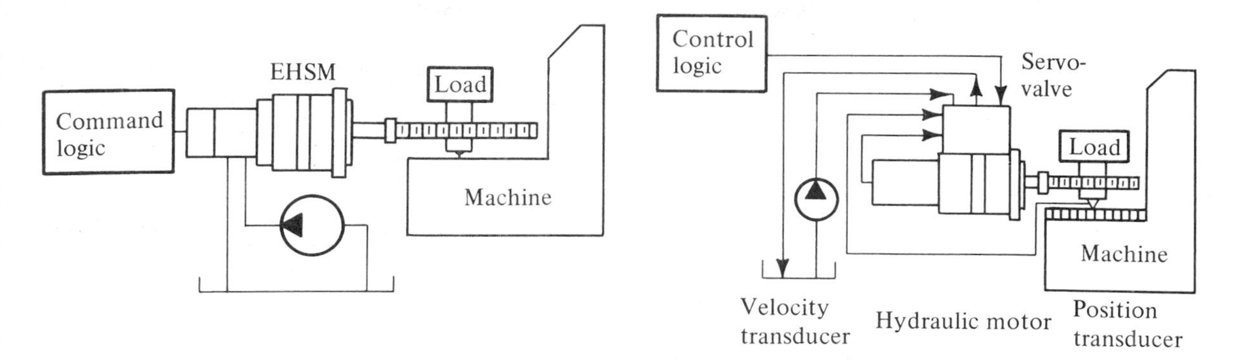

Figure 17-16. Comparison of open-loop versus closed-loop.

popular in the United States and the open-loop system in Japan and Europe. The only advantage of the closed-loop concept over the open-loop concept is the increase in reliability of the system. This is achieved at the price of installing a number of additional devices on the machine like tachometer, generators, resolvers, LVDT's, inductive or optical scales etc. This also increases the number of circuits that comprise the servo-system. The complexity of the whole system is increased, the attachment to the machine is complicated and the servo-system requires periodic readjustments. Overshoot and hunting can occur depending upon the gain in the servo-loop and low resonant frequency of the machine. The open-loop system on the other hand is simple, easy to install and maintain, has no hunting or overshoot, and has a non-cumulative final point error.

17-9 ELECTRO-HYDRAULICS - A NATURAL FOR CONTROLLED POWER

The various sources of power available to most designers include D.C. motors, A.C. motors, Air Motors, I.C. Engines, Hydraulic Motors etc. All these systems have some unique advantages but the range of optimum performance overlap so much that for any application, maybe three different sources of power could be utilized. D.C. variable speed drives have had considerable success in the last few years. A.C. drives, though expensive, have superior mechanical and low maintenance characteristics. I. C. engines are very reliable, but produce air pollution, noise and vibration. Pneumatic power has light weight, low maintenance and is well suited for frequent overloads and hazardous environments. Though all these systems have their individual advantages, the advantages obtained by using hybrid systems cannot be over-emphasized. The superior characteristics of electronic control far surpass the control achievable by alternate systems. Electronic controls are reliable, inexpensive and highly flexible. When these controls are applied to other sources of power, a reliable, accurate and easily programmable form of power is achieved.

The advantages of fluid power systems as a source of power are compactness, ruggedness and high efficiency. Operation of systems controlled by fluid power components is smooth, flexible and readily coordinated. Because of these and other desirable characteristics the trend has shifted away from pure mechanical and electro-mechanical systems towards electro-hydo-mechanical systems. Hybrid systems are also gaining ground with other sources of power. Ingersoll Rand has recently developed a electro-pneumatic control for accurate torquing of nuts and bolts.

Electro-hydraulic systems can provide infinitely variable regulation of output speed and torque and have simple controls requiring a minimum of effort. Mr. Von Hoene, Director, Research and Product Development, Vickers Division of Sperry Rand Corporation, has this viewpoint, "Important developments are taking place in the field of electro-hydraulics. Electro-hydraulic pulse motors, modulating valves and servo systems are gaining in popularity. Utilizing the best of both worlds, they provide for remote control, simplify plumbing systems and enhance economics of modular packaging."

All this when coupled with the fact that mini-computers keep dropping in price, would open up many applications now considered economically unfeasible for direct digital control. The concept of electronic control and hydraulic torque amplification, is in my opinion, the most promising new concept to have appeared on the hydraulic scene in recent years.

REFERENCES

1. Design News Specifier's Annual 1973 Directory.

2. Product Engineering Magazine, December 1972.

3. Hydraulics and Pneumatic Magazine, June 10, 1971.

INDEX

†